Spread Spectrum and CDMA

Spread Spectrum and CDMA

Principles and Applications

Valery P. Ipatov

University of Turku, Finland

and

St. Petersburg Electrotechnical University 'LETI', Russia

John Wiley & Sons, Ltd

Telephone (+44) 1243 779777

Email (for orders and customer service enquiries): cs-books@wiley.co.uk
Visit our Home Page on www.wiley.com

Other Wiley Editorial Offices

John Wiley & Sons Inc., 111 River Street, Hoboken, NJ 07030, USA

Jossey-Bass, 989 Market Street, San Francisco, CA 94103-1741, USA

Wiley-VCH Verlag GmbH, Boschstr. 12, D-69469 Weinheim, Germany

John Wiley & Sons Australia Ltd, 33 Park Road, Milton, Queensland 4064, Australia

John Wiley & Sons (Asia) Pte Ltd, 2 Clementi Loop #02-01, Jin Xing Distripark, Singapore 129809

John Wiley & Sons Canada Ltd, 22 Worcester Road, Etobicoke, Ontario, Canada M9W 1L1

Wiley also publishes its books in a variety of electronic formats. Some content that appears in print may not be available in electronic books.

British Library Cataloguing in Publication Data

A catalogue record for this book is available from the British Library

ISBN 0-470-09178-9 (HB)

Typeset in 10/12pt Times by Integra Software Services Pvt. Ltd, Pondicherry, India.

This book is printed on acid-free paper responsibly manufactured from sustainable forestry in which at least two trees are planted for each one used for paper production.

Contents

Preface

Spread spectrum and CDMA (code division multiple access) are up-to-date technologies widely used in operational radar, navigation and telecommunication systems and playing a dominant role in the philosophy of the forthcoming generations of systems and networks. The amount of interest and effort invested in this encouraging area by research institutions and industry is gigantic and constantly growing, especially after the prominent commercial success of CDMA mobile telephone IS-95 and the use of CDMA as the basic platform of 3G (and beyond) mobile radio. No wonder that the fundamentals of spread spectrum theory have assumed a solid place in the basic university disciplines, while the detailed issues form the contents of numerous advanced courses.

This book was conceived as a textbook for postgraduate and undergraduate students, and is also expected to be useful in training industry personnel and in the daily work of researchers. It is based on the experience and knowledge gained by the author during more than three decades of research activity in the area, as well as on his lecture courses. The original version of such a course started in the late 1970s at the Saint Petersburg Electrotechnical University 'LETI' and has since been continually developed and modernized, absorbing many state-of-the-art achievements and being presented to audiences from Russia, the UK, Australia, China, Finland and other countries.

The intention of the author in preparing this book was to present the key ideas of spread spectrum in the most general form equally applicable to both systems of collecting and recovering information (such as radar and navigation) and telecommunication systems or networks. The author's second concern was to link the material as tightly as possible to classical signal and communication theory, which gives Chapter 2 a special role. The goal pursued everywhere was harmony between mathematical rigour and physical transparency of some or other issue under discussion and the reader's deep understanding of the reasons underlying the preference for spread spectrum and CDMA. The main question the author tried to answer in considering this or that problem was 'Why?'—i.e. why a designer may or should prefer one solution over others.

A particular emphasis of the book is designing spread spectrum signals. Many popular books, although deservedly reputable, do not go into this problem beyond presenting a brief survey of m-sequences and Gold codes. A reader may thereby get a false idea that nothing valuable exists outside this narrow range of attractive signal families. In Chapters 6 and 7 we try to show that the designer's freedom and the multitude of alternatives are much broader and comprise many solutions potentially competitive or clearly superior to those mentioned above.

In no way is this book intended to be looked upon as a manual introducing concrete operational or projected systems and standards. However, some such systems give a rich soil to illustrate the theory and for this reason are frequently mentioned in the text as examples of practical realization of spread spectrum principles. Another aid for better adoption of the contents is offered by the problems at the end of every theoretical chapter. Especially recommended are the Matlab-based problems, since their running involves and develops investigatory skills and allows execution of an extensive experimental study.

The book is supported by the companion website on which instructors and lecturers can find a solutions manual for the problems and matlab programming within the book, electronic versions of some of the figures and other useful resources such as a list of abbreviations etc. Please go to ftp://ftp.wiley.co.uk/pub/books/ipatov. If you have any comments regarding the book please feel free to contact the author directly at valery.ipatov@utu.fi.

The author is sceptical enough to realize that no book—including this one—can be totally free of shortcomings. In our case the difficulties were greatly intensified by the necessity of writing in a non-mother tongue. Nevertheless, the author is entirely responsible for all of the statements as well as the drawbacks of the book and is ready to accept any constructive remarks or criticism.

I would like to express my sincere gratitude to the Department of Information Technology of the University of Turku for the friendly and creative atmosphere during my work in Finland. I address my special appreciation to Professor Jouni Isoaho and Dr Esa Tjukanoff for their daily support and cooperation.

Many thanks to my colleagues Dr Nastooh Avessta and Dr Igor Samoilov, who kindly and carefully read the manuscript and, by way of innumerable discussions, helped in my endeavour to streamline it. The assistance of Jarkko Paavola and Alexey Dudkov in rectifying and debugging the manuscript can hardly be overestimated, too.

This is a good opportunity to emphasize my deepest gratitude to my dear teachers Professor Yu. A. Kolomensky, Professor Yu. M. Kazarinov and Professor Yu. D. Ulianitsky, who introduced me to the fascinating world of signals and noise, and were for decades my advisors in many professional as well as personal matters.

Warmest thanks to all my colleagues at the Department of Radio Systems of Saint Petersburg State Electrotechnical University 'LETI' for a long-standing collaboration.

I bring my gratitude also to Sarah Hinton and her colleagues at John Wiley & Sons, Ltd for initiating this project and inspiring me in the course of writing, and my special thanks to the Nokia Foundation for the grant awarded to me at the final stage of preparing the manuscript.

And finally I cannot help mentioning my wife's patience and care during the year of my working on this book.

Valery P. Ipatov

1

Spread spectrum signals and systems

1.1 Basic definition

The term *spread spectrum* is today one of the most popular in the radio engineering and communication community. At the same time, it may appear difficult to formulate an unequivocal and precise definition distinctively separating the spread spectrum philosophy from a 'non-spread spectrum' one. Certainly, every expert in system design and every experienced researcher has an intuitive understanding of the core of the issue, but—unlike a newcomer—such a person does not need to think about definitions in order to respond successfully to his or her professional challenges. From the point of view of the target audience of the book it seems worthwhile to dedicate some space to elaborating an appropriate explanation of what is implied in the following text under the spread spectrum concept.

Let us start with a reminder of the basics of spectral analysis. Every signal $s(t)$ of finite energy can be synthesized as a sum of an uncountable number of harmonics whose amplitudes and phases within the infinitesimal frequency range $[f, f + \mathrm{d}f]$ are determined by a spectral density or spectrum $\tilde{s}(f)$. It is the pair of inverse and direct Fourier transforms that expresses this fact mathematically:

$$s(t) = \int_{-\infty}^{\infty} \tilde{s}(f) \exp(j2\pi ft)\, \mathrm{d}f \qquad \tilde{s}(f) = \int_{-\infty}^{\infty} s(t) \exp(-j2\pi ft)\, \mathrm{d}t \tag{1.1}$$

Due to the one-to-one correspondence between the signal representation in the time domain $s(t)$ and in the frequency domain $\tilde{s}(f)$, we are able to switch arbitrarily between these two tools, selecting the more convenient one for any specific task. To characterize the size of the zones occupied by signal energy in the time and frequency domains we use the notions of signal duration T and bandwidth W, respectively. A signal whose energy

Spread Spectrum and CDMA: Principles and Applications Valery P. Ipatov

is concentrated within strictly limited space in the time domain cannot have finite (i.e. non-zero in only limited frequency interval) spectrum and vice versa. Because of this, to define at least one of the parameters T, W, or both, some agreement is necessary about what is meant by duration or bandwidth. In this way effective, root mean square, etc. duration and bandwidth came into existence, showing the size of a zone spanned by a substantial part of signal energy in the time and frequency domains, respectively [1].

It is absolutely obvious that one way or another, the word 'spread' is indicative of wide spectrum, i.e. broad bandwidth W of a signal. But against what is the spectrum wide? Where is the reference for comparison? To demonstrate how a definition of spread spectrum may provoke debate, let us consult with several excellent and world-renowned books.

A rather frequent way to explain the concept consists in the statement that a system or a signal is of spread spectrum type if its bandwidth significantly exceeds the minimum bandwidth necessary to send the information [1–6]. What may seem mentally problematic in this definition is the very idea of minimum bandwidth of information or message. According to the fundamental Shannon's bound, spectral efficiency (the ratio between the data rate R and the signal bandwidth W) of a communication system operating over the Gaussian channel obeys the inequality:

$$\frac{R}{W} < \log_2\left(1 + \frac{E_b}{N_0}\frac{R}{W}\right) \quad \text{or} \quad \frac{E_b}{N_0} > \frac{2^{\frac{R}{W}} - 1}{\frac{R}{W}} \tag{1.2}$$

where E_b is signal energy per bit of information and N_0 is the one-side power spectral density of a Gaussian noise. Figure 1.1 represents bound (1.2) graphically, showing that any combinations of R/W and E_b/N_0 falling below the curve are possible, at least in principle. But this means that the theoretical 'minimum bandwidth necessary to send the information' is zero and therefore any real system—which, of course, occupies some

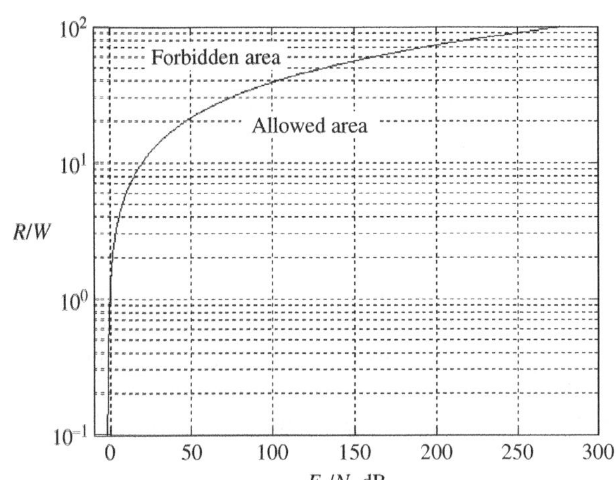

Figure 1.1 Shannon's bound

non-zero bandwidth—should be treated as a spread spectrum one! Undeniably, any attempt to use near-zero data transmission bandwidth would be rather demanding for signal energy. For one thing, to operate with $R = 100W$ one would need to provide bit signal-to-noise ratio E_b/N_0 around 280 dB, which is quite unrealistic. However, data transmission within bandwidth, for instance, up to ten times smaller than data rate is quite typical and is practised in many digital communication links (radio-relay lines, modem communications etc.). This shows the vagueness of the very idea of the 'minimal bandwidth' and the arguable character of taking it as a starting point for explaining the notion of spread spectrum.

As an attempt to eliminate ambiguity we can try to use rate of data in bits per second as a substitute for the above-mentioned minimal necessary bandwidth [7,8]. It is not very logical, however, that one of many possible and, in principle, equal in rights units of measurement of data rate is rendered some conceptually prominent role. Besides, defining spread spectrum in terms of bandwidth significantly exceeding data rate in bits per second is risky of comprising systems which are in no way of spread-spectrum type. Take, for example, the uplink between a single user and a base station in a GSM mobile telephone. With the rate of primary digitized speech data of 9.6 kbits/s, the user's signal has bandwidth around 200 kHz, which may mislead someone to classify GSM as a spread spectrum system. However, no genuine features of spread spectrum are involved in the band broadening in the GSM uplink: the only reason why bandwidth exceeds the data rate is time-division multiple access (TDMA) forcing operation with much shorter transmitted symbols in comparison with the actual average time interval per information bit.

There is still one more reason to look for alternative definitions. Even ignoring the troubles discussed earlier, linking a definition to data rate or 'message bandwidth' can serve only data transmission systems, whereas spread spectrum is widely employed in many others, like radar, sonar, navigation or remote control for time and distance measuring, signal resolution etc. Actually, these systems were among the first to adopt the advantages of the technology under discussion. In those applications such categories as 'information rate' or 'data bandwidth' are hardly meaningful or, at least, have nothing to do with the aims of spreading spectrum. In the wake of the endeavour to define ideas of spread spectrum in some universal way, matching not only communication aspects but the needs of other application areas as well, the following definition of spread spectrum seems more relevant.

Let us turn to the Gabor uncertainty principle, according to which the product of signal duration and bandwidth (*time–frequency product*) satisfies inequality $WT \geq a$, where constant a depends on the exact way in which duration and bandwidth are specified; however, it is always of the order of 1. A signal for which $WT \approx 1$, and therefore duration and bandwidth are tightly linked to each other can be called *plain* (non-spread spectrum). The only way to widen the bandwidth of a plain signal is to reduce its duration, i.e. to shorten it. On the other hand, a deterministic signal for which $WT \gg 1$ and bandwidth can be governed independently of duration is a *spread spectrum* one. Putting it in other words, we may say that any spread spectrum signal occupies a rectangle in the time–frequency plane whose square is much greater than 1. This definition automatically defines a spread spectrum system, too: a system employing spread spectrum signals is a spread spectrum system.

Note that in this definition the independence of duration and bandwidth is particularly emphasized, meaning that one can broaden the bandwidth (duration) without shortening the signal in time (frequency). This has a further implication for the critical role of angle (phase or frequency) modulation in all spread spectrum technology. Indeed, how can amplitude modulation help in widening the spectrum? The answer is: only by reducing the area over which signal energy is effectively spread in the time domain, i.e. by actually reducing the effective signal duration. It is only angle modulation that is capable of widening the signal spectrum with no influence on the time-distribution of signal energy.

As an illustration, Figure 1.2 gives the example of two rectangular pulses having the same duration T and carrier frequency f_0: (a) a signal with no internal modulation and (b) a linearly frequency-modulated (LFM) signal with deviation $W_d = 20/T$. The lower curves show the spectra of these signals. As is seen for signal (a), bandwidth W has the order $W \approx 1/T$, meaning that the signal energy spans in the frequency domain an interval approximately equal to inverse pulse duration. Thereby, duration and bandwidth are strictly tied, the time–frequency product is fixed and widening the spectrum can be achieved only in exchange for pulse shortening. At the same time the bandwidth of pulse (b) is close to frequency deviation ($W \approx W_d$) and much greater than the inverse duration. As a result bandwidth can be easily controlled independently of signal duration by just varying the deviation. Accordingly, we classify the first signal as plain and the second as of spread spectrum type.

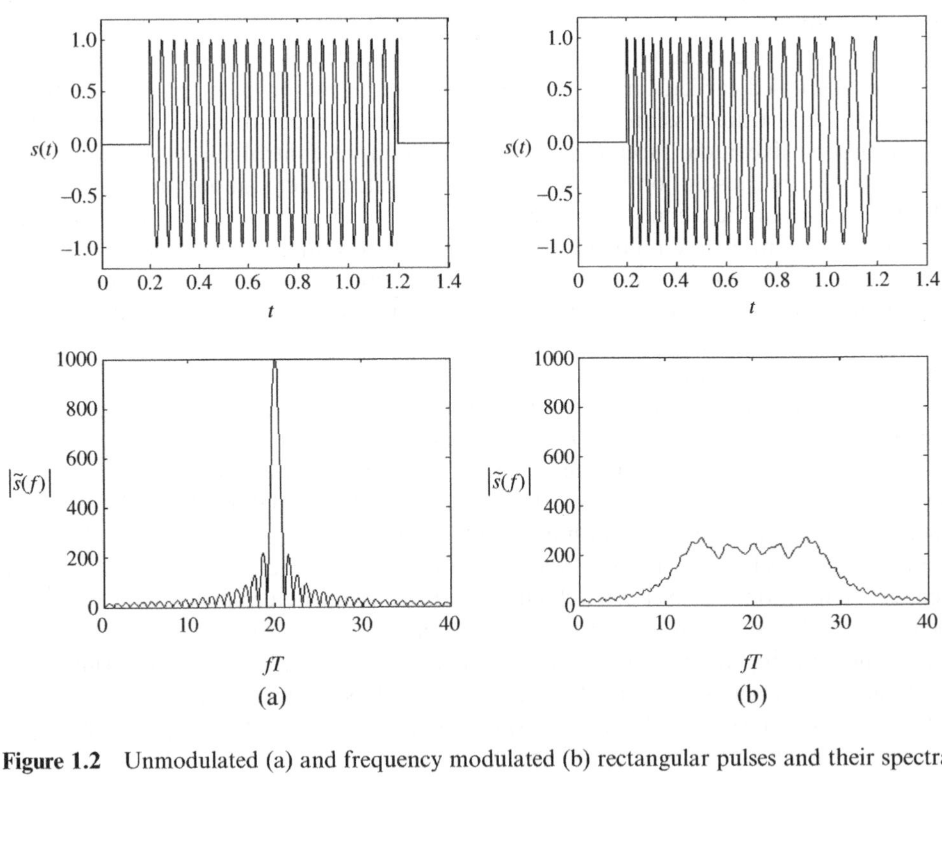

Figure 1.2 Unmodulated (a) and frequency modulated (b) rectangular pulses and their spectra

The definition given is in fact the one which has been widely and long since adopted in the systems of radar-akin philosophy, but it is also consistent with data communication problems. That is why we will rely on it in the following text.

1.2 Historical sketch

The history of spread spectrum covers over six decades and may serve as a topic of separate study. The reader interested in learning the chronology of the key events can address in-depth (albeit focused almost totally on US developments) surveys in [9,10]. Here we limit ourselves to only a very brief mention of the main historical landmarks.

Probably the first patent on the radar, which in modern terminology may be without doubt treated as spread spectrum, was obtained by G. Guanella in 1938. During and after World War II, intensive research in radar spread spectrum systems had been undertaken in Germany, the USA, the UK and the USSR. In parallel with technological and technical advancements, numerous solid theoretical investigations had been conducted into the precision and signal resolution of radar. The most influential and deep results in this regard were published by P. M. Woodward in his 1953 book. It should be noted in passing that many of these results were explainable based on fundamental works by C. Shannon and V. A. Kotelnikov between 1946 and 1948, the role of which thereby goes far beyond only 'pure' data communication applications.

Certainly, for a long time a great deal of information on new practical developments in spread spectrum radar and navigation was classified, because military and intelligence services supervised the great majority of projects. However, many ideas were getting widely known as soon as they were realized in systems of mass-scale usage. A good example of this is the world-wide navigation system Loran-C deployed in the early 1960s in which ground-based longwave radio beacons transmitted 'genuine' spread spectrum (PSK) signals having time–frequency product $WT = 16$. To imagine how viable this system appeared to be, it is enough to stress that with continual modernization and numerous improvements it has managed to remain in operation to see the third millennium.

Another giant step in the practical implementation of the spread spectrum concept in time–distance measuring systems was taken with the creation of the 2G space-based global navigation networks GPS (USA) and GLONASS (USSR/Russia) in the 1980s and early 1990s. Signals with very large time–frequency products, measured in the thousands, are at the heart of these systems, which today constitute an integral part of human civilization as satellite television and mobile radio.

The earliest works in spread spectrum applications to data transmission were primarily aimed at speech masking and communication protection. They started again before World War II in Germany and were soon taken up in the USA, the USSR and elsewhere. An intriguing action of the novel *The First Circle* by Alexander Solzhenitsyn unfolds in the special Soviet jail where convicted scholars and engineers are collected together to elaborate the noise-masked speech transmission system.

Among the turning points in spread spectrum communication, the RAKE algorithm proposed by R. Price and P. Green (1957) should be pointed to, which marked the beginning of the direction later called multipath diversity. Works in the 1960s by

S. Golomb, N. Zierler, R. Gold, T. Kasami and others in the field of discrete sequences with special correlation properties played a crucial role in the formation of spread spectrum technology and numerous practical achievements.

The commercial spread spectrum era started around the late 1970s, at the time when the mobile telephone began its triumphant conquest of the world. The first proposals for CDMA cellular networks in the USA and Europe (1978–1980) yielded to alternative projects, which later evolved into the GSM and DAMPS standards. However, in the mid 1990s the 2G standard IS-95 was put forward, resting on a fully spread spectrum/CDMA platform. At a cosmic pace, networks of this standard (later named cdmaOne) gained wide recognition in America, Asia and the former Soviet Union countries. The great success of IS-95, as well as careful analysis and further experiments, had led to acceptance of the spread spectrum/CDMA philosophy as the basic platform for the major 3G mobile radio specifications: UMTS and cdma2000. Both of them are now in the pre-operational stage and undoubtedly will become the main mobile communication instruments for the next decades.

To conclude this introductory chapter, there are a few words about the development of spread spectrum technology in the Soviet Union and later in Russia. Surveys published in the West usually report only a little on Soviet research in this area. There are a number of objective reasons for this, characteristic of the cold war period: the country's self-isolation, strict limits on the contacts of Soviet specialists with their foreign colleagues and publications abroad, excessive and often needless secrecy etc. The language barrier has also been a serious impediment. But as a matter of fact, Soviet advancement in the spread spectrum field between the 1950s and the 1990s was very up-to-date and quite competitive with developments in the USA and Europe. Works by D. E. Vackman, Ya. D. Shirman, M. B. Sverdlick (spread spectrum radar signal design and processing), I. N. Amiantov and L. E. Varakin (spread spectrum communications) were pioneering in many respects and recruited generations of young professionals into this attractive and absorbing research area.

2

Classical reception problems and signal design

It is typical of communication theory to start analysing a system from the receiving end. The aim is usually to design an optimal receiver, which retrieves the information contained in the observed waveform with the best possible quality. Knowing optimal reception processing algorithms depending on a specific transmitted signal structure, it is possible afterwards to design an optimal transmitted signal, i.e. to choose the best means of encoding and modulation. In this chapter we investigate how classical reception problems appeal to the spread spectrum, or, in other words, which of the classical reception problems demand (or not) the involvement of spread spectrum signals. We call reception problems 'classical' if they are based on the traditional Gaussian channel model.

2.1 Gaussian channel, general reception problem and optimal decision rules

The following abstract model can describe any information system in which data is transmitted from one point in space to another. There is some source that can generate one of M possible messages. This source may be governed or at least created by some human being, but it may also have a human-independent nature. In any case, each of the M competitive messages is carried by a specific signal so that there is a set S of M possible signals: $S = \{s_k(t): k = 1, 2, \ldots, M\}$. There is no limitation in principle on the cardinality of S, i.e. the number of signals M, and, if necessary, the set S may even be assumed uncountable. The source selects some specific signal $s_k(t) \in S$ and applies it to the channel input (see Figure 2.1). At the receiving side (channel output) the observation waveform $y(t)$ is received, which is not an accurate copy of the sent signal $s_k(t)$ but, instead, is the result of $s_k(t)$ being corrupted by noise and interference intrinsic to any real channel. For the receiver there are M competitive hypotheses H_k on which one of M possible signals was actually transmitted and turned by the channel into this specific observation $y(t)$, and only one of these hypotheses should be chosen as true. Denote the

Spread Spectrum and CDMA: Principles and Applications Valery P. Ipatov

Figure 2.1 General system model

result of this choice, i.e. the decision, as $\hat{H}_j$, read as 'the decision is made in favour of signal number j'. With this the classical reception problem emerges: what is the best strategy to decide *which one of the possible messages (or signals) was sent, based on the observation* $y(t)$?

To answer this question it is necessary to know the channel model. The channel is mathematically described by its *transition probability* $p[y(t)|s(t)]$, which shows how probable it is for the given input signal to be transformed by the channel into one or another output observation $y(t)$. When the transition probability $p[y(t)|s(t)]$ is known for all possible pairs $s(t)$ and $y(t)$, the channel is characterized exhaustively.

When all source messages are equiprobable (which is typically the case in a properly designed system) the optimum observer's strategy, securing minimum risk of mistaking an actually sent signal for some other, is the *maximum likelihood* (ML) rule. According to this rule, after the waveform $y(t)$ is observed the decision should be made in favour of the signal which has the greatest (as compared to the rest of the signals) probability of being transformed by the channel into *this very* observation $y(t)$.

The primary channel model in communication theory is the additive white Gaussian noise (AWGN) or, more simply, the Gaussian channel in which the transition probability drops exponentially with the growth of the squared *Euclidean distance* between a sent signal and output observation:

$$p[y(t)|s(t)] = k \exp\left(-\frac{1}{N_0} d^2(\mathbf{s}, \mathbf{y})\right) \quad (2.1)$$

where k is a constant independent of $s(t)$ and $y(t)$, N_0 is white noise one-side power spectral density, and the Euclidean distance from $s(t)$ to $y(t)$ is defined as:

$$d(\mathbf{s}, \mathbf{y}) = \sqrt{\int_0^T [y(t) - s(t)]^2 \mathrm{d}t} \quad (2.2)$$

Explanation of the particular importance of the Gaussian model lies in the physical origin of many real noises. According to the central limit theorem of probability theory, the probability distribution of a sum of a great number of elementary random components, which are neither strongly dependent on each other nor prevailing over the others, approaches the Gaussian law whenever the number of addends goes to infinity. But thermal noise and many other types of noise, typical of real channels, are produced precisely as the result of summation of a great many elementary random currents or voltages caused by chaotic motion of charged particles (electrons, ions etc.).

When talking about the distance between signals or waveforms, we interpret them as vectors, which is universally accepted in all information-related disciplines. If the reader finds it difficult to imagine the association between signals and vectors, a very simple mental trick may be a useful aid. Imagine discretization of a continuous signal in time, i.e. representing $s(t)$ by samples $s_i = s(iT_s)$, $i = 0, 1, \ldots$, taken with a sampling period T_s. If the total signal energy is concentrated within the bandwidth W and $T_s \leq 1/2W$ (ignoring that theoretically no signal is finite in both the time and the frequency domains), samples s_i represent exhaustively the original continuous-time signal $s(t)$. With signal duration T there are $n = T/T_s$ such samples altogether, and therefore the n-dimensional vector $\mathbf{s} = (s_0, s_1, \ldots, s_{n-1})$ describes the signal entirely. Having done the same with observation $y(t)$, we come to its n-dimensional vector equivalent $\mathbf{y} = (y_0, y_1, \ldots, y_{n-1})$ and find the Euclidean distance between vectors $\mathbf{s}$ and $\mathbf{y}$ by Pythagorean theorem for the n-dimensional vector space:

$$d(\mathbf{s}, \mathbf{y}) = \sqrt{\sum_{i=0}^{n-1} (y_i - s_i)^2}$$

One possible way of finishing the game is letting T_s go to zero. Then vectors $\mathbf{s}$, $\mathbf{y}$, remaining signal and observation equivalents, become of infinite dimension (actually repeat $s(t)$, $y(t)$ since there is no longer any discretization in the limit). At the same time, the sum above (ignoring the cofactor) turns into the integral in the right-hand side of equality (2.2). The latter, thereby, is the definition of Euclidean distance for continuous time waveforms.

Now come back to the ML rule for the Gaussian channel. According to equations (2.1) and (2.2), signal likelihood (the probability of being transformed by the channel into the observed $y(t)$) falls with Euclidean distance between $s(t)$ and $y(t)$. Therefore, the ML decision in the Gaussian channel can be restated as the *minimum distance* rule:

$$d(\mathbf{s}_j, \mathbf{y}) = \min_k d(\mathbf{s}_k, \mathbf{y}) \Rightarrow \hat{H}_j \text{ is taken} \tag{2.3}$$

i.e. the decision is made in favour of signal $s_j(t)$ if it is closest (in terms of Euclidean distance) to observation $y(t)$ among all M competitive signals (Figure 2.2). Another, more direct, notation of (2.3) is:

$$\hat{\mathbf{s}} = \arg \min_{\mathbf{s} \in S} d(\mathbf{s}, \mathbf{y})$$

where $\hat{\mathbf{s}}$ is an estimation of the received signal (i.e. the signal declared received).

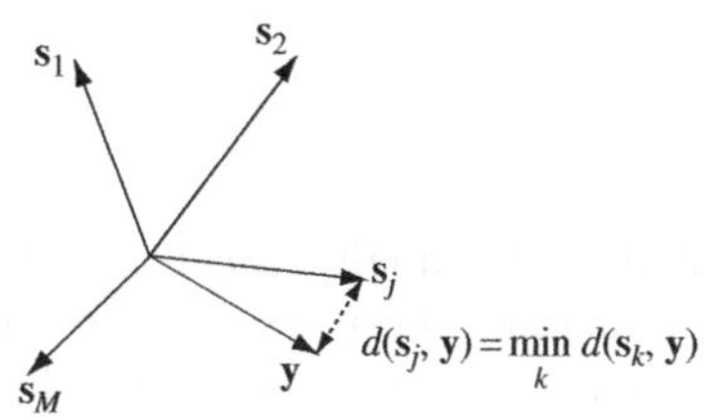

Figure 2.2 Illustration of minimum distance rule

Continuing the geometrical interpretation of signals, we can introduce signal *geometric length* (*norm*) $\|\mathbf{s}\|$ as its distance from the origin. Then from (2.2) it follows that $\|\mathbf{s}\| = d(\mathbf{s}, \mathbf{0}) = \sqrt{E}$, where:

$$E = \int_0^T s^2(t)\mathrm{d}t \tag{2.4}$$

is signal energy. Another important geometrical characteristic is the inner (scalar) product $(\mathbf{u}, \mathbf{v})$ of two signals $u(t)$, $v(t)$:

$$(\mathbf{u}, \mathbf{v}) = \int_0^T u(t)v(t)\mathrm{d}t \tag{2.5}$$

which again can be thought of as a limit form of an inner product of two n-dimensional vectors. The same entity may also be calculated through the lengths of the vectors and the cosine of the angle α between them: $(\mathbf{u}, \mathbf{v}) = \|\mathbf{u}\|\|\mathbf{v}\| \cos\alpha$, and thus the inner product describes the closeness or *resemblance* between signals, since the closer the signals are to each other, with lengths (energies) fixed, the closer to one is $\cos\alpha$ and the greater is the inner product. Because of this the inner product is also called the *correlation* of signals.

In order to outline the special role of this entity, let us now give a slightly different version of the minimum distance rule. Opening the brackets in (2.2) leads to:

$$d^2(\mathbf{s}_k, \mathbf{y}) = \int_0^T y^2(t)\,\mathrm{d}t - 2\int_0^T y(t)s_k(t)\,\mathrm{d}t + \int_0^T s_k^2(t)\,\mathrm{d}t = \|\mathbf{y}\|^2 - 2z_k + \|\mathbf{s}_k\|^2 \tag{2.6}$$

where z_k stands for correlation of observation $y(t)$ with kth signal $s_k(t)$:

$$z_k = (\mathbf{y}, \mathbf{s}_k) = \int_0^T y(t)s_k(t)\,\mathrm{d}t \tag{2.7}$$

The first summand in the right-hand side of equation (2.6) is fixed for a given observation, and therefore does not affect comparing distances and the decision on which signal is received. The last term is just the kth signal energy E_k. With this in mind, distance rule (2.3) can be reformulated as the following *correlation* decision rule:

$$z_j - \frac{E_j}{2} = \max_k\left(z_k - \frac{E_k}{2}\right) \Rightarrow \hat{H}_j \text{ is taken} \tag{2.8}$$

meaning, in particular, that it is maximally correlated with observation $y(t)$ signal, which is announced as having actually been received among all M competitive signals of equal energies. The last case is very well explainable physically: preference is simply given to the signal which has stronger resemblance to $y(t)$ than all the rest, correlation (inner product) being accepted as a criterion of resemblance.

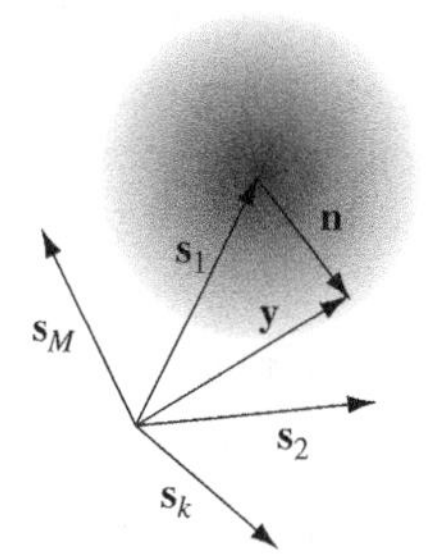

Figure 2.3 Observation scattering and signal design problem

It is interesting to note in passing that these deliberations, although very preliminary, already give a rather clear idea of good signal set design. Look at Figure 2.3, where the signal vectors are depicted. Suppose signal $\mathbf{s}_l$ is transmitted and corrupted by the AWGN channel, which adds to $\mathbf{s}_l$ noise vector $\mathbf{n}$. A Gaussian vector $\mathbf{n}$ has symmetrical (spherical) probability distribution dropping exponentially with the length of the vector $\mathbf{n}$, which is readily seen from (2.1) after removing the signal from it (substituting $s(t) = 0$). Hence, observation vector $\mathbf{y} = \mathbf{s}_l + \mathbf{n}$ proves to be scattered around $\mathbf{s}_l$, as is shown by the figure, and, according to the minimum distance rule (2.3), as soon as $\mathbf{y}$ comes closer to some other signal than to $\mathbf{s}_l$ a wrong decision will happen. To minimize the risk of such an error we should have all the other signals as distant from $\mathbf{s}_l$ as possible. Because any one of M signals may be transmitted equiprobably, i.e. be in place of $\mathbf{s}_l$, it is clear that all distances $d(\mathbf{s}_k, \mathbf{s}_l)$, $1 \leq k < l \leq M$ should be as large as possible. When M is large enough it is not a simple task to maximize all the distances simultaneously, since they can conflict with each other: moving one vector from another may make the first closer to a third one. Due to this, the problem of designing a maximally distant signal set (entering a wide class of so-called *packing* problems) is in many cases rather complicated and has found no general solution so far.

Note that in what preceded all M signals were by default treated as fully *deterministic*, i.e. all their parameters are assumed to be known a priori at the receiving end, the observer being unaware only of which of the competitive M signals is received. This model is adequate to many situations in baseband or coherent bandpass signal reception. However, the general thread, with some adjustments, also retains its validity in more complicated scenarios, such as noncoherent reception (Section 2.5).

Having refreshed these basic ideas of optimal reception, we are now ready to get down to specific problems, putting particular emphasis on aspects of signal design and analysing the potential advantages of spread spectrum—or their absence—in various classical reception scenarios.

2.2 Binary data transmission (deterministic signals)

To demonstrate the strong dependence of reception quality on the distances between signals, let us start with the simplest but very typical communication problem of *binary data transmission*, where one of only $M = 2$ competitive messages is sent over the channel. Practically, this may correspond to the transmission of one data bit in a system where no channel coding is used, or one symbol of binary code in a system with error-correcting code and hard decisions, etc. Numbering the messages 0 and 1 and assuming that signals $s_0(t)$ and $s_1(t)$ (again deterministic!) are used for their transmission, we can represent the minimum distance decision rule (2.3) as:

$$d(\mathbf{s}_0, \mathbf{y}) \underset{\hat{H}_1}{\overset{\hat{H}_0}{\lessgtr}} d(\mathbf{s}_1, \mathbf{y}) \tag{2.9}$$

where placement of the decision symbols points directly to when one or the other of two decisions is made. The same can be rewritten in correlation-based form following from rule (2.8):

$$z = z_0 - z_1 \underset{\hat{H}_1}{\overset{\hat{H}_0}{\gtrless}} \frac{E_0 - E_1}{2}, \tag{2.10}$$

with correlations $z_k, k = 0, 1$ of each signal and observation $y(t)$ defined by equation (2.7) and $E_k = \|\mathbf{s}_k\|^2, k = 0, 1$, being the kth signal energy given by (2.4). Optimal rules (2.9) and (2.10) of distinguishing between two signals can be explained geometrically in a very clear way. Two signal vectors $\mathbf{s}_0$ and $\mathbf{s}_1$ always lie in a signal plane SP. Observation vector $\mathbf{y}$ does not necessarily fall onto this plane but the closeness of it to one or the other signal is determined by the closeness to them of the projection $\mathbf{y}'$ of $\mathbf{y}$ onto SP (see Figure 2.4a). Therefore, we can divide SP into two half-planes by the straight-line bound passing strictly perpendicular to the straight line connecting the signal vectors, and base decisions $\hat{H}_0, \hat{H}_1$ on $\mathbf{y}'$ hitting the corresponding half-plane (Figure 2.4b). It is also seen from Figure 2.4b that the probability of mistaking one signal for the other (error probability) depends on the distance between vectors $\mathbf{s}_0$ and $\mathbf{s}_1$ in comparison with the range of random fluctuations of $\mathbf{y}'$ caused by channel noise. According to (2.10), the actually received signal $s_0(t)$ will be erroneously taken for the wrong one $s_1(t)$ if and only if the correlation difference is lower than the threshold $(E_0 - E_1)/2$. Therefore, the probability p_{01} of such an error is found as:

$$p_{01} = \Pr\left(z < \frac{E_0 - E_1}{2} | s_0(t)\right) = \int_{-\infty}^{\frac{E_0 - E_1}{2}} W(z|s_0(t))\, dz \tag{2.11}$$

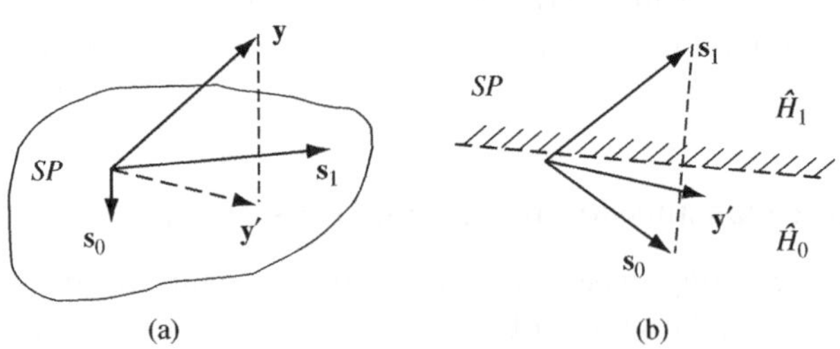

Figure 2.4 Signal plane and decision half-planes

where $\Pr(A|B)$ stands for the conditional probability of event A given that event B occurred, and $W(z|s_0(t))$ is the conditional probability density function (PDF) of correlation difference z in (2.10), given that the signal $s_0(t)$ is actually received. One of the remarkable features of the Gaussian process is that any linear transform of it again produces a Gaussian process. Therefore z, as a result of a linear transform of the Gaussian observation $y(t)$ (see (2.7) and (2.10)), has the Gaussian PDF:

$$W(z|s_0(t)) = \frac{1}{\sqrt{2\pi}\sigma} \exp\left[-\frac{(z-\bar{z})^2}{2\sigma^2}\right]$$

integration of which according to (2.11) results in:

$$p_{01} = Q\left(\frac{2\bar{z} - E_0 + E_1}{2\sigma}\right) \tag{2.12}$$

where

$$Q(x) = \frac{1}{\sqrt{2\pi}} \int_x^\infty \exp\left(-\frac{t^2}{2}\right) \mathrm{d}t$$

is the *complementary error function.*

The expectation of $\bar{z}$ conditioned in the received signal (the bar above will be used from now on to symbolize expectation) and variance $\sigma^2 = \operatorname{var}\{z\}$ of z can be found directly from equations (2.7) and (2.10). When the signal $s_0(t)$ is assumed true, i.e. $\overline{y(t)} = s_0(t)$, expectation of z:

$$\bar{z} = \int_0^T \overline{y(t)}[s_0(t) - s_1(t)]\, \mathrm{d}t = E_0 - \rho_{01}\sqrt{E_0 E_1} \tag{2.13}$$

where:

$$\rho_{kl} = \frac{(\mathbf{s}_k, \mathbf{s}_l)}{\|\mathbf{s}_k\|\|\mathbf{s}_l\|} = \frac{1}{\sqrt{E_k E_l}} \int_0^T s_k(t) s_l(t) \mathrm{d}t \tag{2.14}$$

is called the *correlation coefficient* of the signals $s_k(t), s_l(t), E_k, E_l$ being their energies. As is seen, geometrically ρ_{01} is simply the cosine of the angle between the signals $s_0(t), s_1(t)$ (or the signal vectors $\mathbf{s}_0, \mathbf{s}_1$), and hence characterizes closeness or *resemblance* of the signals.

To find $\sigma^2 = \operatorname{var}\{z\}$ we rely on the fact that it is not affected by a deterministic component of the observation $y(t)$, i.e. in the situation in question the signal $s_0(t)$, since the noise is additive. Therefore we can virtually remove the signal from $y(t)$, putting

$y(t) = n(t)$, where $n(t)$ is white noise with two-sided power spectral density $N_0/2$. After this let us calculate the variance of correlation (2.7) of $y(t)$ and some arbitrary signal $s(t)$:

$$\sigma^2 = \mathrm{var}\{z\} = \overline{\left\{\int_0^T n(t)s(t)\,\mathrm{d}t\right\}^2} = \int_0^T\int_0^T \overline{n(t)n(t')}s(t)s(t')\,\mathrm{d}t\,\mathrm{d}t'$$

where the squared integral is presented as a double integral with separable variables, order of integration and averaging is changed (expectation of sum is sum of expectations!) and, finally, averaging is applied to the only random cofactor in the integrand.

Recall now, that due to uniformity of the spectrum of white noise over the entire frequency range, its *autocorrelation function* (statistical average of product of samples at two time moments) is the Dirac delta function: $\overline{n(t)n(t')} = (N_0/2)\delta(t - t')$. In other words, any two samples of white noise, notwithstanding how close in time, are uncorrelated. Using this result in the integral above along with the *sifting* property of the delta function:

$$\int_0^T s(t')\delta(t' - t)\,\mathrm{d}t' = s(t)$$

leads to:

$$\sigma^2 = \frac{N_0}{2}\int_0^T s^2(t)\,\mathrm{d}t = \frac{N_0 E}{2} \tag{2.15}$$

where E is the energy of the signal $s(t)$.

In the case under consideration, as (2.10) and (2.7) show, substitution $s(t) = s_0(t) - s_1(t)$ should be made in (2.15), i.e. E is energy E_d of the signal difference $s_0(t) - s_1(t)$. Deriving it gives:

$$E_d = \int_0^T [s_0(t) - s_1(t)]^2\,\mathrm{d}t = d^2(\mathbf{s}_0, \mathbf{s}_1) = E_0 + E_1 - 2\rho_{01}\sqrt{E_0 E_1} \tag{2.16}$$

Taking into account the geometrical content of the correlation coefficient and energy, this is just the cosine theorem from 'school' mathematics.

Now, substitute (2.13), (2.15) and (2.16) into (2.12), arriving at:

$$p_{01} = Q\left(\sqrt{\frac{d^2(\mathbf{s}_0, \mathbf{s}_1)}{2N_0}}\right) \tag{2.17}$$

Since the problem is absolutely symmetrical, the same result will be obtained for the probability of mistaking $s_1(t)$ for $s_0(t)$. With this in mind, the complete (unconditional) error probability P_e does not depend on a priori probability w of sending the signal $s_0(t)$ and is expressed by the equation:

$$P_e = wp_{01} + (1-w)p_{10} = Q\left(\sqrt{\frac{d^2(\mathbf{s}_0, \mathbf{s}_1)}{2N_0}}\right) \tag{2.18}$$

It is quite obvious now that the only way to achieve high data transmission fidelity is to make the distance between the two signals $d(\mathbf{s}_0, \mathbf{s}_1)$ large enough. Certainly, $d(\mathbf{s}_0, \mathbf{s}_1)$ can be increased at the cost of large signal energies or vector lengths, as is seen from (2.16). But what is the *optimal* signal pair when recourse to this 'brute force' approach is limited, i.e. signal energies are fixed beforehand? Consider first the very typical case of signals of equal energies: $E_0 = E_1 = E$, meaning that signal intensity is not utilized as a message indicator. Decision rule (2.10) is now reduced to just comparison between z_0 and z_1 or, equivalently, to testing the polarity of their difference:

$$z = z_0 - z_1 \underset{\hat{H}_1}{\overset{\hat{H}_0}{\gtrless}} 0$$

Of course, to maximize the distance between two vectors of fixed lengths, one should make them *antipodal*, as shown in Figure 2.5a. Then the angle between $\mathbf{s}_0, \mathbf{s}_1$ $\alpha = \pi$, $\cos\alpha = \rho_{01} = -1$ and $d(\mathbf{s}_0, \mathbf{s}_1) = 2\sqrt{E}$, which turns (2.18) into the following:

$$P_{e,a} = Q\left(\sqrt{\frac{2E}{N_0}}\right) \tag{2.19}$$

representing the minimal achievable error probability in binary data transmission, signal energy E being fixed. The *correlator* and *matched filter*, which will often be referred to in this book, are devices typically used to physically calculate correlation z, and parameter $q = \sqrt{2E/N_0}$ is nothing but *signal-to-noise ratio* (SNR) at the correlator or matched filter output.

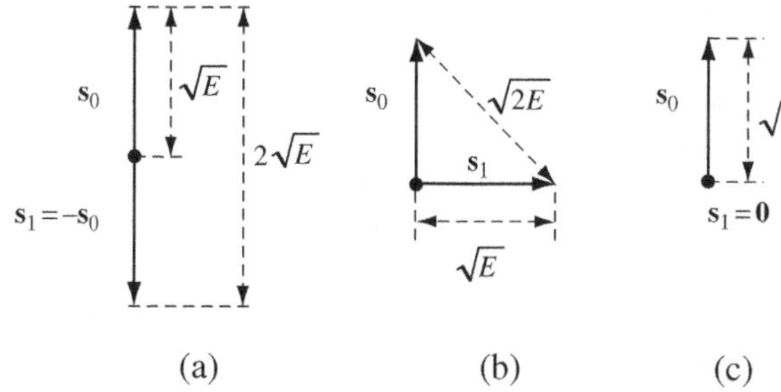

Figure 2.5 Signal pairs in various binary transmission modes

The optimal signal pair is therefore the antipodal pair $s_1(t) = -s_0(t)$. *Binary phase shift keying* (BPSK) is the implementation that is widespread in digital data transmission systems, in which 0 is transmitted by some bandpass signal with phase zero while the same signal but with phase π serves to transmit message 1.

To find out how critical adequate choice of signals may prove to be, compare BPSK with another popular binary transmission mode. Although BPSK is the best possible method of binary signalling, it is based on a phase discrepancy of two signals carrying data, and thus requires accurate knowledge of the frequency carrier current phase at the receiver end. To realize it, a special frequency recovery loop should be used in the receiver, which is sometimes regarded as an undesirable complication. One way to avoid it is to employ *frequency shift keying* (FSK), in which messages 0 and 1 are carried by signals of different frequencies. Typically frequencies are chosen so that the signals are *orthogonal*: $\cos\alpha = \rho_{01} = 0$, $d(\mathbf{s}_0, \mathbf{s}_1) = \sqrt{2E}$ (see Figure 2.5b). Substitution of it into (2.18) gives:

$$P_{e,o} = Q\left(\sqrt{\frac{E}{N_0}}\right) \tag{2.20}$$

Comparing this result with (2.19), one can see that for an orthogonal pair (FSK) twice the signal energy is needed to provide the same error reception fidelity as is secured by the antipodal signals (BPSK). To put it another way, the orthogonal signals rank 3 dB worse than the antipodal ones in necessary energy.

There is one more very old mode of binary transmission still in use: *amplitude shift keying* (ASK), in which the bit '1' is transmitted by the signal $s_1(t) = s(t)$ of energy $E_1 = E$, and '0' is transmitted by a pause: $s_0(t) = 0, E_0 = 0$. In this case (see Figure 2.5c) $d(\mathbf{s}_0, \mathbf{s}_1) = \sqrt{E}$ and the error probability (2.18) becomes:

$$P_{e,as} = Q\left(\sqrt{\frac{E}{2N_0}}\right) \tag{2.21}$$

Comparing the last result with (2.19), one may conclude that ASK requires four times (6 dB) higher energy than BPSK for the same reception quality. This is true when peak energy is a limiting issue. More practical is usually the limitation on average energy. Since in ASK no energy is spent at all when '0' is transmitted, for equiprobable messages 0 and 1 the average energy is $(E_0 + E_1)/2 = E/2$. Thus, average energy only two times higher than BPSK will give the same error probability with the ASK mode, and loss of ASK to BPSK is the same as in the case of FSK, i.e. 3 dB.

Now we have to make a final remark on the potential impact of the binary data transmission problem upon signal pair design. We should conclude that there is *no hint of any special benefits of spread spectrum* in this case, since widening signal bandwidth versus its minimum $1/T$ promises no improvements in error probability. Indeed, to provide the necessary reception quality it is enough to have two maximally distant signals, which automatically involves two antipodal signals with no additional demand as to their shape or modulation. If for some reason an antipodal pair is rejected, orthogonal (say, frequency shifted) or ASK pairs can be used and, again, this in no way calls for the use of spread spectrum.

2.3 *M*-ary data transmission: deterministic signals

In the case of $M > 2$, the probability $p_{1,e}$ of mistaking the actually received signal $s_1(t)$ for one of $M-1$ wrong signals $s_l(t), l = 2, 3, \ldots, M$ in accordance with rules (2.3) and (2.8) is:

$$p_{1,e} = \Pr\left(d^2(\mathbf{s}_1, \mathbf{y}) \neq \min_k d^2(\mathbf{s}_k, \mathbf{y}) | s_1(t)\right) = 1 - \Pr\left(z_1 - \frac{E_1}{2} = \max_k\left(z_k - \frac{E_k}{2}\right) | s_1(t)\right)$$

Accurate evaluation of this probability consists in integration of the conditional joint PDF of all M correlations given that $s_1(t)$ is received over the whole area, where $z_1 \geq z_k - (E_k - E_1)/2$ for all $k = 1, 2, \ldots, M$. This M-fold integral in the general case, i.e. with no special assumption about signal set properties, can not be simplified in any way. However, a very productive and straightforward upper border for $p_{1,e}$ can be derived using the *union* bound. Let the event A_l mean that observation $y(t)$ is closer to some wrong signal $s_l(t)$ with a specific number l from the range [2, M] than to $s_1(t)$. Then the confusion of $s_1(t)$ with some other signal will obviously be a union of all A_l. Let us recollect now that according to the union bound, the probability of a union of events is never greater than the sum of their probabilities:

$$p_{1,e} = \Pr(A_2 \cup A_3 \cup \ldots \cup A_M) \leq \sum_{l=2}^{M} \Pr(A_l)$$

On the other hand, $\Pr(A_l)$, following definition of A_l, is exactly the probability of confusion between only two signals, $s_1(t)$ and $s_l(t)$. This probability is determined by (2.17) after a proper change of the signal numbers:

$$\Pr(A_l) = p_{1l} = Q\left(\sqrt{\frac{d^2(\mathbf{s}_1, \mathbf{s}_l)}{2N_0}}\right)$$

Substituting this into the previous inequality leads to the desired estimate:

$$p_{1,e} \leq \sum_{l=2}^{M} Q\left(\sqrt{\frac{d^2(\mathbf{s}_1, \mathbf{s}_l)}{2N_0}}\right)$$

A similar result (with necessary substitutions of new numbers) will be valid under the assumption that signal $s_k(t)$ is actually received instead of $s_1(t)$, so that with a priori equiprobable M signals the final upper union bound on the complete (unconditional) error probability takes the form:

$$P_e = \frac{1}{M}\sum_{k=1}^{M} p_{k,e} \leq \frac{1}{M}\sum_{k=1}^{M}\sum_{\substack{l=1 \\ l \neq k}}^{M} Q\left(\sqrt{\frac{d^2(\mathbf{s}_k, \mathbf{s}_l)}{2N_0}}\right) \quad (2.22)$$

The first noteworthy fact about (2.22) seen directly is that it becomes a precise equality when $M = 2$. Another observation is linked to its asymptotic behaviour with growth of SNR. As a matter of fact, the complementary error function $Q(x)$ drops approximately as $\exp(-x^2/2)$ when x is sufficiently large, and even a small increment of large x may reduce $Q(x)$ to a negligible level in comparison to its initial value. Due to this, when SNR is large enough only the closest signal pairs may contribute perceptibly to the sum in (2.22), and if $d_{\min}$ is the minimum distance over all signal pairs occurring $n_{\min}$ times among them, estimation (2.22) transforms asymptotically into the following:

$$P_e \sim \frac{n_{\min}}{M} Q\left(\sqrt{\frac{d_{\min}^2}{2N_0}}\right) \qquad \frac{d_{\min}^2}{2N_0} \gg 1 \tag{2.23}$$

Approximation (2.23) points, above all, to an asymptotic convergence of the union bound to the genuine value of the error probability, SNR increasing. To explain it physically, return to Figure 2.3 and note that when the noise level is very low only the signal vectors which are nearest to the true one are at risk of being erroneously mistaken for the latter. This means that asymptotically only signal pairs with distance $d_{\min}$ determine the true error probability P_e itself (not only its upper bound!), which entails closeness between P_e and its union bound.

Result (2.23) underlies one of the possible and most important formulations of the signal design problem: *maximization of minimum distance* between M signals. As was already mentioned in Section 2.1, such a task is geometrically equivalent to packing M vectors in such a manner that the closest pairs of them have maximal achievable distance: $d_{\min} = \max$. Various limitations can be imposed on a signal (vector) constellation. First of all, some energy constraint should be prescribed, allowing for practical power/energy limits. If only the average energy of signals is fixed $\overline{E} = \sum_{k=1}^{M} E_k/M = const$, then signal vectors can have different lengths and their selection procedure may be called *volume packing*. Very frequently, however, it is required that no energy be involved in mapping messages onto signals, i.e. that all energies be the same $E_k = E = const, k = 1, 2, \ldots, M$. In this case all signal vectors have equal lengths, i.e. lie on the sphere surface, hence the name *spherical packing*.

The other typical limitation in signal design is the dimension n_s of signal space, inside which signal vectors are packed. The physical content of this constraint is again associated with a very practical limit on the bandwidth resource. To explain the interconnection between them, consider first the case of baseband signals and suppose that the total (two-sided) bandwidth and time interval which can be allocated to all M signals together are limited to W_t and T_t, respectively. The first of these restrictions allows for bandwidth saving, while the second reflects the desire to transmit necessary data during an acceptable time period, i.e. with acceptable transmission rate $R = \log M/T_t$. Then, according to the sampling theorem, only about W_tT_t independent samples are at our disposal to synthesize M signals, each signal being thereby treated as a vector in the space of dimension $n_s = W_tT_t$. Some caution in estimation of the number of independent samples is caused by the impossibility of the energy of any signal being concentrated within finite intervals of both the time and frequency domains simultaneously. But in the first-approximation estimates this theoretical fact can be ignored.

To cover also the case of bandpass signals, let us turn to a general model of such a signal:

$$s(t) = S(t)\cos[2\pi f_0 t + \gamma(t)] \tag{2.24}$$

in which $S(t)$ is the real envelope (amplitude modulation law), $\gamma(t)$ is the phase modulation law and f_0 is the carrier frequency. Using trigonometry identity for the cosine of the sum of angles, we can represent this equation as:

$$s(t) = S_I(t)\cos 2\pi f_0 t - S_Q(t)\sin 2\pi f_0 t \tag{2.25}$$

where $S_I(t) = S(t)\cos\gamma(t)$ and $S_Q(t) = S(t)\sin\gamma(t)$ are the signal *quadrature components*. Since both $S(t)$ and $\gamma(t)$ are baseband waveforms, the same is true as for $S_I(t), S_Q(t)$, meaning that, given the carrier frequency, every bandpass signal is exhaustively characterized by two independent baseband quadrature components. Therefore, twice the number of independent coordinates (samples) can be used to design bandpass signals compared to the case of baseband signals with the same total time frequency product and $n_s = 2W_t T_t$.

Now the general problem of signal set design can be formulated as follows: find the constellation of M points or vectors in space of a given dimension n_s satisfying energy limitations and having maximal possible minimum distance between points $d_{\min} = \max$. This can also be restated in a dual form: find the constellation of M points in space of a given dimension n_s with pre-assigned minimum distance $d_{\min}$ minimizing energy expenditure in terms of either average energy $\overline{E} = \min$ (volume packing) or the same energy of all signals $E = \min$ (spherical packing).

The simplest version of this problem ($n_s = 1$) corresponds to ASK (the simplest version of which with $M = 2$—a binary one—was touched upon in Section 2.2). Another name for ASK is pulse amplitude modulation (PAM). In this case all signal points lie on the same straight line and with $M > 2$ only the 'volume' packing is tractable. It is not hard to see that the optimal constellation minimizing average energy, $d_{\min}$ pre-assigned, is uniform and symmetrical with space between neighbouring signal points exactly equal to $d_{\min}$ (see Figure 2.6a for the example $M = 4$).

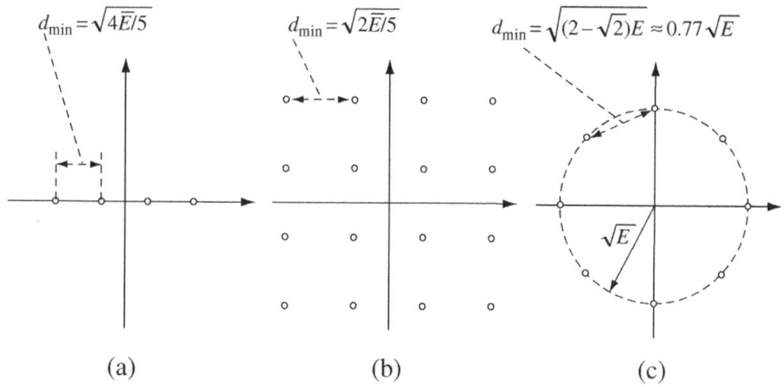

Figure 2.6 One- and two-dimensional constellations: 4-ASK (a), 16-QAM (b) and 8-PSK (c)

When $n_s = 2$, finding the optimal constellation with volume packing becomes more difficult and may even lead to asymmetrical patterns, while spherical packing is trivial and is performed by a uniform placing of M points on the circle of radius $\sqrt{E}$. Widely practised in modern digital communications, M-ary *quadrature amplitude modulation* (QAM) gives an example of symmetrical volume-packed two-dimensional constellations which are not necessarily theoretically optimal but convenient from a hardware implementation point of view (Figure 2.6b, $M = 16$). On the other hand conventional M-ary *phase shift keying* (PSK) constellations have uniformly spaced points on the circle and are optimal in terms of spherical packing (Figure 2.6c, $M = 8$).

The problem of optimal packing in spaces of higher dimension $n_s > 2$ is very complex and has no general mathematical solution so far. Many useful particular results are scattered over the range of books and papers (see, e.g. the bibliography in [11] and web site [12]).

Let us now try to find an upper limit on the minimum distance in the loosest statement imposing no preliminary binding on the signal space dimension n_s, and estimate the minimal value of n_s, which allows this limit to be achieved. Restricting our attention to the spherical packing ($E_k = E, k = 1, 2, \ldots, M$), calculate the sum of all M^2 squared distances, including trivial ones (from any signal to itself). Cosine theorem (2.16) gives:

$$\sum_{k,l=1}^{M} d^2(\mathbf{s}_k, \mathbf{s}_l) = 2M^2E - 2E\sum_{k,l=1}^{M} \rho_{kl} \tag{2.26}$$

with ρ_{kl} being the correlation coefficient of the kth and lth signals. To estimate the sum of all correlation coefficients use definition (2.14) for ρ_{kl}, change the order of integration and summation, and note that the double sum in the integrand has separable summation indexes k and l, which transforms it into a product of two identical sums:

$$E\sum_{k,l=1}^{M} \rho_{kl} = \int_0^T \left(\sum_{k,l=1}^{M} s_k(t)s_l(t)\right) \mathrm{d}t = \int_0^T \left(\sum_{k=1}^{M} s_k(t)\right)^2 \mathrm{d}t$$

Since the integral of a square is never negative, it follows from (2.26):

$$\sum_{k,l=1}^{M} d^2(\mathbf{s}_k, \mathbf{s}_l) \leq 2M^2E$$

At the same time, the sum above is no smaller than $M(M-1)d^2_{\min}$. Combining this inequality with the preceding one results in the upper border on the minimum distance:

$$d^2_{\min} \leq \frac{2M}{M-1}E \tag{2.27}$$

If M signals achieving this upper bound existed they would be quite fairly called optimal in terms of the minimum distance criterion. To show that they do exist, take M vectors $\mathbf{u}_k, k = 1, 2, \ldots, M$ having zero pairwise inner products and unit lengths: $(\mathbf{u}_k, \mathbf{u}_l) = \delta_{kl}, k, l = 1, 2, \ldots, M$, where $\delta_{kl} = 0, k \neq l$; $\delta_{kk} = 1$ is the Kronecker delta function. Such vectors, called *orthonormal*, exist in any vector space whose dimension is no smaller than M. Now form M new vectors $\mathbf{v}_k, k = 1, 2, \ldots, M$, subtracting from each of $\mathbf{u}_k$ the sum $\mathbf{u} = \sum_{k=1}^{M} \mathbf{u}_k$ weighted by a coefficient $1/M$: $\mathbf{v}_k = \mathbf{u}_k - \mathbf{u}/M$. Calculate the inner product of $\mathbf{v}_k$ and $\mathbf{v}_l$. Due to its linearity:

$$(\mathbf{v}_k, \mathbf{v}_l) = (\mathbf{u}_k, \mathbf{u}_l) - \frac{1}{M}(\mathbf{u}_k, \mathbf{u}) - \frac{1}{M}(\mathbf{u}, \mathbf{u}_l) + \frac{1}{M^2}(\mathbf{u}, \mathbf{u}) = \delta_{kl} - \frac{1}{M} \tag{2.28}$$

where use is made of the orthonormality of vectors $\mathbf{u}_k$. Let us change the lengths of vectors $\mathbf{v}_k$, multiplying them by $\sqrt{ME/(M-1)}$, and take the resulting vectors as signal ones:

$$\mathbf{s}_k = \sqrt{\frac{ME}{M-1}}\mathbf{v}_k, \; k = 1, 2, \ldots, M \tag{2.29}$$

Then the squared distance between two signals according to cosine theorem (2.16) and equation (2.29) is:

$$\begin{aligned} d^2(\mathbf{s}_k, \mathbf{s}_l) &= \frac{ME}{M-1} d^2(\mathbf{v}_k, \mathbf{v}_l) \\ &= \frac{ME}{M-1}\left[\|\mathbf{v}_k\|^2 + \|\mathbf{v}_l\|^2 - 2(\mathbf{v}_k, \mathbf{v}_l)\right] = \frac{2ME}{M-1}, \; k \neq l \end{aligned} \tag{2.30}$$

which coincides with the right-hand side of (2.27). Hence, signals lying on the bound (2.27) really exist. More than this, the distances between any two of them are the same, i.e. these signals fall into the category of *equidistant* ones. They are widely known under the special name of *simplex* signals. Directly from their definition it follows that:

$$\sum_{k=1}^{M} \mathbf{s}_k = \sqrt{ME/(M-1)} \sum_{k=1}^{M} \mathbf{v}_k = \sqrt{ME/(M-1)}(\mathbf{u} - \mathbf{u}) = \mathbf{0}$$

meaning that simplex signals are linearly dependent, unlike initial orthonormal vectors $\mathbf{u}_k$. It is easily verified that the dimension $n_s = M - 1$, i.e. smaller by one than the number of signals, is necessary and sufficient for constructing M simplex signals.

The property of equidistance of simplex signals also entails equality of correlation coefficients ρ_{kl} for any pair. Evaluation of ρ_{kl} with the help of (2.14), (2.29) and (2.28) results in:

$$\rho_{kl} = \frac{(\mathbf{v}_k, \mathbf{v}_l)}{\|\mathbf{v}_k\|\|\mathbf{v}_l\|} = -\frac{1}{M-1}, \; k \neq l, \; k, l = 1, 2, \ldots, M$$

showing that the angles between any two simplex signals are the same and greater than $\pi/2$. For the simplest sets of $M = 2, 3, 4$ simplex signals (see Figure 2.7), the values of the correlation coefficient equal -1 (antipodal signals), $-1/2$ and $-1/3$, respectively, which in turn correspond to angles 180°, 120° and approximately 110°. When $M = 4$, simplex vectors form the simplest regular polyhedron (tetrahedron), which explains the name of the signals: *simplex* is Latin for 'simple'.

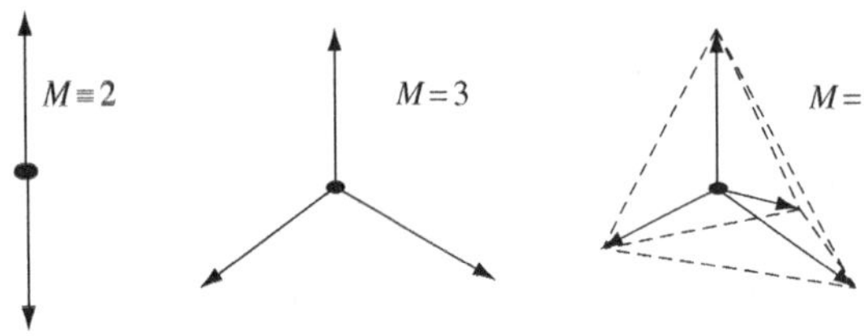

Figure 2.7 Examples of simplex signals

For any equidistant signal set $d(\mathbf{s}_k, \mathbf{s}_l) = d_{\min}$ for all pairs of distinct vectors so that in (2.23) $n_{\min} = M(M-1)$, and this is also the number of summands in (2.22). Substitution of this along with (2.30) in equation (2.23) gives an approximation of the asymptotic error probability achievable with simplex signals, which according to (2.22) is at the same time the upper bound of the error probability:

$$P_{e,\min} \leq (M-1)Q\left(\sqrt{\frac{ME}{(M-1)N_0}}\right) \tag{2.31}$$

Since simplex signals are optimal as for minimum distance, the right-hand side of the latter expression presents simultaneously the minimum possible asymptotic error probability for M signals of fixed and equal energies E.

The *orthogonal* signals, which are another example of equidistant signals, are practically as effective as the simplex ones when the number of signals M is sufficiently large. Indeed, the correlation coefficient of orthogonal signals is zero and the distance between any two of them $d(\mathbf{s}_k, \mathbf{s}_l) = d_{\min} = \sqrt{2E}$. This, used in (2.23), produces an asymptotic error probability for M orthogonal signals, which again borders the exact error probability from above:

$$P_{e,ort} \leq (M-1)Q\left(\sqrt{\frac{E}{N_0}}\right) \tag{2.32}$$

Comparing (2.32) and (2.31) shows that to equalize the error probabilities in both cases, orthogonal signals should be of $M/(M-1)$ times higher energy than simplex signals, i.e. energy loss γ of the first to the second ones is defined as $\gamma = M/(M-1)$. When $M \gg 1$ this loss is negligible and orthogonal signals can be considered optimal; e.g. for $M = 64$ $\gamma = 64/63$, which corresponds to an increase of energy of orthogonal signals against simplex ones by less than 0.07 dB (or 2%). This discrepancy is certainly of no practical significance, and whenever M is large enough orthogonal and simplex signals can be used interchangeably depending on implementation or other reasons.

Talking about M-ary *orthogonal signalling* (in the literature the terms *orthogonal modulation* and *orthogonal coding* are also used), let us remember that the maximal number of orthogonal signals is exactly equal to the signal space dimension: $M = n_s$. Therefore, within the fixed total bandwidth W_t and duration T_t, up to W_tT_t baseband or $2W_tT_t$ bandpass orthogonal signals can be accommodated. Additional physical reasoning for a doubling of the number of orthogonal bandpass signals against baseband ones follows directly from equations (2.24) and (2.25): building up n_s orthogonal signals of the form (2.24), we can add to them n_s more, obtained by just shifting the carrier frequency phase by angle $\pi/2$. This possibility is practical only when all signals are deterministic or *coherent*, which means that their carrier phases are controllable and can actually be used for message identification. In reality, however, this may often not be the case because either a transmitter itself or a channel may destroy the coherence of signals in such a manner that their phases become random and as a consequence cannot be used for distinguishing messages. This case is addressed in Section 2.5.

2.4 Complex envelope of a bandpass signal

Before extending our discussion to the more complicated models of the M-ary transmission, it is reasonable to diverge from the main line in order to recollect some more facts about handling bandpass signals.

Let us begin with the observation that the real envelope $S(t)$ in equation (2.24) is fictitious, i.e. is just a suitable artificial instrument, whereas only the signal $s(t)$ itself is an observable physical reality. More than this, equation (2.24) does not give any unique definition of the envelope of $s(t)$. In fact, it follows from (2.24) that one may take an arbitrary 'phase modulation' law $\gamma(t)$, and then 'envelope' $S(t) = s(t)/\cos[2\pi f_0 t + \gamma(t)]$ will produce a given signal $s(t)$. Therefore, some special agreement is needed on how to interpret the notion of an envelope or amplitude modulation $S(t)$.

A universally adopted basis for determining the envelope is the Hilbert transform. By its physical content, the Hilbert transform is just filtering which rotates the phases of all harmonic components independently of frequency through the same angle, $-\pi/2$, and does not change the amplitudes of the harmonics. In the frequency domain such a transform simply means multiplication of the signal spectrum by $-j\pi/2$ for positive frequencies and by $j\pi/2$ for negative ones, and, therefore, the transfer function of a Hilbert filter is $\tilde{h}_g(f) = -j(\pi/2)\text{sign}f$, where $\text{sign}x = 1, x \geq 0$; $\text{sign}x = -1, x < 0$. Straightforward calculation of the inverse Fourier transform of this leads to a filter pulse response $h_g(t) = 1/\pi t$. Hence, in the time domain the Hilbert transform $s_\perp(t)$ of signal $s(t)$ may be presented via the convolution integral:

$$s_\perp(t) = \frac{1}{\pi}\int_0^\infty \frac{s(\theta)}{t-\theta}\,d\theta \tag{2.33}$$

Using the definition of the Hilbert transform and the Parseval theorem, the reader may easily prove the following relations:

$$s(t) = -\frac{1}{\pi}\int_0^\infty \frac{s_\perp(\theta)}{t-\theta}\,d\theta$$

which is nothing but the inverse Hilbert transform, and:

$$(\mathbf{u}, \mathbf{v}) = (\mathbf{u}_\perp, \mathbf{v}_\perp) \qquad (\mathbf{u}, \mathbf{v}_\perp) = -(\mathbf{u}_\perp, \mathbf{v}) \tag{2.34}$$

The first equation in (2.34) shows that the Hilbert transform preserves the inner product of signals $u(t), v(t)$, while the second establishes the interrelation between the inner products of one of the signals and the Hilbert transform of the other.

Returning now to the issue of the signal envelope definition, we put:

$$S(t) = \sqrt{s^2(t) + s_{\perp}^2(t)} \tag{2.35}$$

At first glance, this definition of the envelope looks somewhat artificial; however, a deeper insight uncovers its complete naturalness. Indeed, how would we calculate the unknown constant amplitude A of the unmodulated continuous wave (CW) i.e. the observed signal $u(t) = A\cos(2\pi f_0 t + \gamma)$? One way is to take the signal itself and its copy $v(t)$ rotated through the angle $-\pi/2$ and then make use of Pythagorean theorem: $A = \sqrt{u^2(t) + v^2(t)}$. But it is seen at once that for the unmodulated signal $u(t)$ its phase-shifted copy $v(t)$ is nothing but the Hilbert transform: $v(t) = u_{\perp}(t)$. Thus, we have a result absolutely consistent with (2.35). Now take a modulated signal $s(t)$. Its envelope $S(t)$ is just instant amplitude at time moment t. For a bandpass signal it changes slowly as compared to the CW $\cos 2\pi f_0 t$, and we can treat $s(t)$ within a sufficiently small time interval around the moment t as though it is the unmodulated harmonic with amplitude $S(t)$. Then how do we find this amplitude $S(t)$? Exactly as it is done for the unmodulated signal, i.e. by $-\pi/2$ phase shifting (Hilbert transform) and application of Pythagorean theorem (2.35). Figure 2.8 illustrates this. Thus the problem of unambiguous understanding of a bandpass signal envelope is solved and definition (2.35) may be used universally.

Analysing bandpass signals becomes much easier with the introduction of one more very convenient tool—the *complex envelope* $\dot{S}(t)$, which is a complex-valued function of time defined immediately by equation (2.24) or (2.25) once the definition of the real envelope is specified:

$$\dot{S}(t) = S_I(t) + jS_Q(t) = S(t)[\cos\gamma(t) + j\sin\gamma(t)] = S(t)\exp[j\gamma(t)] \tag{2.36}$$

where $j = \sqrt{-1}$ and the Euler formula is used. As is seen, the complex envelope integrates in itself both amplitude and angle modulation of the signal. If several signals are considered, given the common frequency carrier, their distinction consists only in modulation laws, and hence complex envelopes give an exhaustive description of them.

Certainly, a complex envelope, along with a real one, is just a suitable mathematical fiction and the 'true' signal (2.24) is expressed in terms of the complex envelope as:

$$s(t) = \mathrm{Re}[\dot{S}(t)\exp(j2\pi f_0 t)] \tag{2.37}$$

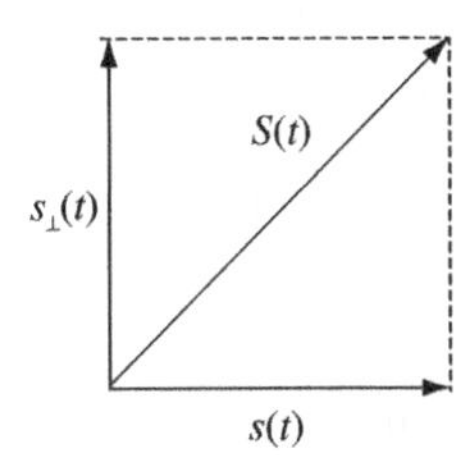

Figure 2.8 The definition of envelope

where Re stands for taking the real part of the complex entity, and the second cofactor in the square brackets is a complex notation of the CW of a carrier frequency f_0 through the Euler formula. Turning again to Figure 2.8, we can see that with $s(t)$ treated according to (2.37) as a real part of the complex signal $\dot{S}(t)\exp(j2\pi f_0 t)$, the imaginary part of the latter is the Hilbert transform of $s(t)$:

$$s_\perp(t) = \text{Im}[\dot{S}(t)\exp(j2\pi f_0 t)]$$

This leads to one more complex substitute of the real signal, called the *analytic signal*:

$$\dot{s}(t) = \dot{S}(t)\exp(j2\pi f_0 t) = s(t) + js_\perp(t) \tag{2.38}$$

Formally, the analytic signal uses complex notation to advance factorization model (2.24) of a bandpass signal so that the first factor covers all modulation (not only amplitude modulation) and the second is responsible for only the unmodulated CW of a carrier frequency f_0.

Using the basic rule of spectral analysis, it may be easily proved that the spectrum of the complex envelope of a bandpass signal (2.37) is located around zero frequency. Therefore—since, given the carrier frequency, the signal is entirely presented by its complex envelope—the latter is a baseband equivalent of a bandpass signal, simplifying the analytical and computational work by getting rid of the carrier frequency dependence.

In what follows we will need a generalized version of the inner product (2.5), which is applicable not only to real signals $u(t)$, $v(t)$ but also to their complex substitutes—analytic signals $\dot{u}(t)$, $\dot{v}(t)$ or complex envelopes $\dot{U}(t)$, $\dot{V}(t)$. This modified inner product is defined as:

$$(\dot{\mathbf{u}}, \dot{\mathbf{v}}) = \int_0^T \dot{u}(t)\dot{v}^*(t)\mathrm{d}t = \int_0^T \dot{U}(t)\dot{V}^*(t)\,\mathrm{d}t = (\dot{\mathbf{U}}, \dot{\mathbf{V}}) \tag{2.39}$$

where a complex conjugation is used to preserve the equality between the inner product of the vector by itself and the vector squared length (always real and non-negative!), while coincidence of the inner products of analytic signals and complex envelopes follows from definition (2.38). Specifically, for a signal $s(t)$ of energy E, according to equations (2.36) and (2.35):

$$(\dot{\mathbf{S}}, \dot{\mathbf{S}}) = \|\dot{\mathbf{S}}\|^2 = \int_0^T |\dot{S}(t)|^2 \mathrm{d}t = \int_0^T S^2(t)\,\mathrm{d}t = \int_0^T s^2(t)\mathrm{d}t + \int_0^T s_\perp^2(t)\mathrm{d}t = 2E \tag{2.40}$$

since the Hilbert transform does not affect the amplitude–frequency spectrum and therefore the energies of $s(t)$ and $s_\perp(t)$ are always the same.

Let us now take two signals $s_k(t)$, $s_l(t)$ and calculate the squared distance between their complex envelopes $\dot{S}_k(t)$, $\dot{S}_l(t)$:

$$\begin{aligned} d^2(\dot{\mathbf{S}}_k, \dot{\mathbf{S}}_l) &= \|\dot{\mathbf{S}}_k - \dot{\mathbf{S}}_l\|^2 = (\dot{\mathbf{S}}_k - \dot{\mathbf{S}}_l, \dot{\mathbf{S}}_k - \dot{\mathbf{S}}_l) \\ &= 2E_k + 2E_l - 4\text{Re}\left[\dot{\rho}_{kl}\sqrt{E_k E_l}\right] \end{aligned} \tag{2.41}$$

where use is made of the linearity of inner product (2.39) and equation (2.40); while $\dot{\rho}_{kl}$, as in (2.16), is again the correlation coefficient but adapted to complex-valued signals, e.g. complex envelopes:

$$\dot{\rho}_{kl} = \frac{(\dot{\mathbf{S}}_k, \dot{\mathbf{S}}_l)}{\|\dot{\mathbf{S}}_k\|\|\dot{\mathbf{S}}_l\|} = \frac{1}{2\sqrt{E_k E_l}} \int_0^T \dot{S}_k(t)\dot{S}_l^*(t)\,dt \tag{2.42}$$

Equation (2.41) may be treated as the cosine theorem for complex vectors and $\mathrm{Re}[\dot{\rho}_{kl}]$ is an adequate measure of resemblance between the complex envelopes of signals $s_k(t)$ and $s_l(t)$. Using the equality between inner products of analytical signals and complex envelopes (2.39) and equations (2.38) and (2.34), the integral in (2.42) can be reduced as follows:

$$\int_0^T \dot{S}_k(t)\dot{S}_l^*(t)\mathrm{d}t = \int_0^T (s_k(t) + js_{k\perp}(t))(s_l(t) - js_{l\perp}(t))\,\mathrm{d}t = 2(\mathbf{s}_k, \mathbf{s}_l) + 2j(\mathbf{s}_{k\perp}, \mathbf{s}_l)$$

so that $\mathrm{Re}(\dot{\rho}_{kl}) = \rho_{kl}$, i.e. coincides with the ordinary correlation coefficient of signals $s_k(t), s_l(t)$ defined by (2.14). This being allowed for in (2.41) ties together the distances between the complex envelopes and the signals themselves:

$$d^2(\dot{\mathbf{S}}_k, \dot{\mathbf{S}}_l) = 2d^2(\mathbf{s}_k, \mathbf{s}_l) \tag{2.43}$$

The last result is one of many examples of the productiveness of the notion of complex envelope: manipulations with complex envelopes are very often much more compact and feasible than those with bandpass signals themselves, being free from the bulky trigonometric functions of carrier frequency terms.

2.5 *M*-ary data transmission: noncoherent signals

Let us now proceed to a problem of M-ary data transmission, but this time, unlike Sections 2.2 and 2.3, assuming that the signals are not fully deterministic. As has already been pointed out, in real life situations are very likely when either the transmitter or a channel can not preserve the coherence of bandpass signals and the latter acquire random phases at the receiving side. In this case initial phases cannot take part in message distinguishing, and the distinctness of signals should go beyond only phase shifts. Scenarios of this sort are termed *noncoherent* reception.

Suppose that the kth bandpass signal $s_k(t)$ has the modulation law described by a deterministic complex envelope $\dot{S}_k(t)$ and a random time-constant initial phase ϕ_k. Then it can be presented according to model (2.37) in the form:

$$s_k(t; \phi_k) = \mathrm{Re}[\dot{S}_k(t; \phi_k)\exp(j2\pi f_0 t)]$$

where a 'complete' complex envelope consists of the deterministic part and a part allowing for a random initial phase: $\dot{S}_k(t;\ \phi_k) = \dot{S}_k(t)\exp(j\phi_k)$.

To calculate the distance between the signals $s_k(t;\,\phi_k), s_l(t;\,\phi_l)$ we can make use of equation (2.43) and evaluate the distance between the complex envelopes $\dot{S}_k(t;\,\phi_k), \dot{S}_l(t;\,\phi_l)$ instead of the signals themselves. Doing this, consistent with the generalization of the cosine theorem (2.41) and under the assumption that all signals have identical energies E, we get:

$$d^2(\dot{\mathbf{S}}_{k\phi}, \dot{\mathbf{S}}_{l\phi}) = 4E[1 - \mathrm{Re}\dot{\rho}_{kl}(\phi)]$$

where the additional subscript ϕ underlines the correspondence of the vector $\dot{\mathbf{S}}_{k\phi}$ to the complete complex envelope $\dot{S}_k(t;\,\phi_k)$, independence of signal energy E of initial phase ϕ_k is taken into account, and:

$$\dot{\rho}_{kl}(\phi) = \frac{(\dot{\mathbf{S}}_{k\phi}, \dot{\mathbf{S}}_{l\phi})}{2E} = \frac{1}{2E}\int_0^T \dot{S}_k(t;\phi_k)\dot{S}_l^*(t;\phi_l)\,\mathrm{d}t$$

is the correlation coefficient of the complete complex envelopes $\dot{S}_k(t;\,\phi_k), \dot{S}_l(t;\,\phi_l)$. Phases ϕ_k, ϕ_l being independent of time, the latter quantity can be rewritten as $\dot{\rho}_{kl}(\phi) = \dot{\rho}_{kl}\exp[j(\phi_k - \phi_l)]$, where the correlation coefficient $\dot{\rho}_{kl}$ covering only the deterministic (random-phase-free) complex envelopes $\dot{S}_k(t), \dot{S}_l(t)$ of the signals is introduced:

$$\dot{\rho}_{kl} = \frac{(\dot{\mathbf{S}}_k, \dot{\mathbf{S}}_l)}{2E} = \frac{1}{2E}\int_0^T \dot{S}_k(t)\dot{S}_l^*(t)\,\mathrm{d}t = |\dot{\rho}_{kl}|\exp(j\phi_{kl}) \tag{2.44}$$

where $\phi_{kl} = \arg(\dot{\rho}_{kl})$.

Now, the squared distance above takes the form:

$$d^2(\dot{\mathbf{S}}_{k\phi}, \dot{\mathbf{S}}_{l\phi}) = 4E[1 - |\rho_{kl}|\cos(\phi_{kl} + \phi_k - \phi_l)] \tag{2.45}$$

The trouble about this distance is its dependence on unknown signal phases ϕ_k, ϕ_l. Due to this, there are a lot of distances for the fixed deterministic modulation laws $\dot{S}_k(t), \dot{S}_l(t)$ governed by a random phase difference $\phi_k - \phi_l$. With a perceptible level of the correlation modulus $|\rho_{kl}|$ the hazard is always present that due to unfavourable combination of phases (that is, when $\phi_{kl} + \phi_k - \phi_l$ is small enough) the distance (2.45) may prove very small. To fully ensure against it, the correlation modulus should be as low as possible and the best signal set should obey the condition:

$$\dot{\rho}_{kl} = \frac{(\dot{\mathbf{S}}_k, \dot{\mathbf{S}}_l)}{2E} = 0, \;\; k \neq l, \quad k,\; l = 1, 2, \ldots, M \tag{2.46}$$

As is seen, we again, as in Section 2.3, come to orthogonal signals. This time, however, the orthogonality condition is much more binding, forcing complex envelopes, or in other words modulation laws, of signals to be orthogonal, not just the signals themselves. Because of this, bandpass signals retain orthogonality under any combinations of their phases, since $\dot{\rho}_{kl} = 0$ entails $\dot{\rho}_{kl}(\phi) = 0$. On the other hand, condition (2.46) excludes the opportunity to provide orthogonality by means of a quadrature phase shift of the carrier, which was available for coherent bandpass signals. The direct implication of this is a halving of the signal space dimension n_s, given the total bandwidth W_t, from $2W_tT_t$, as was the case for deterministic bandpass signals, to only W_tT_t. This can also be explained in a slightly different way: the orthogonality should now be observed between complex signals, i.e. in the complex vector space. Each complex vector is actually a pair of real ones (real and imaginary parts), and hence requires for representation two-dimensional vector space, i.e. a plane. All of these planes should be orthogonal to each other. Since the entire available dimension is $2W_tT_t$, it may accommodate only W_tT_t orthogonal planes, i.e. W_tT_t orthogonal complex signals.

It is noteworthy that, in contrast to what has taken place for M deterministic signals, in the noncoherent case orthogonal signals are strictly optimal regardless of their number. For instance, the optimal noncoherent pair of signals is orthogonal, the antipodal one making no sense since phase randomness erases any distinction between the antipodal signals.

One more comment is appropriate as clarification of the optimal decision strategy for noncoherent reception. Observation $y(t)$, being some bandpass waveform, may itself be expressed in terms of its complex envelope or modulation law $\dot{Y}(t)$: $y(t) = \mathrm{Re}[\dot{Y}(t)\exp(j2\pi f_0 t)]$. Naturally, $\dot{Y}(t)$ is a random process. To declare one of M competitive signals to be the received one, the distances between signals and observation should be compared, which according to equation (2.43) is equivalent to comparing the (squared) distances between signal complete complex envelopes and the observed complex envelope $d^2(\dot{\mathbf{S}}_{k\phi}, \dot{\mathbf{Y}}) = \|\dot{\mathbf{Y}}\|^2 + 2E - 2\mathrm{Re}[\dot{z}_k(\phi_k)]$, where $\dot{z}(\phi_k)$ is the correlation (inner product) of the complex envelopes of the observation $y(t)$ and the kth random-phase signal $s_k(t;\phi_k)$. To get rid of the dependence on a random phase, it is logical to pick for comparison only the minimum distance over all the range of ϕ_k for each k. Dropping details reproducing almost literally those above, the resulting decision rule is expressed in terms of a correlation modulus:

$$Z_k = \left| \int_0^T \dot{Y}(t)\dot{S}_k^*(t)\,\mathrm{d}t \right|, k = 1, 2, \ldots, M \tag{2.47}$$

meaning that the jth signal is declared received if its deterministic complex envelope has the highest correlation modulus with the observation complex envelope. In other words, the old and well-tried idea is exploited: the signal whose modulation law has maximal resemblance with the modulation law of the observation is picked up.

2.6 Trade-off between orthogonal-coding gain and bandwidth

The previous discussion clearly illuminated the special role played by orthogonal signals: they are practically ($M \gg 1$, coherent signals) and even theoretically (noncoherent case) optimal in M-ary data transmission. Let us now evaluate the benefit which accompanies employing orthogonal signals against uncoded, i.e. direct source bit stream transmission. Suppose the energy resource allows the energy E_b, to be put into each data bit, no special coding of the bit stream is used and an individual bit is transmitted by optimal antipodal pair, in other words by BPSK. In a block of m successive bits any bit patterns are possible, including those which differ from each other in only one bit. Therefore, the minimum squared distance between signals corresponding to non-identical m-bit blocks is the same as the squared distance between the one-bit antipodal signals, i.e. according to Figure 2.5a, $d^2_{\min,u} = 4E_b$, where the second subscript stands for 'uncoded'.

Now consider another system, where all different m-bit blocks are transmitted by the orthogonal signals. Clearly, every such signal is allowed to have energy mE_b, preserving the fixed energy per bit E_b. Then the squared distance (all of them are now the same, because orthogonal signals are equidistant) between signals is again easily found referring to Figure 2.5b: $d^2_{\min,ort} = 2mE_b$. It is obvious now that orthogonal signals have a gain G_a in minimum squared distance versus uncoded transmission equal to $m/2$. Putting it another way, to provide the same minimum distance, uncoded transmission requires $G_a = m/2$ times higher energy than orthogonal signalling. In view of equation (2.23), minimum distance asymptotically determines error probability for whatever signals are used in M-ary transmission. Hence, when required reception fidelity is high, which automatically entails high value of required SNR, identical reliabilities of the two considered systems are possible with G_a times higher energy for the uncoded one. Thus, the *asymptotic* orthogonal coding gain G_a is an adequate indicator of the benefits of orthogonal signalling in the limit, i.e. under SNR tending to infinity.

In order to get an idea about the order of orthogonal coding gain under finite SNR as well as the rate of its convergence to the asymptotic limit, Figure 2.9 presents families of curves for two lengths of m-bit blocks: $m = 6$ and $m = 20$. The first curve (solid) shows the probability of wrong reception of an uncoded block, and the other two are calculated for the case when m-bit blocks are encoded into $M = 2^m$ orthogonal

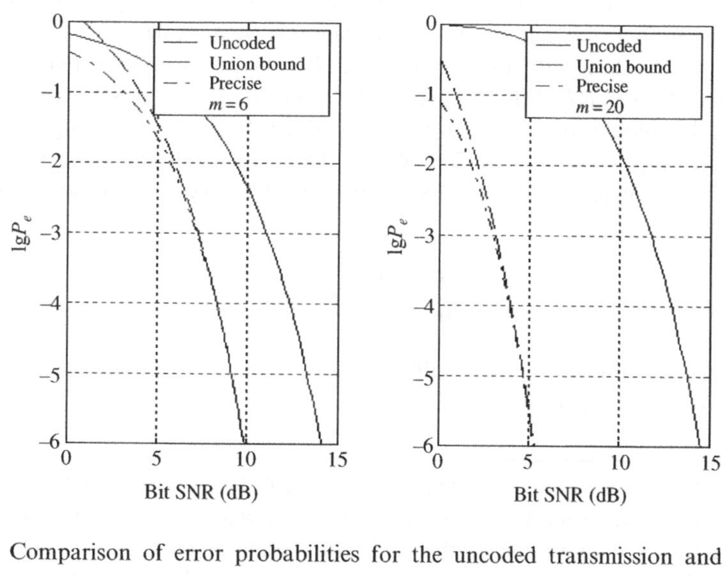

Figure 2.9 Comparison of error probabilities for the uncoded transmission and orthogonal signalling

signals processed coherently. The dashed curve is built up according to the union bound (2.32), while the dash-dot one corresponds to the accurate formula for error probability of coherent reception of M orthogonal signals, derivation of which the reader can find in many popular books on communication theory fundamentals (e.g. [5,7]):

$$P_{e,ort} = 1 - \int_{-\infty}^{\infty} \exp\left[-\frac{(x - q_b\sqrt{m})^2}{2}\right] \Phi^{M-1}(x)\,dx$$

where $q_b = \sqrt{2E_b/N_0}$ is SNR per bit and $\Phi(x) = 1 - Q(x)$ is the error function.

The by-product conclusion, which can be drawn from Figure 2.9, is the very high trustworthiness of the union bound: practical applications typically demand small error probabilities and it is seen that in the area $P_e \leq 10^{-2}$ the gap between the value of bit SNR estimated from the union bound and that calculated precisely appears to be less than 0.5 dB. It drops rapidly when the requirements for transmission fidelity get tougher (becoming no greater than 0.2 dB when $P_e \leq 10^{-3}$). In the range of P_e from 10^{-2} to 10^{-6} actual gain G of orthogonal signalling increases from 3.5 to 4.2 dB ($m = 6$) and from 8.5 to 8.9 dB ($m = 20$). Comparing these figures with the asymptotic ones (4.8 and 10 dB, respectively) one can see rather good consistency between them, justifying usage of asymptotic coding gain as a first approximation of the orthogonal coding efficiency.

The very optimistic judgement on orthogonal signalling prospects which may arise in the light of the above results is significantly shadowed by the real cost of the coding gain. The latter is bought at the expense of bandwidth widening because, as was established in Section 2.3, the signal space dimension n_s, i.e. the number of orthogonal signals $M = n_s$, is immediately governed by the total system time–frequency resource W_tT_t. Being mainly interested in finding the order of quantities and ignoring trivial way of doubling the number of coherent orthogonal signals, we have $M = W_tT_t$ or $W_t = M/T_t$. Let the necessary data rate in a system be R bits per second (bps), which means transmission of $m = RT_t$ bits over the time period T_t. It is seen that orthogonal encoding of bit-blocks of this length will produce $M = 2^m = 2^{RT_t}$ signals, providing asymptotic coding gain $G_a = m/2 = RT_t/2$. Then the spectral efficiency R/W_t, i.e. the rate per 1 Hz of bandwidth (see Section 1.1), of the system employing orthogonal signals:

$$\frac{R}{W_t} = \frac{RT_t}{2^{RT_t}} = \frac{2G_a}{2^{2G_a}} \tag{2.48}$$

falls quite steeply (almost exponentially) with the desired coding gain.

Let us turn to a typical numerical example.

Example 2.6.1. The data rate $R = 9.6$ kbps is very typical of digital speech transmission (mobile telephone, multimedia systems etc.). Suppose someone wants to reduce by three (4.8 dB) the transmitted power without sacrificing the data rate. If that person intends to achieve this goal with the help of orthogonal signals, it is possible in exchange for lowering the spectral efficiency from 1 to 6/64, as predicted by equation (2.48). In other words, to maintain a rate of 9.6 kbps bandwidth wider than 100 kHz will be involved. This is not a prohibitive figure for many applications and, for instance, the cdmaOne cellular telephone standard exploits exactly this principle in the uplink (for more details see Section 11.3).

Let us now imagine a designer who is quite impressed by the figure above and plans to go further along the same way, targeting a 10-times (10 dB) reduction of transmitted power. To realize that, blocks of $m = 20$ bits should be converted into $M = 2^{20} > 10^6$ orthogonal signals. This will lead to spectral efficiency less than 2×10^{-5} or bandwidth occupied wider than 480 MHz, which looks wholly impractical versus the data rate 9.6 kbps.

The discussion undertaken illustrates the very tough character of the trade-off between energy efficiency and spectral efficiency inherent in orthogonal signalling. At the same time it is pertinent to note that, although big energy gains are unattainable in practice with orthogonal signals due to the enormous demand for bandwidth, asymptotic orthogonal coding gain may serve as a good reference point, being the upper border of theoretical efficiency of any m-bit-block coding.

Return now to equation $M = W_t T_t$ and consider the question: when the number of orthogonal signals and thus the product $W_t T_t$ is measured in the tens or more, does it point to a spread spectrum? In other words, is a system exploiting numerous orthogonal signals always a spread spectrum one? As the discussion in the next section shows, the answer to this question is in general negative.

2.7 Examples of orthogonal signal sets

Throughout this section we will again ignore the opportunity of doubling the number of orthogonal signals by quadrature carrier shifts, which is always present in a coherent bandpass system, and concentrate only on the dependence between M and equivalent baseband system time–frequency resource $W_t T_t$. We will demonstrate first how to build the simplest orthogonal sets based on fragmentation of an available resource.

2.7.1 Time-shift coding

It is obvious that the inner product of any two non-overlapping time-shifted signals is zero. Consider M signals shown in Figure 2.10a, which occupy jointly time period T_t. With signal duration no greater than $T = T_t/M$ and the time shift between successive signals no smaller than the signal duration, this *time-shift coding* produces orthogonal signals. The estimated bandwidth W of each of these signals is inverse to its duration and all the signals are permitted to occupy the same bandwidth with no violation of orthogonality: $W = W_t$. Hence, the maximal number of orthogonal signals of this sort which can be accommodated within a given total time–frequency resource T_t, W_t is $M = T_t/T = W_t T_t$, i.e. as is easily foreseen, it is equal to the signal space dimension $n_s = W_t T_t$. A high necessary number of signals $M \gg 1$ implies a large product $W_t T_t = M$, which may seem to point to spread spectrum. However, for any individual signal the time–frequency product is $WT = W_t T = W_t T_t/M = 1$, so that the signals are *not* of spread spectrum type. In the

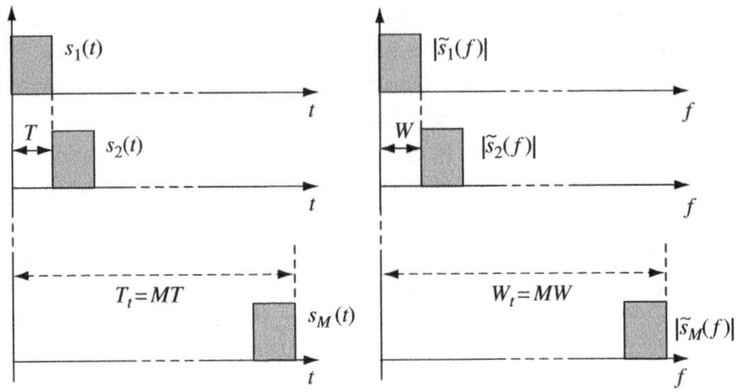

(a) (b)

Figure 2.10 Orthogonal time-shift coded (a) and frequency-shift coded (b) signals

wake of the agreement to call a system 'spread spectrum' only if it uses spread spectrum signals (see Section 1.1), orthogonal time-shift coding has nothing to do with spread spectrum.

Let the total time–frequency resource be identified with a rectangle having sides T_t, W_t in the t, f coordinate plane. Then time-shift coding just means slicing this resource into M vertical strips, each being assigned to some individual signal (see Figure 2.11a). Orthogonality in this transmission mode is provided by a rigorous distribution of the time resource between signals, each exploiting the total spectral resource.

The orthogonal signalling scheme just introduced may seem attractive from an implementation point of view due to its apparent simplicity. Its weaknesses, however, are also conspicuous and should be kept in focus. First, accurate synchronization is necessary, any potential fluctuations of signal time positions being capable of destroying orthogonality. This requires secure safety margins between signals, which reduces the number of signals compared to the theoretical maximum, i.e. worsens spectral efficiency. Another issue is the

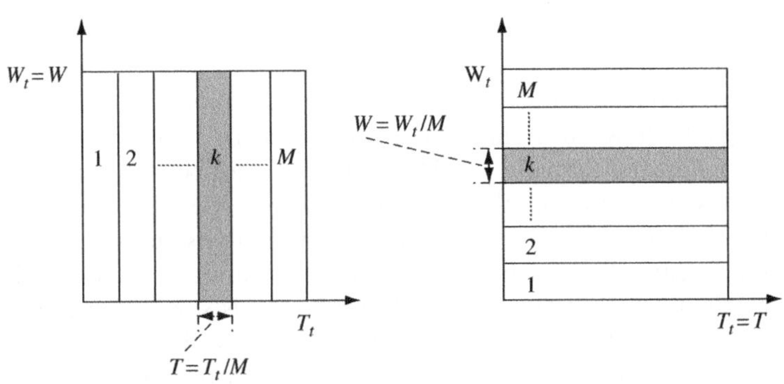

(a) (b)

Figure 2.11 Resource distribution in orthogonal time-shift (a) and frequency-shift (b) coding

value of the peak-factor ν, which is the ratio between peak and average powers. Because an individual signal occupies only an Mth part of the available time resource, average power is M times smaller than peak power and $\nu = M \gg 1$. At the same time, in designing a transmitter power amplifier, a small value of ν is crucial: the closer it is to 1, the softer are the demands on the linearity of an amplifier and the better is its power performance.

2.7.2 Frequency-shift coding

The other straightforward way to provide orthogonality is *frequency-shift coding*. Due to time–frequency duality or Parseval theorem, the inner products of signals $u(t)$, $v(t)$ and of their spectra $\tilde{u}(f)$, $\tilde{v}(f)$ coincide:

$$(\mathbf{u}, \mathbf{v}) = \int_{-\infty}^{\infty} u(t)v(t)\,dt = \int_{-\infty}^{\infty} \tilde{u}(f)\tilde{v}^*(f)df = (\tilde{\mathbf{u}}, \tilde{\mathbf{v}}) \tag{2.49}$$

which allows transfer of the idea discussed above into the frequency domain (see Figure 2.10b). With an entire overlap of the signals in time ($T = T_t$) each of them has bandwidth $W = 1/T_t$ at the least. Thus the maximum number of orthogonal signals formed by shifting the spectra is again $M = W_t/W = W_tT_t = n_s$. As in the previous case, the total resource is again 'sliced', but differently: the strips are horizontal, meaning that the total time resource T_t but only an Mth part of the entire frequency resource W_t are utilized by every signal (Figure 2.11b). Clearly, each individual signal is again *non-spread-spectrum* since its time–frequency product $WT = (W_t/M)T_t = 1$, and any system with however large a number of orthogonal signals of this sort is certainly not a spread spectrum one.

The peak-factor of this mode of orthogonal signalling, unlike time-shift coding, is $\nu = 1$ and synchronization errors are not that dramatic because orthogonality is provided by signal non-overlap in the frequency domain. Instead, spectra drifts (e.g. because of Doppler shifts) may sometimes be destructive. Still, this transmission mode is extremely popular and the conventional M-ary FSK modulation is its direct embodiment.

The examples considered explain why employing even a great number of orthogonal signals and, hence, the necessity for a total resource $W_tT_t \gg 1$ does not automatically mean the involvement of spread spectrum technology.

2.7.3 Spread spectrum orthogonal coding

Fragmentation of the total time–frequency resource inherent to the two discussed modes of orthogonal signalling may in some cases be a preferable solution in connection with hardware implementation aspects. However, with M increasing reasons of this sort are getting more doubtful since, as mentioned above, time-shift coding demands a high peak-factor while frequency-shift coding implies optimal processing with a bank of numerous parallel frequency-detuned filters.

Under such circumstances spread spectrum orthogonal signalling can prove very competitive, allowing all signals to share a total time–frequency resource with no distribution or slicing of the latter. Consider a simple example of realization of the idea in the form of discrete BPSK signals. Compose each of M signals of N consecutive contiguous elementary pulses or *chips*, each having the same rectangular shape and duration Δ. Let the chip polarities of the signal number k be manipulated by a *code sequence* (or simply *code*) of binary symbols $a_{k,i} = \pm 1$, where $k = 1, 2, \ldots, M$ and the second subscript is chip number (discrete time): $i = 0, 1, \ldots, N-1$. Then the baseband version of such a signal may be written as:

$$s_k(t) = \sum_{i=0}^{N-1} a_{k,i} s_0(t - i\Delta) \qquad (2.50)$$

with $s_0(t)$ symbolizing the rectangular chip of duration Δ.

Calculate now the inner product or correlation (2.5) of the kth and lth signals. After changing the order of summation and integration:

$$(\mathbf{s}_k, \mathbf{s}_l) = \sum_{i=0}^{N-1} \sum_{j=0}^{N-1} a_{k,i} a_{l,j} \int_0^T s_0(t - i\Delta) s_0(t - j\Delta) dt \qquad (2.51)$$

The integral here is the inner product of two chips time-shifted to each other by $(i - j)\Delta$. When $i \neq j$ it equals zero since chips in the integral have no overlap in time. Thus:

$$\int_0^T s_0(t - i\Delta) s_0(t - j\Delta) dt = E_0 \delta_{ij}$$

where E_0 is the chip energy. Using this in equation (2.51) produces:

$$(\mathbf{s}_k, \mathbf{s}_l) = E_0 \sum_{i=0}^{N-1} a_{k,i} a_{l,i} = E_0 (\mathbf{a}_k, \mathbf{a}_l) \qquad (2.52)$$

Equation (2.52) relates the inner product of the signals (2.50) with an inner product of N-dimensional vectors of the corresponding code sequences $\mathbf{a}_k = (a_{k,0}, a_{k,1}, \ldots, a_{k,N-1})$. As can be seen, M orthogonal code sequences automatically generate M orthogonal signals of the type (2.50). With $M \leq N$ there are many ways to construct such sequences because the case in point is simply finding $M \leq N$ orthogonal N-dimensional vectors. In our discussion those vectors are binary, i.e. with components taking values of ± 1 only. $M = N$ orthogonal binary vectors used as rows form a square matrix called the *Hadamard* matrix. It is not difficult to prove (the reader may try attempt it; see Problem 7.14) that only Hadamard matrices of size divisible by 4 can exist: $M \equiv 0 \bmod 4$, where the symbol of congruence $a \equiv b \bmod c$ is used, meaning equal residuals of dividing integers a, b by the integer c. No answer has been found as yet as for the sufficiency of this necessary condition.

A number of algorithms are known for building Hadamard matrices of the special (not sparse) lengths. One is the very popular Sylvester rule, which doubles the matrix size recursively. To explain its content let us suppose that Hadamard matrix $\mathbf{H}_M$ of size M has been somehow found. Then the double-sized Hadamard matrix $\mathbf{H}_{2M}$ can be constructed of four repetitions of $\mathbf{H}_M$, taken as blocks, one of them being sign-changed:

$$\mathbf{H}_{2M} = \begin{bmatrix} \mathbf{H}_M & \mathbf{H}_M \\ \mathbf{H}_M & -\mathbf{H}_M \end{bmatrix} = \begin{bmatrix} 1 & 1 \\ 1 & -1 \end{bmatrix} \otimes \mathbf{H}_M \tag{2.53}$$

where the second equality expresses the rule in terms of matrix Kronecker product $\otimes$. The orthogonality of rows of $\mathbf{H}_{2M}$ is obvious: if two rows have numbers differing by any integer but M, they have zero inner product, since their two M-element halves are orthogonal. Otherwise, the first M components of the rows coincide, while the rest of the components are opposite, which again gives zero inner product.

To make use of the Sylvester algorithm one can start with matrix

$$\mathbf{H}_2 = \begin{bmatrix} 1 & 1 \\ 1 & -1 \end{bmatrix}$$

which is evidently a Hadamard one, and construct $\mathbf{H}_4$ (using the symbols '+' and '−' in place of +1 and −1 for brevity), then from $\mathbf{H}_4$ produce $\mathbf{H}_8$, and so forth:

$$\mathbf{H}_4 = \begin{bmatrix} \mathbf{H}_2 & \mathbf{H}_2 \\ \mathbf{H}_2 & -\mathbf{H}_2 \end{bmatrix} = \begin{bmatrix} + & + & + & + \\ + & - & + & - \\ + & + & - & - \\ + & - & - & + \end{bmatrix}$$

$$\mathbf{H}_8 = \begin{bmatrix} \mathbf{H}_4 & \mathbf{H}_4 \\ \mathbf{H}_4 & -\mathbf{H}_4 \end{bmatrix} = \begin{bmatrix} + & + & + & + & + & + & + & + \\ + & - & + & - & + & - & + & - \\ + & + & - & - & + & + & - & - \\ + & - & - & + & + & - & - & + \\ + & + & + & + & - & - & - & - \\ + & - & + & - & - & + & - & + \\ + & + & - & - & - & - & + & + \\ + & - & - & + & - & + & + & - \end{bmatrix}$$

Thereby a Hadamard matrix of any order $M = 2^m$ (2, 4, 8, 16, 32, ...) can be built up. Rows of Hadamard matrices of this kind are also known as *Walsh functions*.

Figure 2.12 shows baseband orthogonal BPSK signals (2.50)—Walsh functions—generated with the aid of Hadamard matrix $\mathbf{H}_8$.

Figure 2.13 illustrates that within this signalling mode there is no resource distribution: all signals share the common resource, fully overlapping in both the time and frequency domains. Indeed, the bandwidth of each signal is estimated as $W = 1/\Delta$ while duration $T = M\Delta$, thus producing $WT = M = W_t T_t$. Orthogonality is now achieved at the cost of an appropriate signal modulation, rather than either time interval or bandwidth fragmentation.

Analysing the benefits of spread spectrum orthogonality, one can note that methods of generation and processing of signals (2.50) are quite well matched to modern digital microchip circuitry (ASIC, VLSI, microprocessors). Another factor is the automatic

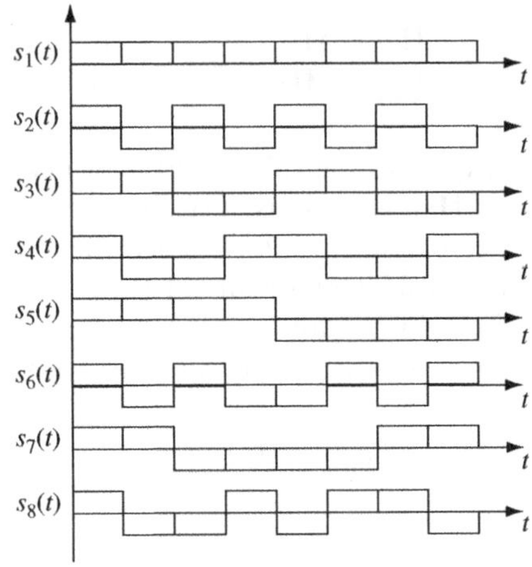

Figure 2.12 Baseband Walsh functions

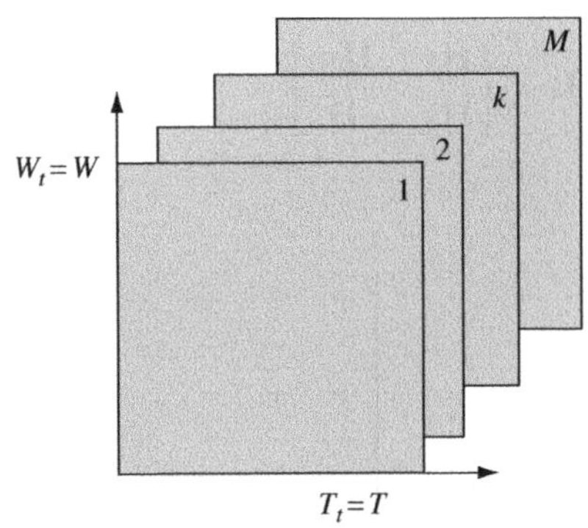

Figure 2.13 Resource allocation in orthogonal spread-spectrum signalling

acquiring of those merits of the spread spectrum which cannot be seen directly within the classical reception framework but are numerous and very valuable in practice (for details see Chapter 3). This gives an explanation of the great popularity of orthogonal signalling of this sort in advanced telecommunication systems (e.g. cdmaOne, UMTS, cdma2000; see Chapter 11).

Now the moment has come to draw an overall conclusion on the results of Sections 2.5–2.7. As one may see, theoretically the classical M-ary transmission problem *does not lean* implicitly towards the spread spectrum, and in principle optimal signals can be realized as plain ones. On the other hand, there are implementation reasons, along with the desire to gain the numerous advantages pertaining to spread spectrum beyond the classical reception model. Because the latter opportunity is potentially promised by a large total necessary time–frequency resource $W_t T_t \gg 1$, this can incline a system designer to prefer spread spectrum signals to plain ones.

2.8 Signal parameter estimation

2.8.1 Problem statement and estimation rule

Everywhere in radio systems we encounter the problem of signal *parameter measurement* or *estimation.* It describes any situation when information which is interesting to an observer is carried by a current value of some signal parameter (e.g. amplitude, frequency, initial phase, time delay etc.). Therefore, to extract necessary information the observer needs to measure or estimate the corresponding parameters.

Let us turn to some transparent examples. In conventional AM (FM) broadcasting, dependence of amplitude (frequency) on time carries audio information: the volume and pitch of a tone. To restore the audio message and give it out to the listener, instant values of the amplitude (frequency) should first be measured and reproduced as a continuous-time waveform. Another similar example is conventional analog TV, where both amplitude and frequency are involved in information transmission. To restore colour moving images amplitude measurement should be performed, since the luminance and chrominance components are broadcast via AM, while audio transmission is accomplished via FM and, hence, in any TV receiver frequency measurement is present.

Another parameter estimation task is found in the synchronization or timing problem, where time–frequency mismatch between the received signal and a local reference clock should be measured to synchronize the second with the first. This procedure is characteristic of a great number of systems, ranging from TV horizontal and vertical synchronization channels to pilot channels of 2G and 3G mobile radio.

Numerous estimation problems are typical of radar and navigation: measurement of time delay and signal arrival direction provides knowledge of the mutual distance and angle coordinates between a receiver and a target; if knowledge of a target velocity and manoeuvre is necessary Doppler frequency shift should be measured etc.

The list of examples could easily be continued, since parameter measurement is an integral part of practically any system in which information is transmitted, recovered and processed.

In terms convenient for our context, the parameter estimation problem may be stated in the following manner. The observation $y(t)$ along with noise contains the signal $s(t;\lambda)$, which is deterministic except for the unknown constant value of the parameter λ. The latter may be a vector or a scalar, depending on the specific situation. The observer, based on the analysis of $y(t)$, should produce a *decision* on what value, within the range of possible ones, is taken by the signal parameter in question. This decision, in association with the problem itself also called *estimate*, is denoted by $\hat{\lambda}$. Since noise is always present in $y(t)$, in any separate session of reception $\hat{\lambda}$ differs from the unknown true value of the parameter λ, and the question is how to make the optimal decision, which guarantees the smallest harm caused by this discrepancy.

The simplest clue to this issue may be found in understanding that in principle the estimation problem is not anything radically new with respect to the problem of distinguishing M signals studied in Section 2.3. In fact, suppose at first that a parameter λ to be measured is discrete and takes one of M competitive values $\lambda_1, \lambda_2, \ldots, \lambda_M$. Then the decision about which of these possible values is assumed by a signal parameter in this specific observation is nothing more than just determining between M hypotheses on which of M competitive signals $s_1(t), s_2(t), \ldots, s_M(t)$ is being received, where the signals are just copies of $s(t;\lambda)$, differing from each other only in value of parameter λ: $s_k(t) = s(t;\lambda_k)$. To cover the case of a continuous parameter λ with this reasoning too, one can just imagine an infinite (up to uncountable) number M of the parameter values and, consequently, of the signals to be recognized.

The conclusion from these arguments is that the already well-known optimal decision strategy, the ML rule, remains applicable to the parameter estimation. This means that among all competitive values of λ the one should be picked as the estimate $\hat{\lambda}$ which maximizes the probability of transforming the sent signal $s(t;\lambda)$ into the observed waveform $y(t)$ at the channel output. For the AWGN channel this rule is equivalent to the minimum distance one, which, rewritten in the current designations, looks as follows:

$$d(\mathbf{s}_{\hat{\lambda}}, \mathbf{y}) = \min_{\lambda} d(\mathbf{s}_\lambda, \mathbf{y}) \Rightarrow \hat{\lambda} \text{ is given out} \tag{2.54}$$

where $\mathbf{s}_\lambda$ is vector notation of signal $s(t;\lambda)$. This rule produces the *maximum likelihood* estimate $\hat{\lambda}$ by finding the value of λ under which signal $s(t;\lambda)$ is closest to the observation $y(t)$ by the Euclidean distance. Figure 2.14 gives an illustration of it. The signal $s(t;\lambda)$ may be thought of as a vector $\mathbf{s}_\lambda$, which moves, tracing the changes of the parameter λ. Its extreme point travels along some trajectory, points of which map one-to-one to specific values of λ (Figure 2.14a). The point of the trajectory closest to the observation vector $\mathbf{y}$ is found according to rule (2.54) and the corresponding value of λ is announced as the estimate, which is also seen in Figure 2.14b, showing the dependence of the distance between the observation and the signal copy on the value of λ. The ML estimate $\hat{\lambda}$ is the parameter value minimizing this distance.

All signal parameters can be categorized as *energy* or *non-energy* ones, the terms reflecting whether the parameter affects signal energy. If λ is of the second type, the energy of the signal $s(t;\lambda)$ does not depend on the value of λ:

$$E(\lambda) = \int_{-\infty}^{\infty} s^2(t;\lambda)\mathrm{d}t = E$$

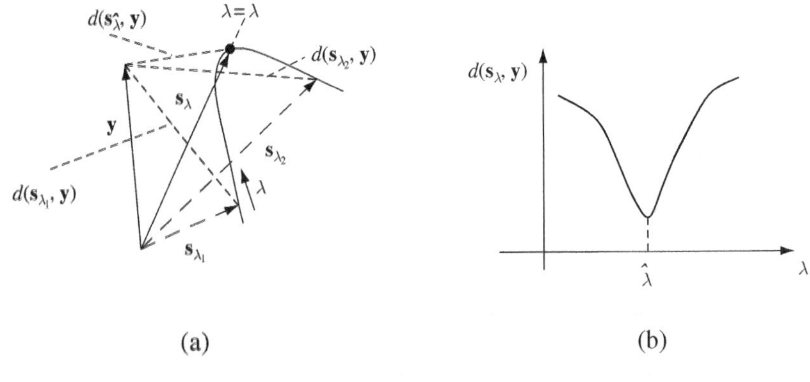

Figure 2.14 Illustration of ML estimation

Amplitude and duration, for example, are energy parameters, whereas time delay, frequency and initial phase are non-energy ones. It should be clear now that estimation of a non-energy parameter is the particular case of the problem of distinguishing between competitive signals of equal energies, for which correlation rule (2.8), in the new designations, can be presented as:

$$z(\hat{\lambda}) = \max z(\lambda) \Rightarrow \hat{\lambda} \text{ is given out} \tag{2.55}$$

or

$$\hat{\lambda} = \arg\max_{\lambda} z(\lambda)$$

where according to (2.7):

$$z(\lambda) = (\mathbf{y}, \mathbf{s}_\lambda) = \int_0^T y(t)s(t;\lambda)\mathrm{d}t \tag{2.56}$$

is a correlation between the observed waveform $y(t)$ and the signal $s(t;\lambda)$ in dependence on the value of the measured parameter λ.

In the light of the physical content of the correlation, the estimation rule (2.55) has quite a transparent interpretation: the ML estimate $\hat{\lambda}$ is the value of λ under which signal $s(t;\lambda)$ has maximal resemblance with the observed waveform $y(t)$.

2.8.2 *Estimation accuracy*

Let us recollect now the fact discussed in depth in Sections 2.2–2.3 that the fidelity of signal distinguishing is critically governed by the correlation coefficient (2.14). In the case of parameter estimation, signals to be distinguished are just copies of $s(t;\lambda)$ with different values of λ. In many practical situations the correlation of any two such copies $s(t;\lambda_1), s(t;\lambda_2)$ depends only on their mismatch in λ, i.e. the difference $\lambda_2 - \lambda_1$, rather than on values λ_1, λ_2 separately, so that putting $\lambda_1 = 0, \lambda_2 = \lambda$ leads to the following representation of the correlation coefficient (2.14) for the case of non-energy parameter λ:

$$\rho(\lambda) = \frac{(\mathbf{s}_0, \mathbf{s}_\lambda)}{E} = \frac{1}{E}\int_{-\infty}^{\infty} s(t;0)s(t;\lambda)\,\mathrm{d}t \tag{2.57}$$

As is customary for a correlation coefficient, this entity characterizes the resemblance of two signal copies depending on their mismatch in the parameter λ. It is evident that $\rho(\lambda) \leq \rho(0) = 1$, which has an instructive implication: signal copies mismatched in λ can not be more similar to each other than fully identical ones, which have, in their turn, unity correlation. Another property of the quantity (2.57) induced by its dependence on only $\lambda = \lambda_2 - \lambda_1$ is evenness: $\rho(\lambda) = \rho(-\lambda)$.

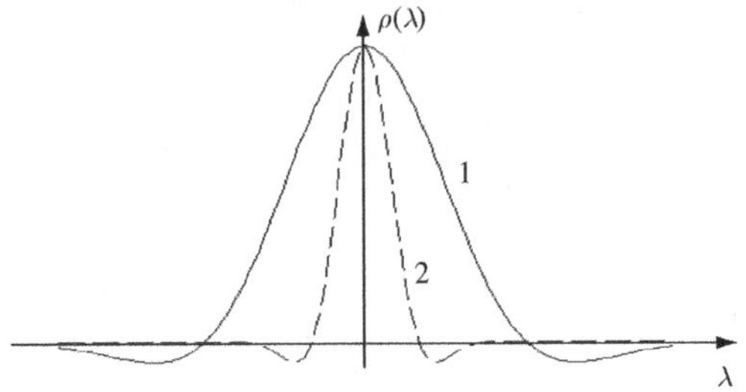

Figure 2.15 Typical curves of $\rho(\lambda)$ in dependence on λ

Figure 2.15 gives example curves of $\rho(\lambda)$ in dependence on λ for two hypothetical signals. The solid curve is flatter than the dashed one, which means that the first signal is less sensitive to changing λ: the resemblance between its copies mismatched in λ is higher than that between copies of the second with the same mismatch.

Now it is necessary to stress one detail, which was deliberately omitted above. As a matter of fact, one or another decision rule is always optimal only in some strict sense specified by an optimality criterion. The ML rule referred to here is optimal in the sense of the *minimum probability of error* criterion, which is quite natural and adequate when discrete signals are distinguished or a discrete parameter is measured. But it does not look that adequate when a continuous parameter is measured. It seems much more reasonable in this case to characterize estimation fidelity by a precision, i.e. magnitude of deviation $\varepsilon = \hat{\lambda} - \lambda$ of estimate $\hat{\lambda}$ from the true value of λ. First of all, it looks quite normal to require that expectation of error ε over all possible observations $y(t)$, true value λ fixed, be zero for any λ, i.e. that estimate $\hat{\lambda}$ be equal to the true λ on average:

$$\bar{\varepsilon} = \overline{\hat{\lambda} - \lambda} = 0 \Leftrightarrow \bar{\hat{\lambda}} = \lambda, \forall \lambda \tag{2.58}$$

An estimate meeting this condition is called *unbiased*. But fulfilment of (2.58) does not yet allow us to consider the estimate good, since the magnitude of random scattering of the estimate around the true value is of critical importance. The variance of error $\mathrm{var}\{\varepsilon\} = \overline{(\hat{\lambda} - \lambda)^2}$ is a traditional and very adequate measure of this scattering, and seeking for the rule providing unbiased estimate with minimal $\mathrm{var}\{\varepsilon\}$ over all true λ:

$$\mathrm{var}\{\varepsilon\} = \overline{(\hat{\lambda} - \lambda)^2} = \min \qquad \forall \lambda$$

would be highly justified. Thus, minimization of variance of unbiased estimate is a natural way of pursuing greatest measurement accuracy.

In estimation theory the fundamental Cramer–Rao bound is proved bordering the variance of any unbiased estimate from below. An estimate whose variance lies on this bound is called an *efficient* one. 'Purely' efficient estimates are rather infrequent but this is not a big problem from the application standpoint. The matter is that the ML estimate is *asymptotically* unbiased and efficient as it is again established in classical estimation theory. Physically the term 'asymptotically' means 'in situations where high measurement accuracy is necessary', or, to put it even more practically, when sufficiently high

SNR or observation time is provided. Therefore, in any task where high accuracy is wanted, the ML rule is optimal not only by the error-probability criterion but also by the estimate accuracy criterion. Certainly, in the real world, high measurement precision is a typical demand and this is why the ML estimates are extensively used.

For the case of non-energy parameter λ the Cramer–Rao bound acquires on especially simple form and provides a practical tool to calculate the ML estimate variance:

$$\mathrm{var}\{\hat{\lambda}\} = \mathrm{var}\{\varepsilon\} \approx -\frac{1}{\rho''(0)q^2} \qquad q \gg 1 \tag{2.59}$$

The presence of SNR $q^2 = 2E/N_0$ in the denominator of the right-hand side of (2.59) is no surprise: naturally, for any reasonable estimation rule the greater is SNR, the smaller is the error and the higher is the measurement precision. At the same time, the dependence on the second derivative of a correlation coefficient deserves a more extensive comment. As is well known from mathematical analysis, the second derivative describes the curvature or sharpness of a function at the examined point and for a convex curve is negative. The sharpness of $\rho(\lambda)$ at zero point, in its turn, shows the sensitivity of a signal towards mismatch in λ: the sharper it is, the faster a signal copy mismatched in λ loses its resemblance to the initial copy. Recollect now that estimation is a particular case of signal distinguishing and is the more reliable the smaller the signal similarity is. This gives a complete explanation as to why $\rho''(0)$ may affect the precision of measuring λ: when copies of $s(t; \lambda)$ have low resemblance even with close values of λ, they are more easily distinguished in comparison with the case of their stronger similarity.

The latter fact points at the general trend of signal design in problems of non-energy parameter λ estimation. To achieve the desired result not at the cost of just 'brute force', i.e. energy increase, one may try to find signals with a steep dependence of correlation coefficient $\rho(\lambda)$ on λ.

In the following sections we turn to concrete estimation problems, among which examples of measuring both energy and non-energy parameters are considered. The main idea remains as before: to find out where estimation problems may call for the use of spread spectrum.

2.9 Amplitude estimation

The problem of measuring signal intensity (level, power) may be encountered in numerous applications, from TV broadcasting to digital PAM or QAM data transmission and mobile radio. Let us set it as the problem of measuring unknown amplitude A remaining constant during the observation interval $[0, T]$. In this statement the following signal model can be assumed:

$$s(t; A) = As(t)$$

where $s(t)$ is some deterministic reference signal whose amplitude is equal to one by convention. Then signal $s(t; A)$ is the result of scaling the reference signal by an unknown factor A. Let E be the energy of the reference signal. Then the energy $E(A)$ of the signal with amplitude A and its correlation $z(A)$ with observation $y(t)$ are:

$$E(A) = \int_0^T s^2(t; A)\, dt = A^2 \int_0^T s^2(t)\, dt = A^2 E \qquad z(A) = \int_0^T y(t)s(t; A)\, dt = Az$$

where

$$z = \int_0^T y(t)s(t)\, dt \tag{2.60}$$

is the correlation of the observation with the reference signal.

Turning to the correlation version of minimum distance rule (2.8) and noticing that the roles of E_j and z_j are now played by $E(A)$ and $z(A)$, respectively, the ML estimation procedure may be treated as maximization of the difference $z(A) - E(A)/2$ in A. Using the equations above, this difference takes the form of a quadratic binomial $Az - A^2E/2$ in A with the known coefficients. Its maximum is easily found, producing the ML estimate of amplitude:

$$\hat{A} = \frac{z}{E}$$

Thus, calculating a correlation of the observed waveform with the reference signal and scaling the result by the constant $1/E$ is exactly the desired optimal amplitude estimate. After finding the expectation of z from equation (2.60):

$$\bar{z} = \int_0^T \overline{y(t)}s(t)\, dt = \int_0^T s(t; A)s(t)\, dt = A \int_0^T s^2(t)\, dt = AE$$

it is readily seen that on average $\hat{A}$ strictly coincides with the true value of the amplitude, $\bar{\hat{A}} = \bar{z}/E = A$, meaning that the ML estimate of amplitude is *rigorously* (not only asymptotically) *unbiased*. No more difficult is the evaluation of variance of $\hat{A}$:

$$\text{var}\{\hat{A}\} = \frac{\text{var}\{z\}}{E^2} = \frac{N_0}{2E} = \frac{1}{q^2} \tag{2.61}$$

where result (2.15) is used and q^2 is SNR for the reference signal (i.e. for a signal having a unit amplitude). It may be shown that (2.61) strictly reproduces the Cramer–Rao bound, proving rigorous (not only asymptotic) efficiency of the ML estimate of signal amplitude. This rare case of an estimate's rigorous optimality is associated with the energy nature of amplitude and will not be met later when non-energy parameters are considered.

Now, what sort of demand does amplitude measuring impose on signal design? As (2.61) demonstrates, nothing but sufficient energy exhaustively determining estimation precision. No complications of the signal modulation law are able to improve amplitude measurement accuracy if they do not govern the signal energy. Consequently, *no momentum* to the involvement of spread spectrum appears in connection with this classical reception problem.

2.10 Phase estimation

We now address the situation where the parameter carrying useful information is the initial phase of the signal. This case is typical of coherent radar and navigation receivers, carrier reference recovery loops of PSK/QAM data transmission links, demodulators of 2G and 3G mobile radio receiver, chrominance channels of TV and many more applications.

We modify the bandpass signal model (2.24), separating the constant (during observation interval) initial phase φ, which is unknown and to be measured:

$$s(t;\varphi) = S(t)\cos(2\pi f_0 t + \gamma(t) + \varphi)$$

Since φ is a non-energy parameter, $E(\varphi) = E$ and ML estimation of phase consists in maximizing $z(\varphi) = \int_0^T y(t)s(t;\varphi)\,\mathrm{d}t$ over all $\varphi \in [-\pi, \pi]$. To make use of equation (2.59), note that by definition $\rho(\varphi)$ is the cosine of the angle between two signal copies phase-shifted by φ; that is, between two vectors separated by the angle φ. Therefore $\rho(\varphi) = \cos\varphi, \rho''(0) = -1$ and variance of the ML estimate $\hat{\varphi}$ is:

$$\mathrm{var}\{\hat{\varphi}\} \approx \frac{1}{q^2}, q \gg 1$$

Again, in common with amplitude measuring, precision of phase estimation is governed only by SNR. Thus, this classical problem is also indifferent to the signal modulation law, whenever signal energy is maintained constant, and *does not stimulate* spreading a signal spectrum.

2.11 Autocorrelation function and matched filter response

Spread spectrum theory rests to a very large extent on the notion of a signal *autocorrelation function* (ACF), which is defined as the inner product of two copies of the same signal time-shifted to each other by τ seconds:

$$R(\tau) = (\mathbf{s}_0, \mathbf{s}_\tau) = \int_{-\infty}^{\infty} s(t)s(t-\tau)\,\mathrm{d}t \tag{2.62}$$

Signal time delay τ is a non-energy parameter ($E(\tau) = E$) and scaling (2.62) by E^{-1} produces normalized ACF, which is simply a correlation coefficient of the time-shifted signal copies:

$$\rho(\tau) = \frac{(\mathbf{s}_0, \mathbf{s}_\tau)}{\|\mathbf{s}\|^2} = \frac{(\mathbf{s}_0, \mathbf{s}_\tau)}{E} = \frac{1}{E}\int_{-\infty}^{\infty} s(t)s(t-\tau)\,\mathrm{d}t \tag{2.63}$$

Clearly, the latter shows how rapidly the likeness of the signal time-spaced copies dies away with the delay mismatch τ. According to the general properties of a correlation coefficient $\rho(\lambda)$ indicated in Section 2.8, ACF is an even function of τ, attaining its maximum at zero point:

$$R(\tau) \le R(0) = E, R(\tau) = R(-\tau) \Leftrightarrow \rho(\tau) \le \rho(0) = 1, \rho(\tau) = \rho(-\tau) \tag{2.64}$$

Using equations (2.39) and (2.34) it is not difficult to verify that for any bandpass signal (2.37) ACF:

$$R(\tau) = \mathrm{Re}\left[\frac{\dot{R}(\tau)}{2}\exp(j2\pi f_0 t)\right], \rho(\tau) = \mathrm{Re}[\dot{\rho}(\tau)\exp(j2\pi f_0 t)] \tag{2.65}$$

where

$$\dot{R}(\tau) = (\dot{\mathbf{S}}_0, \dot{\mathbf{S}}_\tau) = \int_{-\infty}^{\infty} \dot{S}(t)\dot{S}^*(t-\tau)\,\mathrm{d}t \tag{2.66}$$

is the ACF of the complex envelope $\dot{S}(t)$, or, in other words, the modulation law. The normalized version of ACF (2.66):

$$\dot{\rho}(\tau) = \frac{(\dot{\mathbf{S}}_0, \dot{\mathbf{S}}_\tau)}{\|\dot{\mathbf{S}}\|^2} = \frac{(\dot{\mathbf{S}}_0, \dot{\mathbf{S}}_\tau)}{2E} = \frac{1}{2E}\int_{-\infty}^{\infty} \dot{S}(t)\dot{S}^*(t-\tau)\mathrm{d}t \tag{2.67}$$

being the correlation coefficient of two time-shifted copies of the complex envelope $\dot{S}(t)$, serves (after taking the modulus) as a measure of the rapidness with which the time-shifted modulation law loses similarity with the initial one when the mismatch τ grows. As may be seen from (2.65), ACF $\rho(\tau)$ of a bandpass signal $s(t)$ is a bandpass signal itself whose modulation law is ACF $\dot{\rho}(\tau)$ of the complex envelope of $s(t)$. In particular, the real envelope $\rho_0(\tau)$ of ACF of $s(t)$ is the modulus of $\dot{\rho}(\tau)$: $\rho_0(\tau) = |\dot{\rho}(\tau)|$, which is illustrated by Figure 2.16.

Any ACF can be obtained physically as an output of the *correlator*, i.e. the device running the straightforward operations set up by definitions (2.62) or (2.66). In this case calculations for a range of τ are fulfilled point by point, i.e. repeatedly in time or in hardware. An alternative solution is the *matched filter*, i.e. a linear system with the pulse response reproducing a mirror image of the signal: $h(t) = s(T - t)$, where T is, as usual, signal duration and an immaterial proportionality factor is set to be one. This filter emerges every now and again as an integral element of an optimal receiver in the

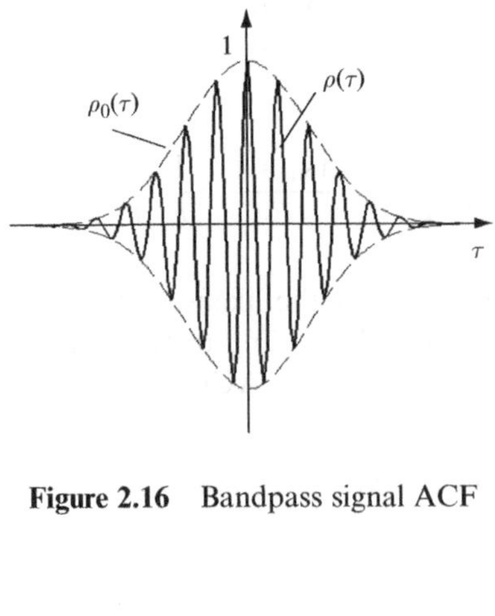

Figure 2.16 Bandpass signal ACF

AWGN channel, but its optimality often goes far beyond this specific channel model. In particular, it maximizes output SNR among all linear systems, signal given. In our current context the matched filter is important due to its ability to calculate and reproduce ACF as a real-time output waveform. To examine this, apply the signal $s(t)$ to the input of the filter matched to $s(t)$. The filter response $r(t)$ may then be calculated as the convolution integral:

$$r(t) = \int_{-\infty}^{\infty} s(\theta)h(t-\theta)\,d\theta = \int_{-\infty}^{\infty} s(\theta)s(T-t+\theta)\,d\theta$$
$$= \int_{-\infty}^{\infty} s(\theta)s[\theta-(t-T)]\,dt = R(t-T) \qquad (2.68)$$

and duplicates the ACF in real time with predictable delay equal to the signal duration.

To elucidate what was said, consider Figure 2.17. The rectangular baseband pulse $s(t)$ of duration T (Figure 2.17a) has a triangular ACF $R(\tau)$ of duration $2T$ with a maximum at the zero point (b, dashed line). In accordance with equation (2.68) the matched filter response reproduces a copy of this ACF delayed by the signal duration T so that maximal voltage at the filter output occurs at the moment when the input signal ends (b, solid line). If the pulse were bandpass with a rectangular envelope $s(t)$, its ACF would be a triangular bandpass pulse (c, dashed bold line) and its T-delayed copy would appear at the bandpass matched filter output (c, solid bold line). The maximum of the filter response to the signal the filter is matched to always occurs at the moment of signal ending (at least no earlier), since this filter processes the whole signal. It is very instructive to note that for a bandpass signal moments of maximal envelope and maximal value of carrier cosine at the matched filter output always coincide, since ACF always assumes its maximum at the zero point (see also Figure 2.16).

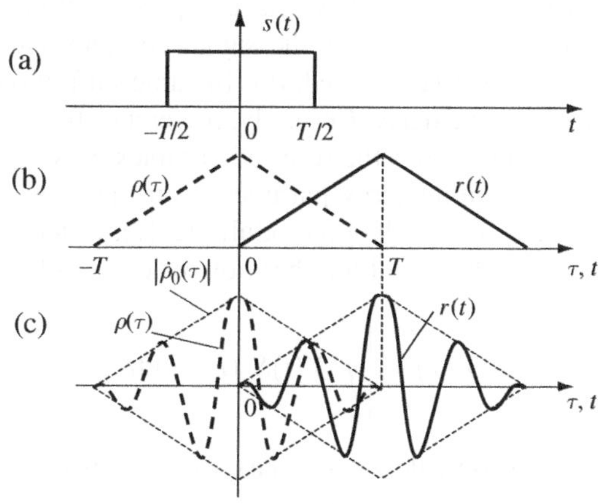

Figure 2.17 Illustration to the definition of ACF and its forming by the matched filter

2.12 Estimation of the bandpass signal time delay

2.12.1 Estimation algorithm

The problem we address in this section is among the most frequently encountered. It is typical of TV broadcasting (synchronization channels), digital mobile radio (pilot channels, timing recovery loops), radar (target distance measurement), space-based and ground-based navigation (beacon distance measurement) and so forth. To operate adequately, practically any modern information processing system needs to retrieve timing data from the received waveform, and this is exactly what is meant by *time delay estimation.*

Suppose that a bandpass signal (2.37) $s(t) = \text{Re}[\dot{S}(t)\exp(j2\pi f_0 t)]$ passing through the channel acquires unknown time delay τ and initial phase φ_0, i.e. takes the form:

$$s(t;\tau;\varphi_0) = s(t-\tau;\varphi_0) = \text{Re}\{\dot{S}(t-\tau)\exp[j2\pi f_0(t-\tau)+j\varphi_0]\}$$

In many situations phase φ_0 is random and uniformly distributed over the interval $[-\pi,\pi]$, i.e. it has no bearing on the only object of interest—the time delay τ. Let us incorporate the phase component caused by delay τ into the integral initial phase $\varphi = -2\pi f_0\tau + \varphi_0$. The latter, remaining random and uniformly distributed over $[-\pi,\pi]$, is again independent of τ, i.e. contains no information on it due to the destructive contribution of φ_0. Then the received signal may be represented by the following:

$$s(t-\tau;\varphi) = \text{Re}[\dot{S}(t-\tau)\exp(j2\pi f_0 t)\exp(j\varphi)] \tag{2.69}$$

where time delay τ is an unknown useful parameter to be measured and φ is a useless initial phase, whose uncertainty may only complicate the measurement of τ.

As was stated in Section 2.8, any estimation procedure is a particular case of signal distinguishing. In the case in question, we need to distinguish between multiple copies of signal (2.69) that differ from each other by value of the time shift τ, a *nuisance* parameter—initial phase φ—being an extra care. Fortunately, there is a straightforward way to overcome the signal indeterminacy related to the randomness of φ: in Section 2.5 it was shown that optimal choice between noncoherent signals is performed by distinguishing between their deterministic modulation laws, i.e. complex envelopes. In the delay estimation, consequently, time-shifted copies of the signal complex envelope $\dot{S}(t)$ should be compared and one of them declared received. It is the time shift of the latter which is given out as the ML estimate $\hat{\tau}$ of time delay. Certainly, preference for this copy over the rest is based on its minimum distance from the received complex envelope $\dot{Y}(t)$ or, taking into account that time delay is a non-energy parameter, on maximal correlation with $\dot{Y}(t)$. This correlation is evaluated by correlation modulus (2.47), which can be rewritten taking into account that the role of the signal number now belongs to the value of τ:

$$Z(\tau) = \left| \int_0^T \dot{Y}(t)\dot{S}^*(t-\tau)\,dt \right| \tag{2.70}$$

Based on this entity estimation rule $Z(\hat{\tau}) = \max_{\tau} Z(\tau)$ is absolutely transparent physically: the ML estimate $\hat{\tau}$ is just the time delay under which the signal modulation law has maximal resemblance with the observed one.

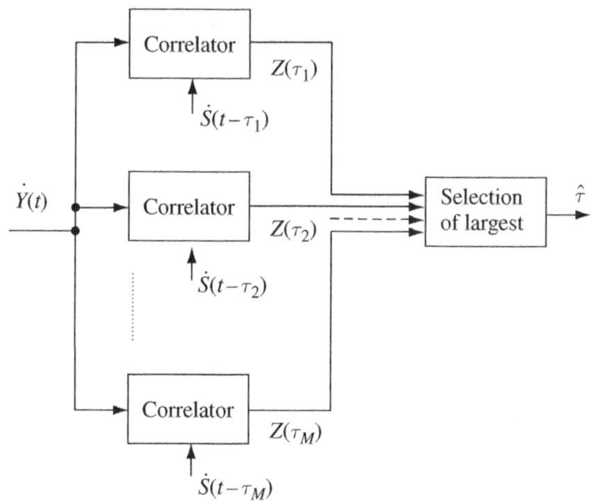

Figure 2.18 Correlator-bank implementation of the ML estimate of time delay

One possible solution to implement this estimation rule is a bank of correlators, shown in Figure 2.18. The observed complex envelope is processed in parallel by M correlators, whose reference signals are time-shifted copies of the signal complex envelope. At the correlator outputs values of $Z(\tau_i), i = 1, 2, \ldots, M$ are present, and the rightmost block compares them to select the largest. The ML estimate is the delay of the reference in the correlator whose output is maximal.

Of course, this structure treats the time delay as though it takes only discrete values. When this is not the case it simply quantizes continuous τ and the number of correlators (or, which is the same, discrete values τ_i) should be chosen sufficient to make the quantization error tolerably low.

The matched filter offers an alternative version of delay estimator. Let the observation $y(t)$ be applied to the filter matched to the signal $s(t)$. Find the output waveform $r(t)$ using the convolution integral and filter pulse response $h(t) = s(T - t)$:

$$r(t) = \int_{-\infty}^{\infty} y(\theta)h(t-\theta)\,\mathrm{d}\theta = \int_{-\infty}^{\infty} y(\theta)s(T-t+\theta)\,\mathrm{d}\theta.$$

This integral, being the inner product of $y(t)$ and $s(T - t + \theta)$, may be calculated with the aid of equations (2.39) and (2.34):

$$\begin{aligned} r(t) &= \int_{-\infty}^{\infty} y(\theta)s(T-t+\theta)\,\mathrm{d}\theta = \frac{1}{2}\mathrm{Re}\left[\int_{0}^{T} \dot{y}(\theta)\dot{s}^*(T-t+\theta)\,\mathrm{d}\theta\right] \\ &= \mathrm{Re}\left\{\left[\frac{1}{2}\int_{0}^{T} \dot{Y}(\theta)\dot{S}^*(T-t+\theta)\,\mathrm{d}\theta \cdot \exp(-j2\pi f_0 T)\right]\exp(j2\pi f_0 t)\right\} \end{aligned}$$

Comparing this result with the general bandpass signal model (2.37), one can see that the square brackets single out nothing but the complex envelope at the filter output.

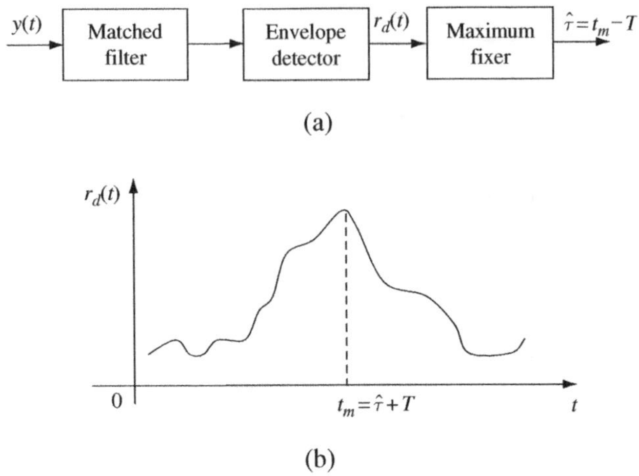

Figure 2.19 Matched filter implementation of the ML estimate of time delay

Therefore, the real output envelope (amplitude modulation) defined as the modulus of the expression within the square brackets:

$$\left| \frac{1}{2} \int_{-\infty}^{\infty} \dot{Y}(\theta)\dot{S}^*[\theta - (t - T)]\, d\theta \right| = \frac{1}{2} Z(t - T) \tag{2.71}$$

i.e. replicates in real time (with immaterial scaling by a factor 1/2) a copy of the correlation modulus (2.70) time-shifted by the known signal duration. This points directly to the possible structure of ML estimator of delay τ shown in Figure 2.19a. The observation $y(t)$ is first filtered by a matched filter and then passes through the envelope detector. The last unit in the structure registers the time moment t_m when the detector output waveform $r_d(t)$ takes a maximum value and the desired ML estimate $\hat{\tau}$ is obtained after subtraction of the known constant T from t_m (Figure 2.19b).

The scheme of Figure 2.19 seems more transparent for explaining the idea, but many practical software-based ML estimators may appear not so directly identifiable with either of the two structures just discussed.

2.12.2 Estimation accuracy

According to equation (2.59), the variance of estimation of τ depends on the steepness with which the time-shifted signal copy loses its similarity with the initial one. But in the case of random-phase signals only the deterministic complex envelopes are being compared to carry out the ML estimate. The similarity of time-mismatched copies of the complex envelope is characterized by the envelope of the signal ACF (2.65):

$$\rho_0(\tau) = |\dot{\rho}(\tau)| = \left| \frac{1}{2E} \int_{-\infty}^{\infty} \dot{S}(t)\dot{S}^*(t - \tau)\, d\tau \right| \tag{2.72}$$

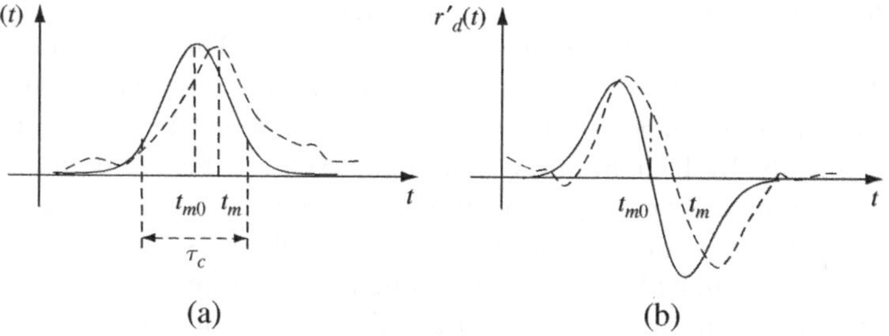

Figure 2.20 Illustration to the time measuring accuracy

and therefore its steepness should affect the error variance of measuring time delay, which can be strongly argued on physical grounds. Indeed, as is seen from equations (2.70) and (2.71), ACF modulus (2.72) is the noise-free envelope at the matched filter output (neglecting constant factor) taking a maximum value at some 'true' time-position t_{m0} (Figure 2.20a, solid line). When noise is added the time-position t_m of this maximum fluctuates with respect to the true point (Figure 2.20a, dashed line) spanning the range depending on the sharpness of the signal envelope at the filter output, i.e. of the ACF modulus (2.72).

To make the latter more evident, note that fixing the moment of a maximum of the detector output $r_d(t)$ is tantamount to registering the time-point where its derivative $r'_d(t)$ crosses the zero level (provided it is selected properly against spurious 'zero-points' caused by possible side maximums). This is illustrated by Figure 2.20b. When SNR is high enough, deviation $\varepsilon = t_m - t_{m0}$ is small and we may assume that the noisy (dashed) curve $r'_d(t)$ is linear within the range $[t_{m0}, t_m]$, having the same slope as the noiseless (solid) curve has at the point t_{m0}. Hence, solving the right triangle seen in Figure 2.20b, ε can be found as a result of division of its dashed-dotted leg by the steepness of the noiseless curve $r'_d(t)$ at the point t_{m0}, i.e. by the second derivative $r''_d(t_{m0})$. The latter, in its turn, is exactly $\rho''_0(0)$, so that $\varepsilon \approx r'_d(t_{m0})/\rho''_0(0)$. On the other hand, variance of scattering of $r'_d(t_{m0})$ around the noiseless zero value is greater the smaller is SNR and the higher is the rate of random change of the noisy detector output $r_d(t)$ (dashed line in Figure 2.20a). The sharpness of ACF of a random process tells us about the rate of its change, and ACF of the detector output random process under high SNR repeats the envelope of ACF at the detector input. The latter envelope, when the filter is matched to the signal, is nothing but the envelope $\rho_0(\tau)$ of the signal ACF. Since the sharpness of any ACF is measured by its second derivative at zero point with a minus sign, the variance of $r'_d(t_{m0})$ is proportional to $-\rho''_0(0)$. Then:

$$\text{var}\{\varepsilon\} = \frac{\text{var}\{r'_d(t_{m0})\}}{|\rho''_0(0)|^2} \equiv \frac{1}{-\rho''_0(0)}$$

is inversely proportional to the sharpness of the signal ACF $-\rho''_0(0)$ as was predicted earlier and as follows from the Cramer–Rao bound (2.59).

We thereby came to quite an important conclusion: time-measurement accuracy is critically governed by the signal ACF *sharpness*, and the sharper is ACF, the smaller is the variance of the ML estimate of time delay τ.

In parallel with $-\rho_0''(0)$, another indicator of signal ACF sharpness can be introduced, which we will call the *correlation spread* and denote τ_c. This parameter characterizes signal ACF width (see Figure 2.20a) and—like duration or bandwidth—should be defined by a convention, since ACF can be of rather complicated shape and fall to zero only asymptotically. In the light of the meaning of ACF, we believe that signal copies (or copies of the complex envelope) with mutual time-shift $\tau < \tau_c$ have significant resemblance while with $\tau > \tau_c$ their resemblance is negligible. It is clear that the conclusion above may be reformulated in terms of this new entity: signals with *narrow* ACF, i.e. *small* correlation spread, are generally preferable for the high precision of time-delay estimate.

Continuing, we may recall one of the basic facts of spectrum analysis following directly from result (2.72) after applying the Parseval theorem to it: signal ACF and energy spectrum are related to each other by the Fourier transform. In terms of the complex envelope:

$$\rho_0(\tau) = \frac{1}{2E}\left|\int_{-\infty}^{\infty} \left|\dot{\tilde{S}}(f)\right|^2 \exp(j2\pi f\tau)\, \mathrm{d}f\right|$$

where the complex envelope spectrum $\dot{\tilde{S}}(t)$ is physically (neglecting a proportionality coefficient) a bandpass signal spectrum moved to the baseband area. Then according to the general Fourier transform property, $\tau_c \approx 1/W$, or in other words, the narrower is ACF the wider is the signal spectrum and vice versa. The implication of this is straightforward: a possible way of improving time-measurement accuracy consists in employing signals with a wide spectrum. To come to the same conclusion formally, one may differentiate the last expression of the ACF envelope and substitute the result into the Cramer–Rao bound (2.59). With some tedious work, this leads to the equation often referred to as the Woodward formula:

$$\mathrm{var}\{\hat{\tau}\} \approx \frac{1}{(2\pi W_{rms})^2 q^2}, \quad q \gg 1$$

where the signal *root-mean-square* (rms) bandwidth is introduced:

$$W_{rms} = \frac{1}{2E}\int_{-\infty}^{\infty} f^2 \left|\dot{\tilde{S}}(f)\right|^2 \mathrm{d}f$$

It may appear difficult to understand why a measure like this reports about the spectrum width. In this case analogy with the more customary probabilistic scattering parameter is advisable. Variance of the random variable x with zero mean and PDF $W(x)$ is by definition $\mathrm{var}\{x\} = \int_{-\infty}^{\infty} x^2 W(x)\mathrm{d}x$, characterizing a scattering range of x around its expectation, or, which is equivalent, width of PDF $W(x)$. But the normalized energy spectrum $\left|\dot{\tilde{S}}(f)\right|^2/2E$ is non-negative and satisfies the condition:

$$\int_{-\infty}^{\infty} \left[\left| \tilde{\dot{S}}(f) \right|^2 / 2E \right] \mathrm{d}f = 1$$

i.e. may be treated as a PDF of some appropriate random variable. Then W_{rms} is a measure of scattering of this 'dummy' random variable, and by that characterizes the width of an energy spectrum $\left| \tilde{\dot{S}}(f) \right|^2$.

Thus, roughly speaking, the criterion for good signals in time-measurement tasks is: signals with short ACF (small correlation spread τ_c) or, equivalently, wide bandwidths W, are of the primary interest. Compared with the previously considered two estimation problems the situation looks pretty new: there exists a sustainable resource to improve estimation fidelity beyond a brute-force course, i.e. just extra energy expenditure.

It should be stressed now that 'wide band' and 'spread spectrum' are not synonyms. As a matter of fact, if conventions adopted on the correlation spread τ_c and the signal duration T definitions are sound enough, signal copies shifted by more than T are practically non-overlapping, i.e. have negligible inner product or resemblance. Therefore, $\tau_c \leq T$, showing that shortening signal ACF and, consequently, widening its bandwidth can be achieved trivially by shortening the signal itself.

However, going this way one should not forget that SNR depends on signal energy $E = k_p PT$ where P is the signal peak power and k_p is a coefficient determined by the signal shape. It has to be clear that preserving SNR when the signal is being shortened requires a proportional increase of peak power. Consequently, in the pursuit of higher and higher precision, it is possible to arrive at the situation where the necessary peak power becomes prohibitively large. Generally, very high transmitted power entails large mass and dimension of transmitter equipment and an energy source. In addition to that, short high-power pulses may appear substantially damaging for neighbouring systems and the surrounding ecology.

A more elegant way of improving the time-measurement accuracy is prompted by the fact that signal shortening is not the only method of widening the spectrum or, which is the same, reducing the correlation spread. Consider a signal whose duration T is large enough to provide necessary energy, i.e. SNR, in combination with acceptably low peak power P. Suppose that the internal angle modulation law is found, making a signal correlation spread much smaller than signal duration: $\tau_c \ll T$. Then, signal ACF is sharp, providing high-accuracy estimation of time delay, despite the long duration of the signal itself. But in the light of the dependence between the correlation spread and the bandwidth $\tau_c \approx 1/W$ inequality $\tau_c \ll T$ means that the signal has a large time–frequency product $WT \gg 1$, i.e. it is a spread spectrum one. Putting it another way, involvement of the spread spectrum allows the contradiction between peak power and estimation precision to be removed: necessary energy is put into a signal at the cost of duration, not power, while high measurement accuracy is achieved by designing an appropriate modulation law.

Physically, fulfilment of the condition $\tau_c \ll T$ means that a long signal becomes short after processing in the matched filter of the estimator shown in Figure 2.19a. It should be clear that this matched filter *time-compression* phenomenon is achievable only with spread spectrum signals. In principle, any signal may be time-shortened in some purposely designed (generally mismatched) filter, e.g. an equalizer, but for plain signals the price of this is loss in SNR, and only spread spectrum signals promise best 'noise-clearance' simultaneously with the time compression. At the same time it has to be understood that the spread spectrum condition $WT \gg 1$ is just the necessary one, and finding signals combining long proper duration with a sharp ACF is a rather sophisticated task. There is further discussion of this in Chapter 5.

Consider Figure 2.21 where waveforms are shown simulated by Matlab. The plots in column (a) present, respectively, a plain bandpass bell-shaped pulse, matched filter response to it (delayed ACF) and 10 superimposed noisy realizations at the detector output. The plots of column (b) show similar waveforms but for the spread spectrum signal, which is a linear-frequency-modulated pulse having the same shape, duration and energy as that of column (a). The time-compression effect for case (b) manifests itself clearly and results in a noticeably narrower range of fluctuations of the time position of a maximum at the detector output in comparison with case (a). This demonstrates convincingly how spread spectrum can improve time-measurement accuracy without compromising peak power.

We now formulate the following conclusion. When no peak power limitation is imposed, the classical problem of time-delay estimation *does not appeal* strongly to the spread spectrum. However, spread spectrum *is an imperative demand* when tough peak power constraints need to be obeyed. Note in passing that the latter situation is quite typical of the pulse radar, which explains why this application area has for decades been stimulating the development of spread spectrum technology.

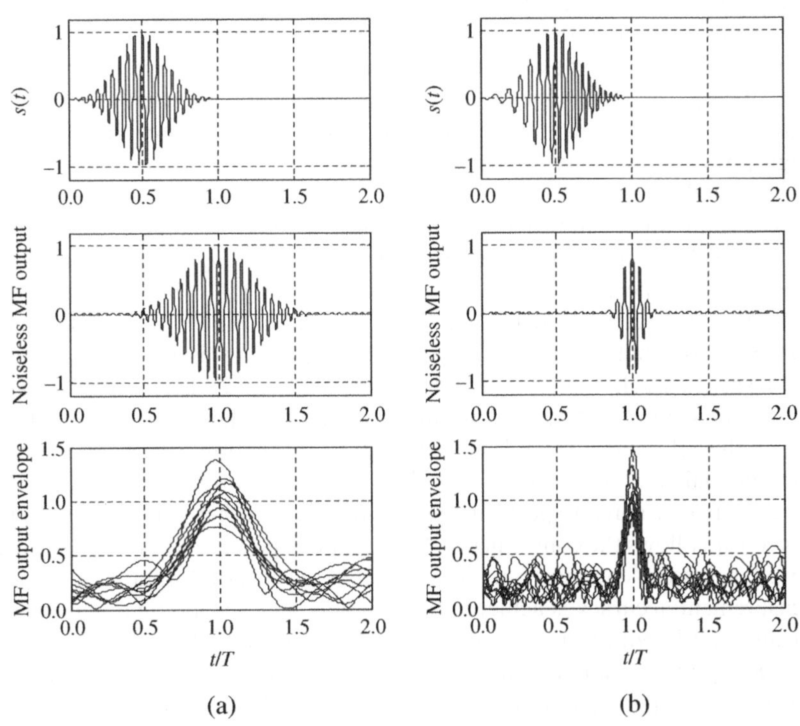

Figure 2.21 Illustration of time estimation and matched filter time-compression effect

2.13 Estimation of carrier frequency

Consider now the situation where signal carrier frequency is an unknown informative parameter. This problem is as widespread as the previous one. It is met in radar measuring target velocity via Doppler frequency shift (e.g. police radar for road traffic monitoring), reference recovery loops of 2G and 3G mobile telephone receivers, automatic frequency control system in FM and TV receivers and so on.

In actual practice it is typical that only bias F versus the carrier frequency nominal value f_0 has to be measured, whereupon a signal model can be written as:

$$s(t; F; \varphi) = \mathrm{Re}\{\dot{S}(t)\exp[j2\pi(f_0 + F)t + j\varphi]\} = \mathrm{Re}[\dot{S}(t; F)\exp(j2\pi f_0 t)\exp(j\varphi)]$$

where $\dot{S}(t; F) = \dot{S}(t)\exp(j2\pi Ft)$ is the signal complex envelope including the linear phase drift due to the frequency shift F, and φ is again a nuisance parameter—random initial phase containing no information on F.

Because of the non-energy nature of F, the correlation rule $Z(\hat{F}) = \max_F Z(F)$ is relevant, where the correlation modulus $Z(F)$ is again produced by a slight adaptation of (2.47) with F playing the role of k:

$$Z(F) = \left| \int_0^T \dot{Y}(t)\dot{S}^*(t; F)\,dt \right| = \left| \int_0^T \dot{Y}(t)\dot{S}^*(t)\exp(-j2\pi Ft)\,dt \right|$$

The scheme of Figure 2.18 may be used to realize this optimal estimation if instead of time-shifted copies of the complex envelope, its frequency shifts $\dot{S}(t; F_k) = \dot{S}(t)\exp(j2\pi F_k t)$, $k = 1, 2, \ldots, M$ are used as correlator references. On the other hand, it is seen from equation (2.71) that the amplitude at the output of the matched filter, which is tuned to the frequency $f_0 + F_k$, takes a value $Z(F_k)$ (neglecting a coefficient) at the moment of signal ending T. Thus, an alternative structure (see Figure 2.22) may be used containing a bank of M matched filters, the kth one being detuned with respect to the nominal frequency f_0 by F_k Hz. After amplitude detecting and sampling of detector outputs at moment T, the set of $Z(F_k)$ is obtained, the largest of

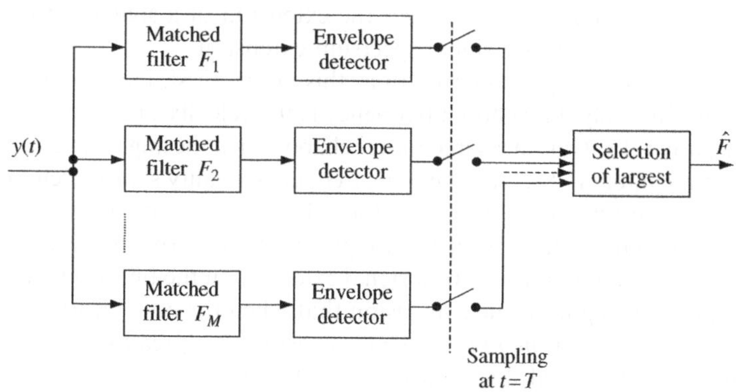

Figure 2.22 Matched filter implementation of the ML estimate of frequency

which is then selected to derive the ML estimate $\hat{F}$ as a frequency mismatch of the filter outputting the largest $Z(F_k)$. Clearly, continuous F here is again quantized and the number of filters M should be sufficient to tolerate the quantization error.

Precision of the frequency measurement in accordance with equation (2.59) is governed not only by SNR but also by the sharpness of the correlation between frequency-shifted signal copies in dependence on their mutual frequency mismatch. Due to the phase randomness, we operate with the complex envelope, and the resemblance between its frequency-detuned copies $\dot{S}(t;0)$ and $\dot{S}(t;F)$ is characterized by the modulus of the correlation coefficient (2.44):

$$\rho_0(F) = |\dot{\rho}(F)| = \left|\frac{1}{2E}\int_{-\infty}^{\infty} \dot{S}(t;0)\dot{S}^*(t;F)\,\mathrm{d}t\right| = \left|\frac{1}{2E}\int_{-\infty}^{\infty} \left|\dot{S}(t)\right|^2 \exp(-j2\pi Ft)\,\mathrm{d}t\right|$$
$$= \left|\frac{1}{2E}\int_{-\infty}^{\infty} S^2(t)\exp(j2\pi Ft)\,\mathrm{d}t\right| \qquad (2.73)$$

As is seen, this correlation coefficient as a function of F duplicates in its shape the amplitude spectrum of the signal squared real envelope. Since it is a real envelope, i.e. the amplitude modulation law is always some baseband signal, it follows from the Fourier transform properties that the longer is the signal, the sharper is $\rho_0(F)$. An appropriate name for the extension of $\rho_0(F)$ along the frequency axes is the 'envelope frequency spread'. Denoting it by F_e we have from the fact just mentioned $F_e \approx 1/T$. Thus, beyond the 'brute force' way of increasing the signal energy, it is possible to improve the accuracy of the frequency measurement by employing a signal having a rather compact spectrum of the envelope (small F_e), i.e. sufficiently long duration T. Formally, this conclusion may be deduced again by evaluation of the second derivative of $\rho(F)$ and substituting it into the Cramer–Rao bound (2.59):

$$\mathrm{var}\{\hat{F}\} = \frac{1}{(2\pi T_{rms})^2 q^2}$$

where the rms duration T_{rms} characterizes the extension of a signal in time just as rms bandwidth W_{rms} characterizes a signal spectrum extension.

Two physical aspects may be referred to at this point. First, increase of precision with signal duration T is easily understood: frequency is the velocity of a signal complete phase angle and, like any velocity, is measured more reliably when the angle increment is observed over a longer time interval. Second, the time–frequency duality shows itself in comparison of the frequency and time-delay estimates. Indeed, while the time-measurement precision is governed by a signal extension in the frequency domain (bandwidth W), frequency-measurement precision is controlled by a signal extension in the time domain (duration T).

The result above bring us to the conclusion that when the only informative parameter is a signal frequency there is *no momentum* to resort to the spread spectrum, since, apart from the energy, only signal duration is influential as to the estimate accuracy. This explains why the Doppler speed-monitoring radar very often operates with the plainest unmodulated CW harmonic signal.

2.14 Simultaneous estimation of time delay and frequency

Our next subject is the situation where both time τ and frequency F shifts of the received signal are unknown and informative, i.e. to be measured. This matches many practical scenarios. In digital communications, e.g. 2G and 3G mobile radio, a receiving session typically starts with synchronizing the local reference with the received signal. This procedure consists of measuring time–frequency misalignment of the local clock against the received signal and subsequent time–frequency adjustment of the former towards its synchronism with the latter. In radar technology to measure simultaneously the distance and the velocity of the target relative to the receiver the signal time delay and Doppler frequency are estimated. In navigation, e.g. GPS, much the same sort of estimation serves to measure the user's own location and velocity. There are many similar examples.

In distinction to the material of Sections 2.9–2.13 the parameter to be estimated is now the two-dimensional vector $\lambda = (\tau, F)$ rather than a scalar. Accordingly, the model of the received signal combines the two of Sections 2.12 and 2.13:

$$s(t; \tau, F; \varphi) = \mathrm{Re}\left[\dot{S}(t; \tau, F)\exp(j2\pi f_0 t)\exp(j\varphi)\right]$$

where $\dot{S}(t;\ \tau, F) = \dot{S}(t-\tau)\exp(j2\pi F t)$ is the complex envelope incorporating both the time delay and the frequency shift; and φ is, as before, a non-informative initial phase. It is obvious that now the correlation rule of estimating τ, F operates with the correlation modulus:

$$Z(\tau, F) = \left|\int_0^T \dot{Y}(t)\dot{S}^*(t; \tau, F)\,\mathrm{d}t\right| = \left|\int_0^T \dot{Y}(t)\dot{S}^*(t-\tau)\exp(-j2\pi F t)\mathrm{d}t\right| \tag{2.74}$$

which indicates how closely the time and frequency biased copy of the signal complex envelope $\dot{S}(t;\ \tau, F)$ resembles the observed complex envelope $\dot{Y}(t)$. Finding τ and F, which maximize such a resemblance, provides the pair of the ML estimates $\hat{\tau}, \hat{F}$: $Z(\hat{\tau}, \hat{F}) = \max\limits_{\tau, F} Z(\tau, F)$.

As previously, one can imagine the correlator-bank structure of the ML estimator where references of different correlators are the copies $\dot{S}(t;\tau, F) = \dot{S}(t-\tau)\exp(j2\pi F t)$ with different values τ, F. More instructive, however, is the matched-filter-based scheme combining those of Figures 2.22 and 2.19a and illustrated by Figure 2.23. It follows directly from the comparison of equations (2.74) and (2.71) showing that (neglecting a constant factor) $Z(\tau, F)$ is duplicated by the real envelope at the output of the matched filter, the filter being frequency-detuned by F. M filter-detector branches are taken and each of them is tuned to its specific frequency, securing thereby estimation of F, while estimation of τ is obtained by fixing the point of a maximum value at the detector output. The 'Selection of largest' unit implements both these operations together, fixing the time moment where the global maximum among the output values of all the branches occurs. Then this moment itself (after subtracting the signal duration) gives $\hat{\tau}$, while the frequency tuning of the branch at whose output the global maximum is registered provides $\hat{F}$.

The estimation precision depends on the rapidity with which the resemblance between the time–frequency mismatched copies $\dot{S}(t; 0, 0)$ and $\dot{S}(t; \tau, F)$ of the signal complex

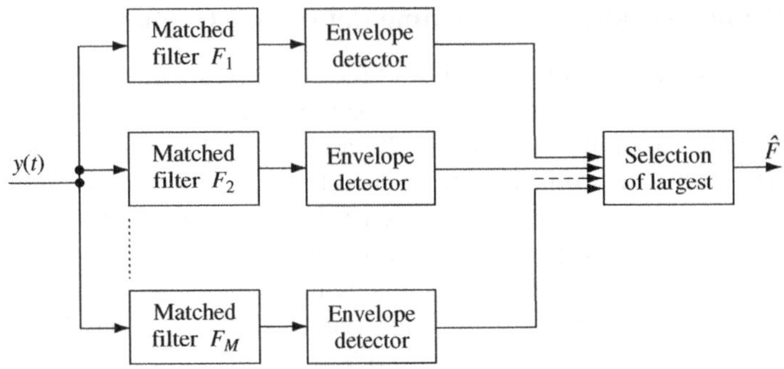

Figure 2.23 Matched filter implementation of the time delay and frequency ML estimate

envelope decays with growth of τ, F. In other words, the precision is governed by the sharpness of the modulus of the correlation coefficient (2.44):

$$\rho_0(\tau, F) = |\dot{\rho}(\tau, F)| = \left| \frac{1}{2E} \int_{-\infty}^{\infty} \dot{S}(t; 0, 0)\dot{S}^*(t; \tau, F)\,\mathrm{d}t \right|$$
$$= \left| \frac{1}{2E} \int_{\infty}^{\infty} \dot{S}(t)\dot{S}^*(t - \tau)\exp(-j2\pi Ft)\mathrm{d}t \right| \tag{2.75}$$

as a function of the two variables τ, F. This function, often called the Woodward *ambiguity function*, is crucially important in signal theory. Geometrically it may be plotted as a three-dimensional surface over the plane τ, F having a maximum equal to one at the origin: $\rho_0(\tau, F) \leq \rho_0(0, 0) = 1$. Figure 2.24 gives an example view of the ambiguity function.

When $F = 0$ signal copies are time-detuned only, and in accordance to it, the ambiguity function (2.75) turns into the ACF modulus (2.72): $\rho_0(\tau, 0) = \rho_0(\tau)$. On the other

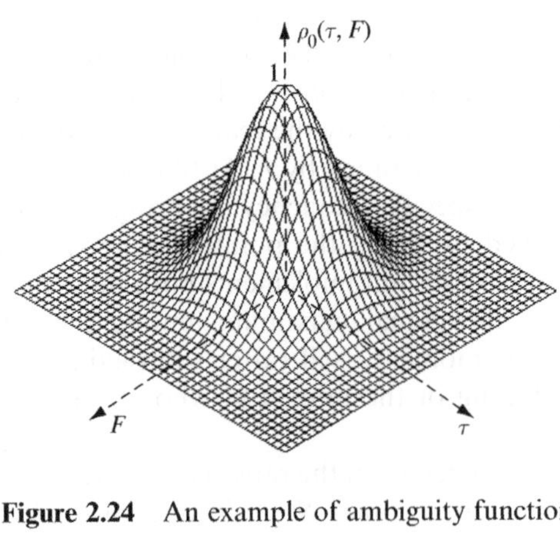

Figure 2.24 An example of ambiguity function

hand, with no time shift $\tau = 0$ signal copies are mismatched in frequency only, and the ambiguity function becomes the spectrum (2.73) of the squared signal amplitude modulation law: $\rho_0(0, F) = \rho_0(F)$. In other words, the signal ACF and squared envelope spectrum represent profiles of the ambiguity function along the vertical planes $\tau = 0, F = 0$, respectively. Returning to the structure of Figure 2.23, one may see that in the case of noiseless observation the kth channel of it reproduces in real-time the profile of the ambiguity function along the vertical plane $F = F_i$.

As follows from the reasoning above, the only extra-energy-free resource of increasing the accuracy of the time–frequency estimation is sharpening the ambiguity function. The latter should die rapidly enough along any direction in the plane τ, F if no material a priori information about possible values of τ, F is involved. One relevant way to characterize the sharpness of the $\rho_0(\tau, F)$ is to consider its horizontal section (called *ambiguity diagram*) at some predetermined level, e.g. 0.5. Figure 2.25 exemplifies such a section. Since the extensions of the two basic profiles $\rho_0(\tau)$ and $\rho_0(F)$ are measured by the correlation spread ($\tau_c \approx 1/W$) and envelope frequency spread ($F_e \approx 1/T$), those spreads define automatically the sizes of the section along the axes τ, F. Certainly, the sharper the ambiguity function, the smaller is the square of the ambiguity diagram. The latter is proportional to the product $\tau_c F_e \approx 1/WT$, which entails the statement: *only spread spectrum signals allow increasing the accuracy of time delay (frequency) estimation without compromising the accuracy of measuring frequency (time delay)*. Indeed, for any plain signal $WT \approx 1$ and, consequently, $\tau_c F_e \approx 1$, so that making an ambiguity function sharper along one direction (e.g. τ) is not possible without simultaneously stretching it along the other direction (F).

With spread spectrum signals the opportunity exists to escape the contradiction between the extensions τ_c, F_e of an ambiguity function, or equivalently between the signal duration and bandwidth, similarly to what was studied in Section 2.12. Specifically, an appropriate internal angle modulation secures signal bandwidth W (correlation spread τ_c) and thereby time-measurement accuracy, while the choice of an appropriate duration T can be carried out independently to guarantee the necessary frequency-measurement accuracy.

Comparing the statement we arrived at with the previous ones, we see that among all the classical reception problems examined so far, simultaneous time–frequency

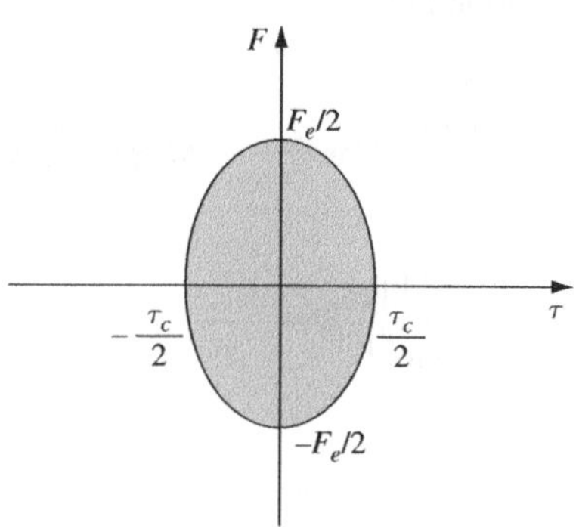

Figure 2.25 Ambiguity diagram: horizontal section of the ambiguity function

estimation is the first *appealing* to the spread spectrum philosophy in an absolutely categorical way. There are *no other means* to make the requirements for high precision in both time and frequency estimations non-conflicting than the involvement of spread spectrum signals.

2.15 Signal resolution

It is extremely typical of many practical systems that the received signal is actually a superposition of many copies of the original 'pure' signal, each copy having its specific amplitude, phase, time delay and frequency shift. Time-overlapped, these copies interfere with each other, creating a rather complicated resulting signal and often hampering the extraction of the necessary information. The procedure directed to separating the interfering signal copies or neutralizing their mutual harmful effects is called signal *resolution*. There may be different ways of treating this problem in terms of the classical reception approach depending on what the final objectives are. All of them, however, are again based critically on the category of the signal copies' distance or resemblance.

To explain the idea better, consider the very indicative case of the *time resolution*. Start with the bandpass signal $s(t) = \mathrm{Re}[\dot{S}(t)\exp(2\pi f_0 t)]$ and assume that two of its copies having equal amplitudes and different time delays $\tau_1 = 0, \tau_2 = \tau$ arrive at the receiver input. Then the superposition waveform:

$$s_r(t) = s(t) + s(t-\tau) = \mathrm{Re}\{[\dot{S}(t) + \dot{S}(t-\tau)]\exp(j2\pi f_0 t)\}$$

is strongly determined by the mutual time shift τ in comparison to both the signal duration T and carrier frequency as well as by the 'fine' details of the signal modulation. Figure 2.26 illustrates several situations for the example of the plain bell-shaped pulse, plot a corresponding to the 'pure' signal itself, while value τ for each of the three other plots is given in terms of the carrier frequency f_0.

The case when the time difference τ is greater than the signal duration T (plot b) causes no problem, since the signal copies are completely separated in time and an

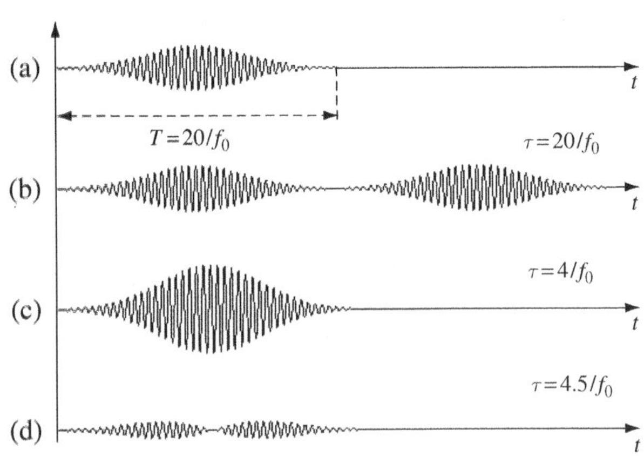

Figure 2.26 Illustration of the time-resolution problem

observer is able to recognize that there are two of them. Obviously, no interference effect is present and if, apart from time position, some other signal parameters are of interest this information may be retrieved from every copy without any major difficulties.

Another matter is the situation where signals are strongly overlapping. Then interference may lead either to amplifying (if the copies are nearly in-phase, plot c) or attenuation of the signal (the copies' phase difference is around π, plot d). The likely problems of an observer are clearly seen, considering that the received waveform is always corrupted by channel noise. He can hardly recognize the presence of more than one copy (and be fully confident that the interference does not damage the information retrieved, plot c), or meets prohibitively low SNR (d).

Scenarios of this kind are widespread throughout systems related to transporting and collecting information. The main physical phenomena underlying them are the limited channel bandwidth and multipath wave propagation. The latter will be discussed in more detail in Section 3.5. For now it is enough to look briefly at some characteristic examples. In digital telecommunications two sorts of channel distortion are often among the most troublesome: *intersymbol interference* (ISI) and *fading*. They are somewhat akin, both being the result of the linear summation of the multitude of the weighted and delayed signal copies. When the range of delays surpasses the signal correlation spread, or, equivalently, the channel transfer function is explicitly nonuniform within the signal spectrum, the received superposition is substantially distorted against the original signal, which is typical of ISI or *frequency-selective fading*. Non-selective or *flat* fading appears when delays of signal copies differ insignificantly in terms of the signal correlation spread (the channel transfer function is nearly uniform) but are sufficient to create destructive mutual phase shifts between copies, causing risk of a perceptible loss in the resulting signal power. Clearly, the problem of resolution, i.e. effective separation of the signal time-shifted copies, is directly relevant to the task of ISI suppression.

Analogously, in radar a set of time-shifted signal copies may enter the received waveform due to reflection of the emitted pulse by multiple targets. One of the primary necessary capabilities of radar is to tell how many targets are present on some definite direction and to measure the distances to all of them. It is immediately seen that the scenarios described by Figure 2.26b are easy from this standpoint, while the situations where inter-target distances are small enough to cause overlapping of their echoed signals (Figure 2.26c, d) demonstrate the essence of the resolution problem.

Multipath propagation is also typical of navigation problems. In many ground-based systems its nature is attributed to the ionosphere reflecting strongly long and medium waves. Due to this the earth's surface and the ionosphere create a waveguide, along which many alternative propagation modes, i.e. paths, may exist. In space-based systems the same problem arises, since along with the line-of-sight signal from the satellite multipath signals caused by reflections from some surrounding objects (e.g. ship masts or deck erections) may also reach a receiver input. All these situations fall well within the realms of the resolution problem.

Analysing Figure 2.26, it would seem reasonable to jump to the conclusion that the appropriate way of achieving good time-resolution consists in employing short signals. Certainly, shortening a signal may be a solution to the problem; however, the limitations discussed thoroughly in Section 2.12 remain in force. When pursuing high-resolution capability one cannot simply reduce the signal duration; a proportional increase of the peak-power should also be provided to preserve the necessary SNR. Thus, peak-power constraints make the potency of this simple strategy not very encouraging.

Fortunately, an alternative and much more elegant way of solving the time-resolution problem still exists, actually described previously. Just as in the case of time-measurement, it is the duration of the signal ACF but not of the signal itself that affects resolution capability, since clearing the signal off the noise by the matched filter may be looked upon as an integral part of any reception procedure. Consequently, the requirement for a small correlation spread τ_c or sharp ACF is dictated by the time-resolution, duplicating the one formulated in the problem of time-delay estimation. Putting it differently, the time-shifted signal copies may strongly overlap and interfere with each other severely, but if their mutual time shift τ exceeds the correlation spread τ_c they will be observed as non-overlapping at the matched filter output, i.e. will be resolved. Clearly, we are arriving again at the idea of Section 2.12: necessary signal energy (SNR) is secured by a high signal duration T allowing compliance with the peak-power limitation, while the proper choice of the internal angle modulation is responsible for a sufficiently small correlation spread $\tau_c \ll T$. Then the time-compression effect in the matched filter will be able to provide separation of the overlapping signal copies; that is, their successful resolution. This is potentially achievable only with spread spectrum signals, whose time–frequency product $WT \approx T/\tau_c \gg 1$.

Therefore, we may literally cite the conclusion of Section 2.12 in application to time resolution. When no peak-power limitation is imposed the problem *does not appeal* strongly to spread spectrum. However, spread spectrum *is a vital demand* when tough peak-power constraints must be obeyed.

Figure 2.27 presents two indicative situations simulated in Matlab. Column (a) shows the bell-shaped plain pulse $s(t)$, its τ-shifted copy $s(t-\tau)$ strongly overlapping with $s(t)$, superposition of $s(t)$ and $s(t-\tau)$ at the matched filter input and the filter response to this superposition. Column (b) contains similar plots for the spread spectrum (linear-frequency modulated) bell-shaped pulse of the same duration and energy. With identical time-shifts of the copies in both cases the second signal shows excellent resolution performance—the two signal copies are fully separated—while the pattern with the first one shows no resolution at all.

The concept developed above for the case of time-resolution may be easily extended to resolution in other parameters. When several signal replicas with only frequency shifts differing are overlapping the frequency resolution problem arises, where performance is affected by the correlation coefficient (2.73) similarly to the frequency estimation. Naturally, nothing urges the use of spread spectrum in this case (see Section 2.13). If both time and frequency shifts are characteristic of the superimposed signal replicas, the procedure is the time–frequency resolution and its quality is governed by the ambiguity function (2.75). As in the case of time–frequency measurement (Section 2.14), this situation *appeals* to the spread spectrum crucially: *no other ways* offer the promise of obtaining a sharp ambiguity function along any directions in the τ, F plane.

To conclude this section, there are also numerous scenarios of space resolution where signal copies reach the receiving antenna from different directions and the task of an observer consists in separate processing of each copy. It is the antenna itself (or antennas, including the transmit one) that then plays the role of a space signal,

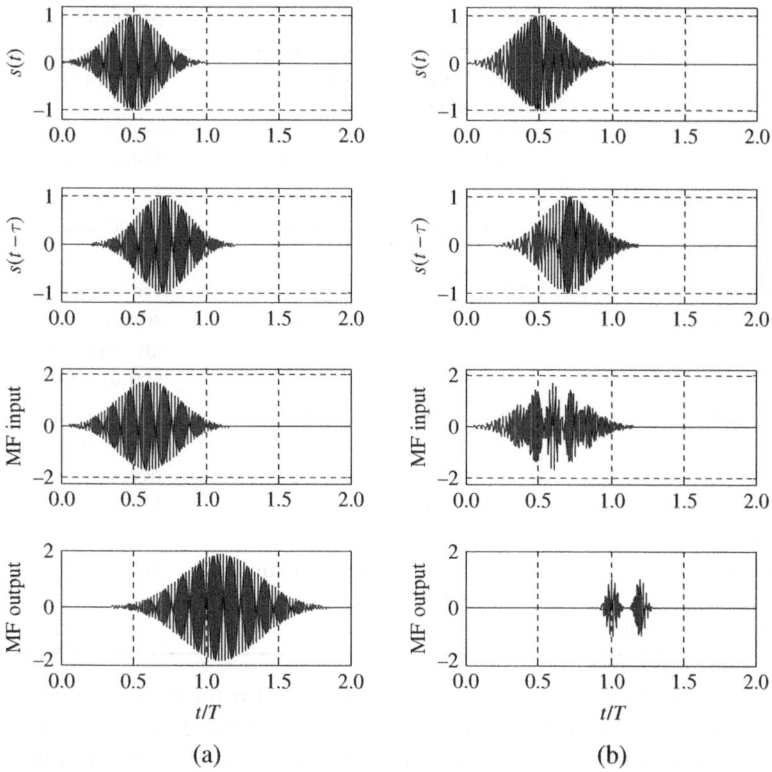

Figure 2.27 Time-resolution problem: the plain (a) and spread spectrum (b) signals

and the 'signal' design is directed towards the most effective combined processing of the waveforms received by antenna elements. Many ideas from time–frequency resolution can be applied to space resolution problems.

2.16 Summary

In this chapter we visited briefly the basic procedures of signal reception: detection, recognition, parameter estimation and resolution. Throughout the chapter the Gaussian background noise model was accepted, conforming to the classical problem formulation, and the goal pursued was to see whether this somewhat idealized approach appeals strongly to the spread spectrum technology. The conclusions are summarized in Table 2.1, which lists the signal parameters that affect the performance in every specific procedure, and indicates the motivation towards involving the spread spectrum when non-energy-consuming performance improvement is wanted.

As the table shows, it would not be fair to say that the classical reception problems appeal much to the spread spectrum philosophy. Only the simultaneous time–frequency estimation and resolution give unequivocal momentum to its involvement. This may

Table 2.1 Role of spread spectrum signals in the classical reception problems

Problem	Performance-influencing signal parameters	Spread spectrum signals
Detection, amplitude and phase measurement	SNR (signal energy only)	Unnecessary
Binary data transmission ($M = 2$)	SNR, correlation coefficient	Unnecessary
M-ary data transmission, $M > 2$	SNR, correlation coefficients of all signals	Unnecessary but sometimes attractive from an implementation standpoint
Time delay measurement, time-resolution	SNR, signal bandwidth	Unnecessary when power-limit-free, necessary otherwise
Frequency measurement, frequency resolution	SNR, signal duration	Unnecessary
Time–frequency measurement, time–frequency resolution	SNR, signal bandwidth and duration	Necessary

seem very odd and lead to questioning of the grounds for the wide popularity of spread spectrum nowadays. As the following chapter shows, however, these grounds are quite solid and manifest themselves clearly whenever we base our study on a more realistic channel model than the sometimes 'academic' Gaussian one, or need to draw in some additional performance criteria.

Problems

2.1. Three signals $s_1(t), s_2(t), s_3(t)$ are given in Figure 2.28.
By how many times is the maximum distance in this signal set greater than the minimum one?

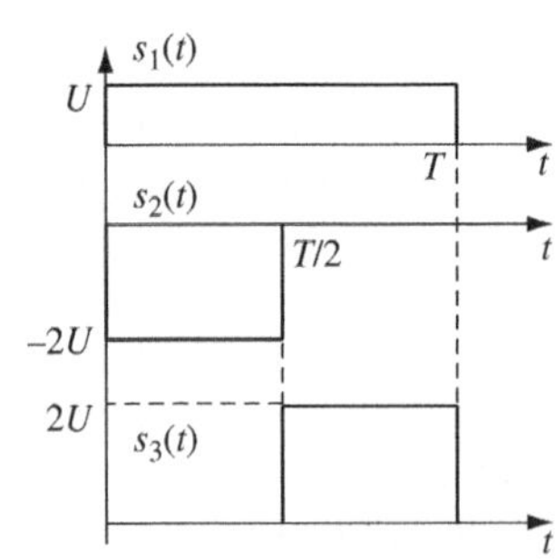

Figure 2.28 Set of three signals

2.2. Observation $y(t)$ at the AWGN channel output is given in Figure 2.29, and the input signals are the same as in Problem 2.1. What will the decision of the optimal receiver be?

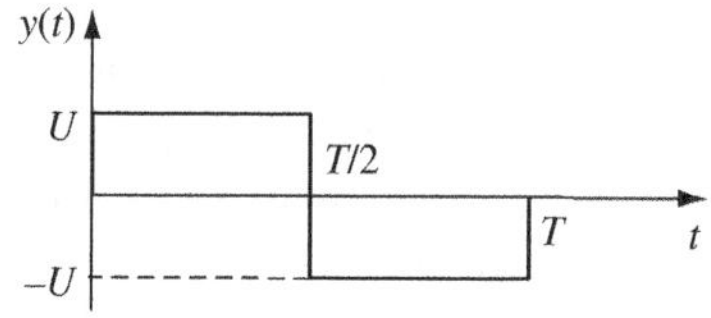

Figure 2.29 Waveform at the channel output

2.3. A source generates data at the rate $R = 10$ kbps. Each bit of the source data is transmitted over the AWGN channel by BPSK. Bandwidth $W = 10$ MHz is available in principle. Is it reasonable to use signals with bandwidth $W = 10$ MHz?

2.4. What is better for BPSK signalling over a Gaussian channel?

(a) Rectangular pulses of peak power 1000 W with bandwidth 100 kHz.
(b) Rectangular pulses of the same duration with peak power 900 W and bandwidth 10 MHz?

2.5. Calculate the energy losses for the pairs of signals of Figure 2.30 used for binary transmission over the AWGN channel with respect to the optimal pair.

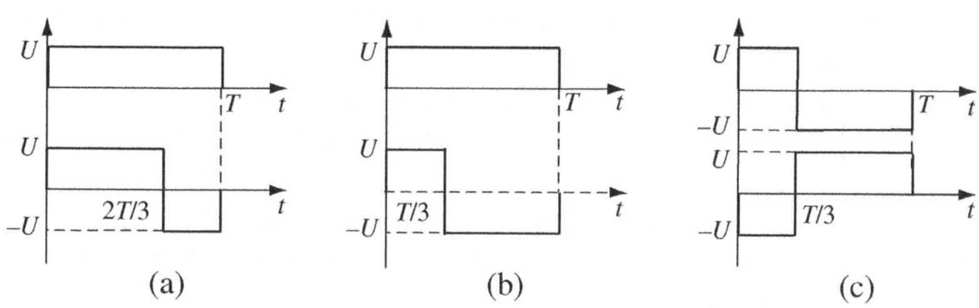

Figure 2.30 Three signal pairs

2.6. In differential BPSK (DBPSK) a bit content is transmitted by alternation or nonalternation of the polarities of two consecutive pulses, identical polarities corresponding to zero and different polarities to one. Compare to the first approximation DBPSK and BPSK in energy consumption (based on minimum distances only) and the bandwidth occupied, transmission rates being the same.

2.7. In quadrature PSK (4-PSK or QPSK) two bits (four messages) are transmitted by four signal phases: 0, π, $\pm\pi/2$. Is this an optimal transmission mode for two bits? If not, describe a better technique and its asymptotic coding gain against QPSK.

2.8. Is it possible to have 10 equidistant signals with correlation coefficient between any two of them equal to $-1/7$? What is the maximal possible number of signals with this correlation coefficient?

2.9. Calculate and sketch versus M the energy loss (in dB) of the set of M orthogonal signals as compared to the set of M optimal signals (AWGN channel). Find the asymptotic loss when M grows.
2.10. Asymptotic energy gain of orthogonal coding against uncoded binary transmission for the case of $M = 2$ messages turns to be 1/2 or -3 dB, i.e. negative, showing loss rather than gain. What does this mean physically?
2.11. Asymptotic energy gain of orthogonal coding against uncoded binary transmission for the case of $M = 4$ messages turns to be 1 (0 dB), i.e. no gain at all. Give the physical reasoning for the result.
2.12. In 8-PSK $M = 8$ messages are transmitted by identical bandpass pulses having 8 different equidistant initial phases. Is this way of transmitting 8 messages over the Gaussian channel the best possible one if no bandwidth limitation is imposed? If not, what is the energy loss of 8-PSK against the optimal set of M signals?
2.13. Compare asymptotic (determined by minimum distance) efficiency of M-PSK against orthogonal coding in energy consumption (given the error probability) and bandwidth occupied.
2.14. In the IS-95 mobile phone uplink blocks of 6 bits are transmitted by orthogonal signals. The transmission rate is 28.8 kbps. Estimate the bandwidth occupied by the encoded signals (ignoring further spreading by the long code).
2.15. In a digital communication system the allowed bandwidth $W = 1.2288$ MHz. What maximal number of orthogonal signals M may be used for data transmission at the rate of 38.4 kbps?
2.16. In a system data is transmitted over the Gaussian channel at the rate of 10 kbps. The system designer tries to provide 6 dB energy gain against the uncoded transmission. Is this achievable on the basis of orthogonal signals if only bandwidth within 320 kHz is tolerable?
2.17. Some system is allowed to use bandwidth of 10.24 MHz while the necessary transmission rate is 100 kbps. What potential asymptotic coding gain is achievable in the system?
2.18. It is necessary to transmit data over the Gaussian channel at the rate of 100 kbps using carrier frequency 2 GHz. Is it realistic to count on energy gain $G = 10$ dB on the basis of orthogonal signals?
2.19. Build up the Hadamard matrix of size 16.
2.20. Which of the following transformations preserve/violate the main property (orthogonality of rows) of a Hadamard matrix?

(a) Permutation of rows.
(b) Permutation of columns.
(c) Simultaneous alternation of signs of all elements.
(d) Alternation of signs of all elements of several rows.
(e) Alternation of signs of all elements of several columns.
(f) Alternation of sign of upper leftmost element only.

2.21. Hadamard matrix H_M of size $M = 2^m$ is built up by the Sylvester rule starting with $H_2 = \begin{bmatrix} 1 & 1 \\ 1 & -1 \end{bmatrix}$. The first column of H_M is discarded and the rows of the matrix thus obtained are used to generate signals for M-ary transmission. What sort of signal ensemble do we have this way? What is the bandwidth they occupy in comparison with orthogonal ones?

2.22. LFM signal (with linear frequency modulation):

$$s(t) = \begin{cases} A\cos\left[2\pi f_0 t + \dfrac{\pi W_d (t-\tau)^2}{T} + \phi\right], |t-\tau| \le T/2 \\ 0, |t-\tau| > T/2 \end{cases}$$

has amplitude A, carrier frequency f_0, frequency deviation W_d, duration T, time delay τ and initial phase φ. Classify these six parameters as energy or non-energy ones. (For any bandpass signal $f_0 T \gg 1$, $W \ll f_0$.)

2.23. Non-energy signal parameter λ needs to be measured. The correlation coefficient $\rho(\lambda)$ of the signal copies detuned in λ is shown in Figure 2.31 for three different cases. In which of these will the accuracy of measuring λ be highest?

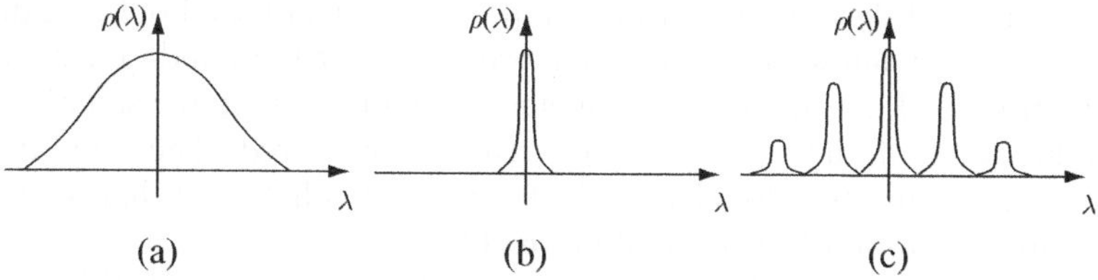

Figure 2.31 Dependence of correlation coefficient on a measured parameter

2.24. Two non-energy scalar parameters λ_1, λ_2 need to be measured simultaneously. The correlation coefficient of two signal copies with different pairs of values of λ_1, λ_2 is $\rho(\lambda_1, \lambda_2)$ and geometrically is represented by some surface in three-dimensional space. A cross-section of this surface by a horizontal plane at some level (e.g. 0.5) is given in Figure 2.32 for three typical cases. In which of these will the accuracy of simultaneous estimation of λ_1, λ_2 be highest if no a priori knowledge about their values is available?

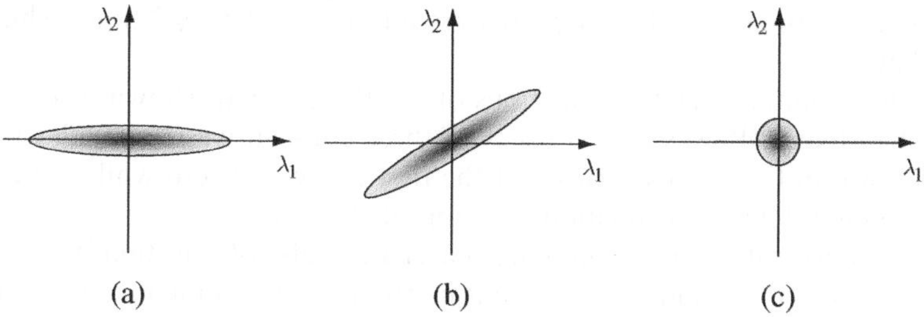

Figure 2.32 Horizontal sections of the surface $\rho(\lambda_1, \lambda_2)$

2.25. Amplitude A of the signal $s(t; A) = As(t)$ needs to be measured. The reference signal $s(t)$ for three cases is given in Figure 2.33. In which of these cases will the accuracy of measuring A be highest?

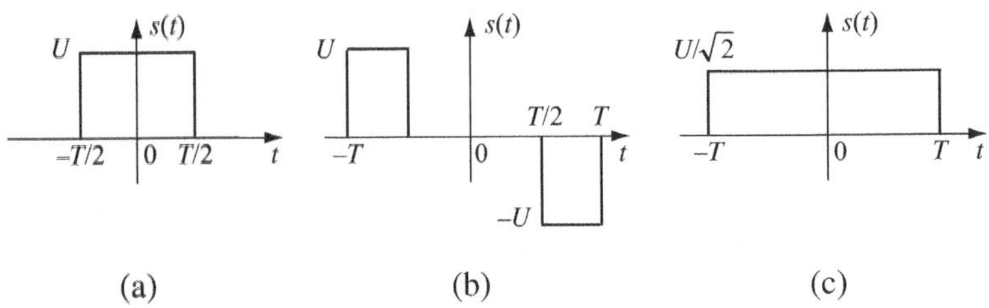

Figure 2.33 Three forms of a reference signal

2.26. Amplitude A of the signal $s(t; A) = As(t)$ is measured. Someone is dissatisfied with the accuracy of the amplitude estimation. How many times should the duration of the reference signal $s(t)$ be increased in order to halve the standard deviation of the estimation of A, all the other parameters of $s(t)$ remaining the same?

2.27. Amplitude A of the signal $s(t; A) = As(t)$ is measured. The amplitude of the reference signal $s(t)$ is doubled while its duration is halved. What happens to the standard deviation of the estimation of A?

2.28. The initial phase of the bandpass signal needs to be measured. What happens to the standard deviation of the phase estimation when:

(a) Carrier frequency of the signal is doubled?
(b) Signal duration is doubled?
(c) Signal amplitude is halved?
(d) Signal amplitude is doubled and duration is reduced by four times?

2.29. The initial phase of the bandpass signal needs to be measured. Three variants of the signal envelope are given in Figure 2.33. In which of these cases will the accuracy of phase measurement be highest?

2.30. The initial phase φ of the LFM signal of Problem 2.22 is measured. Variation of which of the parameters A, f_0, W_d, T, τ and in which direction will affect the precision of the phase estimation? What happens to the standard deviation of the phase estimation when A, f_0, W_d are all increased by $\sqrt{2}$ times while T and τ are halved?

2.31. Sketch the autocorrelation functions of the three signals shown in Figure 2.34.

2.32. The bandpass BPSK signal consists of three consecutive rectangular pulses each being of duration Δ. The phases of the first two equal zero while the third phase is π. Sketch the autocorrelation function of the signal.

2.33. The matched filter for a rectangular baseband pulse of duration Δ is given. What sort of circuitry should be added to it in order to obtain the matched filter for the signals of Figure 2.34? Sketch the response of the filter matched to signal (c) when this very signal inputs the filter.

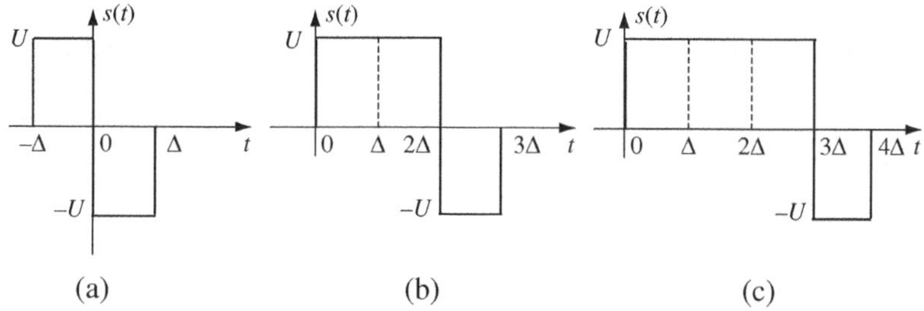

Figure 2.34 Three forms of a signal

2.34. The matched filter for a rectangular bandpass pulse of duration Δ is given. What circuitry added to it will produce a matched filter for the signal of Problem 2.32? Sketch the response of the matched filter to the signal of Problem 2.32.

2.35. The autocorrelation functions of three alternative signals are given in Figure 2.35. Which of these is best for measuring time delay?

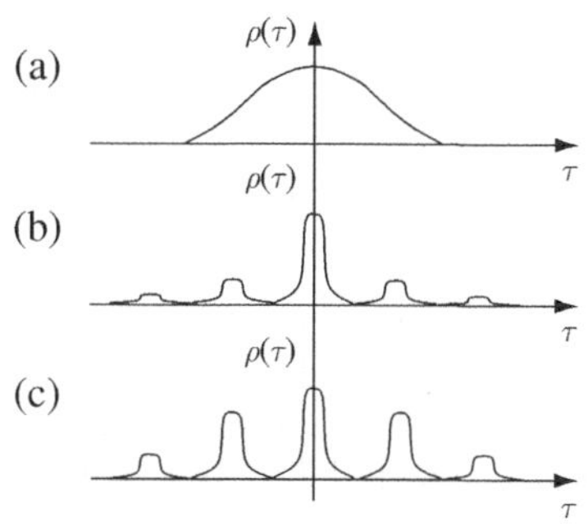

Figure 2.35 Examples of autocorrelation function

2.36. Parameters of the plain pulse signal provide standard deviation of time-delay measurement 0.5 μs and SNR at the matched filter output $q = \sqrt{2E/N_0} = 10$. Estimate roughly the signal duration.

2.37. In some radar a plain pulse signal is used. The system designer wants to reduce the peak-power by 100 times without sacrificing SNR and at the same time to reduce by 10 times the standard deviation of measuring time delay. What should the time–frequency product of the signal in the improved radar be?

2.38. In some radar distance is measured through emitting LFM pulse (see Problem 2.22) with time–frequency product $WT = 10^3$. Due to a breakage of the modulator the radar started emitting unmodulated pulses of the same peak-power and duration. What happens to the standard deviation of distance measuring?

2.39. In some system it is necessary to reduce by 10 times the standard deviation of frequency measurement. The signal power can be increased only 25/16 times (1.94 dB). How should the signal duration be changed?

2.40. In some system signal duration was quadrupled, signal power remaining unchanged. What happens to the standard deviation of the frequency estimation?

2.41. Horizontal sections of the ambiguity function for four signals are given in Figure 2.36, all sizes marked by the same symbols (τ_{c1} etc.) being equal. Which of the signals is best for:

(a) Measuring only time delay?
(b) Measuring only frequency shift?
(c) Measuring both time delay and frequency shift?

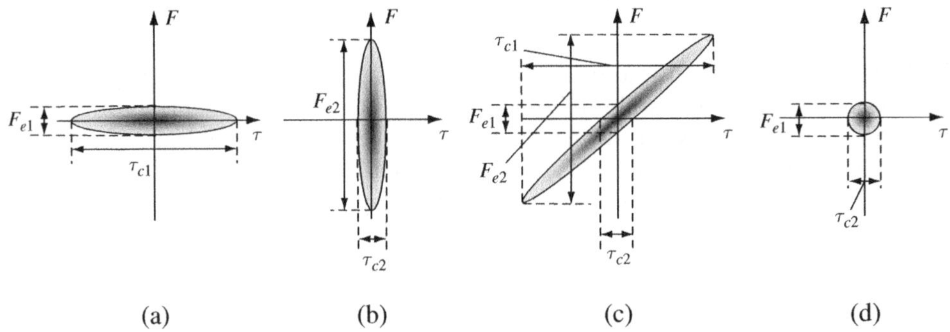

Figure 2.36 Horizontal sections of the ambiguity function

2.42. In the course of modernizing some system operating originally with a plain signal, emitted power was reduced by 6 dB. At the same time signal duration was increased by four times and the plain signal was replaced by a spread spectrum one with time frequency product $WT = 100$. What happens to the standard deviations of time delay and frequency estimations as compared to the initial ones?

2.43. Which of the signals having the ambiguity diagrams of Figure 2.36 is best for:

(a) Time resolution?
(b) Frequency resolution?
(c) Time–frequency resolution?

2.44. Horizontal sections of the ambiguity function for four signals are given in Figure 2.37, all sizes marked by identical symbols being equal. Which of the signals is best for:

(a) Time resolution?
(b) Frequency resolution?
(c) Time–frequency resolution?

Matlab-based problems

2.45. Write a program illustrating the receiver decision on which of two competitive signals is received. Do the following:

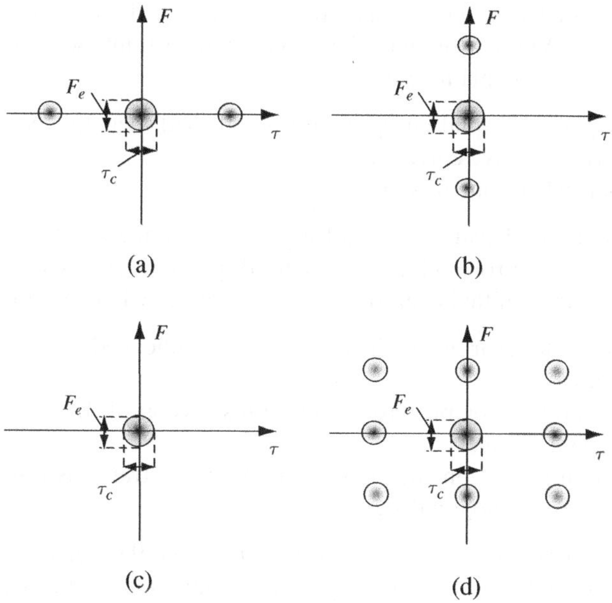

Figure 2.37 Horizontal sections of the ambiguity function

(a) Form and plot two 100-dimensional antipodal signal vectors with elements $\{\pm 1\}$.
(b) Choose one of them as a transmitted signal.
(c) Form the 1000×100 matrix of Gaussian noise, putting standard deviation σ 6–10 times higher than the square root of the signal energy.
(d) Form the 1000×100 observation matrix, each row being the sum of signal of point (b) and noise; plot observations.
(e) For every observation calculate two distances (or another sufficient statistic) from the observation to each of the two signals.
(f) For every observation take the minimum-distance decision.
(g) Compare the decision with a preset signal of point (b) for each observation.
(h) Calculate the number of errors in all 1000 observations.
(i) Retaining the signal vector lengths and noise deviation, change the signals into an orthogonal pair and run the program again.
(j) Do the same as in (i) for the pair of signals with positive correlation coefficient and for the pair with the second signal equal to zero.
(k) Compare the results of items (i)–(k), both with each other and with theoretical predictions, and give your comments.

2.46. Write a program calculating and plotting binary transmission error probability versus SNR for an arbitrary pair of signals preset as vectors. Use power SNR defined for average energy of the signals:

$$q^2 = \frac{E_1 + E_2}{N_0}$$

E_1, E_2 and N_0 being the energies of the two signals and the one-sided power spectrum of AWGN. Plot all of the curves in logarithmic scale. Run the program, and compare and explain results for:

(a) Signal pairs of equal energies: antipodal, with negative correlation, orthogonal, with positive correlation.
(b) The pair where one of signals is zero.

2.47. Write a program calculating and plotting (in logarithmic scale) curves of the message recognition error probabilities versus SNR per bit (see Figure 2.9) when m-bit messages are transmitted over an AWGN channel, including for each value of m:

(a) The accurate error probability for bit-by-bit uncoded transmission of $M = 2^m$ messages.
(b) The accurate error probability for transmission of $M = 2^m$ messages by the orthogonal signals.
(c) The union bound on the error probability for transmission of $M = 2^m$ messages by the orthogonal signals.

Run the program for $m = 1$ to 10 and interpret the behaviour of the curves. Explain why for $m = 1, 2$ the uncoded transmission curves go lower than the orthogonal signalling ones. For $m = 3$ to 10, find the values of the orthogonal coding gain corresponding to the error probabilities $10^{-3}, 10^{-5}$ and compare them with the asymptotic ones.

2.48. Write a program demonstrating experimentally energy gain of orthogonal coding as compared to uncoded transmission of a 6-bit message. The following operations are relevant:

(a) Form and plot an uncoded 6-bit pattern corresponding to BPSK mode oversampled with N_s samples per bit (it is advisable to take $N_s = 64$, i.e. 384 samples over 6 bits).
(b) Form the $1000 \times 6N_s$ matrix of the Gaussian noise samples with standard deviation equal to a quadrupled bit amplitude.
(c) Form and plot the $1000 \times 6N_s$ matrix of observations, adding the signal to the noise matrix.
(d) Demodulate all observations, calculate and output the message-wise error rate.
(e) Form the Hadamard matrix of order 64, take one of its rows as an encoded message and oversample it into the signal vector of dimension $6N_s$.
(f) Repeat items (b)–(d).
(g) Compare and treat the error rates for the two explored transmission modes.
(h) Run the program for the range of SNR, changing the noise level; record the results and compare them with theoretical ones (see Figure 2.9).

2.49. Write a program demonstrating experimentally the trade-off between energy gain and spectral efficiency of the orthogonal signalling. The following operations are relevant:

(a) Set the number of bits $m = 8$.

(b) Form all possible $2^m = 256$ bit patterns and oversample them so that each bit occupies 160 samples.
(c) Calculate the power spectra of all bit patterns and the average power spectrum under uncoded transmission.
(d) Form the Hadamard matrix of order 256 and oversample it to come to the same dimension of signal vectors as previously.
(e) Calculate the power spectra of all 256 orthogonal signals and the average power spectrum under orthogonal signalling.
(f) Plot the average power spectra, estimate and compare the bandwidths occupied for both investigated cases and compare their ratio to the theoretical prediction.

2.50. Write and run a program illustrating optimal measurement of the amplitude of the triangle pulse of duration T: $s(t; A) = As(t) = \frac{At}{T}, 0 \le t \le T$ (see Figure 2.38). The following steps should be fulfilled:

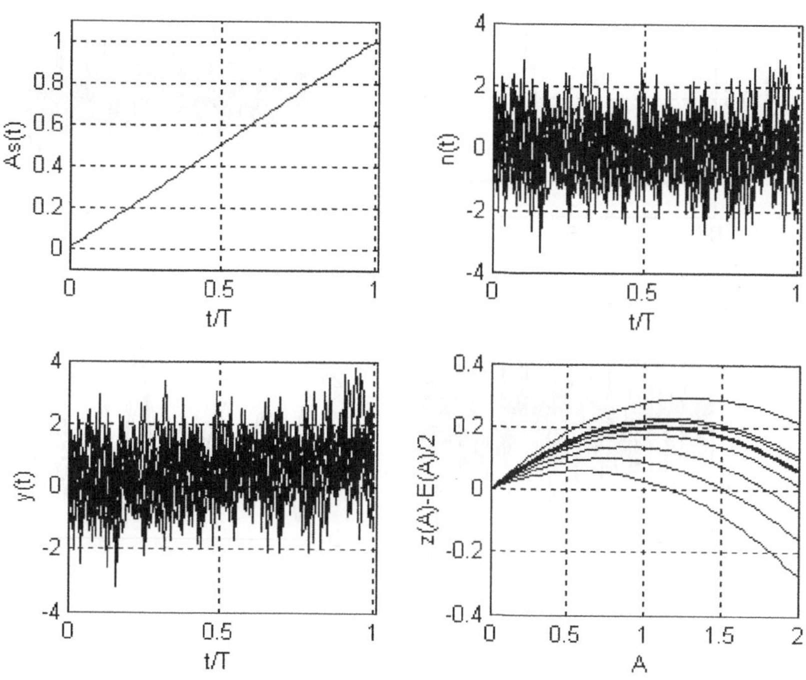

Figure 2.38 Simulation of measuring amplitude

(a) Specify the value of amplitude A on your own and form a signal vector of dimension around 100.
(b) Form 10 vectors of the Gaussian noise $n(t)$ with standard deviation approximately equal to the preset signal amplitude.
(c) Form 10 vectors of the observation $y(t)$ by adding noise vectors to the signal vector.

(d) Plot waveforms of the observations. Is the signal clearly visible in them?
(e) Calculate and plot the curves of the decision statistics (distances or squared distances or differences $Az - A^2E/2$) for all 10 observations versus amplitude estimation.
(f) Find optimal estimations of A for all 10 observations.
(g) Calculate average and variance of the optimal estimations over all observations.
(h) Run the program with several preset values of A and compare the results of measuring with the preset values. Give a theoretical explanation for the results.

2.51. Write and run a program illustrating the optimal measurement of time delay τ of the bell-shaped baseband pulse of duration T (by the level 0.01). Assume $\tau \in [0, 9T]$. The following steps illustrated by Figure 2.39 should be completed:

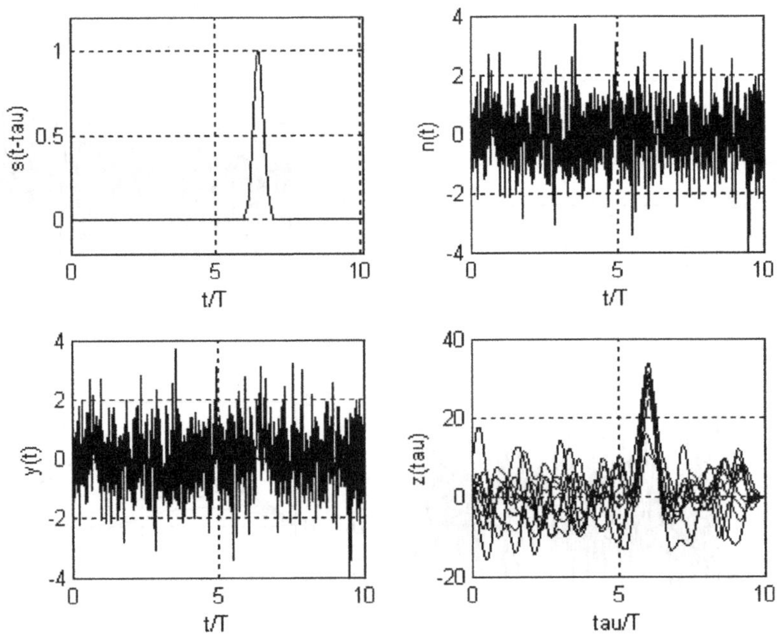

Figure 2.39 Simulation of measuring time delay

(a) Use around 1000 sample points over the whole observation interval $\tau \in [0,10T]$. Preset the value of τ (in number of sampling points) on your own and form the time-shifted signal vector;
(b) Form 100 vectors of Gaussian noise $n(t)$ with standard deviation within the range 1 to 2 times the signal amplitude.
(c) Form 100 observation vectors $y(t)$ by adding noise vectors to the signal vector.
(d) Plot waveforms of the observations (see Figure 2.39). Is the signal clearly visible in them?

(e) Calculate and plot curves of the correlations between observations and time-shifted signal copies versus time shift. Is the signal effect now more visible?
(f) Find optimal estimations of τ for all 100 observations.
(g) Calculate the average and the variance of optimal estimations over all observations.
(h) Run the program with several preset values of τ and in the range of SNR. Compare the results of measurements with the preset values. Give a theoretical explanation for the results.

2.52. The bandpass rectangular pulse of duration T is applied to a bandpass filter with rectangular pulse response of duration $2T$, which is frequency-detuned against the signal by F. Based on the complex envelope, calculate and plot signal shapes (real envelopes) at the filter output for $F = a/T$, where $a = 0, 0.5, 1.0$.

2.53. The bandpass rectangular pulse of duration T and carrier frequency $f_0 = 10/T$ is applied to the bandpass filter with rectangular pulse response of the same duration which is frequency-detuned against the signal by $F \in \{0, \frac{1}{T}, \frac{2}{T}\}$. Based on the complex envelope, calculate and plot the bandpass signals at the filter output for all three values of F.

2.54. Write a program calculating and plotting the autocorrelation function of both baseband and bandpass signals of the same arbitrary form and duration T. Take the carrier frequency of the baseband signal equal to $10/T$. See the example in Figure 2.40. Run the program for:

(a) Rectangular pulse.
(b) Triangular symmetric pulse.

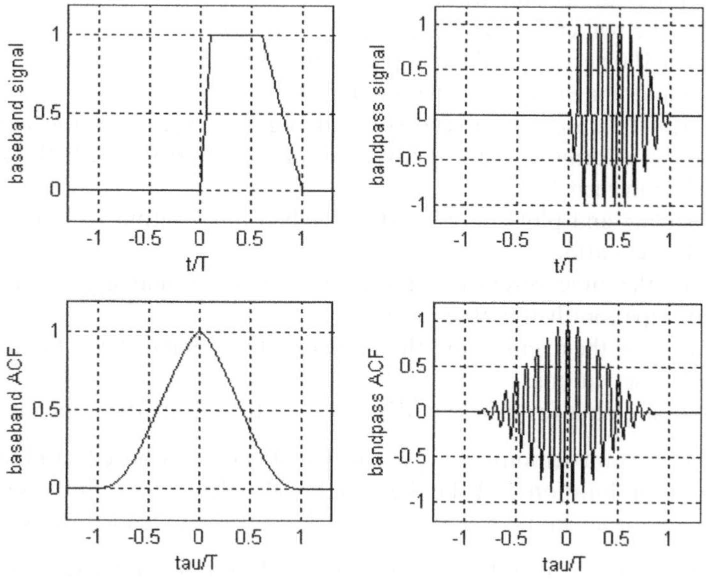

Figure 2.40 Autocorrelation functions of baseband and bandpass signals

(c) Triangular pulse rising within $[0, T)$.
(d) Pulse of your own choice.

2.55. Write a program demonstrating the time-compression effect in a matched filter. Take a rectangular linearly-frequency-modulated (LFM) signal of duration T with complex envelope:

$$\dot{S}(t) = \begin{cases} \exp\left(j\dfrac{\pi W_d t^2}{T}\right), & 0 \le t \le T \\ 0, \ t < 0 \text{ or } t > T \end{cases}$$

Take the initial carrier frequency $f_0 = 50/T$ and five values of frequency deviation $W_d = a/T$, $a \in \{0, 10, 20, 30, 40\}$. For each value of deviation show bandpass signals at the matched filter input and output.

2.56. Write a program illustrating the dependence of the accuracy of time-delay estimation on the signal bandwidth (see Figure 2.21). Take an LFM signal with the bell-shaped envelope, i.e. the complex envelope:

$$\dot{S}(t) = \begin{cases} \exp\left[-20\left(\dfrac{t}{T} - \dfrac{1}{2}\right)^2\right] \exp\left(j\dfrac{\pi W_d t^2}{T}\right), & 0 \le t \le T \\ 0, \ t < 0 \text{ or } t > T \end{cases}$$

Perform calculation and plotting for the values of the deviation $W_d = a/T$, $a = 0, 10, 25, 50$. The following steps should be completed:

(a) Specify the value of the deviation W_d and form the signal (complex envelope) vector of dimension around 100.
(b) Plot the real envelope of the signal.
(c) Form 100 vectors of the complex Gaussian noise with standard deviations of real and imaginary parts within [0.5, 4.0].
(d) Form 100 observation vectors by adding noise vectors to the signal vector.
(e) Plot waveforms of the real envelopes of the observations. Is the signal clearly visible in them?
(f) Calculate and plot curves of the matched filter output real envelopes for all 100 observations.
(g) Find the time positions of the maximums of matched filter output real envelope for all 100 observations.
(h) Calculate the average and the variance of the time-delay estimations over all 100 observations.
(i) Compare data for different W_d and treat the results theoretically.

2.57. Write a program demonstrating resolution of two time-shifted bandpass rectangular pulses of duration T. Take the signal of Problem 2.55 with the same values of carrier frequency and deviation (Figure 2.27). Recommended steps:

(a) Specify the value of the deviation W_d and form the signal (complex envelope) vector of necessary dimension.

(b) Form and plot a bandpass signal.
(c) Specify the value of delay $0 < \tau < T$, form and plot a delayed signal copy.
(d) Calculate and plot superposition of the two signals.
(e) Calculate and plot matched filter response to it.
(f) Run the program, varying the frequency deviation and delay, and give a theoretical treatment of the results.

2.58. Write a program calculating and plotting the time–frequency ambiguity function of a plain bandpass pulse with arbitrary real envelope and duration T. Provide for plotting also the basic sections of the ambiguity function (along the τ and F axes) as well as the horizontal section at the level 0.5. Run the program for:

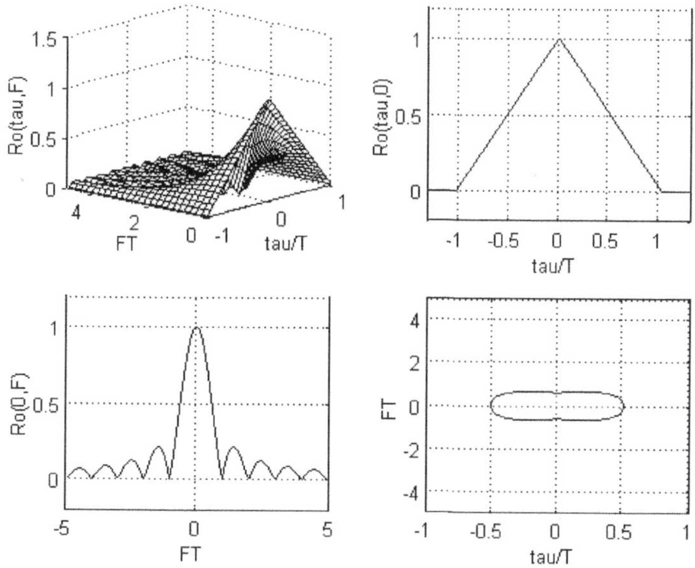

Figure 2.41 Ambiguity function of a rectangular pulse

(a) Rectangular pulse (see Figure 2.41).
(b) Triangular symmetric pulse.
(c) Triangular pulse rising within $[0,T)$.
(d) Pulse of your own choice.

3

Merits of spread spectrum

3.1 Jamming immunity

The surrounding environment in which a specific system transmits and retrieves information is not perfectly friendly towards it. Along with thermal noise, interferences of varying physical nature may accompany the useful signal at the receiver input. In particular, interference can emerge as a result of the presence of some other system on the air if its operating frequency is close enough to that of the first system. Following the universally adopted terminology, we will call this sort of interference a *jammer*.

A great variety of jammer types may be encountered in practice and special measures are as a rule necessary to counter their destructive effect. In this section we will show that the spread spectrum is quite a powerful instrument of jammer neutralization. An exhaustive investigation into system behaviour subject to the combined effects of both jammer and thermal noise would require the calculation of integral performance characteristics such as error probability or estimate precision. The concrete results of this challenging work may be found in books (e.g. [5,6]) and mainly in numerous specialized papers (see, for example, the bibliography in [3]). However, our aim is much more modest and directed at getting a general idea of why and how spread spectrum helps in combating a jammer. For this reason we limit ourselves to only the simplest assessments based on the power ratio between the signal and overall interference. Putting it another way, we choose for the analysis here only the plainest, but still very characteristic, type of jammer, which is approximated by the Gaussian random process whose spectrum overlaps with that of the signal. Sometimes such a simplified approach is fully adequate, as it is in the case of BPSK or ASK data transmission, where the error probability depends only on the ratio referred to above, whenever the overall interference may be assumed Gaussian. Other situations (M-ary transmission, parameter measuring) are not that straightforward and the power ratio does not contain all the necessary information on the performance quality, but still remains indicative as a 'rule of thumb' to judge the potential advantages of spread spectrum. We consider two basic models of a jammer, starting with a narrowband one.

Spread Spectrum and CDMA: Principles and Applications Valery P. Ipatov

3.1.1 Narrowband jammer

This type of jammer is more typical of situations where some side system or systems have no hostile intentions with respect to the useful system and create a jammer just as a result of normal functioning. Figure 3.1 shows the amplitude spectrum $|\tilde{s}(f)|$ of the useful signal and the jammer power spectrum $\tilde{J}(f)$ approximated by rectangles against the uniform background AWGN power spectrum $N_0/2$. We call a jammer *narrowband* only to stress that its bandwidth W_j is smaller than the signal bandwidth W and there are areas where the signal spectrum is not corrupted by the jammer. The narrowband jammer may be further classified to partial-band, tone, etc. [3,5,6]; however, for our study the specific value of its bandwidth is immaterial.

First suppose that the useful system undertakes no special measures to combat a jammer except for, maybe, just appropriate signal design. This kind of scenario means that a system designer may foresee the risk of the presence of a jammer and allow for it in the signal choice, but the system is not adaptive, and makes no adjustment of either signal modulation or processing algorithm to the current interference environment. In other words, it uses always only the filter, which is matched to AWGN regardless of the presence or absence of a jammer at the receiver input.

To find the power signal-to-interference ratio (SIR) q_I^2 at the matched filter output note that with a rectangular signal spectrum (where $|\tilde{s}(f)|$ equals the constant $\tilde{s}$ within signal bandwidth W and is zero elsewhere), the filter amplitude transfer function is also uniform within the signal bandwidth W and zero outside it. Without loss of generality we can put its non-zero level equal to 1. Therefore the filter passes a jammer (treated as a random process) to the output without any change of its power J, the filtered AWGN power being $N_0 W$. On the other hand, the filter is matched to the signal and sums all signal harmonics coherently to produce the output peak $A_{out} = \int_{-\infty}^{\infty} |\tilde{s}(f)|\,df = 2W\tilde{s}$, where spectrum uniformity within the bandwidth W is used and doubling is responsible for 'negative frequencies'. By the same token signal energy calculated through the Parseval theorem $E = \int_{-\infty}^{\infty} |\tilde{s}(f)|^2\,dt = 2W\tilde{s}^2$. Consequently:

$$q_I^2 = \frac{A_{out}^2}{J + N_0 W} = \frac{4W^2\tilde{s}^2}{J + N_0 W} = \frac{2E}{N_0 + J/W} \tag{3.1}$$

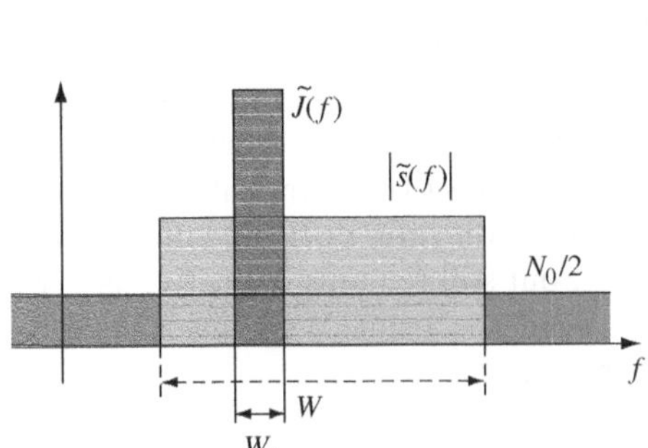

Figure 3.1 Spectra of signal, jammer and background AWGN

From the last equality it can be seen that, regardless of a specific jammer bandwidth W_j, the matched filter output SIR behaves as if the jammer power were uniformly spread over the signal (not jammer!) bandwidth W, creating an additional 'AWGN' with the power spectrum J/W.

Let us turn to the other scenario where the useful system adapts its receiver to the current interference pattern. The optimal processing procedure would be filtering matched to the overall interference, including a narrowband jammer. It is obvious physically that when a jammer is very strong against the background AWGN such processing is equivalent to a full cutting off of the frequency band damaged by the jammer. Figure 3.2 shows spectra at the band-elimination filter output. No jammer is present there but the signal frequency components within the jammer bandwidth are also forced to zero as well as the noise components. The spectral pattern may be treated as though the signal originally occupied only the part of bandwidth W free of the jammer, having the energy $E(1 - W_j/W)$. Accordingly, the matched filter clearing this residual signal off the AWGN will provide output power SNR (the subscript J stands for jammer):

$$q_J^2 = \frac{2E(1 - W_j/W)}{N_0} = q^2(1 - W_j/W) \tag{3.2}$$

with $q^2 = 2E/N_0$ being a 'pure' matched filter power SNR in the absence of jammer.[1]

Analysing equations (3.1) and (3.2) we note that they both clearly point to the benefits of wideband signals for the anti-jamming capability: the wider the signal bandwidth W versus the jammer bandwidth W_j, the smaller is the additional power spectrum in the first case and the energy loss in the second (jammer power J constant), and the greater are q_I^2 and q_J^2. But when the signal peak power P is limited and not allowed to increase, widening the bandwidth cannot be realized by a trivial signal

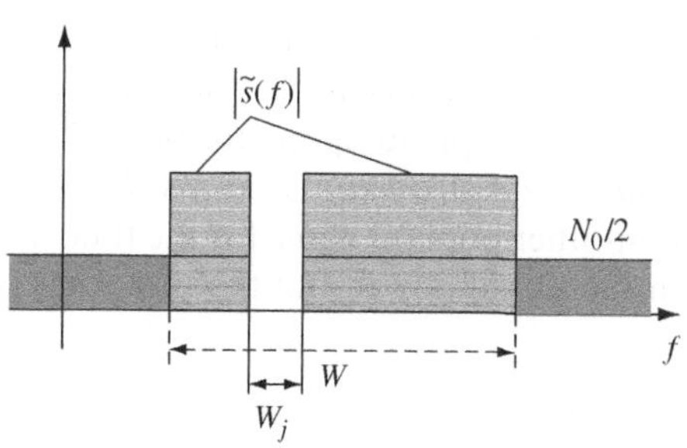

Figure 3.2 Spectral pattern after the band-elimination filtering

[1] We again stress that the SNR is not a universal characteristic of performance. It is appropriate for BPSK or ASK data transmission but in the general case band elimination affects the correlation properties of signals along with their energies. E.g. orthogonal signals may lose orthogonality under cutting off partial bandwidth. More detailed analysis is necessary to allow for this sort of effect.

shortening, since otherwise signal energy, SIR and SNR will suffer. Thus, we arrive at the following conclusion: to achieve a higher narrowband jamming immunity with no mobilization of brute-force resource (increase of signal energy or peak power) the only way is to widen the spectrum independently of signal duration, i.e. *to use spread spectrum technology*.

3.1.2 Barrage jammer

In many military scenarios and intelligence games, a jammer is frequently created deliberately as an electronic countermeasure. In such cases the jammer transmitter may suspect that the system under counteraction will appear smart enough to register the presence of the jammer and properly adapt itself to it. In particular, when the jammer is narrowband the system can resort to band-elimination filtering or even to changing a signal to shift its spectrum to the jammer-free zones. To prevent this, the *barrage noise jammer* can be engaged whose power spectrum covers the signal spectrum with no gaps (Figure 3.3). It is clear that the barrage jammer corrupts the signal in the same way as an additional AWGN with power spectrum density $N_J = J/W$. Therefore, the power SNR at the matched filter output of the useful system:

$$q_J^2 = \frac{2E}{N_0 + N_J} = \frac{2E}{N_0 + J/W}$$

coincides with SIR (3.1). In this case, however, the malignant intent urges the jammer transmitter to provide a much more damaging effect compared to the natural AWGN. It is possible only if $J/W \gg N_0$, resulting in:

$$q_J^2 \approx \frac{2EW}{J} = \frac{2P(WT)}{J} \tag{3.3}$$

We again clearly see that when the peak power of the useful system is limited, and so is the power resource of the jammer transmitter, the involvement of signals with high time–frequency product WT, i.e. spread spectrum ones, *is the only instrument at the disposal of the system to improve its immunity towards the barrage jammer*.

Formula (3.3) explains another popular name for the time–frequency product WT. As is seen, the ratio between the signal power and the power of a uniform-spectrum noise

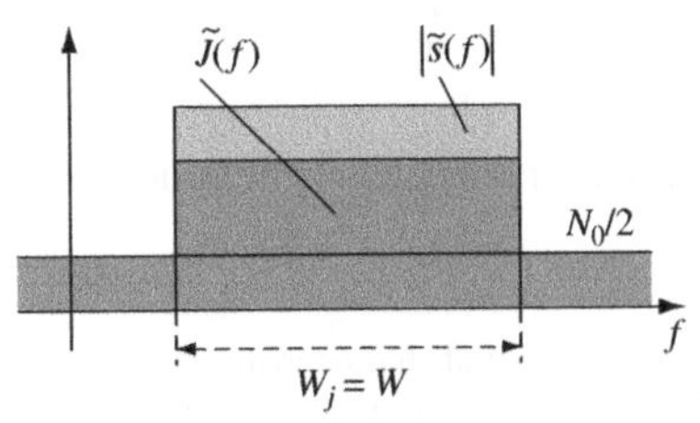

Figure 3.3 Spectra of signal, barrage jammer and background AWGN

within the signal bandwidth increases at the matched filter output $2WT$ times against the input value P/J. Thus, it is natural to call WT the *processing gain*.

The conclusions above are well supported visually by two figures, obtained by simulation in Matlab. Figure 3.4 illustrates matched filtering of the plain rectangular signal (column a) and a spread spectrum signal (LFM, $WT \approx 50$) of the same duration and energy (column b). The same jamming CW interference is added to both signals (second row). While the plain signal is fully masked by the jammer and is not clearly seen at the filter output, the spread spectrum one, time-compressed by the matched filter, is observable distinctively.

In Figure 3.5, where the columns correspond to the same two signals, the upper plots show the signal power spectra. The plots of the second row give the power spectra of two random realizations of different barrage jammers, the same fixed average jammer power being distributed over the signal bandwidth. Because of this the average level of the spectrum in column (b) is about 50 times lower than that in column (a). The third row shows example observed waveforms, where the intensity of the jammer is approximately the same for both signals, which are well hidden under the jammer. As for the lower plots, they again confirm explicitly the superiority of a spread spectrum signal in resistance to a barrage jammer.

In closing, note once again that this section is in no way aimed to answer questions on what sort of jammer is most dangerous in a concrete scenario and what the system should

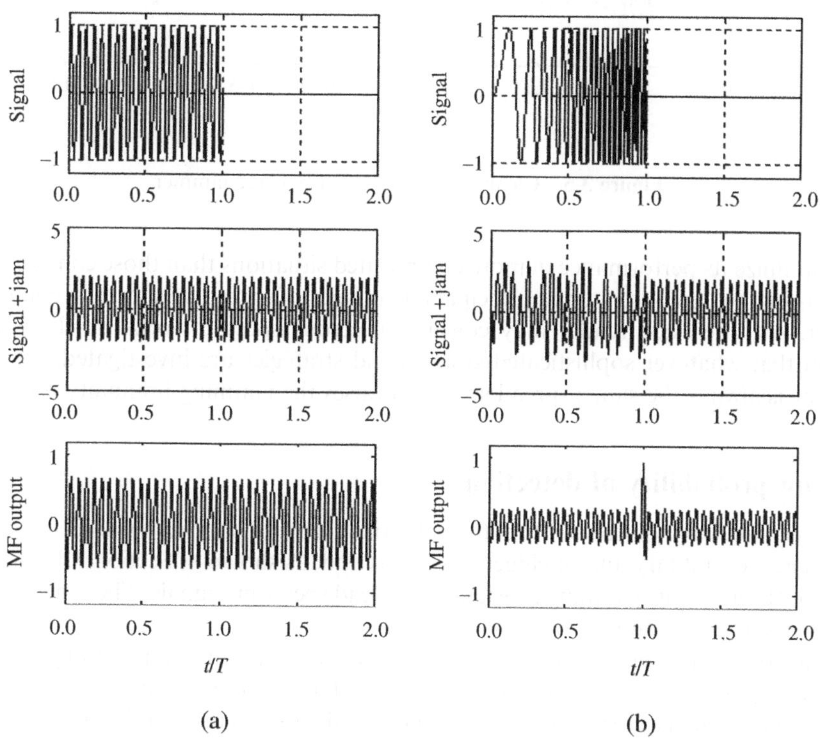

Figure 3.4 Examples of matched filtered jammer and signal

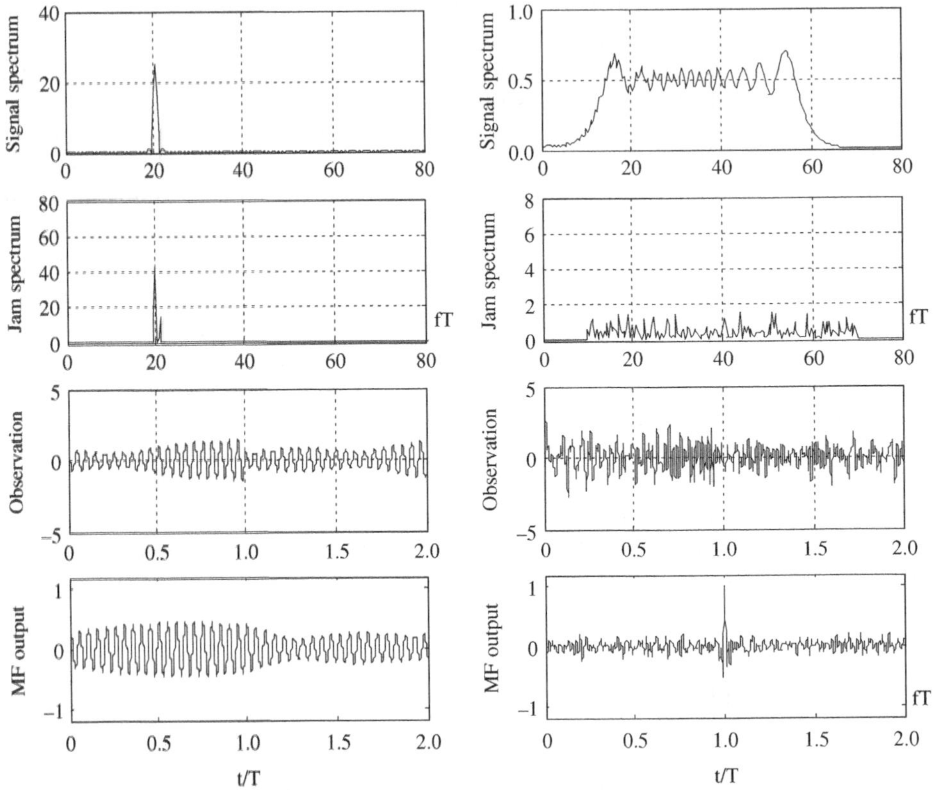

Figure 3.5 Clearing a signal off a barrage jammer

do to optimize its performance in more complicated situations than those considered. The idea was just to demonstrate the principal advantages of spread spectrum in countering a jammer. An interested reader may consult numerous specific works on the issue and confirm that whatever sophisticated systems and strategies are investigated, the general tendency is always the same: spread spectrum raises the jamming immunity potential.

3.2 Low probability of detection

It had been already pointed out that the potential of spread spectrum was first recognized by designers of military and intelligence systems and, as shown in Section 3.1, one reason for this is the high anti-jamming resistance of spread spectrum signals. The other reason we will discuss in this section.

In the confrontation of electronic systems effective jamming may be organized only after detecting the presence of an adversary system on the air and estimation of its parameters, such as carrier frequency and bandwidth. This entails a very popular scenario of the confrontation of two systems, when the first (call it *intended*) tries to operate as covertly as possible and escape unintended detection of its signal, while the second (*interceptor* or *eavesdropper*) is on the alert constantly, doing all in its power to discover an active state of the first. From the perspective of the intended system, let us explore how the spread spectrum can help in its conflict with an interceptor.

There are lots of strategies and techniques, which may be hypothesized to be at the interceptor's disposal potentially. They may be rather sophisticated and hard to analyse (see [6,9] and their bibliographies). We are again pursuing the goal of getting the general idea of why a spread spectrum appears to be a good option in this case. Let us assume that the intended system uses a signal with some non-trivial modulation law, details of which are not known to the interceptor, depriving the latter of the chance to use a matched filter or a correlator for signal detection. It is natural to believe, then, that the eavesdropper has no other choice but to treat the intercepted signal as random and base its detection on just the presence or absence of some extra energy in the suspicious frequency band. Thus, an *energy detector*, also called a *radiometer*, which is optimal for detecting a band-limited noise signal against the AWGN background, is accepted as the operational instrument at the intercepting side. Figure 3.6 gives the structure of an energy detector. A bandpass filter, whose bandwidth W_i spans the whole signal spectrum or only part of it, filters the observation to remove any off-band noise. Then a square-law amplitude detector forms an estimate of instant power, which is further integrated to produce an estimate of energy $\hat{E}$ within an observation interval T_{ob}. The energy estimate is then compared with the threshold E_t, and inequality $\hat{E} \geq E_t$ entails the decision that the observation contains a signal along with the 'natural' background noise, while in the opposite case no signal presence is declared. In practice, an interceptor may not know beforehand the frequency zone and time interval occupied by a signal. In such circumstances he tries various combinations of these parameters, implementing the whole procedure with the aid of either scanning the time–frequency area or a bank of parallel channels, each analysing its specific time–frequency zone. In any case the performance of the interceptor receiver will depend radically on the performance of the energy detector tuned to the true signal time–frequency zone. This allows us to idealize an interceptor's prior knowledge and believe that he is well aware of where on the time–frequency plane the signal energy may manifest itself. As long as observation outside the signal duration carries no information about the signal presence we may put $T_{ob} = T$, as is done in Figure 3.6.

Figure 3.7 shows a rectangular approximation of the signal spectrum along with the uniform natural background AWGN power spectrum (a) and the amplitude–frequency transfer function of the radiometer bandpass filter (b). From the interceptor's viewpoint, an indication of the signal presence is an extra (signal) power spectrum density $N_s/2 = P/2W$ added to that of the background thermal noise $N_0/2$. The radiometer bandpass filter has on its output the noise process with power $\sigma_n^2 = N_0 W_i$ in the absence of the signal and $\sigma_n^2 + \sigma_s^2 = (N_s + N_0)W_i = (P/W + N_0)W_i$ in its presence.

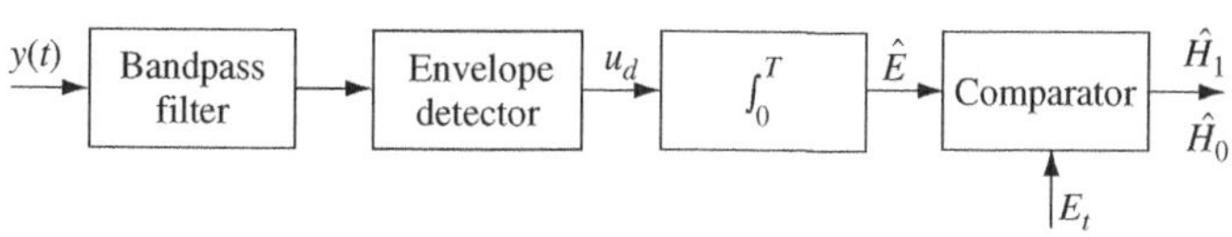

Figure 3.6 Energy detector (radiometer)

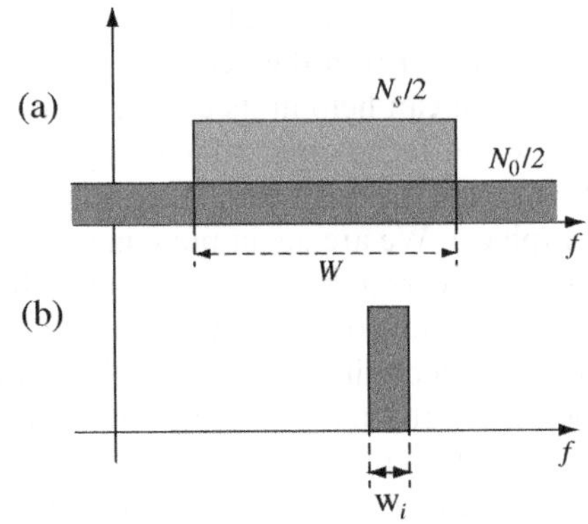

Figure 3.7 Spectra at the radiometer input and the bandpass filter bandwidth

Let us find the mean and variance at the envelope detector output. First of all, an output voltage u_d of the square-law detector equals the input instant power. Therefore, expectation $\overline{u}_d$ of u_d in the absence of a signal is simply the average power of the filtered noise $\overline{u}_{d0} = \sigma_n^2$, while when the signal is present it goes up to $\overline{u}_{d1} = \sigma_n^2 + \sigma_s^2$. It is exactly the increment of $\overline{u}_d$ caused by the signal:

$$\overline{\Delta u_d} = \overline{u}_{d1} - \overline{u}_{d0} = \sigma_s^2 = N_s W_i = PW_i/W \tag{3.4}$$

which allows the interceptor to hope to detect the presence of a signal. On the other hand, this useful (from the interceptor's viewpoint) effect is masked by the random fluctuations of u_d, measured by its variance var$\{u_d\}$. The latter can be found if it is remembered that the instant power of a bandpass process is its instant envelope Y squared and halved, and thus $u_d = Y^2/2$, meaning that var$\{u_d\}$ = var$\{Y^2\}/4$. Variance of any random variable may be calculated as the mean square minus the mean squared [1,13,14]:

$$\mathrm{var}\{Y^2\} = \overline{Y^4} - \left(\overline{Y^2}\right)^2 \tag{3.5}$$

Now make use of the fact that the envelope Y of a Gaussian bandpass random process with variance σ^2 has Rayleigh PDF [1]:

$$W(Y) = \begin{cases} \dfrac{Y}{\sigma^2}\exp\left(-\dfrac{Y^2}{2\sigma^2}\right), Y \geq 0 \\ 0, Y < 0 \end{cases}$$

and its even-order moments are found via elementary integration [13]:

$$\overline{Y^{2n}} = \int_0^\infty Y^{2n} W(Y)\,\mathrm{d}Y = (2\sigma^2)^n \int_0^\infty \left(\frac{Y^2}{2\sigma^2}\right)^n \exp\left(-\frac{Y^2}{2\sigma^2}\right) \mathrm{d}\left(\frac{Y^2}{2\sigma^2}\right) = n!(2\sigma^2)^n$$

Substituting this into (3.5) gives $\mathrm{var}\{Y^2\} = 8\sigma^4 - 4\sigma^4 = 4\sigma^4$. When the signal is not present at the interceptor receiver input we should use in this expression $\sigma^2 = \sigma_n^2$. Strictly speaking, with the signal advent the filtered observation may differ from a Gaussian process, making applicability of the result just obtained doubtful. This detail, however, has no importance to the case in question, since the intended system does its best to hide its signal under the thermal noise and we have every reason to assume that a signal has a negligible effect on the variance of instant power and thus on the variance of the detector response. Therefore, independently of signal presence, variance of u_d can be assumed the same:

$$\mathrm{var}\{u_d\} = \mathrm{var}\{Y^2\}/4 = \sigma_n^4 = (N_0 W_i)^2 \tag{3.6}$$

In order to estimate the mean $\bar{u}_d$ and register its growth due to a signal presence, the integrator in the scheme of Figure 3.6 runs time averaging of the detector response over the observation period T. To make a constant component at the detector output distinctive enough against the random fluctuations the latter should be smoothed as a result of integration. This is possible only if fluctuations of the detector response around $\bar{u}_d$ are sufficiently fast and change their polarity many times during the period T to compensate each other and produce an averaging effect. In other words, the number of statistically independent samples n_s of the detector response within T should be large enough. Extension in time (correlation spread τ_c) of the autocorrelation function of a random process is a trustworthy first approximation of a minimal time interval between samples, starting with which samples may be treated as independent. Since the filtered observation has the bandpass W_i its correlation spread is estimated as $\tau_c \approx 1/W_i$, which gives the number of independent samples $n_s \approx W_i T$.

Although practically integration may be implemented as continuous, its result is rather close to that of just summation of n_s independent samples [6,9], which is even a more practicable technique, especially in digital circuitry. To perform an accurate analysis, the PDF of the integrator output value $\hat{E}$ should be found for both hypotheses (signal absence and presence) and then integrated over the decision regions to obtain two probabilities, of *false alarm* and *detection*. These PDFs are subject to the chi-square law, which is a bit bulky and not quite transparent enough for a physical treatment. However, we may again exploit the fact that a signal is weak and its reliable detection requires a large number of integrated samples n_s. Then the central limit theorem allows assuming Gaussian PDF for the integrator output $\hat{E}$, therefore finding the mean and variance of $\hat{E}$ is enough to accomplish calculation of the probabilities just mentioned.

When only AWGN is observed the mean and variance of $\hat{E}$ are $\bar{\hat{E}} = n_s\bar{u}_{d0}$, $\mathrm{var}\{\hat{E}\} = n_s \mathrm{var}\{u_d\} = n_s\sigma_n^4$, where the second result follows from (3.6) and statistical independence of the integrated samples. Similarly, when AWGN plus signal is observed, $\bar{\hat{E}} = n_s\bar{u}_{d1}$ but the variance remains unchanged because of signal weakness. Thus PDF at the integrator output corresponding to the hypotheses H_0 (signal absence) and H_1 (signal presence) are:

$$W(\hat{E}|H_i) = \frac{1}{\sqrt{2\pi n_s \sigma_n^2}} \exp\left[-\frac{(\hat{E} - n_s\bar{u}_{di})^2}{2n_s\sigma_n^4}\right], \quad i = 0, 1$$

When $\hat{E}$ exceeds the threshold E_t the false alarm happens if a signal is actually absent and detection if a signal does arrive at the receiver. Hence, the probabilities p_f, p_d of these events are respectively:

$$p_f = P(\hat{E} \geq E_t|H_0) = \int_{E_t}^{\infty} W(\hat{E}|H_0)d\hat{E} = Q\left(\frac{E_t - n_s\bar{u}_{d0}}{\sqrt{n_s\sigma_n^2}}\right)$$

and

$$p_d = P(\hat{E} \geq E_t|H_1) = \int_{E_t}^{\infty} W(\hat{E}|H_1)d\hat{E} = Q\left(\frac{E_t - n_s\bar{u}_{d1}}{\sqrt{n_s\sigma_n^2}}\right)$$

Having rewritten the second of these equations as:

$$p_d = Q\left(\frac{E_t - n_s\bar{u}_{d0}}{\sqrt{n_s\sigma_n^2}} - \frac{n_s(\bar{u}_{d1} - \bar{u}_{d0})}{\sqrt{n_s\sigma_n^2}}\right) \tag{3.7}$$

one may see that if a tolerable level of the false alarm probability is predetermined, the first fraction in the brackets of (3.7) is fixed and the detection probability is completely defined by the ratio:

$$q_i = \frac{n_s(\bar{u}_{d1} - \bar{u}_{d0})}{\sqrt{n_s\sigma_n^2}} = \frac{n_s\overline{\Delta u_d}}{\sqrt{n_s\mathrm{var}\{u_d\}}} \tag{3.8}$$

The physical content of the latter is obvious: it is a voltage SNR at the integrator output showing the proportion between the useful (increment of expectation due to the signal advent) and hampering (standard deviation of random fluctuations) components of $\hat{E}$.

Making use of (3.4) and (3.6) together with the equation $n_s = W_iT$, we can represent (3.8) as:

$$q_i = \sqrt{n_s}\frac{P}{WN_0} = \sqrt{W_iT}\frac{P}{WN_0}$$

This equation allows us to see that from the interceptor's point of view, the maximal possible filter bandwidth, i.e. equal to the signal one ($W_i = W$), is optimal, providing the greatest output SNR:

$$q_i = \frac{P\sqrt{T}}{\sqrt{W}N_0} = \frac{q^2}{2\sqrt{WT}} \tag{3.9}$$

where $q^2 = 2E/N_0 = 2PT/N_0$ is, as always, SNR at the matched filter output of the intended receiver.

Certainly, q^2 should be maintained large enough otherwise the intended system will not be able to do its main job. It is quite clear, then, that the intended system has the only way to prevent detection of its signal by a potential interceptor: *use a spread spectrum signal with as large a processing gain WT as possible*. Coming back to Figure 3.7 uncovers the physical basis for this conclusion. Widening the spectrum of the signal of a constant energy and duration reduces the signal power spectrum density, hiding it under the background spectrum of the natural thermal noise.

Example 3.2.1. Consider the system transmitting sporadically and rather infrequently one of 64 messages using orthogonal signals. To provide error probability no worse than 10^{-3} it needs to use SNR around 7 dB per one bit, or 15 dB per 6-bit message (see Figure 2.9). Thus q^2 is 15 dB and transforming the interceptor's (voltage!) SNR (3.9) into decibels we have:

$$(q_i)_{\text{dB}} = 2(q^2)_{\text{dB}} - 20\lg 2 - 10\lg WT$$

If the system employs spread spectrum signals with processing gain $WT = 1000$ the interceptor's SNR turns to be $(q_i)_{\text{dB}} = -6$ dB or $q_i = 1/2$, which is not at all sufficient for reliable detection of the intended system's presence on the air in one session. If, for instance, the interceptor tolerates a probability of false alarm of $p_f = 10^{-3}$ then according to (3.7) the detection probability $p_d \leq 5 \times 10^{-3}$, i.e. is extremely small and exposes no serious threat to the intended system.

Finishing this section, note that the discussed advantage of spread spectrum is widely utilized today not only by the military or special services. The fact that a spread spectrum signal may be practically unnoticeable for the equipment that monitors the state of the radio air has serious implications for licensing policy. In particular, the range of commercial systems that may actively operate on the air without applying for a licence becomes broader, and in some regions special spectral zones are currently allocated for such licence-free use.

3.3 Signal structure secrecy

Continuing the line of the previous section, let us remember once more that the only reason for an interceptor to resort to such an ineffective detection instrument as an energy receiver is lack of information about the structure of the detected signal, i.e. its modulation law. As a result, the interceptor cannot process the signal in the manner used by the intended receiver (matched filtering). Of course, if the signal structure is not complicated enough and the interceptor is aware that it was chosen from only a few alternatives he may try them all. Appropriate equipment for doing so may be a bank of parallel matched filters or a single filter (several filters) reconfigured to fit the candidate signal structures serially in time, if the signal is known to be received for an adequate duration. Therefore, another aspect of the strategy of the intended system in its conflict with an interceptor consists in making a signal structure practically unbreakable.

A similar task is characteristic of military or commercial systems that do not tend to make the fact of their operation a mystery, e.g. if they function continuously, but are very keen to avoid unauthorized access to services addressed only to classified consumers, or forging of the transmitted information. The satellite-based global navigation system GPS is a convincing example of this kind. It has two positioning channels (see Section 11.2): open (or clear access, abbreviated C/A) and special (or protected, P). The signal transmitted over the second channel allows super-high precision of positioning, and the US government, which runs the system, does not permit unconditional access to this channel. In order to protect it from unauthorized use some special measures are undertaken concerning the signal modulation.

In disciplines dealing with information security, the extent of data protection is measured by a number of competitive equiprobable *keys*, which an enemy cryptanalyst (eavesdropper) should try to crack the ciphertext, i.e. encrypted data. In application to the signal structure, each of those keys is just a modulation law, which is typically repeated with some period T. Suppose that a signal is built of chips (see the example in Section 2.7.3) on the basis of an M-ary alphabet, i.e. using M different symbols to manipulate chips. If the bandwidth allocated to the system is W then the total signal space has dimension measured as WT (ignoring bandpass doubling; see Sections 2.3–2.5), i.e. a modulation law may be thought of as being constructed of WT chips. It is clear, then, that M^{WT} is the total number of possible modulation laws, i.e. competitive keys, and the system designer concerned with secrecy of the modulation format in the developed system should employ signals with rather large processing gain WT.

Example 3.3.1. The signal of the P-channel (P-code) in GPS is binary ($M = 2$) with the bandwidth $W \approx 10$ MHz. Its structure is quite regular and repeated with the period $T = 7$ days. Being hidden under the thermal noise, this signal cannot be retrieved by symbol-wise reception and only knowledge of its fine structure permits it to be cleared with highest efficiency off the AWGN. To prevent an unauthorized interceptor from accessing the P-code, a secret binary key (W-code) is modulo 2 added to it, masking the structure of the resulting Y-code. A single symbol of the W-code spans 20 symbols of the P-code; therefore, to break this mask by a trial and error method, up to $2^{WT/20}$ alternatives should be tested. Since $WT = 7 \times 86\,400 \times 10^7 > 10^{12}$, the number of tried keys is greater than 2 to the power of ten billion, which is fully beyond any imagination. For this reason the Y-code is believed to be unbreakable and no reports have emerged in nearly 10 years of its history on any successful cryptanalytical attack on it.

We conclude the section with another declaration on the advantages of spread spectrum: this technology *is very conducive to cryptographic protection of a signal structure*.

3.4 Electromagnetic compatibility

The problem of electromagnetic compatibility (EMC) is one of the most topical in modern wireless engineering. EMC implies friendly co-existence of different systems on the air despite each of them receiving not only its proper signal but also the signals of the other systems. Certainly, it is impossible to root out entirely mutual disturbance when several systems are operating simultaneously within a relatively small area. Any active system, i.e. emitting electromagnetic waves, inevitably affects all neighbouring ones and a system designer should try to minimize this potentially harmful influence.

There are two parties playing the EMC game. The first, which may be called 'emanating', tries to minimize the interference created by its emitted power to other nearby, so-called 'susceptible', systems [15]. The motivation for this is not only ethical but is also enforced by strict international and domestic regulations, compliance with which is carefully monitored by the services authorized to impose relevant sanctions. Any system of the second, 'susceptible', party takes its own measures aimed at neutralization of alien signals falling in its receiver front end.

Among the traditional ways of providing EMC are stringent frequency allocation controlled by national and world institutions, employing antennas with high directivity, careful design of RF circuitry etc. Here we briefly show why spread spectrum technology may also be included in this list.

From the point of view of the emanating system, the following logic is justified. As long as it is possible to make the emitted signal almost imperceptible for a special monitoring receiver (see Section 3.2) at the cost of complicating the modulation law, i.e. spreading the spectrum, such a signal will all the more be less harmful to an ordinary outside system operating in the same band. The issue is only in the choice of processing gain guaranteeing that the signal power spectral density appears to be sufficiently low compared to the natural noise spectrum intensity at the input of an outside receiver. As a rule of thumb, assume that 'sufficiently low' means −7 dB, i.e. $N_s/N_0 \leq 0.2$. Substituting $N_s = P/W = E/WT$ into this proportion leads to the criterion of EMC $E/WTN_0 \leq 0.2$ or $q^2/WT \leq 0.4$, where again the targeted parameter is expressed in terms of the intended receiver SNR q^2 and processing gain WT. If, for instance, intended SNR at the point of an outside receiver were 20 dB, then $WT \geq 400$ might be considered satisfactory in respect of the EMC problem. In a real design estimates like this have to be coordinated with distance so that some circle around the emanating system exists outside of which the signal of the latter is practically harmless for other systems [16].

From the position of a susceptible system, any outside signal at its receiver output may be treated as a narrowband or broadband jammer and all the reasoning behind the benefit of spread spectrum in anti-jamming (see Section 3.1) is applicable. Therefore we see that the *spread spectrum philosophy fits well with the issue of EMC*.

3.5 Propagation effects in wireless systems

To explore the next merit of spread spectrum we need to collect some supplementary knowledge on wave propagation effects in wireless channels, and this section will be a sort of excursion into this area. First of all, a key parameter affecting performance in any reception problem is signal intensity or SNR. Certainly, signal energy and power in all preceding formulas expressing error probability, variance of estimate etc. characterize signal level at the receiver input. Hence, it is important to be able to predict signal intensity at some point in space remote from the transmitting antenna, allowing for effects accompanying electromagnetic wave propagation.

The issue of wave propagation is quite complicated and hard to analyse theoretically. There are a great variety of factors causing both deterministic and random attenuation of a signal reaching the receiver input. Due to them the received signal is corrupted not only by additive noise (AWGN) but also by *multiplicative* interference, whose name stems from the fact that it changes signal intensity, or putting it another way, scales signal amplitude.

3.5.1 Free-space propagation

To begin with consider an idealized free-space model (see Figure 3.8), where there are no obstacles between the transmitter and receiver antennas and the transmitted wave propagates along the single possible path, called *line-of-sight* (LOS).

Let D denote the distance between the transmitter and the receiver. If the transmitter antenna were omnidirectional the transmitted power P_t would be uniformly distributed over the inner surface of the sphere of radius D and power $P_t/4\pi D^2$ would fall on every unit of area of this surface. The receiving antenna with an effective area A_r would then capture a received power $P_r = P_t A_r/4\pi D^2$. If the transmitting antenna is directional it emanates in the receiver direction power which is G_t higher than that of the omnidirectional one, and G_t is called a transmit antenna *power gain*. In this case the received power becomes G_t times higher as well. To represent the received power in a symmetrical form, we make use of the relation between A_r and a receiving antenna power gain $G_r = 4\pi A_r/\lambda_w^2$, where λ_w is wavelength. We will come to the Friis free-space formula [4]:

$$P_r = P_t G_t G_r \left(\frac{\lambda_w}{4\pi D}\right)^2 \tag{3.10}$$

showing that the attenuation of the signal power along the LOS is inversely proportional to the squared distance.

The free-space model may be directly applied to communication links whose environment is well described as an open space, e.g. between space vehicles or aircrafts, ground control centre and space vehicle etc. The propagation medium of terrestrial systems is much less favourable and in its influence on the signal intensity two main components are typically categorized: *shadowing* and *multipath fading*.

3.5.2 Shadowing

Shadowing is caused by landscape details obstructing LOS: hills, forests, bushes, buildings and so forth. Due to them the signal intensity drops with distance much faster than equation (3.10) predicts. Of course, the irregular nature of terrestrial patterns makes attempts at creating some universal theoretical model of shadowing impossible or worthless. A great deal of field testing has been carried out to collect knowledge about the general character of the dependence between the received power and the length of the propagation path and a number of empirical models have been proposed [17–19]. One of the most popular with mobile communication specialists is the Okumura–Hata model, according to which the behaviour of an average received power $\overline{P}_r$ obeys the law $\overline{P}_r = kP_t/D^e$ where the specific value of the exponent e depends on the kind of landscape, typically ranging from 3 (rural area) to 5 (dense urban area) and the coefficient k is determined by the frequency band and

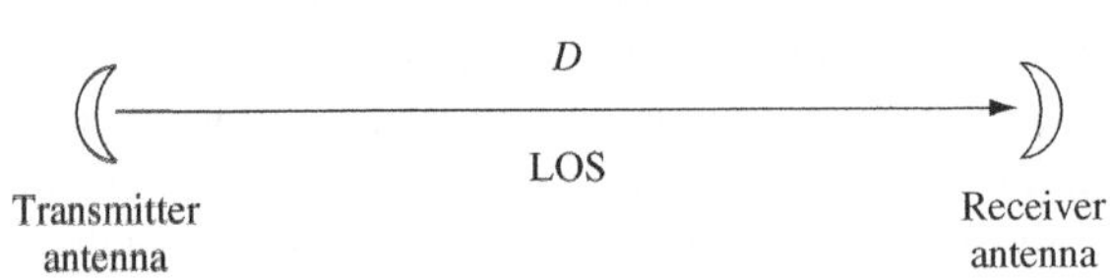

Figure 3.8 Free-space propagation model

the heights of the antennas [2,6,15,19]. The received power predicted by this model gives only a very rough reference point, being the result of averaging over the different positions of the receiver with the same distance D from the transmitter. Fluctuations of P_r along the arc of radius D centred at the transmitter location are significant and approximated by the lognormal PDF, meaning that the distribution of the decibel content of the received power $x = 10\lg P_r$ is Gaussian (normal):

$$W(x) = \frac{1}{\sqrt{2\pi}\sigma_x}\exp\left[-\frac{(x-\overline{x})^2}{2\sigma_x^2}\right]$$

The standard deviation σ_x of $10\lg P_r$ in the last expression is commonly accepted in the literature to be between 6 and 12 dB.

Attenuation of power caused by shadowing has a static character and, even when the receiver is in motion, usually changes in time comparatively slowly due to the large scale of landscape components (tens or hundreds of metres). For this reason shadowing is also often referred to as *large-scale* or *long-term fading*.

3.5.3 Multipath fading

Let us turn now to the second factor affecting the received signal intensity: *multipath propagation*. As a matter of fact, the transmitted signal can reach the receiving antenna travelling by many paths. The LOS may appear as one of them or be utterly obstructed, all the other paths emerging as a result of the transmitted wave being reflected by various objects. Examples of such reflectors are buildings, towers, cars, aircrafts, the earth's surface and many more (see Figure 3.9).

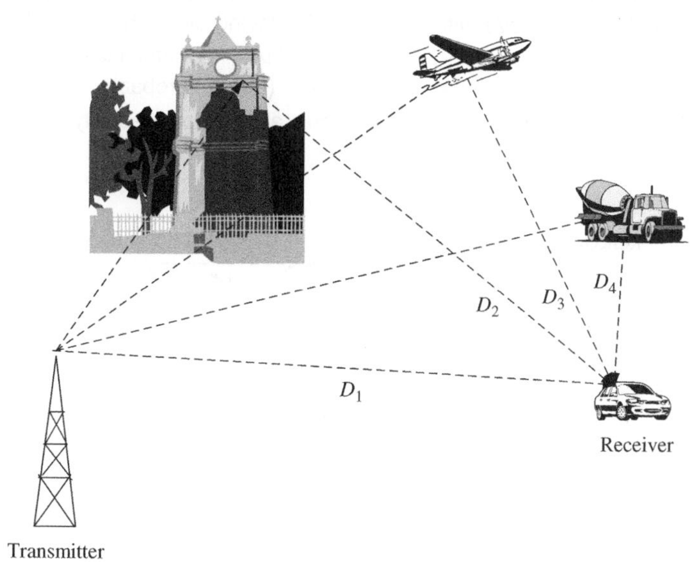

Figure 3.9 Illustration of multipath propagation

Suppose that travelling by the ith path, the transmitted signal with the reference complex envelope $\dot{S}(t)$ assumes amplitude A_i, delay τ_i and initial phase ϕ_i. Then the received complex envelope will be found as:

$$\dot{S}_r(t) = \sum_i A_i \dot{S}(t - \tau_i) \exp(j\phi_i) \tag{3.11}$$

where without loss of generality the real amplitude of the reference signal is set equal to one. When the *delay spread* τ_{ds}, i.e. maximal mutual delay between signals of different paths, is within the signal duration all of the multipath signals will overlap and interfere with each other. To better understand the phenomenon, consider first one simple scenario, which may take place in mobile communications, TV broadcasting or elsewhere.

Example 3.5.1. Figure 3.10 describes the situation where the transmitted signal reaches the receiver through the paths created by two reflectors (buildings, vehicles etc.), the LOS path being totally blocked by an obstacle (e.g. a building) and (3.11) including only two summands. Both reflectors are oriented so that they emit the secondary wave towards each other. The receiver placed on the line connecting the reflectors will observe superposition of the two interfering waves whose phase difference ϕ is governed by the ratio of propagation difference $\delta = D_1' + D_1'' - D_2' - D_2''$ to the wavelength λ_w: $\phi = 2\pi\delta/\lambda_w$. With amplitudes of the reflected signals at the receiver location A_1, A_2, the resulting amplitude A_r may be found by the cosine theorem as $A_r = \sqrt{A_1^2 + A_2^2 + 2A_1A_2\cos\phi}$ (see phasor diagram in Figure 3.11a). The periodicity of A_r as a function of ϕ means its periodicity in dependence on the propagation difference δ. When the receiver moves along the line connecting the reflectors its displacement by $\lambda_w/2$ in any direction changes δ by one wavelength λ_w so that ϕ changes by 2π and values of A_r at the points separated by $\lambda_w/2$ are identical. In other words, interference of two impinging waves creates a stationary wave with period $\lambda_w/2$. Moving along the tested line the receiver will alternate observing maximal $A_1 + A_2$ and minimal $|A_1 - A_2|$ amplitudes each $\lambda_w/2$ m. If the amplitudes of the reflected signals are close (rather probable case) the resulting power $P_r = A_r^2/2 = (A_1^2 + A_2^2)/2 + A_1A_2\cos\phi$ drops almost (or precisely) to zero when the

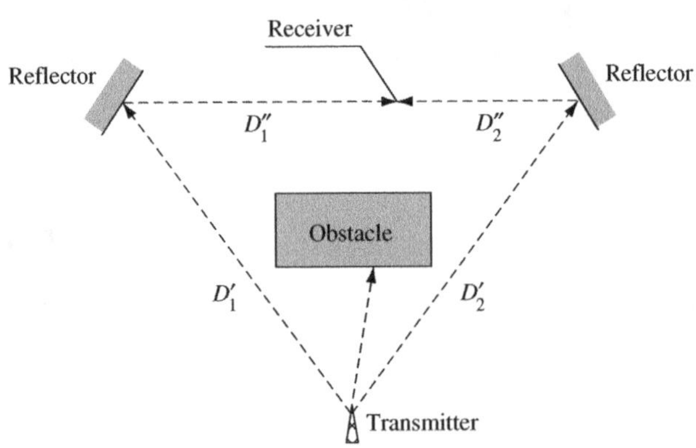

Figure 3.10 The two-path case

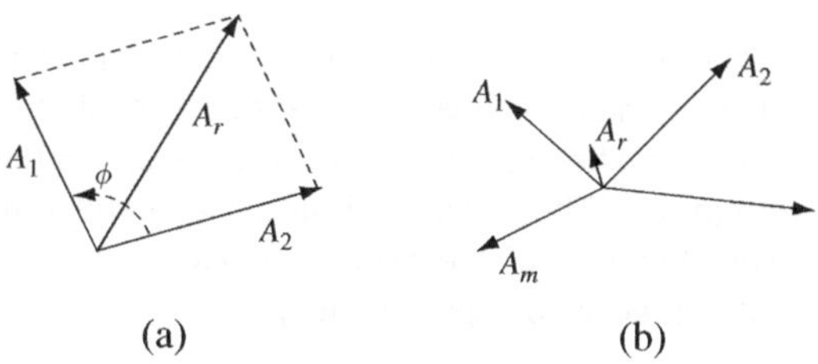

Figure 3.11 Interpretation of multipath effects with phasor diagrams

receiver passes the stationary wave nodes. This is exactly the phenomenon called multipath fading. Since the space distance between adjacent peaks of P_r is comparable with a wavelength, for the system operating in metre or decimetre band the time cycles of changing P_r at the moving receiver input will be rather short (typically split seconds).

The plot of P_r in dependence on time in Figure 3.12 gives an example for parameter values typical of mobile communications: $\lambda_w \approx 0.3$ m and the receiver carrier speed $V_r = 60$ km/h. As is seen, even with rather small speed of movement, changes of the received power are very rapid. This explains why multipath fading is also called *short-term fading* or *small-scale fading*.

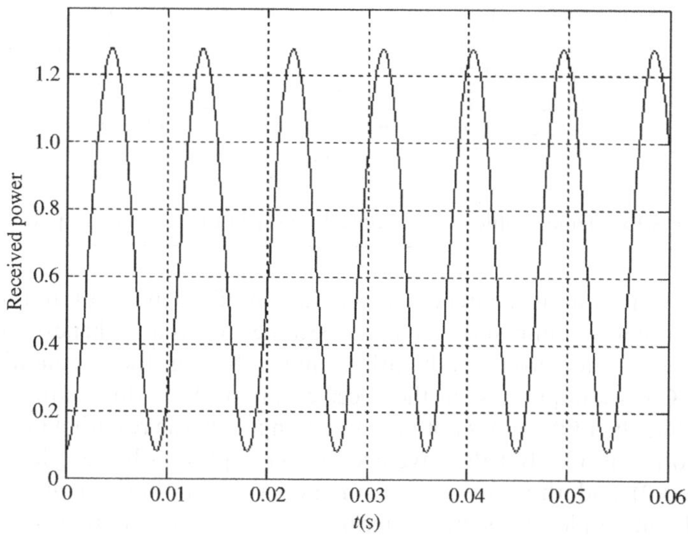

Figure 3.12 Time profile of the received power in the case of two-path fading

Clearly, this example is artificially simplified in order to present the phenomenon in the most explicit way. In practice, the number of multipath signals L received simultaneously may be very large and as a result the interference pattern becomes more complicated. The phasor diagram in Figure 3.11b illustrates a situation of this kind. The chaotic character of the distribution of reflectors or scatterers in the receiver environment makes the interference pattern unpredictable and its statistical description most appropriate.

Example 3.5.2. Figure 3.13 shows the time profile of the received power obtained by modelling in Matlab of the propagation environment with five reflectors located equiprobably within the square of side equal to the transmitter–receiver initial distance $D = 30$ km. The received power is normalized to the average one. The wavelength and user's speed are 0.3 m and 60 km/h, respectively. The irregular character of the power change is fairly explicit as well as the presence of deep drops of the received signal intensity.

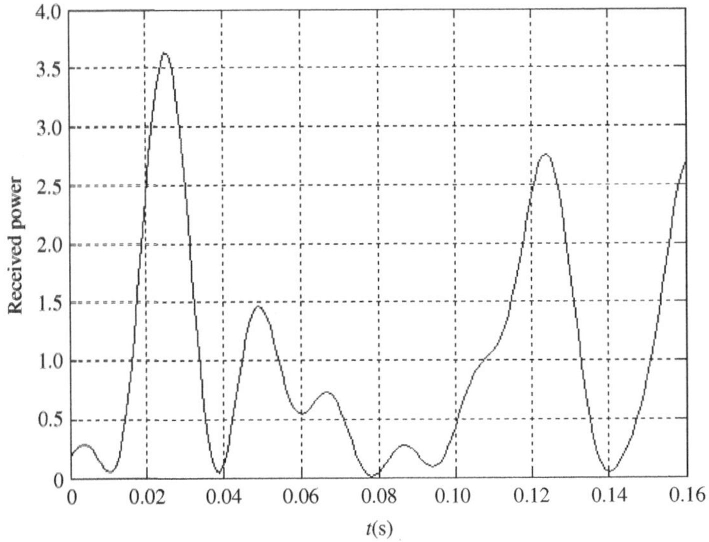

Figure 3.13 Time profile of the received power in the case of five-path fading

Due to the central limit theorem, a superposition of independent and nearly equally contributing random summands tends to become Gaussian whenever their number grows. Therefore, numerous multipath signals obeying these conditions produce a bandpass Gaussian process at the receiver input. If a dominating deterministic component (like the LOS one) is not present among them the resulting Gaussian process will be a zero-mean one. But the envelope of such a process has Rayleigh distribution (see Section 3.2) and thus we come to a model of *Rayleigh fading channel.* Now, the received amplitude A_r is not deterministic but, instead, random, meeting the Rayleigh PDF:

$$W(A_r) = \begin{cases} 2A_r \exp(-A_r^2), A_r \geq 0 \\ 0, \quad A_r < 0 \end{cases} \tag{3.12}$$

Since in the product $A_r\dot{S}(t)$, the 'genuine', actually measurable signal amplitude is split between two cofactors, and this may be done arbitrarily, a convenient normalization is assumed in (3.12) setting the mean square of A_r equal to one: $\overline{A_r^2} = 1$. A plot of PDF (3.12) is shown in Figure 3.14.

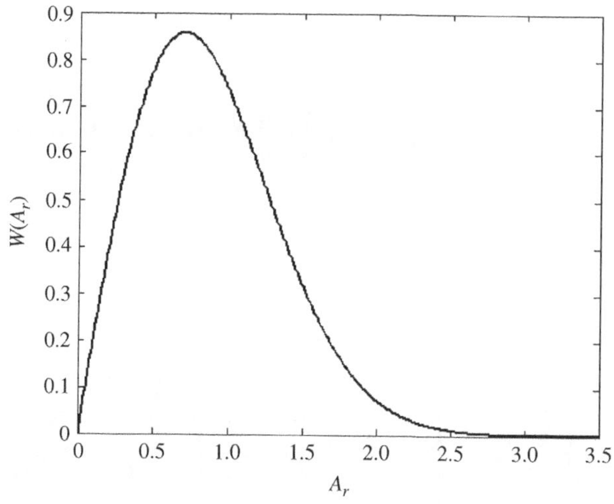

Figure 3.14 Rayleigh PDF

Rayleigh fading is characteristic of numerous systems, including, along with communications, radar, navigation etc. The deep falls in signal intensity inherent in it are not as a rule neutralized by sporadic rises of A_r when multipath signals arrive nearly in phase. As a result, the overall effect of Rayleigh fading on the system performance appears to be pretty destructive, as the analysis below corroborates.

3.5.4 Performance analysis

Consider as an example binary data transmission over the Rayleigh channel with a *slow* and *flat* fading. The first of the attributes means that the interference pattern remains stable during many symbols and the current reference phase may be retrieved from the received signal by averaging over an appropriate time interval. In other words, signal randomness does not exclude the BPSK from the available options. The second term stresses that the delay spread of multipath signals τ_{ds} is small enough compared to the duration T_b of the individual BPSK bit: $\tau_{ds} \ll T_b$. As a result, successive BPSK symbols do not overlap with each other, i.e. ISI (see Section 2.15) does not emerge. To understand why the word 'flat' is relevant, return to (3.11) and note that an adequate model of a multipath channel is a delay line with taps having delays τ_i and complex weights $A_i \exp(j\phi_i)$. The transfer function of such a system strongly depends on tap delays, and when $\tau_{ds} \ll T_b$ is rather uniform (flat) within the signal bandwidth so that all signal frequency components are distorted identically and signal shape remains unchanged. The only sort of corruption which the signal undergoes due to multipath propagation in such a case is Rayleigh amplitude fluctuations described by (3.12).

Figure 3.15 enlarges on these definitions. Plots simulated in Matlab show a slow flat fading (a) as opposed to the *fast* flat one (b) for the case of the bell-shaped symbol pulses. The second of the fading types is characterized by a rapid change of the interference pattern in time so that distortions of successive symbols are practically independent.

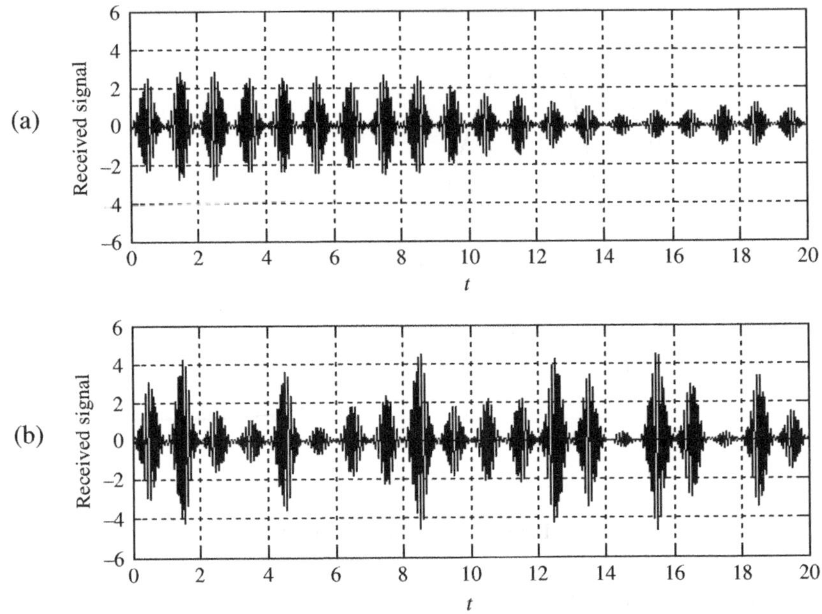

Figure 3.15 Slow (a) and fast (b) fading

Let the energy of the received signal corresponding to $A_r = 1$ be E. Then the energy of a signal with another value of amplitude is $E(A_r) = A_r^2 E$ and the average energy again equals E due to the normalization adopted above: $\overline{E(A_r)} = \overline{A_r^2}E = E$. Equation (2.19) may be used to calculate a conditional error probability $P_e(A_r)$, when the received amplitude is assumed fixed and equal to A_r:

$$P_e(A_r) = Q\left(\sqrt{\frac{2E(A_r)}{N_0}}\right) = Q\left(A_r\sqrt{\frac{2E}{N_0}}\right) = Q(A_r q_b)$$

where SNR $q_b = \sqrt{2E/N_0}$ corresponds to the signal of energy $E(A) = E$, i.e. of amplitude $A_r = 1$. Actual amplitude A_r is random and fluctuates from one receiving session to another according to the Rayleigh PDF (3.12). It is natural, then, to characterize the performance of data transmission by the value of $P_e(A_r)$ averaged over all A_r. Reserving now the term 'error probability', with designation P_e, for this expectation we have:

$$P_e = \overline{P_e(A_r)} = \int_0^{\infty} P_e(A_r)W(A_r)\mathrm{d}A_r = 2\int_0^{\infty} Q(A_r q_b)A_r \exp(-A_r^2)\mathrm{d}A_r$$

$$= \frac{2}{\sqrt{2\pi}}\int_0^{\infty}\int_{A_r q_b}^{\infty} A_r \exp\left(-\frac{x^2}{2}\right)\exp(-A_r^2)\mathrm{d}x\mathrm{d}A_r$$

where definition of the complementary error function $Q(\cdot)$ is used. Reversing the order of the integrals gives:

$$P_e = \frac{1}{\sqrt{2\pi}} \int_0^{\infty} \left[\int_0^{x/q_b} 2A_r \exp(-A_r^2) \mathrm{d}A_r \right] \exp\left(-\frac{x^2}{2}\right) \mathrm{d}x$$
$$= \frac{1}{\sqrt{2\pi}} \int_0^{\infty} \left[1 - \exp\left(-\frac{x^2}{q_b^2}\right) \right] \exp\left(-\frac{x^2}{2}\right) \mathrm{d}x$$
$$= \frac{1}{\sqrt{2\pi}} \int_0^{\infty} \exp\left(-\frac{x^2}{2}\right) \mathrm{d}x - \frac{1}{\sqrt{2\pi}} \int_0^{\infty} \exp\left[-\frac{(q_b^2+2)x^2}{2q_b^2}\right] \mathrm{d}x$$

where the first term equals $Q(0) = 1/2$ and the second is brought to the same form if multiplied by $\left(\sqrt{q_b^2+2}\right)/q_b$. Finally:

$$P_e = \frac{1}{2}\left(1 - \frac{q_b}{\sqrt{q_b^2+2}}\right) \tag{3.13}$$

To assess quantitatively the extent of the harmful influence of the fading, look at Figure 3.16, which presents error probabilities of the BPSK transmission over the AWGN and Rayleigh channels. As is seen, $P_e = 10^{-3}$ in the AWGN channel may be guaranteed with the bit SNR around 10 dB, while the Rayleigh channel requires bit SNR of at least 27 dB,

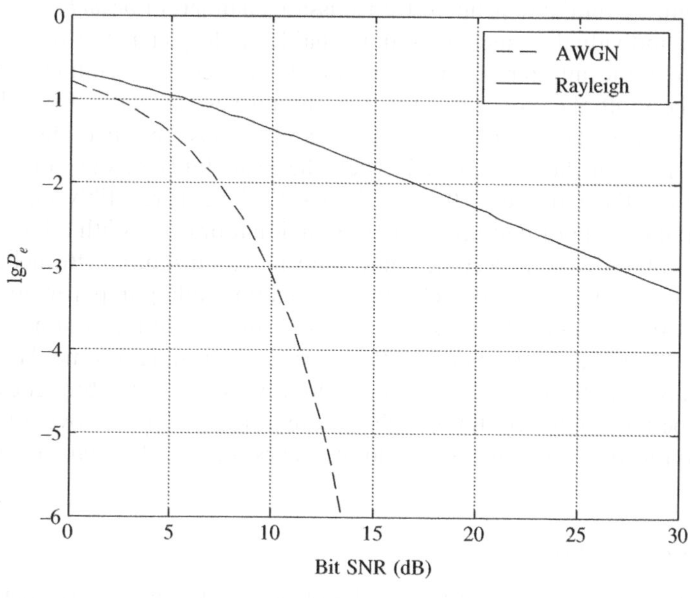

Figure 3.16 Bit error probability for AWGN and Rayleigh channels

i.e. 50 times higher. This drastic energy loss due to the fading gets greater when higher transmission reliability is necessary and becomes close to 25 dB (300 times) for $P_e = 10^{-4}$.

The physical explanation of a quite detrimental fading effect is rather straightforward. Sporadic sharp drops in signal intensity due to the multipath interference are rather likely in the Rayleigh channel. Elementary integration of PDF (3.12) shows, for instance, that the probability of A_r falling below the level 0.4 (SNR decreases by 8 dB) is about 0.15. Then, as is seen from Figure 3.16 (dashed line), in those sessions where such drops happen, the error probability can not be lower than 0.1 if reference SNR is 10 dB. Since the share of such sessions is 0.15, their contribution to the total (average) error probability will be no less than $0.1 \times 0.15 = 0.015$, i.e. 15 times bigger than the value corresponding to the reference SNR. This effect cannot in any way be compensated by possible favourable sessions with high SNR, since their contribution to the total error probability is never negative.

The consequence of multipath propagation may potentially be even more dramatic when the fading is *frequency selective*. This term, as opposed to the attribute 'flat', defines the situation where the channel transfer function is not uniform within the signal bandwidth. This happens if the delay spread covers several transmitted bits so that at the channel output the previous bits overlap with the current one. To counter this ISI, special filters (*equalizers*) are used, which rectify the channel transfer function non-uniformity. On the other hand, frequency selectivity when used properly is a good resource for countering fading by arranging the multipath diversity discussed in Section 3.7.

3.6 Diversity

The general idea of combating destructive multipath effects consists in *diversity*, which means arranging several independent transmission channels or *branches*. Thanks to this, despite every individual branch remaining liable to Rayleigh (or other) fading, the probability that the interference patterns in all of them are simultaneously poor is defined by the multiplication rule and thus diminishes radically. Take the figures of the example at the end of the previous section and suppose that two identical independent branches are somehow organized. Then the probability of the same fall of signal power in both of them at once is $0.15^2 = 2.25 \times 10^{-2}$, i.e. perceptibly smaller compared to the probability of poor conditions in an individual branch. With a larger number of branches this diversity gain becomes more and more substantial. Branches operate in parallel, as though they secured each other, mitigating fading impairment.

In other words, we know that the poor performance resulting from multipath fading is entirely due to the deep drops of SNR occurring from time to time. Hence, the final objective of diversity techniques is to process jointly signals of the branches in a manner that makes 'better' (higher SNR) branches more influential on the overall performance in comparison to the worse ones. Such joint processing is called *combining*.

3.6.1 Combining modes

Various strategies for combining the results of processing signals arriving via different branches may be used at the receiver. Suppose that there are n_d diversity branches altogether and let A_i, ϕ_i and σ_i be current signal amplitude, signal phase and noise standard deviation in the ith branch, where $i = 1, 2, \ldots, n_d$. What, then, is the best linear processing producing the maximum possible resultant SNR? Any linear form of the branch responses $u_i, i = 1, 2, \ldots, n_d$ is their weighted sum $\sum_{i=1}^{n_d} w_i^* u_i$ where weights w_i^* are in general complex. Then the resultant power SNR q_r^2 is just the ratio between the magnitude of the deterministic component of this sum and the variance of its noise component. The latter is simply the sum of branch noise variances weighted by $|w_i|^2$, since branches are independent. Therefore:

$$q_r^2 = \frac{\left|\sum_{i=1}^{n_d} w_i^* A_i \exp(j\phi_i)\right|^2}{\sum_{i=1}^{n_d} |w_i|^2 \sigma_i^2} \tag{3.14}$$

The sum in the numerator of (3.14) may be treated as an inner product of two n_d-dimensional vectors whose components are $w_i \sigma_i$ and $(A_i/\sigma_i) \exp(j\phi_i)$. No inner product can have modulus greater than the product of the lengths of vectors, which is well known as Schwarz's inequality [1,2]. Consequently:

$$q_r^2 \leq \frac{\left(\sum_{i=1}^{n_d} |w_i|^2 \sigma_i^2\right)\left(\sum_{i=1}^{n_d} \frac{A_i^2}{\sigma_i^2}\right)}{\sum_{i=1}^{n_d} |w_i|^2 \sigma_i^2} = \sum_{i=1}^{n_d} \frac{A_i^2}{\sigma_i^2} = \sum_{i=1}^{n_d} q_i^2 \tag{3.15}$$

where $q_i = A_i/\sigma_i$ is voltage SNR of the ith diversity branch. When optimal weights:

$$w_i^* = \frac{A_i}{\sigma_i^2} \exp(-j\phi_i)$$

are taken, inequality (3.15) becomes an equality, i.e. maximal possible resultant SNR is achieved. Such weights, as is readily seen, realize joined matched filtering of the responses of the diversity branches. Technically it is possible only when accurate values of all signal amplitudes and phases are known. Then signals can be summed coherently with an appropriate amplitude weighting. This combining technique is often referred to in the literature as the *maximal ratio* technique [5,18].

To assess the efficiency of combining, denote maximal SNR over all the diversity branches by $q_{\max}$ and introduce the diversity gain as $G_d = q_r^2/q_{\max}^2$. Since $\sum_{i=1}^{n_d} q_i^2 \leq n_d q_{\max}^2$ no combining scheme can provide gain greater than n_d, and the latter is achievable only in the maximal ratio scheme under the additional stipulation that all the diversity branches have the same SNR.

In practice, some other combining schemes find application, too, because maximal ratio processing is rather demanding as to the extra arrangements necessary (to measure SNR and phase in a diversity branch some special pilot signal may appear necessary etc.). Alternative combining modes are equal-weight combining and selection of a maximum SNR branch. The first approaches the maximal ratio mode in effectiveness if all the diversity branches have nearly equal SNR. The gain of the second is close to that of the optimal scheme if one of the diversity branches dominates over the rest in value of SNR.

Certainly, these strategies can be combined with each other, e.g. several branches with the best SNR values are selected and then their outputs are summed with equal weights.

Consider now traditional ways of organizing independent diversity branches.

3.6.2 Arranging diversity branches

The traditional ways to set up independent diversity branches may be categorized as follows:

- Space diversity
- Frequency diversity
- Time diversity
- Polarization diversity
- Multipath diversity.

Space diversity implies creating several independent propagation paths at the expense of involving multiple antennas, which explains the other popular name for this technique: *antenna diversity*. Duplicating antennas may be used at the receiving side as well as at the transmitting side. Being spaced from each other by a distance of 7–10 wavelengths or more, they provide practical independence of parallel interference patterns at the receiver input. When used at the receiver (Figure 3.17) (*receive diversity*) antenna diversity is most effective, since additional antennas utilize signal energy which otherwise would not be captured at all. In this case diversity signals are separated automatically since different antennas receive them. Being matched-filtered individually, they may be further combined as described above.

Transmitting antenna diversity (*transmit diversity*) is not that straightforward. First, as is seen from Figure 3.18, a limited total transmitter energy resource should be divided between several transmitting antennas. Second, the receiver antenna receives the mixture of signals emitted by all transmitting antennas. Therefore, some measures should be taken to provide an opportunity for separation and individual processing of those signals by the receiver before combining. These factors make this sort of diversity a sophisticated optimization problem, solving which is the subject of a special branch of communication theory called *space–time coding* (see Section 10.3).

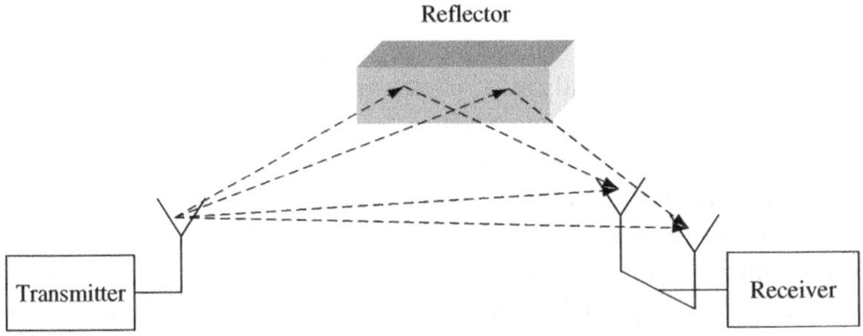

Figure 3.17 Receive antenna diversity

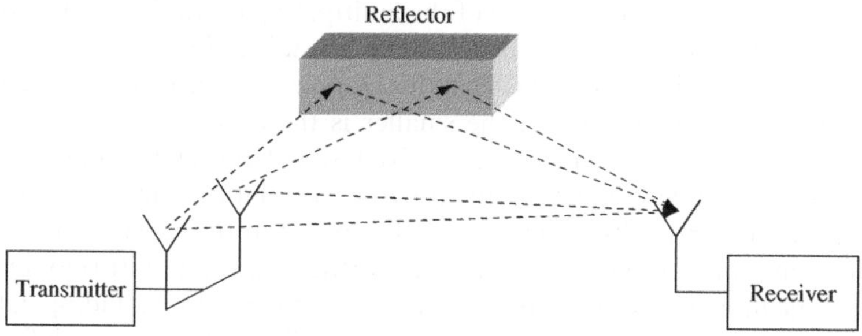

Figure 3.18 Transmit antenna diversity

Certainly, when the technological constraints allow it, the combination of transmit and receive antenna diversity may be used to gain maximal benefits.

The idea of frequency diversity is based on the concept of the channel *coherence bandwidth*. This notion determines the frequency range within which fading is considered as flat, i.e. distortion of signal frequency components is strongly dependent. On the other hand, the harmonics with frequency space beyond the coherence bandwidth may be treated as independently distorted by the channel. As was already underlined in the previous section, the frequency range of the flat fading depends inversely on the delay spread, so the wider the range of dispersing signals in time, the shorter the coherence bandwidth. Evidently, transmitting the same signal simultaneously at n_d carriers whose frequencies are offset by coherence bandwidth or more creates n_d diversity branches. We may say that frequency diversity puts frequency selectivity of fading to good use. Figure 3.19 gives an elementary clarification of the idea. The waves of two wavelengths λ_{w1} and λ_{w2} propagating along the same couple of paths have identical geometrical propagation differences δ. However, the phase differences between the signals of the two paths are individual for each wavelength and equal to $2\pi\delta/\lambda_{w1}$ and $2\pi\delta/\lambda_{w2}$, respectively. When one of these phase differences leads to attenuation of the resultant signal the other may appear less destructive. With many parallel propagation paths present statistical interpretation comes into force, and frequency difference exceeding the channel coherence bandwidth provides the independence of diversity branches in this scheme. An appropriate choice of frequencies in this diversity scheme provides separation of branches at the receiver with the help of bandpass filtering.

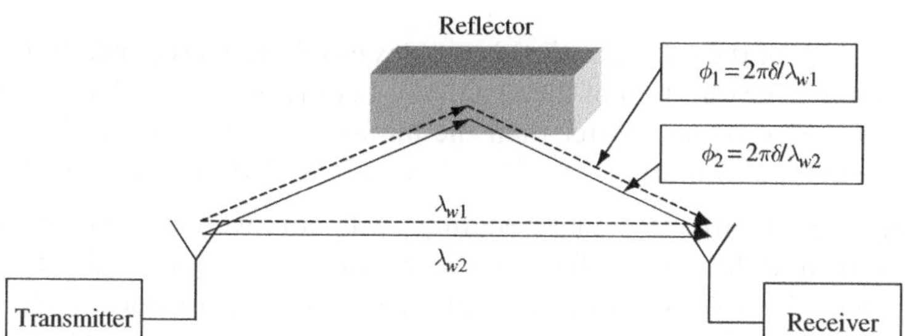

Figure 3.19 Frequency diversity

Time diversity exploits time-variance of the multipath pattern. Even when the receiver does not move, the multipath profile may be unstable due to the motion of the transmitter or surrounding reflectors. Thereby Doppler scattering of the received signal arises, and the greater is its spread, the smaller is the *coherence time* of the channel, i.e. the time interval during which the received signal power remains nearly stable. Consider again the time–frequency duality: the correlation range in the frequency domain (coherence bandwidth) is inverse to the spread in time (delay spread), while in the time domain (coherence time) it is inverse to the frequency shift (Doppler) spread. Since at the time moments spaced apart by coherence time or more, fading patterns may be treated as independent, retransmission of n_d replicas of the same information at appropriate time intervals creates n_d diversity branches. With a slight modification this principle is universally used in telecommunications in the form of *interleaving*.

Polarization diversity, which exploits the difference of multipath profiles of waves with different polarization, has not yet found wide application. As for the last item in the list above, it possesses quite an important role in our context and will be discussed separately in the next section.

3.7 Multipath diversity and RAKE receiver

A conventional frequency diversity touched upon above implies parallel transmission of the same signal on several carriers, the spacing between them being bigger than the channel coherence bandwidth. Typically this technique is used when the signal spectrum is narrow enough compared to the coherence bandwidth to make the fading flat. The other version of frequency diversity is multipath diversity, exploiting signals with spectrum deliberately extended beyond the coherence bandwidth. Thereby fading becomes frequency selective, allowing in principle time resolution of multipath signals. Thus, the multipath diversity scheme is based on the fact that the signals propagating along the different paths reach the receiver with different time delays. Suppose that the resultant received signal with the complex envelope (3.11) is passed through the filter matched to the signal $s(t)$. Then, taking into account filter linearity and the connection between signal ACF and the matched filter response (see Section 2.11), we will have for the complex envelope $\dot{S}_{rf}(t)$ at the filter output:

$$\dot{S}_{rf}(t) = \sum_i A_i \dot{R}(t - \tau_i - T)\exp(j\phi_i) \tag{3.16}$$

where $\dot{R}(\tau)$ is ACF of the reference signal complex envelope $\dot{S}(t)$ defined by (2.66) and T, as usually, denotes the duration of the signal $s(t)$. Let the signal correlation spread τ_c, i.e. ACF time-extension, be no greater than the minimum mutual delay of the successive multipath signals $\tau_{\min} = \min_i\{\tau_i - \tau_{i-1}\}$: $\tau_c \le \tau_{\min}$. It is obvious that in such a situation all multipath signals after the matched filter will not overlap. Since they are fully resolved in time and do not interfere with each other, we may treat them as signals of the independent diversity branches and process according to one of the combining algorithms described above. If, for instance, their time positions, amplitudes and initial phases are known (say, preliminarily measured using a separate pilot channel), maximal ratio combining is the best choice.

It should be clear that to realize this multipath diversity scheme, a signal with a short ACF is necessary. As the discussion of Sections 2.12 and 2.15 showed, a 'brute force' solution of this task exists, consisting in employing short signals. This, however, means transmission of high peak power, which cannot be afforded in numerous cases. Much more attractive is the use of special signals featuring time-compression in the matched filter, i.e. having correlation spread small in comparison to duration: $\tau_c \ll T$. These signals can only be found among spread spectrum ones, so we may add to the list of the merits of spread spectrum one more advantage: *feasibility of organizing the multipath diversity scheme*.

Multipath diversity is unique in the sense that it radically changes the attitude towards multipath effects, which at first sight are taken as implicitly harmful. As the discussion above exhibits, reflection of waves has a fruitful side, too. Actually, any contributing reflector directs to the receiver part of the emitted energy which would otherwise be entirely lost. When these reflected signals may be separated from each other (time-resolved), this energy is utilized, improving the system performance against that in their absence. The channel itself, in effect, creates diversity branches in this scheme, the only problem being adequate signal design allowing the multipath replicas to be resolved.

Example 3.7.1. Consider the illustration of the diversity scheme applied to digital communications given by Figure 3.20. Matlab is used to simulate the BPSK transmission with bit duration *T* over

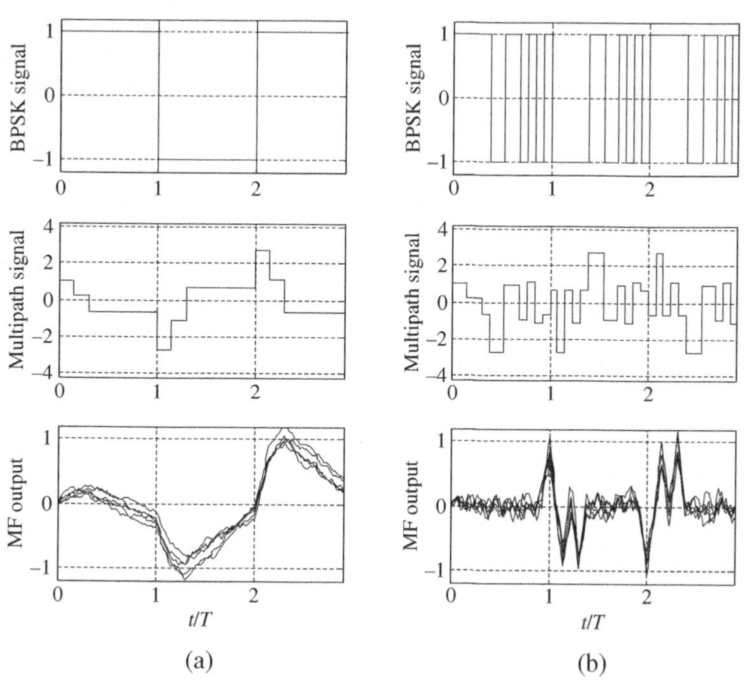

Figure 3.20 Multipath effects in BPSK transmission: (a) plain signals and (b) spread spectrum

the channel whose three paths have mutual delays $\tau_2 - \tau_1 = \tau_3 - \tau_2 \approx 0.15T$. The amplitudes of the multipath signals are $A_1 = 1$, $A_2 = 0.8$, $A_3 = 0.9$, and the phases of the second and third are opposite to the phase of the first. For the sake of transparency, only baseband equivalents (complex envelopes) of all signals are shown. The left column (a) corresponds to the transmission of one bit by a plain rectangular pulse, zero bit content being sent by positive polarity (upper plot). The second plot of the column shows the resultant signal at the channel output, ignoring noise. Although ISI exhibits itself by distortion of the initial part of a bit pulse, the major portion of the pulse undergoes flat fading. The lowest plot demonstrates five superimposed realizations of the matched filter response to the resultant signal, in the presence of noise. The destructive effect of multipath propagation is clearly seen: reliable decisions on the transmitted bits are hardly possible at all.

As an opposite example, column (b) relates to the transmission mode, where every bit manipulates the polarity of a spread spectrum signal, specifically a binary Barker code of length $N = 13$ (see Section 6.4). The upper plot shows three such pulses manipulated by the same bit pattern as earlier. In the second plot a resultant noiseless signal at the channel output is given. The lowest plot demonstrates the matched filter response to the resultant signal corrupted with white noise of the same intensity as in the previous case. The three distinctive peaks per one transmitted bit are all available to retrieve the transmitted data with high confidence. Assuming the channel model is known beforehand, the samples may be taken at the accurate moments of the maximums of each noiseless multipath component at the matched filter output. As is seen, a properly chosen spread spectrum bit pulse provides resolution of all multipath components at the filter output with no mutual interference. The three samples may then be combined optimally, i.e. weighted proportionally to their amplitudes and summed after alternation of the polarity of the second and third components. An equivalent realization of combining is shown in Figure 3.21, where the tapped delay line is used to align in time three noiseless multipath peaks at the filter output. Then the non-delayed waveform weighted by $-A_3 = -0.9$ and the one delayed by $\tau_3 - \tau_2$ weighted by $-A_2 = -0.8$ are summed with the waveform delayed by $\tau_3 - \tau_1$ to produce the optimal combiner output. The voltage SNR of the output is $\sqrt{1 + A_2^2 + A_3^2} \approx 1.56$ times higher compared to that of the first path.

The SNR gain is apparent in Figure 3.22, where the upper plot repeats the matched filter output of Figure 3.20b and the lower plot shows the output voltage of the adder. Sampling the adder output at the moments $\tau_3 + T$, $\tau_3 + 2T$, ... and fixing the polarities of the samples provides the decisions on the transmitted bits.

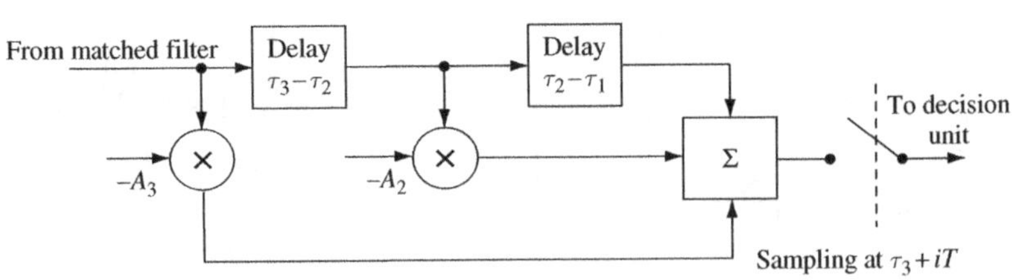

Figure 3.21 Maximal ratio combining for the channel of Figure 3.19

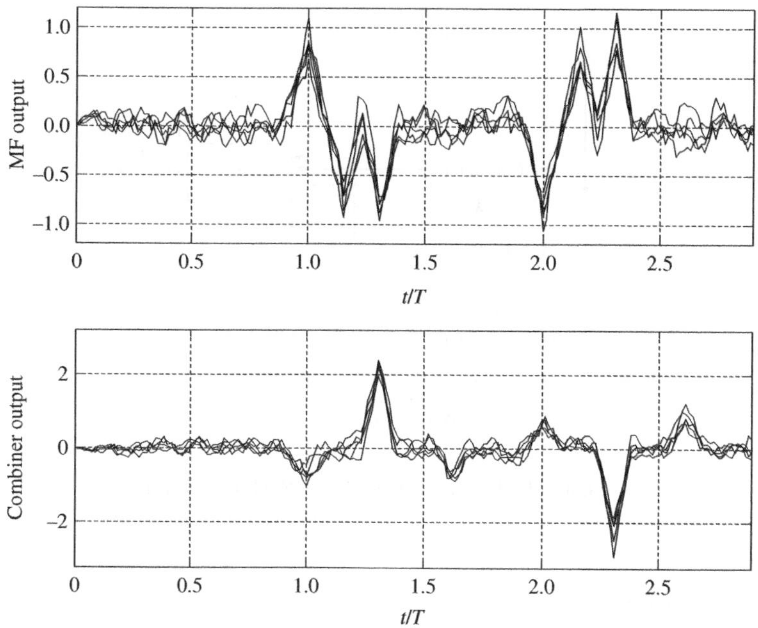

Figure 3.22 Outputs of the matched filter and adder

The multipath diversity principle was proposed by Price and Green in 1958. Since then it has been widely known by the nickname RAKE, because the peaks at the matched filter output (Figures 3.20 and 3.22) resemble the *fingers* of a popular garden implement.

Numerous implementations of the RAKE algorithm are around. One of them uses n_d parallel correlators instead of a matched filter, n_d being the number of fingers, i.e. exploited multipath diversity branches. This structure is most practical when delays of multipath signals are estimated precisely and may be considered as known. A correlator with reference signal delayed by τ outputs the value equal to the sample at the matched filter output at the moment $T + \tau$. Then, since only samples at the moments $T + \tau_i$, $i = 1, 2, \ldots, n_d$ are necessary to make a decision, we may use n_d delayed replicas of the signal as references in the correlators, thereafter combining their outputs in an appropriate manner. The advantage of such a structure compared to the matched filter one is that for a complicated spread spectrum modulation, implementation of the correlator is often much more feasible than that of the matched filter, since, unlike the latter, the former computes only a single sample of the correlation (see Section 2.11). With maximal ratio combining a correlator-based RAKE receiver looks as shown in Figure 3.23 (a complex-envelope-processing version is presented). To emphasize the practical role of the RAKE diversity it is enough to refer to the 2G (IS-95) and 3G (UMTS, cdma2000) cellular CDMA standards, which employ this principle as an integral part of their philosophy.

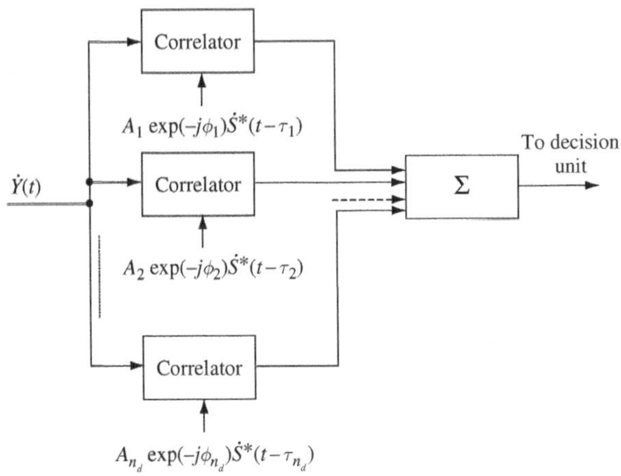

Figure 3.23 Correlator-based RAKE receiver

Problems

General directions

- If the shape of a signal or/and interference spectrum is not specified take it as rectangular.
- The term 'matched filter' is used for the filter matched to signal against AWGN background.
- Where the barrage jammer is considered neglect AWGN.

3.1. A system needs the highest possible ratio of useful signal power to total interference power, including AWGN and a jammer. A narrowband jammer is present. Which of the two strategies is better: ignoring the jammer or band-elimination filtering, if:

(a) Jammer power equals AWGN power within the signal bandwidth and jammer bandwidth is half of the signal bandwidth?
(b) Jammer power is 6 dB below AWGN power within the signal bandwidth and jammer bandwidth is a quarter of the signal bandwidth?
(c) Jammer power is 3 dB above AWGN power within the signal bandwidth and jammer bandwidth is a quarter of the signal bandwidth?

3.2. Spectra of the signal and narrowband jammer are given in Figure 3.24. What is the most harmful central frequency of the jammer when a receiver ignores the jammer and when it employs band-elimination filtering?

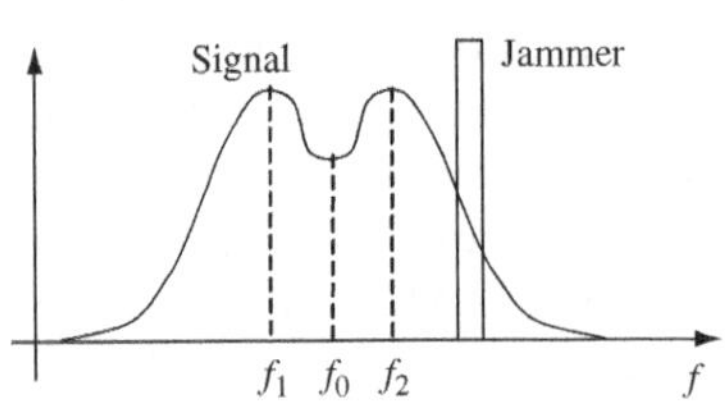

Figure 3.24 Signal and jammer spectra

3.3. In some system power SIR at the matched filter output degrades 101 times as compared to power SNR, while the system can preserve its operation capability with only two-times degrading SIR. What should be changed in the signal, SNR remaining constant, if

(a) Only plain signals are allowed?
(b) Signal peak power cannot be raised (what should the signal processing gain be in this case)?

3.4. In some system a band-elimination filter neutralizes the narrowband jammer. Due to this, the matched filter SNR degrades by 3 dB.

(a) What should be done to the plain signal duration and amplitude if only 2% degradation in power SNR is tolerable (SNR in the absence of band-elimination is fixed)?
(b) Is it possible to push degradation below 2% without increasing signal power? If so, what should the signal processing gain be?

3.5. A system can operate with SNR no smaller than 10 dB. Due to a barrage jammer SNR drops to −3 dB. How could the signal parameters be changed to neutralize the jammer, if:

(a) Only plain signal of the fixed energy is allowed?
(b) Only plain signal of the same peak power can be used?
(c) Peak power and energy of the signal are fixed with no other constraints?
(d) Peak power of the signal is fixed and its bandwidth can be increased only 10 times?

Find the processing gain in cases (c) and (d).

3.6. In conflict with a barrage jammer transmitter, a system increases signal duration by 4 times with simultaneous halving of signal power and widening of the bandwidth by 50 times. The jammer transmitter is capable of increasing its power no more than 13 dB. Who will be the winner in this game?

3.7. A signal occupies two separate sub-bands of identical width W_0. Total signal energy is distributed between them in the proportion 9:16. In the receiver two matched filters process both sub-band parts and their outputs are combined optimally to maximize a resultant SNR. There is a barrage jammer transmitter. What is the most harmful distribution of its total power between the signal sub-bands?

3.8. Matched filter SNR for an intended receiver equals 14 dB. The processing gain of the signal $WT = 400$. Find:

(a) Ratio of power spectrum densities of the signal and background AWGN.
(b) SNR at the integrator output of an interceptor radiometer.

3.9. In the BPSK data transmission system error probability per bit $P_e = 1.5 \times 10^{-3}$ is required. A system designer wants the SNR of the interceptor radiometer to be no greater than −10 dB per one transmitted bit duration. What processing gain per one bit would be satisfactory?

3.10. The interceptor radiometer SNR is -10 dB per signal duration, whereas the intended receiver needs SNR 12 dB. Signal duration $T = 100\,\mu s$. What should its minimal bandwidth be?

3.11. Two spread spectrum BPSK data transmission systems are compared. The first employs binary modulation to widen the bandwidth, providing processing gain of 100 per data bit. The second operates with a ternary modulation resulting in processing gain of 50 per data bit. Which of them has better immunity to cracking a modulation law, the data rate being the same?

3.12. Two spread spectrum BPSK data transmission systems operating at the same data rate are compared. In the first an intended receiver SNR per data bit is 12 dB, while the interceptor radiometer SNR per bit is -12 dB. For the second system these parameters are 16 and -4 dB, respectively. Which of them is more immune to breaking a modulation law?

3.13. A system designer takes care of the EMC of a developed system. The maximal SNR provided by the new system over the entire zone covered by the old ones is 20 dB. Any old system operates with a satisfactory quality if the extra power spectrum density does not exceed -10 dB compared to the background AWGN spectrum. What should the minimum processing gain of the new system be?

3.14. There are two spread spectrum systems occupying the same total bandwidth and operating inside the same geographical area. The maximal SNR over all the intersection of their operational zones are 20 and 17 dB, respectively. For their compatibility, the extra power spectrum density due to the emission of the other system should be at least -7 dB below the natural AWGN level. Find the minimal processing gain of each of the systems.

3.15. For normal functioning, the system operating at the wavelength $\lambda_w = 30$ cm needs signal-to-AWGN ratio $q = 14$ dB. Signal duration $T = 100\,\mu s$ and the noise temperature of the receiver $\theta_n = 1000$ K. The transmitter antenna has gain of 5 dB, while the receiving antenna is omnidirectional. Find the necessary transmitted power for a free-space propagation model, if the system coverage zone should have a radius no smaller than 30 km. How will this power increase for the conditions typical of mobile telephone with the distance-attenuation exponent 3.84, no other adjustment of the free-space model being necessary?

3.16. A system survives if the received voltage SNR drops no more than 4 times below the average predicted level. Find the probability of system failure due to long-term fading with the standard deviation of the decibel power content equal to 9 dB.

3.17. There are two propagation paths: LOS and one through a reflector located 3 km away from the LOS and equidistant from both receiver and transmitter. Find the standing wave period in metres and the time interval between successive power drops at a distance 12 km from the transmitter for a receiver moving at a constant speed of 60 km/h if the wavelength is 30 cm.

3.18. Is it practical to use BPSK in the channel with fast multipath Rayleigh fading? What binary transmission mode is advisable in this case?

3.19. Binary data are transmitted over a channel with lognormal long-term and Rayleigh short-term fading. Due to the long-term fading, the signal power fluctuates around

27 dB with standard deviation 12 dB. Is bit error probability of 10^{-3} achievable in this channel without error-control coding?

3.20. There are two Rayleigh diversity branches with identical average signal energies. Using average power SNR at the combiner output as a criterion, compare the energy gains of two combining techniques: the maximal ratio one and selection of a maximum SNR branch.

3.21. A signal of a system occupies bandwidth of 60 kHz. The delay spread of the channel is 20 μs. The total bandwidth is no greater than 300 kHz. How many frequency diversity branches may be arranged?

3.22. There are four propagation paths with lengths 5 km, 5.4 km, 5.55 km and 6 km. A system transmits data at the rate $R = 20$ kbps. Estimate the approximate signal bandwidth and minimal processing gain necessary for arranging a 4-finger RAKE receiver.

3.23. The minimum difference of lengths of paths in a channel is 300 m. The delay spread of the channel is within 10 μs. The system transmits data using QPSK at the rate 20 kbps. What is the maximal available number of RAKE fingers? Find the necessary bandwidth and processing gain of the signal for arranging the RAKE receiver.

3.24. A RAKE receiver splits the resultant signal at the output of a Rayleigh channel into n_d non-fading signals of equal power using maximal ratio combining. How does the average power SNR at the combiner output differ from the case when the RAKE algorithm is not used? In the light of the answer how can the energy gain of the RAKE technique be explained?

Matlab-based problems

3.25. Write a program illustrating the advantages of spread spectrum in countering a narrowband jammer when the receiver does not use band-elimination filtering (Figure 3.4).

(a) Form and plot two rectangular bandpass signals of the same duration T (vectors of dimension 1000): a plain pulse and an LFM pulse with deviation $(40\text{–}50)/T$ (see Problems 2.55–2.56). Take carrier frequency to have 25–30 periods per signal duration.

(b) Form the matrix of a CW jammer with 10 rows, each row having frequency equal to the signal carrier frequency and random initial phase uniformly distributed over $[-\pi, \pi]$.

(c) Form observation matrices for both signals by adding the signals to the jammer. Set up the jammer level several (2–10) times higher than the signal level. Plot the observations.

(d) Process the observations with corresponding matched filters. Plot the filter output waveforms.

(e) Run the program, varying the jammer intensity and frequency, and explain the results.

3.26. Write a program illustrating the advantages of spread spectrum in countering a narrowband jammer when the receiver uses band-elimination filtering (example plots before and after band elimination are shown in Figure 3.25).

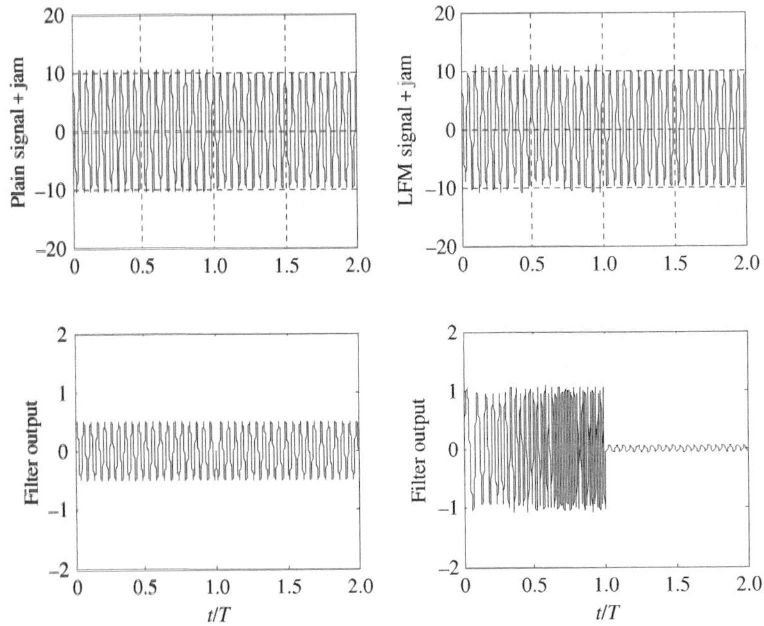

Figure 3.25 Simulation of band-elimination filtering

(a) Repeat items (a)–(c) of the previous problem.
(b) Calculate and plot energy spectra of signals and observations.
(c) Use band-elimination filtering forcing the spectrum component of the observation at the jammer frequency to zero.
(d) Return to the time domain and plot the signals after band-elimination filtering.
(e) Run the program for several combinations of jammer intensity and frequency, and interpret the results.

3.27. Write a program illustrating the advantages of spread spectrum in countering a barrage jammer (see Figure 3.5).

(a) Repeat item (a) of Problem 3.25.
(b) Form 10 jammer spectra for each of two signals so that they are non-zero only within the bandwidth of a proper signal. Take spectral components to be Gaussian complex values with zero mean. Plot power spectra of the jammer;
(c) Convert the jammer into the time domain and sum it with a proper signal to obtain observations. Adjust the power of the jammer so that for both signals voltage signal-to-jammer ratio are identical, lying in the range 0.5–1.
(d) Process the observations with a proper matched filter and plot the filter output.
(e) Run the program, varying the signal and jammer parameters, and interpret the results.

3.28. Write a program illustrating the low-detection-probability feature of spread spectrum (see Figure 3.26).

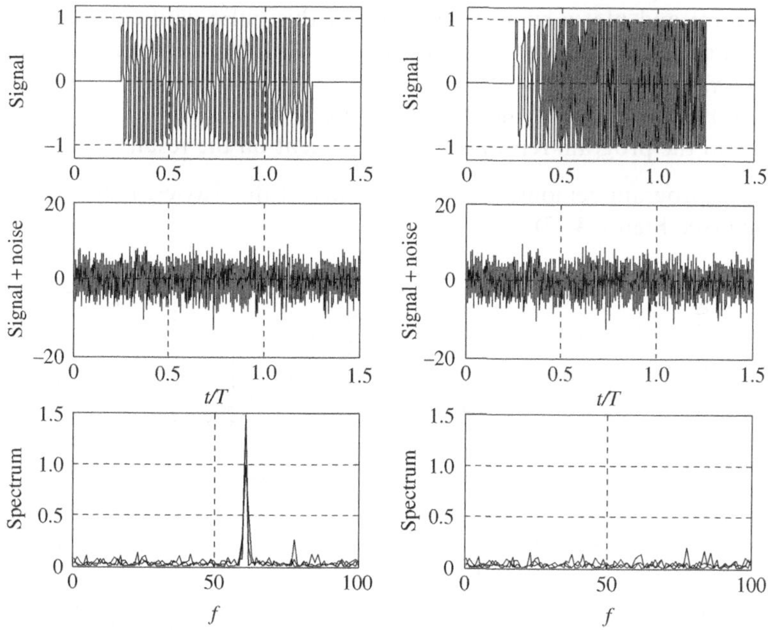

Figure 3.26 Simulation of detecting signals by the radiometer

(a) Repeat item (a) of Problem 3.25.
(b) Form 10 Gaussian noise realizations with zero mean and standard deviation three times higher than signal amplitude.
(c) Sum the noise realizations and signal to produce 10 observations for each signal.
(d) Convert the observations into the frequency domain and plot observation energy spectra for each of the signals. Interpret the plots in terms of the ability of the radiometer to detect the signal.

3.29. Write a program illustrating the good electromagnetic compatibility of spread spectrum systems, both between each other and with conventional systems.

(a) Form complex envelopes of three rectangular pulses of the same energy and duration: a plain one and two LFM pulses with deviation $W_d = 50/T$, the first with growing and the second with dropping frequency.
(b) For all three cases, calculate complex envelopes of a matched filter response to the proper signal and two foreign ones.
(c) Plot the real envelope for each of 9 responses of the previous item.
(d) Interpret the plots in terms of electromagnetic compatibility.

3.30. Plot a logarithmic graph of attenuation of the received power with the distance from a transmitter for the propagation model $\overline{P}_r = k/D^{\epsilon}$, $\epsilon = 2, 3, 3.84, 4$.

3.31. Write a program demonstrating the phenomenon of multipath flat fading.

(a) Form and plot a smooth plain bandpass pulse of duration T and carrier frequency within $(20–30)/T$.
(b) Form and plot a sum of 10 copies of this pulse with equal amplitudes and random initial phases which are independently and uniformly distributed over $[-\pi, \pi]$.
(c) Plot the envelope of the resulting pulse.
(d) Run the program repeatedly and comment on the results.

3.32. Write a program demonstrating the validity of the Rayleigh model of short-term fading (see Figure 3.27).

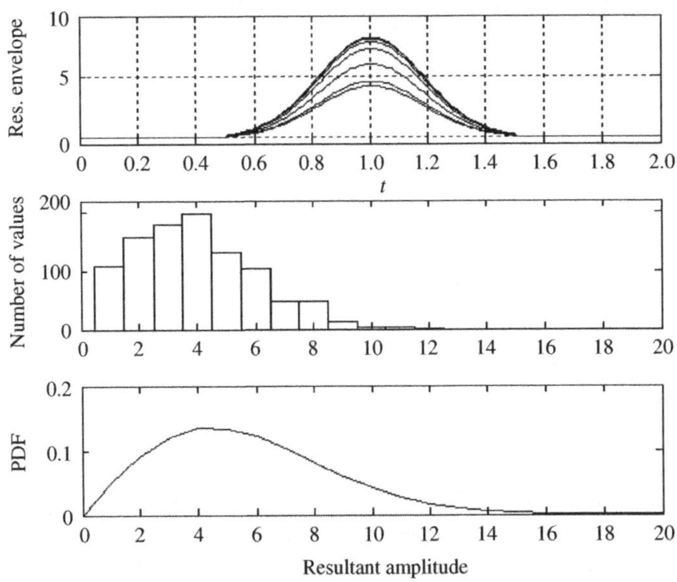

Figure 3.27 Simulation of short-term fading

(a) Repeat item (a) of Problem 3.31.
(b) Form a 1000×1 vector of complex amplitudes, the ith component of which is the sum of 20 exponents $\exp(j\phi_{ij})$, $i = 1, 2, \ldots, 1000$; $j = 1, 2, \ldots, 20$ with all ϕ_{ij} independent and uniformly distributed over $[-\pi, \pi]$.
(c) Plot several (3–7) example copies of the resultant real envelope of the pulse.
(d) Build up a histogram of real amplitudes of the resulting pulse based on item (b).
(e) Plot the Rayleigh distribution, scale it properly and compare it to the histogram of the previous item.

3.33. Write a program illustrating the antenna diversity technique (see Figure 3.28).

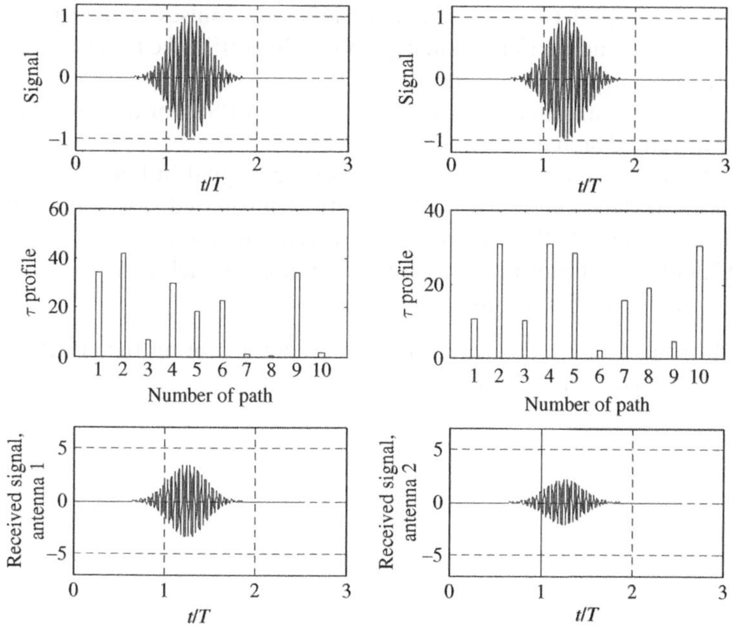

Figure 3.28 Simulation of receive antenna diversity

(a) Form and plot a plain smooth bandpass signal of duration T and carrier frequency $(20\text{–}40)/T$.
(b) Form two delay profiles as vectors of 10 delays with random independent components uniformly distributed over $[0,T/10]$.
(c) Form two complex envelopes of sums of 10 copies of the signal with initial phases calculated in accordance with the two delay profiles.
(d) Plot two delay profiles and the corresponding resultant bandpass signals in a form suitable for comparison.
(e) Run the program repeatedly and explain the results in terms of the diversity gain.

3.34. Write a program illustrating the frequency diversity principle.

(a) Form and plot two plain smooth bandpass pulses of the same duration T and amplitude but with different carrier frequencies, e.g. $20/T$ and $30/T$.
(b) Form a delay profile as a vector of 10 delays independently and uniformly distributed over $[0, T/10]$.
(c) Plot in a form convenient for comparison the delay profile and two sums of 10 copies of each signal with equal amplitudes and the initial phases calculated according to the delay profile and carrier frequency.
(d) Run the program repeatedly, and explain the results in terms of the diversity gain.

3.35. Write a program illustrating the multipath diversity principle. Unlike what is presented in Figure 3.20, use an LFM bandpass pulse.

(a) Form and plot two rectangular bandpass bit pulses of the same duration T, amplitude and carrier frequency about $(30\text{–}40)/T$, the first being plain and the second being LFM with deviation $W_d = (20\text{–}30)/T$.
(b) Set up a bit stream (5 to 6 bits) and arrange BPSK for both bit pulses. Plot the transmitted signals.
(c) Add two delayed copies to the transmitted signal in both cases. Take amplitudes of the copies equal to that of the transmitted signal and delays around $0.15T$ and $0.3T$. Plot the resultant received waveforms.
(d) Calculate and plot matched filter response for each case.
(e) Comment on the results.

3.36. Based on Problem 3.35, write a program demonstrating the effect of combining the path signals in a RAKE receiver.

4

Multiuser environment: code division multiple access

4.1 Multiuser systems and the multiple access problem

Many modern wireless systems are of multiuser type. In a *multiuser* system multiple communication links are arranged within the total time–frequency resource so that every individual user is allowed to transmit or receive his specific data in parallel with the others and independently of them. An instructive example of a multiuser system where a single transmitter transmits data to multiple users is the downlink of a satellite system or of a cellular terrestrial system. Each user receiver in such a system should be able to filter out the data addressed to it individually from the observed signal containing data sent to many users. Another case is the uplink of a satellite or a terrestrial cellular system, where there are a number of parallel transmitters and the only receiver should separate and detect the data of each individual user in the resulting observed signal.

In designing any multiuser system the principal issue is how to provide *multiple access*, i.e. the ability for many subscribers to use the communication channel simultaneously with minimal mutual interference. To describe this problem mathematically, suppose that the kth user's data form a sequence $\mathbf{b}_k = (b_{k,0}, b_{k,1}, \ldots)$, where $b_{k,i}$ stands for the ith symbol in the data stream of the kth user. This sequence in one way or another modulates the specific kth user's signal $s_k(t)$, producing the modulated signal $s_k(t; \mathbf{b}_k)$. Passing through the channel every such signal may acquire amplitude A_k and time delay τ_k and is summed with the signals of other users so that the overall or *group* signal reaching a receiver is:

$$s(t; \mathbf{b}_1, \mathbf{b}_2, \ldots, \mathbf{b}_K) = \sum_{k=1}^{K} A_k s_k(t - \tau_k; \mathbf{b}_k)$$

where K is the number of active, i.e. actively transmitting, users, and arguments after the semicolon in the group signal stress its dependence on data of all active users.

Spread Spectrum and CDMA: Principles and Applications Valery P. Ipatov

Certainly, the group signal is accompanied by channel noise $n(t)$ and the resultant observation is:

$$y(t) = s(t;\mathbf{b}_1,\mathbf{b}_2,\ldots,\mathbf{b}_K) + n(t) = \sum_{k=1}^{K} A_k s_k(t-\tau_k;\mathbf{b}_k) + n(t) \tag{4.1}$$

The receiver should retrieve thc user's data from the observation $y(t)$. According to the general ideas presented in Section 2.1, the key role in decisions on the received data $\mathbf{b}_k$, $k = 1, 2, \ldots, K$ for the AWGN channel belongs to the (squared) Euclidean distance between observation $y(t)$ and various copies of the group signal $s(t;\mathbf{b}_1,\mathbf{b}_2,\ldots,\mathbf{b}_K)$ corresponding to all possible combinations of data of K users:

$$d^2(\mathbf{s},\mathbf{y}) = \int_0^T [y(t) - s(t;\mathbf{b}_1,\mathbf{b}_2,\ldots,\mathbf{b}_K)]^2\,\mathrm{d}t \tag{4.2}$$

Substituting (4.1) into (4.2) and opening the brackets leads to the equation:

$$d^2(\mathbf{s},\mathbf{y}) = \|\mathbf{y}\|^2 - 2\sum_{k=1}^{K} A_k z_k(\mathbf{b}_k) + \sum_{k=1}^{K}\sum_{l=1}^{K} A_k A_l \int_0^T s_k(t-\tau_k;\mathbf{b}_k)s_l(t-\tau_l;\mathbf{b}_l)\,\mathrm{d}t \tag{4.3}$$

where $z_k(\mathbf{b}_k)$ is the correlation (inner product) of the observation $y(t)$ and the kth user's signal modulated by the data sequence $\mathbf{b}_k$ and delayed by τ_k:

$$z_k(\mathbf{b}_k) = \int_0^T y(t)s_k(t-\tau_k;\mathbf{b}_k)\,\mathrm{d}t \tag{4.4}$$

Typically, estimating intensities and delays of all user signals precedes the decision on data sequences so that parameters A_k, τ_k, $k = 1, 2, \ldots, K$ in (4.3) and (4.4) may be assumed known precisely. Then the optimal (ML or minimum distance) strategy of recovering user data consists in substitution of all possible realizations of the sequences $\mathbf{b}_1,\mathbf{b}_2,\ldots,\mathbf{b}_K$ in (4.3) and selecting those of them which jointly minimize the squared distance (4.3).[1] Such a decision rule, called *multiuser detection*, may appear quite impractical in a typical situation where the number of users K is measured in the tens or more. As an example take the simplest case of a synchronous system with zero mutual delays $\tau_k = 0, k = 1, 2, \ldots, K$ and binary data transmission. With the observation interval spanning only one bit, retrieving the individual bits of $K = 40$ users would require testing $2^{40} > 10^{12}$ bit patterns of all users, which looks absolutely infeasible from an implementation standpoint. We will revisit the issue of multiuser detection in Chapters 7 and 10.

[1] This rule remains adequate even if the receiver is intended to recover only the individual (kth) user information sequence, as e.g. takes place in the downlink of a cellular mobile system. After estimating data of all users the receiver just discards the unnecessary data of all users but the kth one.

The so-called *conventional* or *single-user* receiver realizes an alternative decision rule estimating each of the data sequences $\mathbf{b}_k$ separately by maximizing the correlation (4.4). It is evident that this strategy coincides with the optimal (multiuser) one if and only if the third term of (4.3) does not depend on data sequences $\mathbf{b}_k$, $k = 1, 2, \ldots, K$ at all. To meet the latter condition, one can use the modulation scheme possessing the following properties: (a) the user's signal energy does not depend on transmitted data (PSK, FSK); (b) all user signals are orthogonal regardless of transmitted data. Both these requirements are expressed by the equation:

$$\int_0^T s_k(t - \tau_k; \mathbf{b}_k) s_l(t - \tau_l; \mathbf{b}_l)\, \mathrm{d}t = E\delta_{kl} \tag{4.5}$$

Calling this multiple access mode orthogonal and returning to the material of Sections 2.3 and 2.4, we recollect that the maximal number of orthogonal signals is limited by the total signal space dimension, and within the total bandwidth W_t and time resource T_t no more than $2W_tT_t$ bandpass orthogonal signals may exist. To derive from this the maximal available number of users in the orthogonal multiple access scheme, let us limit ourselves to M-ary PSK digital data transmission with fixed rate R bps. Assuming that all user signals should be orthogonal on the time interval equal to the M-ary symbol duration, we arrive at the equation $T_t = (\log_2 M)/R$. Therefore, the maximal signal space dimension is $2W_tT_t = (2W_t \log_2 M)/R$. When $M = 2$ (BPSK), each user occupies only a one-dimensional subspace of the signal space, since only two antipodal pulses (i.e. two collinear vectors) are necessary to transmit one bit (see Figure 2.5a). In this case the maximal number of users coincides with the total signal space dimension. With $M > 2$ each user needs a two-dimensional subspace (i.e. a plane; see Figure 2.6c), and all those subspaces should be orthogonal according to (4.5), so that the maximal number of users becomes two times smaller than the total signal space dimension. Combining these results gives the upper bound of the maximal number of users in the orthogonal multiple access scheme:

$$K = \begin{cases} \dfrac{2W_t}{R}, & M = 2 \\ \dfrac{W_t \log_2 M}{R}, & M > 2 \end{cases} \tag{4.6}$$

In the following three sections we discuss briefly traditional ways of carrying orthogonal multiple access into effect.

4.2 Frequency division multiple access

One of the simplest ways to fulfil requirement (4.5) is employing user signals whose spectra do not overlap. The idea is fully allied to that of frequency-shift orthogonal coding discussed in Section 2.7.2. This multiple access mode called *frequency division multiple access* (FDMA) is illustrated by Figure 4.1a. If M-ary PSK is used for data transmission at the rate R, the data symbol duration is $T_t = (\log_2 M)/R$ so that each

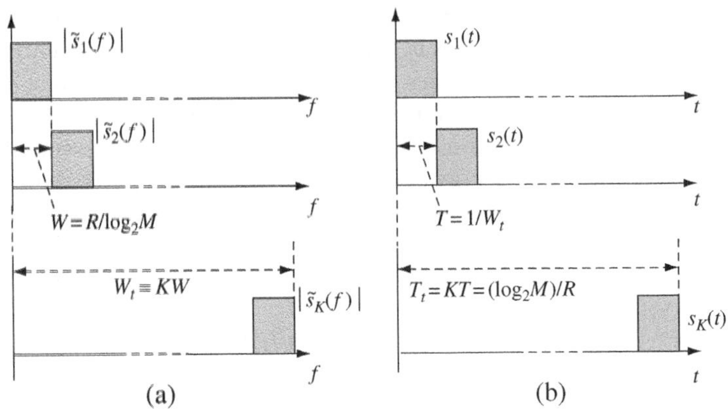

Figure 4.1 Frequency (a) and time (b) division multiple access

user's signal occupies bandwidth no smaller than $W = 1/T_t = R/\log_2 M$. Then the total allowed bandwidth W_t could accommodate no greater than $W_t/W = (W_t \log_2 M)/R$ non-overlapping spectra. This is exactly the maximal number of users if $M > 2$, which is the case shown in Figure 4.1a. When $M = 2$ and phase coherence is available each of those spectra may be utilized by two users whose carriers differ by only quadrature phase shift. As a result, the potential number of users in the FDMA scheme is subject to bound (4.6). In practice, non-ideal filtering, master clock generator drifts and Doppler frequency shifts may cause partial overlapping of the adjacent spectra, i.e. mutual interference between different user signals. To neutralize these effects and preserve separation of user signals a system designer is often forced to introduce guard frequency intervals between adjacent spectra, which decrease the achievable number of users as compared to bound (4.6).

FDMA is the oldest and classical multiple access mode commonly used in both analog and digital wireless systems (radio broadcasting, TV, mobile radio etc.). Non-overlapping spectra secure orthogonality, and hence separability of the user signals, regardless not only of data but also of time delays, thanks to which no synchronization between user signals is required. This fact is frequently referred to as a serious advantage of FDMA (see Section 4.5 for more detail).

4.3 Time division multiple access

Another popular orthogonal multiple access scheme is *time division multiple access* (TDMA), in which user signals do not overlap in the time domain (Figure 4.1b). The idea is again borrowed from time-shift orthogonal coding (see Section 2.7.1). Specifically to the case of *M*-ary PSK, it means that the whole available time resource $T_t = (\log_2 M)/R$ (in TDMA systems it is often called a *frame*) is divided into non-overlapping slots of duration T. If $M > 2$ (this case is shown in Figure 4.1b) every slot may be used by only one user and the duration of the data symbol transmitted by it cannot be smaller than the inverse total bandwidth $1/W_t$. Therefore, the total number of users is limited by the figure $T_t/T = (W_t \log_2 M)/R$. When $M = 2$ in the phase-coherent case two users may use the same slot having quadrature shifted carriers. As a result we again come to the bound (4.6) for the maximal value of K, demonstrating the theoretical equivalence of FDMA and TDMA in terms of the potential number of users accommodated.

TDMA mode has found application in various systems, e.g. in 2G mobile radio (GSM, IS-136 etc.). Although its simplicity is superficially attractive, limiting factors should be mentioned too. First, every user's signal occupies only a Kth (or perhaps a $K/2$th) part of the frame, which entails increasing peak-power K (or $K/2$) times as compared to the case of continuous emission in order to preserve the necessary signal energy, i.e. SNR. Implementation problems related to this have been repeatedly pointed out before. Second, strict synchronization is necessary between user signals at the receiver input, since otherwise they will overlap and create mutual interference. At the same time, in systems with migrating users, like the uplink of a mobile telephone, the lengths of the propagation paths between the user transmitters and the central station receiver are changing continually and over a wide range. Clearly, synchronization of the user signals at the receiver input in such situations, although possible in principle, might appear problematic technologically. The conventional way of getting around these obstructions consists in introducing guard time intervals between adjacent user signals, preventing them from being superimposed on each other within the whole range of variation of their delays. Rather commonly the guard intervals appear significant and may dramatically reduce the number of users compared to the upper bound (4.6). The severity of the problem is typically alleviated when an individual user's slot carries not a single data symbol (e.g. a bit), but, instead, a burst of n_b symbols. Then guard intervals are necessary for separating only the bursts of different users, which entails n_b times smaller guard-time overhead. On the other hand, pauses between the successive bursts of the same user become n_b times longer, too. In many systems (such as mobile telephone, where continuous speech exchange should be maintained) long pauses are not acceptable, and this frequently puts a tough limit on the length of bursts.

For these reasons, 'pure' TDMA is not often encountered in practice. For example, the 2G mobile telephone standards combine TDMA and FDMA.

4.4 Synchronous code division multiple access

Both FDMA and TDMA distribute the total available time–frequency resource between different users so that each user utilizes only his 'personal', user-specific fraction of it and no users share common fractions. In FDMA this fragmentation is done in the frequency domain (Figure 4.2a), and at the kth user's disposal are the whole time resource ($T = T_t$) but only part W of the total frequency resource W_t. When the maximal number of users is a top priority, $W = 1/T \sim W_t/K$. Splitting the time domain in TDMA (Figure 4.2b) makes it possible for the single user to occupy the whole available frequency range ($W = W_t$) but only part of the total time frame ($T = 1/W \sim T_t/K$). If the number of users needs to be maximized the resource fragmentation in both these orthogonal multiple access schemes makes each user signal plain since its time frequency product $WT = 1$.

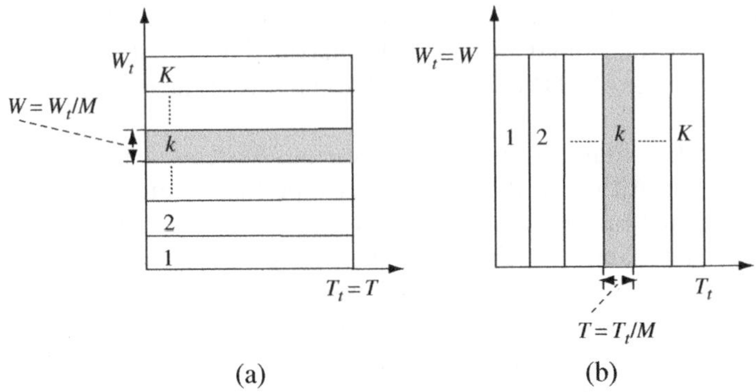

Figure 4.2 Resource distribution in FDMA (a) and TDMA (b)

On the other hand, with a large necessary number of users K the total time–frequency product has to be large too ($W_tT_t \gg 1$; see (4.6)), and if every user's signal occupied both the total available bandwidth ($W = W_t$) and time interval ($T = T_t$) we would have an orthogonal multiple access scheme in which all user signals are spread spectrum ones. Such a multiuser system would enjoy all the advantages of spread spectrum technology studied in the previous chapter.

Let us assume that the transmission may be arranged in a manner providing zero mutual delays between all user signals at the receiver input. Then without sacrificing generality all the absolute delays can be set equal to zero: $\tau_k = 0,\ k = 1, 2, \ldots, K$. Take an arbitrary family of W_tT_t orthogonal spread spectrum signals (see Section 2.7.3), e.g. Walsh functions, and employ each of them as a user signal for M-ary PSK data transmission. An individual spread spectrum signal assigned to the kth user is called the kth *signature*. Every signature occupies the total bandwidth W_t and the total time frame T_t (Figure 4.3), transmitting $\log_2 M$ bits of data over the interval T_t. If $M > 2$ this multiple access mode may serve up to $K = WT = W_tT_t = (W_t \log_2 M)/R$ users, while BPSK allows doubling of K by permitting two different users to exploit quadrature-phase-shifted copies of the same signature. Obviously, we again have the maximal possible number of users determined by (4.6), exactly as for both FDMA and TDMA.

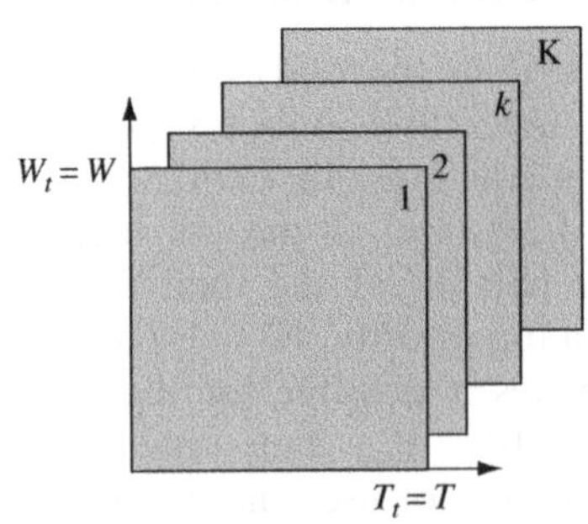

Figure 4.3 Resource utilization in spread-spectrum orthogonal multiple access

In the considered multiple access mode an appropriate signature encoding provides orthogonality of the user signals instead of fragmentation of the time or frequency domain. That is why it has the name *code division multiple access* (CDMA). The advantages of CDMA compared to classical FDMA and TDMA (jamming immunity, low detection probability, opportunity to involve the RAKE algorithm etc) follow automatically from the spread spectrum nature of CDMA signatures. At the same time, signature synchronism is critical for their orthogonality and user separation at the receiving side. In order to distinguish this version of CDMA from the one to be studied in the next section, the term *synchronous* CDMA (S-CDMA) is used. Synchronous mode is easily achievable in systems where only a single transmitter (like the base station of a cellular network) transmits simultaneously individual data streams, each being addressed to a specific user (e.g. mobile station). That is why S-CDMA constitutes the basis of the physical layer of downlinks in 2G (IS-95) and 3G (UMTS, cdma2000) CDMA cellular networks. In parallel, the core idea of S-CDMA is used in both downlink and uplink of 3G standards for arranging the so-called multi-code transmission (see Section 11.3).

4.5 Asynchronous CDMA

A situation typical of numerous applications is where delays τ_k may change over a wide range, making synchronization of signatures at the receiver input problematic or even impossible. An instructive example of this is an uplink of a mobile cellular system where users migrate over the cell, as a result of which the distances between them and the base station change constantly, as do the arrival times of the user signals at the base station receiver. In principle, each user knowing his instantaneous location relative to the base station, and therefore propagation delay τ_k, is capable of transmitting his signal with an advance τ_k. Thereby all path delays will be compensated for and all the user signals will be synchronized at the base station receiver. This operational mode, however, places excessive demands on the complexity of equipment, and in many cases can hardly be regarded as commercially viable.

Let us analyse the consequence of the asynchronous character of the received user signals. First of all, is it possible to preserve the orthogonality of signals in a wide range of mutual time shifts? Take two signals $u(t)$ and $v(t)$ and calculate their cross-correlation function (CCF) $R_{uv}(\tau)$, i.e. the inner product of $u(t)$ and a copy of $v(t)$ time-shifted by τ as a function of the argument τ:

$$R_{uv}(\tau) = \int_{-\infty}^{\infty} u(t)v(t-\tau)\,\mathrm{d}t$$

Applying the Parseval theorem gives:

$$R_{uv}(\tau) = \int_{-\infty}^{\infty} \tilde{u}(f)\tilde{v}^*(f)\exp(-j2\pi f\tau)\,\mathrm{d}f$$

If the goal is signal orthogonality independently of mutual delay τ, the equality $R_{uv}(\tau) = 0$ should hold under all values τ, which due to the Fourier transform linearity is possible if and only if $\tilde{u}(f)\tilde{v}(f) = 0$ everywhere in the frequency domain. This tells us that two signals are orthogonal under arbitrary time shifts if and only if their spectra do not overlap. But the multiple access scheme with non-overlapping spectra is FDMA! Hence, asynchronous orthogonal multiple access is realizable only with FDMA, which is often proclaimed as one of the main advantages of FDMA.

But what sort of penalty will accompany an attempt to realize CDMA when signatures are not synchronous at the receiver input? Since the signatures of the different users within CDMA have overlapping spectra, they cannot remain orthogonal in a wide range of mutual delays, and equality (4.5) cannot be true for arbitrary values of τ_k, τ_l and data sequences $\mathbf{b}_k, \mathbf{b}_l$. As a result inter-user interference emerges, which is a non-zero response of the receiver intended for the kth user to signals of other users.

Consider the kth user conventional receiver. With no loss of generality we may put $\tau_k = 0$, rewriting (4.4) as:

$$z_k(\mathbf{b}_k) = \int_0^T y(t)s_k(t; \mathbf{b}_k)\,\mathrm{d}t \tag{4.7}$$

According to the single-user rule, the estimate $\hat{\mathbf{b}}_k$ of data $\mathbf{b}_k$ should maximize the decision statistic $z_k(\mathbf{b}_k)$ as a function of $\mathbf{b}_k$. Substituting (4.1) (at this step we should replace $\mathbf{b}_k$ of (4.1) by $\mathbf{b}'_k$ in order to mark differently a genuine transmitted data $\mathbf{b}'_k$ from that assumed in the course of decision making $\mathbf{b}_k$) into (4.7) presents $z_k(\mathbf{b}_k)$ in the form:

$$z_k(\mathbf{b}_k) = A_k \int_0^T s_k(t; \mathbf{b}'_k)s_k(t; \mathbf{b}_k)\,\mathrm{d}t + \sum_{\substack{l=1 \\ l \neq k}}^{K} A_l \int_0^T s_l(t - \tau_l; \mathbf{b}'_l)s_k(t; \mathbf{b}_k)\,\mathrm{d}t + \int_0^T n(t)s_k(t; \mathbf{b}_k)\,\mathrm{d}t \tag{4.8}$$

The first and last terms of (4.8) give, respectively, the contribution of the proper, i.e. kth, user signal, and thermal additive noise into the kth user receiver effect. If no side users were present ($K = 1$) the second addend would be zero and the whole problem would be no different to the one considered in Chapter 2. With $K > 1$ and arbitrary signature delays this term differs from zero, expressing the contribution of other user signals to the kth receiver output effect, i.e. mutual or *multiple-access interference* (MAI).

An easy way to assess the influence of MAI is to treat all the alien signals as random noise-like processes similarly to what was frequently done in Chapter 3. In any practical asynchronous CDMA system, measures should be taken to equalize the level of all user signals at the receiver input in order to mitigate the *near–far* problem. The latter implies that MAI created by alien users which are much closer to the receiver than the kth user may significantly overpower the useful signal of the latter due to the strong dependence of the received power on distance (see Section 3.5.2). Therefore, we may assume that, thanks to the efficient *power control*, all A_k, $k = 1, 2, \ldots, K$ are the same; in other words, the powers of all signals are identical and equal to P. Then an lth side-user noise-like signal whose energy is assumed to be uniformly spread over the bandwidth W creates extra noise power spectrum density $N_l = P/W$ added to the thermal noise spectrum. Since there are $K - 1$ independent alien users altogether, the total MAI spectrum density is $N_I = (K-1)N_l = (K-1)P/W$. Now we may determine the power SIR, i.e. signal-to-interference ratio q_I^2 embracing both MAI and thermal noise:

$$q_I^2 = \frac{2E}{N_0 + N_I} = \frac{2E}{N_0 + (K-1)(P/W)} \tag{4.9}$$

Typically the number of users K or/and processing gain of every signature are large enough to enforce the mechanism of the central limit theorem and treat the second sum in (4.8) as a Gaussian random value. This justifies the Gaussian approximation of MAI used universally and meaning that all the results obtained in Chapter 2 for the classical reception problems (error probabilities, estimation precision etc.) are applicable to the similar multiuser problems after replacing SNR q^2 by SIR q_I^2. For example, if data are transmitted by BPSK the bit error probability for any user is calculated via (2.19) where q_I^2 substitutes for $q^2 = 2E/N_0$.

Equation (4.9) makes it possible to estimate the maximal number of users which asynchronous CDMA can accommodate within the total time–frequency resource WT. It is readily seen that in the multiuser environment, absence of thermal noise does not lead to error-free decisions at the receiving side since MAI retains SIR finite and equal to so-called *floor* SIR:

$$q_{If}^2 = \frac{2E}{(K-1)(P/W)} = \frac{2PT}{(K-1)(P/W)} = \frac{2WT}{K-1} \tag{4.10}$$

The last result shows that the floor SIR and, hence, floor reception fidelity is exhaustively determined by the time–frequency product, i.e. spread spectrum processing gain WT and number of users. As long as inequality $q_I^2 \leq q_{If}^2$ is always true the maximal possible number of users may be limited by the relation:

$$K \leq \frac{2WT}{q_I^2} + 1 \tag{4.11}$$

where q_I^2 stands for the required SIR dictated by the necessary fidelity of reception in the analysed system. To be specific, consider a BPSK or QPSK data transmission system where the bit error probability should be provided no worse than $P_e = 10^{-2}$. From (2.19) or Figure 3.16 (dashed line) it may be seen that in the no-fading case SIR of about 7 dB ($q_I^2 = 5$) is necessary to meet this demand. This produces the following estimate of the potential number of users:

$$K \leq \frac{2WT}{5} + 1 \tag{4.12}$$

At the same time, FDMA is capable of accommodating WT users[2] within the same total time–frequency resource ($W_t = W, T_t = T$), which is about 2.5 times greater than the right-hand side of (4.12). This leaves a rather bleak impression about the prospects of asynchronous CDMA in comparison to FDMA. In the next section, however, we will demonstrate that in the systems where the frequency resource needs to be reused in spatially distant areas (e.g. cellular ones), asynchronous CDMA significantly outperforms FDMA in the maximal number of users.

4.6 Asynchronous CDMA in the cellular networks

4.6.1 The resource reuse problem and cellular systems

When creating a new commercial multiuser wireless system the system designer naturally intends to serve as many subscribers as possible, at the same time being tightly bound by some fundamental limitations. The first of these is the power constraint limiting the spatial zone covered by a single transmitter. The curvature of the Earth and fast attenuation of signal intensity with distance, which is characteristic of the UHF band utilized by systems akin to mobile radio (see Section 3.3), rule out a practical opportunity of covering zones whose radius exceeds tens of kilometres. Another tough restriction is imposed by the time–frequency resource, i.e. allocated spectrum band and data rate required. Take, for example, the physical layer bandwidth of cdmaOne (IS-95) $W_t = 1.25\,\text{MHz}$. With rate of encoded voice data $R = 19.2\,\text{bps}$ and BPSK data modulation used in the downlink the potential number of active users according to (4.6) is $K = 130$. This number is obviously too small for coverage of a densely populated urban area and this is all the more true if the service should also include high-speed (e.g. multimedia) data transmission along with a telephone connection.

An effective way of getting around these obstructions is offered by a cellular network topology involving multiple base stations, each servicing its specific zone (cell) and covering collectively the total necessary area. The base station (BS) transmitter of relatively low power sends signals to the users (or mobile stations, MS), which are located within the served cell, and MS receivers form the downlink. The uplink includes the MS transmitters and BS receiver. All BS operate in strong coordination and the whole network has connections with fixed telephone and data transmission networks. When an MS moving across the system coverage zone leaves a current cell, the BS of the adjacent cell automatically takes over servicing this MS: this procedure is called *handover*. Within the framework of the cellular philosophy the wave attenuation manifests its favourable feature, allowing reuse of the same physical sub-channels (e.g. frequency sub-bands in FDMA or time slots in TDMA) by different transmitters, provided they are distant enough to secure a low level of their signals over the foreign coverage zones. As a result, just increasing the number of cells may flexibly solve the problem of raising the number of users and extension of a coverage area. In sparsely populated regions macrocells (measured by kilometres to tens of kilometres) may meet

[2] We ignore here the potential doubling of the number of users by reuse of the same subcarrier frequency with a quadrature phase shift, since this opportunity is not feasible if different users are not time-synchronized.

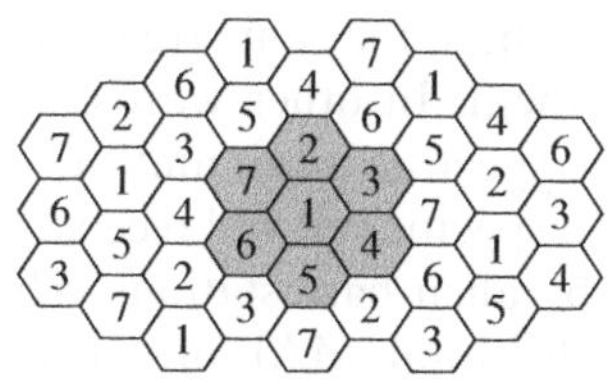

Figure 4.4 Cellular network configuration

coverage demands while in congested zones microcells (hundreds of metres) or even picocells (tens of metres) are likely to be necessary. It is universally accepted that an individual cell of a cellular network be approximated by a hexagon, owing to which the network pattern resembles a honeycomb (Figure 4.4).

Let us estimate the efficiency of utilization of the time–frequency resource in a cellular system employing classical FDMA or TDMA multiple access schemes. To avoid unnecessary repetition and allowing for equivalence of FDMA and TDMA in number of users (Section 4.3), we will use only FDMA terminology. It is obvious that the cell radius cannot be bigger than the radius of total wave attenuation but the latter, as was pointed out, should be at least two times smaller than the distance between the centres of cells utilizing user signals with identical frequencies. If the first condition fails, MS near the edge of a cell will receive too weak a BS signal and contact with the BS will appear unreliable. Violation of the second condition will entail inter-cell interference, since again the MS travelling near the edge of the cell may receive along with the proper signal of its own BS the signal of an alien BS communicating with another MS (serviced by this alien BS) at the same frequency. In other words, the frequency sets of all cells around any specific cell should differ from the set used by the central cell. Thus, a configuration arises called a *cluster* within which no frequency set may be reused. A regular honeycomb structure where frequency alternation between cells meets the condition above may exist for only some specific sizes of cluster. The most typical is the seven-cell cluster highlighted in Figure 4.4. Hence, only one seventh of the total number of physical channels (frequencies) granted by the total time–frequency resource $W_t T_t$ of the system may be utilized by a single cell. This gives the following estimation of the maximal number of users per cell in an FDMA or TDMA system:

$$K_c = \frac{W_t T_t}{7} \tag{4.13}$$

where asynchronous operation typical of uplink is assumed. In the light of this result the pessimistic conclusion about the prospects of asynchronous CDMA made in the previous section needs a serious revision.

4.6.2 Number of users per cell in asynchronous CDMA

Let us recall that asynchronous CDMA is spread spectrum based and every signature occupies the whole available time–frequency resource. Consider the uplink of a CDMA cellular system in which all cells share the same frequency band with no distribution of the spectral resource between them. In other words, signatures of all cells, including adjacent ones, occupy the same spectral band, and the cluster consists of only a single cell. Clearly, the BS receiver of a specific cell will receive MAI not only from users of the cell but also from MS served by alien base stations. The natural question arises: how great is the contribution to the total MAI of the component caused by the MS transmitters of the outer cells? To estimate the intensity of this *inter-cell* MAI, look at Figure 4.5, where two contiguous cells C_1, C_2 are approximated by circles of radius D_c. There are two BS marked as BS1 and BS2 and the MS within the zone of coverage of BS2. Despite the fact that the MS is served by BS2 its signal will also fall at the input of receiver BS1, contributing to the inter-cell MAI. Denote the distances from the MS to BS1 and BS2 as D_1 and D_2, respectively, and recall that precise power control is essential for any asynchronous CDMA system to get around the near–far problem. Thanks to the control loop, the power of the signal received by BS2 from the MS is permanently maintained constant and equal to P. If the power transmitted by MS is P_t then according to the propagation model introduced in Section 3.5 $P = kP_t/D_2^e$. On the other hand, the signal propagating from MS to BS1 will suffer attenuation defined by the distance D_1, so that the power received by BS1 $P_{r1} = kP_t/D_1^e$. We can use the previous equation to express P_{r1} in terms of power P of the useful signal at its 'own' BS receiver input:

$$P_{r1}(D_2, \theta) = \left(\frac{D_2}{D_1}\right)^e P \tag{4.14}$$

where MS coordinates D_2, θ (see Figure 4.5) emphasize the dependence of P_{r1} on the MS position inside C_2.

Now average the result (4.14) over all the cell C_2 assuming that all positions of MS within the cell are equally likely, i.e. joint PDF of the polar coordinates $W(D_2, \theta) = D_2/\pi D_c^2$ inside C_2 and zero outside. Then the mean power $\overline{P_{r1}}$ of MAI created by a single alien MS from the neighbouring cell is:

$$\overline{P_{r1}} = \iint_{C_2} P_{r1}(D_2, \theta) W(D_2, \theta)\, \mathrm{d}D_2\, \mathrm{d}\theta = \frac{P}{\pi D_c^2} \int_0^{D_c} \int_0^{2\pi} \frac{D_2^{e+1}}{D_1^e}\, \mathrm{d}\theta\, \mathrm{d}D_2.$$

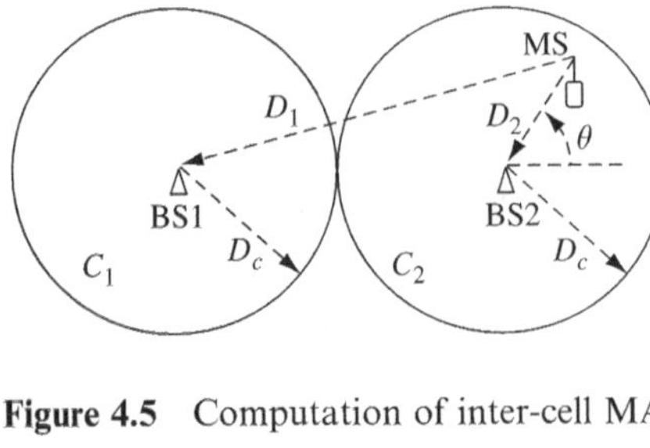

Figure 4.5 Computation of inter-cell MAI

With attenuation exponent $e = 4$ appropriate for many mobile communication scenarios, integration in the last equation may be completed analytically [20]. By the cosine theorem $D_1^2 = (2D_c)^2 + D_2^2 + 2(2D_c)D_2 \cos\theta$ and:

$$\overline{P_{r1}} = \frac{P}{\pi D_c^2} \int_0^{D_c} \int_0^{2\pi} \frac{D_2^5}{(4D_c^2 + D_2^2 + 4D_c D_2 \cos\theta)^2} \, d\theta \, dD_2 = \frac{2P}{\pi} \int_0^1 \int_0^{\pi} \frac{x^5}{(x^2 + 4x\cos\theta + 4)^2} \, d\theta \, dx$$

The internal integral here may be evaluated by trigonometric substitution or found in tables (e.g. [21]), after which the integrand in x becomes $(4x^5 + x^7)/(4 - x^2)^3$. Resorting again to the integral tables [21], we arrive at:

$$\overline{P_{r1}} = P\left(16 \ln\frac{4}{3} - \frac{41}{9}\right) < 0.05P$$

This figure should be multiplied by the number of users per cell K_c and also by the number of adjacent cells surrounding the specific one. With hexagonal cell representation the latter is 6 and the total power of the inter-cell MAI induced by all adjacent cells is no greater than $6 \times 0.05 \times K_c P = 0.3K_c P$. Strictly speaking, this estimation should be further incremented to cover inter-cell MAI from the more remote cells than just the neighbouring ones. However, it is quite predictable from the calculation above that this contribution will be negligible in comparison with the estimation just obtained [20]. Reserving some safety margin, we can therefore say that overall inter-cell MAI power $P_{I,ext} \leq 0.5K_c P$, while internal MAI created by $K_c - 1$ 'own' mobiles has, as earlier, power $P_{I,in} = (K_c - 1)P$. The floor SIR (4.10) may now be modified to allow for both internal and external MAI:

$$q_{If}^2 = \frac{2E}{(K_c - 1)(P/W) + 0.5K_c(P/W)} = \frac{2WT}{1.5K_c - 1} \tag{4.15}$$

This result admits further revision for 'pure' telephoning, since in the dialogue every party does not talk continuously and spends some time reflecting and listening. Certainly, during such pauses the transmitter of a silent speaker may be switched off or at least operate at much lower power. Actually, this opportunity has already been exploited in 2G non-CDMA mobile telephone systems to extend battery lifetime. However, only in CDMA standards does it allow MAI to be reduced simultaneously, thereby, potentially increasing the number of users served by one cell.

A typical figure for the *voice activity factor*, i.e. the fraction of total conversation time during which the phone speaker is talking, is 3/8. Correspondingly, weighting the average MAI power by this factor will transform the floor SIR above as follows:

$$q_{If}^2 = \frac{16WT}{4.5K_c - 3}$$

Solving this with respect to K_c gives a much more encouraging estimation of the number of users in CDMA as compared to the original one (4.12), obtained irrespective of a specific system topology:

$$K_c \leq \frac{32WT}{9q_{If}^2} + \frac{2}{3} \tag{4.16}$$

Substituting in (4.16) the former reference figure of 7 dB for the required SIR gives:

$$K_c \leq \frac{32WT}{45} + \frac{2}{3} \tag{4.17}$$

The estimation produced by this inequality is about five times greater in comparison with that of (4.13), the time–frequency resource being assumed the same. This shows that in cellular systems asynchronous CDMA is significantly more promising than the traditional orthogonal multiple access schemes FDMA and TDMA.

Example 4.6.1. Suppose FDMA is used to provide multiple access within the bandwidth of 5 MHz typical of 3G systems. With the transmission rate of encoded speech 19.2 kbps and BPSK, up to $K_c = W_t/7R = 37$ users per cell may be potentially accommodated. At the same time, the asynchronous CDMA alternative, as follows from (4.17), is significantly better, allowing upto $K_c = 32W/45R + 2/3 \approx 185$ users to be served by a single cell site.

Estimations of the sort of (4.16) and (4.17) may seem excessively optimistic, since they ignore the thermal noise component. A practical situation to which they are more applicable implies that the power P transmitted by every mobile is so great that overall MAI dominates the AWGN noise. On the other hand, a designer may be interested in employing as low transmitted power as possible, e.g. for reasons of battery lifetime or electromagnetic compatibility. Returning to (4.15), adding AWGN spectrum density to the denominator and allowing again for the voice activity factor, it is easy to show that when a 'pure' (i.e. not covering MAI) power SNR is q^2 and SIR q_I^2 covering MAI plus noise is required, (4.16) changes to:

$$K_c \leq \frac{32WT}{9q_I^2}\left(1 - \frac{q_I^2}{q^2}\right) + \frac{2}{3} \tag{4.18}$$

If, for example, an overall MAI power should be of the same level as noise power within the signal bandwidth then $q^2 = 2q_I^2$ and:

$$K_c \leq \frac{16WT}{9q_I^2} + \frac{2}{3} \tag{4.19}$$

that is, about half of that calculated from the floor SIR (see (4.16)). For the required SIR specified as before (7 dB):

$$K_c \leq \frac{16WT}{45} + \frac{2}{3} \tag{4.20}$$

Although with signal power reduction the number of users per cell dropped by two times, it nevertheless remained more than twofold greater than in FDMA or TDMA.

One more merit of asynchronous CDMA is its favourable blocking character. In all real multiuser systems physical channels (frequency sub-bands in FDMA, time slots in TDMA, code signatures in CDMA) are not assigned to the customers once and for all. Instead, the network itself controls the bank of traffic channels and grants one of them to a user only when he requests access to the network. Of course, some system resource in this case should be reserved to arrange the request channel. In FDMA or TDMA systems the number of physical channels is fixed and from time to time blocking may happen, i.e. the situation where the network rejects a user's request since all the channels are busy. A figure of about 2% is often assumed as tolerable probability of blocking and the number of channels should meet this requirement. Sometimes the pattern of distribution of subscribers over the network coverage area may change so seriously that at some cells blocking becomes intolerably probable. Then the network operator may face the challenge of reconfiguring the network, which entails frequency replanning affecting all cells.

Scenarios typical of CDMA are critically different. First, if the number of users already active equals the nominal one calculated by (4.16) or (4.20), and one more request is received, it may be satisfied by assigning a signature which differs from those already employed. This will lead to some (as a rule slight) reduction of SIR and thereby quality of service for all active users. Therefore, instead of an outright denial a smooth blocking happens. Second, when in the course of time traffic in some area increases dramatically the operator may place an additional base station at the 'hot spot' without frequency replanning or any other radical affecting of the other cell sites.

Based on the analysis of this chapter, the conclusion is justified that spread spectrum appears quite a flexible and efficient aid in providing multiple access. In particular, cellular systems are among those where the advantages of CDMA manifest themselves most persuasively.

Problems

4.1. A digital FDMA data transmission system should serve at least 100 users. Estimate the minimal total bandwidth occupied by the system if the necessary data rate per user is 20 kbps and BPSK is used as the data modulation mode. How will the bandwidth change if QPSK replaces BPSK? Answer the same questions if TDMA is preferred to FDMA.

4.2. An FDMA QPSK system is intended for digital data exchange between aircrafts and operates at frequencies around 3 GHz. The maximal vehicle velocity is 1800 km/h, master clock drift is 2×10^{-7} and the guard interval due to non-rectangular filtering is 1 kHz. Find the maximal number of users accommodated within the bandwidth 2.32 MHz if the necessary data rate per one user is 20 kbps.

4.3. A digital TDMA multiuser system should serve at least 100 users. The modulation mode is 8-PSK. Estimate the minimum bandwidth occupied by the system if the necessary transmission rate per user is 20 kbps.

4.4. A single uplink frequency subchannel of a TDMA BPSK digital cellular system is allowed to occupy bandwidth of 200 kHz. The time interval between consecutive data bursts of every user should not be longer than 5 ms, the necessary data transmission rate per user is around 20 kbps and the maximal cell radius is 30 km. Find the maximal number of TDMA channels per frequency subchannel.

4.5. In a synchronous CDMA system 128 physical channels should be arranged. The data transmission mode is 8-PSK and the necessary rate per user is 20 kbps. Estimate the minimum bandwidth demanded.

4.6. A synchronous CDMA system has 50 physical channels operating in 16-PSK mode with a data rate of 20 kbps per user. The total bandwidth occupied is 500 kHz. What is the processing gain of the system? Can the system be free of MAI? What if the bandwidth were four times smaller?

4.7. There are two users within one cell of a cellular CDMA telephone system located at distances 500 m and 5 km from the base station. A more distant mobile emits power of 100 mW. Find the power emitted by the closer mobile under the assumption of perfect power control.

4.8. In the uplink of an IS-95 cellular telephone asynchronous CDMA is used. Data are transmitted at a rate of 28.8 kbps by means of orthogonal signals comprising 6-bit data blocks. The signal bandwidth can be assumed as 1.25 MHz. What is the number of users per cell if the minimum required SIR is 7 dB, voice activity factor is 3/8, external MAI adds 50% to internal interference and thermal noise is neglected. What will change if the SNR for thermal noise only is 9 dB?

4.9. Estimate the number of necessary cell sites under the conditions of Problem 4.8 (AWGN is not neglected) to service an area with 50 000 subscribers if the probability of the active state of a subscriber is 0.02. Compare the result with that for an FDMA system.

4.10. How will the potential number of users change if under the conditions of Problem 4.8 orthogonal 6-bit signalling is replaced by BPSK, the bandwidth and data rate remaining the same?

4.11. Based on the Ocumura–Hata model, prove that asynchronous CDMA is inoperative in a typical macrocell cellular system uplink without an effective power control. Neglect the thermal noise component.

4.12. There is synchronous downlink in a cellular CDMA system. Within one cell the maximal number of orthogonal signatures is used. If no time–frequency resource distribution between cells is involved, how can the effect of the surrounding base stations on the mobile receivers of the given cell be estimated? What would you recommend as a general approach to choosing signatures in such a system?

Matlab-based problems

4.13. Write a program illustrating the principle of FDMA. Example plots are shown in Figure 4.6.

(a) Form the $K \times 100$ matrix of K subcarriers ($K = 2$–10 is recommended). Take frequencies so that for each subcarrier the exact integer number of periods per 100 points is one greater than for the previous one. For the first one 4–6 periods are advised.

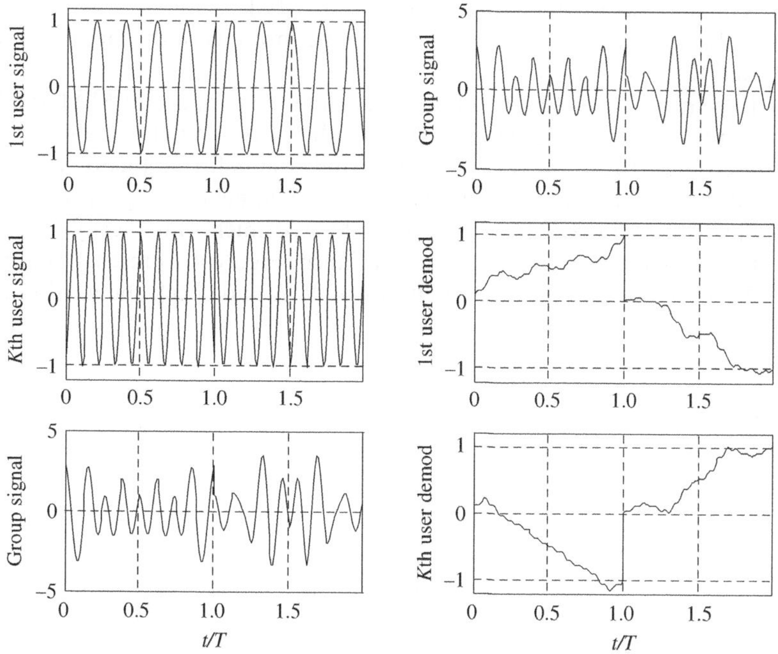

Figure 4.6 Simulation of the principle of FDMA

(b) Take 2–3 random information bits for every subcarrier and perform BPSK modulation of all subcarriers. Plot the modulated signals for two selected users.
(c) Sum all modulated signals to come to a group signal and plot that.
(d) Demodulate each transmitted bit for every user, multiplying the group signal by a relevant subcarrier and then integrating over the bit duration. Plot selected demodulator outputs. Take decisions on the received bits for all users and compare them with the transmitted ones.
(e) Run the program, changing the number of users, and interpret the results.

4.14. Use the program from Problem 4.13 to demonstrate the influence of inter-channel interference accompanying frequency drifts in an FDMA scheme. Enter a frequency shift of +0.25 in the first subchannel and fix identical bit patterns in all subchannels. Reducing the amplitude of the first signal, note the value under which wrong decisions on the channel bits arise. Run the program, varying frequency drift and channel amplitudes, and comment on the results.

4.15. Write a program confirming experimentally equation (4.10) for the floor SIR in asynchronous CDMA.

(a) Set $N = 50$–80 and form the $K \times N$ binary matrix (taking on values ± 1) of independent random numbers.
(b) Use the rows of the matrix as the signatures of K users. Assume the first of them is a useful one, the rest being interfering.

(c) Sum all rows except the first to simulate MAI.
(d) Calculate output receive MAI as the inner product of the input MAI (item (c)) and the first signature.
(e) Repeat items (a)–(d) 5000 times, calculate the variance of MAI and SIR, and compare them with the one evaluated by (4.10). What value of W should be used in (4.10) to make the experimental results fit the theoretical prediction? Give your explanation as to why $WT = N$ appears inappropriate.
(f) Plot a histogram of MAI over all 5000 experiments and verify its closeness to the Gaussian PDF with an equal variance (Figure 4.7 presents example plots).
(g) Run the program for combinations of N, K and check the validity of (4.12).

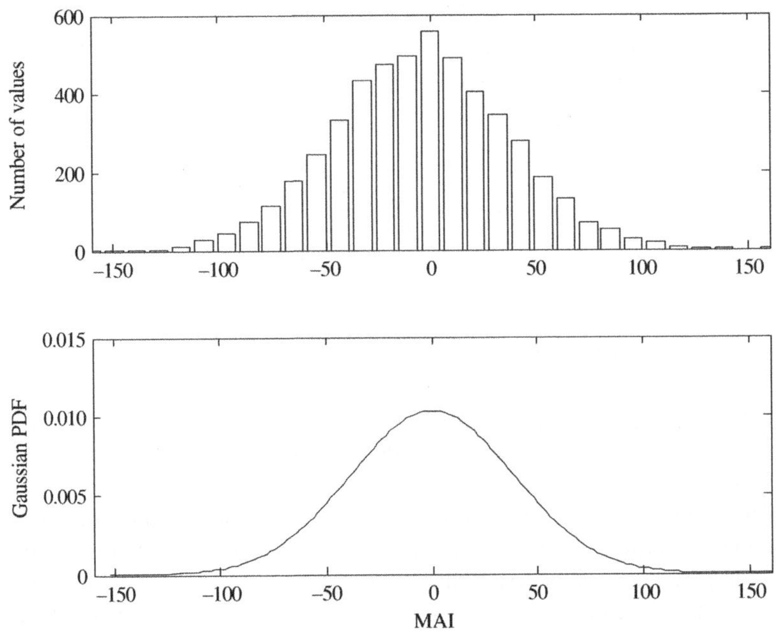

Figure 4.7 Histogram of MAI and its Gaussian approximation

4.16. Write a program illustrating the low contribution of outer mobiles to an overall MAI in an asynchronous cellular CDMA. Advised steps:

(a) Set $N = 80–100$ and number of users per cell $K = 20–25$.
(b) Using uniform random generator, form and plot internal MAI as the sum of $K - 1$ random binary (consisting of elements ± 1) signatures of length N.
(c) Form the $6K \times 2$ matrix of random polar coordinates D_c, θ of outer mobiles (see Figure 4.4) to simulate uniform distribution of mobiles throughout a cell. Note that to realize this the angle should be uniformly distributed over $[-\pi, \pi]$, but the radius has to have linearly rising PDF. The latter is simulated by generation of uniformly distributed random numbers and afterwards taking their square roots.

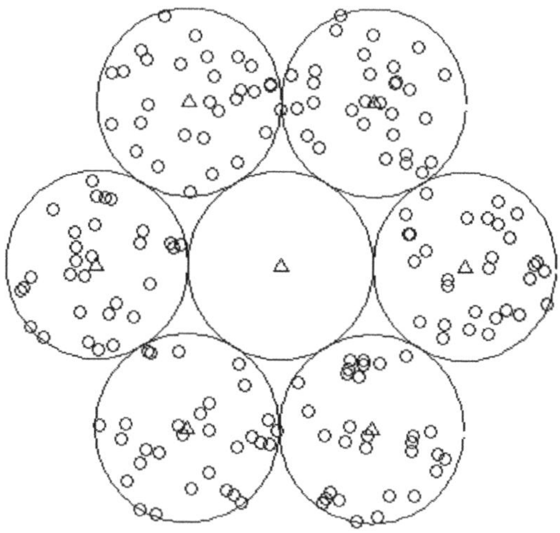

Figure 4.8 Random distribution of users over the surrounding cells

(d) Form the $6K$ dimensional vector of ratios of distances D_2 (from an outer mobile to the BS serving it) and D_1 (to the BS serving internal mobiles):

$$\frac{D_2}{D_1} = \frac{D_2}{\sqrt{4D_c^2 + D_2^2 + 4D_cD_2\cos\theta}}$$

where cell radius D_c may be set equal to 1.

(e) Form and plot the external MAI as sum of all outer signatures, each weighted with an amplitude attenuation factor. The latter is calculated on the basis of previous item with power attenuation exponent $e = 3.8$ (see (4.14)).

(f) Form and plot total MAI as the sum of the internal and external ones, and estimate the increase of the total MAI versus the internal one.

(g) Produce a scatter plot demonstrating the random distribution of mobiles in outer cells (Figure 4.8 shows an example).

(h) Run the program repeatedly for a range of N, K, compare the results with the theoretical prediction and give your comments.

5

Discrete spread spectrum signals

5.1 Spread spectrum modulation

Let us revisit the general model (2.37) of a bandpass signal:

$$s(t) = \mathrm{Re}[\dot{S}(t)\exp(j2\pi f_0 t)], \quad \dot{S}(t) = S(t)\exp[j\gamma(t)]$$

It is quite comprehensible that the spreading of a signal spectrum is accomplished by appropriate controlling of a signal complex envelope $\dot{S}(t)$, i.e. modulation of its instant amplitude $S(t)$ and instant initial phase $\gamma(t)$. As was observed in Chapter 1, 'pure' amplitude modulation cannot be an efficient tool for spectrum spreading, since it may remarkably widen the bandwidth only at the cost of concentrating the signal energy within short time intervals. As a matter of fact, this implies operating with short plain signals. On the contrary, angle (phase or frequency) modulation is capable of unlimited (at least in theory) widening of the spectrum with no effect on the distribution of the signal energy in time, i.e. signal duration, due to which its role in spread spectrum technology is fundamental. Amplitude modulation is just an auxiliary instrument, which sometimes appears to be useful in combination with angle modulation.

Depending on the character of the modulation involved all spread spectrum signals may be classified into *continuous* and *discrete* ones. For the first, the modulation law, i.e. complex envelope $\dot{S}(t)$, is a continuous function of time, while the modulated parameters (amplitude, frequency, initial phase) of the second are piecewise constant and change by hops only at discrete time moments. The example of a continuous spread spectrum signal will be discussed briefly in Section 6.2; however, the main attention will be focused on discrete signals, owing to their predominant role in the majority of modern and forward-looking commercial systems.

Spread Spectrum and CDMA: Principles and Applications Valery P. Ipatov

5.2 General model and categorization of discrete signals

Discrete signals considered in this book may be covered by the following description, generalizing the one already presented in Section 2.7.3: a discrete signal is a sequence of elementary pulses of a fixed form, recurring at some fixed time interval. The elementary pulse is called a *chip*. The complex envelope $\dot{S}_0(t)$ defines its shape and internal angle modulation, if any. The time interval Δ between consecutive chips typically, but not compulsorily, equals or exceeds chip duration Δ_c. Modulation of the whole signal consists in manipulating the amplitudes, phases and, possibly, frequencies of individual chips. Accordingly, formal representation of the complex envelope of a discrete signal is given by the equation:

$$\dot{S}(t) = \sum_{i=-\infty}^{\infty} a_i \dot{S}_0(t - i\Delta) \exp(j2\pi F_i t) \tag{5.1}$$

where, in addition to the designations explained, a_i and F_i are, respectively, complex amplitude and frequency (in terms of shift against the fixed central frequency) of the ith chip. It is obvious that the sequence $\{|a_i|,\ i = \ldots, -1, 0, 1, \ldots\}$ determines real amplitudes of chips, i.e. their amplitude modulation. Similarly, sequences $\{\phi_i = \arg a_i, i = \ldots, -1, 0, 1, \ldots\}$ and $\{F_i, i = \ldots, -1, 0, 1, \ldots\}$ define the laws of modulation of chip phases and frequencies. Figure 5.1 may be helpful in understanding some of the definitions above.

Suppose that in the model (5.1) real amplitudes $|a_i|$ may take on non-zero values only for $0 \le i \le N - 1$ and $|a_i| = 0$ for $i < 0$ and $i \ge N$. Putting it differently, the signal is a burst of a finite number N of manipulated chips. Such a signal we will call *pulse* or *aperiodic*. The duration of an aperiodic signal is $T = (N - 1)\Delta + \Delta_c$. Another important case is a signal for which the modulation law repeats itself with a period of N chips: $a_i = a_{i+N}, F_i = F_{i+N}, i = \ldots, -1, 0, 1, \ldots$. For natural reasons, this sort of discrete signal is called *periodic*. Its real-time period is $T = N\Delta$ and any periodic signal is just a repetition with period $N\Delta$ of an aperiodic one, the latter being a one-period segment of the periodic signal. In both cases we will call parameter N the *length* of a code sequence (see Section 2.7.3).

Within the described general model we distinguish between several categories of discrete signals in accordance with a specific chip modulation mode.

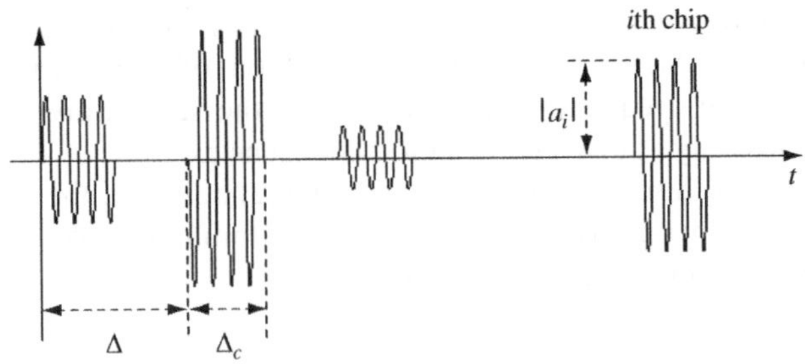

Figure 5.1 Example of a discrete signal

1. If only the complex amplitudes of chips are manipulated, all frequencies remaining the same ($F_i = 0,\ i = 0, 1, \ldots, N-1$), the signal is called an *amplitude-phase shift keying* (APSK) one. The conventional name of the sequence of chip complex amplitudes $\{a_i,\ i = 0, \ldots, N-1\}$ is a *code sequence* or simply *code*.
2. If only the phases of chips in an APSK signal are manipulated, amplitudes being unchanged ($|a_i| = 1,\ i = 0, 1, \ldots, N-1$), the signal is a PSK one. PSK signals are typical of so-called direct sequence spread spectrum systems (see Section 7.1).
3. Within PSK signals further categorization is possible depending on the modulation alphabet. When only binary complex amplitudes are involved ($a_i = \pm 1$, or equivalently, $|a_i| = 1, \phi_i \in \{0, \pi\},\ i = 0, 1, \ldots, N-1$), the signal is a BPSK one; with a quaternary alphabet $a_i = \pm 1, \pm j$, or equivalently, $|a_i| = 1, \phi_i \in \{0, \pi, \pm \pi/2\}$, $i = 0, 1, \ldots, N-1$, the signal is QPSK etc.
4. If only the frequencies of chips are controlled, complex amplitudes remaining constant, the signal is of FSK type. A code sequence of such a signal is just a sequence of frequencies $\{F_i, i = 0, 1, \ldots, N-1\}$. These signals are, in particular, employed in frequency hopping systems (see Section 7.1).

5.3 Correlation functions of APSK signals

Correlation functions showing the likeness of time-shifted copies of signals are of critical importance in problems of time measurement and resolution (see Sections 2.11–2.16). The art of designing spread spectrum systems, as will be seen from the further discussion, is in many aspects the ability to find signals with adequate correlation properties. In this section we are deriving a general expression for the correlation functions of APSK signals. From the definitions above, the complex envelope of an APSK signal has the form:

$$\dot{S}(t) = \sum_{i=-\infty}^{\infty} a_i \dot{S}_0(t - i\Delta) \tag{5.2}$$

Turn to the definition of normalized ACF (2.67), taking into account that for a periodic signal the integrand will also be periodic, and hence, its time-averaging (integration) may be accomplished over one period, normalization being made to the one-period energy. Therefore, with an assumption $\Delta_c \leq \Delta$ mostly typical of applications,[1] we are able to make use of the universal equation:

$$\dot{\rho}(\tau) = \frac{1}{E} \int_0^T \dot{S}(t) \dot{S}^*(t - \tau)\, \mathrm{d}t \tag{5.3}$$

[1] The final results we will come to are valid regardless of whether or not this inequality is true. The assumption only helps to eliminate some secondary details in the derivation below.

for both aperiodic and periodic signals, where $E = \|\mathbf{a}\|^2 E_0$ is complete energy for the first and per-period energy for the second. Here E_0 stands for chip energy and $\|\mathbf{a}\|$ is a geometric length (Euclidean norm) of the *code vector* $\mathbf{a} = (a_0, a_1, \ldots, a_{N-1})$; in other words, $\|\mathbf{a}\|^2 = \sum_{i=0}^{N-1} |a_i|^2$ is the energy of the N-element sequence $\{a_0, a_1, \ldots, a_{N-1}\}$.

Substituting (5.2) into (5.3) gives:

$$\dot{\rho}(\tau) = \frac{1}{E} \sum_{i=-\infty}^{\infty} \sum_{k=-\infty}^{\infty} a_i a_k^* \int_0^T \dot{S}_0(t - i\Delta)\dot{S}_0^*(t - k\Delta - \tau)\,dt$$

$$= \frac{1}{E} \sum_{k=-\infty}^{\infty} \sum_{i=0}^{N-1} a_i a_k^* \int_0^T \dot{S}_0(t - i\Delta)\dot{S}_0^*(t - k\Delta - \tau)\,dt$$

where the last equality follows from vanishing the integral, whenever i is beyond the set $\{0, 1, \ldots, N-1\}$.

Introducing the ACF of a single chip:

$$\dot{\rho}_c(\tau) = \frac{1}{E_0} \int_{-\infty}^{\infty} \dot{S}_0(t)\dot{S}_0^*(t - \tau)\mathrm{d}t \tag{5.4}$$

leads to:

$$\dot{\rho}(\tau) = \sum_{k=-\infty}^{\infty} \left(\frac{1}{\|\mathbf{a}\|^2} \sum_{i=0}^{N-1} a_i a_k^* \right) \dot{\rho}_c[\tau - (i-k)\Delta]$$

Now change the summation index k to $m = i - k$, arriving at:

$$\dot{\rho}(\tau) = \sum_{m=-\infty}^{\infty} \rho(m)\dot{\rho}_c(\tau - m\Delta) \tag{5.5}$$

where:

$$\rho(m) = \frac{1}{\|\mathbf{a}\|^2} \sum_{i=0}^{N-1} a_i a_{i-m}^* \tag{5.6}$$

is the ACF of the code sequence $\{a_0, a_1, \ldots, a_{N-1}\}$ characterizing its resemblance to its replica shifted by m positions.

Equation (5.5) has quite an eloquent implication. Comparing it with the model (5.2) allows the observation that the ACF of an APSK signal is the APSK signal itself! The chip of the latter is the ACF $\dot{\rho}_c(\tau)$ of the original chip, while the code sequence is the ACF (5.6) of the code sequence $\{a_0, a_1, \ldots, a_{N-1}\}$ of the original signal. Therefore, given the chip, the ACF of the APSK signal is entirely determined by the ACF $\rho(m)$ of the code sequence (or code ACF), and designing APSK signals with good autocorrelation properties means searching for sequences with good ACF. Note that, like any ACF, $\rho(m)$ at $m = 0$ equals 1 and is even: $\rho(m) = \rho^*(-m)$.

In multiuser CDMA systems families of discrete signals are necessary with special cross-correlation properties (see Section 4.5 and Chapter 7). Repeating accurately the derivation above but this time for two different (kth and lth) APSK signals having identical chips and lengths results in the following equation for their cross-correlation function (CCF):

$$\dot{\rho}_{kl}(\tau) = \sum_{m=-\infty}^{\infty} \rho_{kl}(m)\dot{\rho}_c(\tau - m\Delta) \tag{5.7}$$

where:

$$\rho_{kl}(m) = \frac{1}{\|\mathbf{a}_k\|\|\mathbf{a}_l\|}\sum_{i=0}^{N-1} a_{k,i}a^*_{l,i-m} \tag{5.8}$$

is the CCF of the code sequences $\{a_{k,0}, a_{k,1}, \ldots, a_{k,N-1}\}, \{a_{l,0}, a_{l,1}, \ldots, a_{l,N-1}\}$ of the two signals, which shows the resemblance of the first to an m-shifted replica of the second. Certainly, CCF (5.7) is again the APSK signal whose code sequence is just the CCF of the two original codes (code CCF). Designing families with the necessary cross-correlation properties means searching for the families of sequences having appropriate CCF. Equations (5.7) and (5.8) are most general, since the ACF of the kth signal is $\dot{\rho}_{kk}(\tau)$ and the same is true for the code sequences.

In what follows we will widely use the results obtained, sometimes omitting front factors in (5.6) and (5.8), i.e. operating with non-normalized correlation functions of code sequences:

$$R(m) = \sum_{i=0}^{N-1} a_i a^*_{i-m}, \quad R_{kl}(m) = \sum_{i=0}^{N-1} a_{k,i}a^*_{l,i-m} \tag{5.9}$$

5.4 Calculating correlation functions of code sequences

Consider the code sequence $\{a_0, a_1, \ldots, a_{N-1}\}$. If it is used to generate a pulse signal, in the general model (5.2) $a_i = 0$ for negative i and $i \geq N$, so that according to (5.6) *aperiodic* or *pulse* ACF is calculated as:

$$\rho_a(m) = \begin{cases} \dfrac{1}{\|\mathbf{a}\|^2}\displaystyle\sum_{i=m}^{N-1} a_i a^*_{i-m}, m \geq 0 \\ \dfrac{1}{\|\mathbf{a}\|^2}\displaystyle\sum_{i=0}^{N-1+m} a_i a^*_{i-m}, m < 0 \end{cases} \tag{5.10}$$

The second row here is somewhat redundant since any ACF features evenness, and in particular, $\rho_a(-m) = \rho^*_a(m)$. As is seen, ignoring the normalizing factor, aperiodic ACF is an inner product of the vector $\mathbf{a} = (a_0, a_1, \ldots, a_{N-1})$ and its m-position *non-cyclically* shifted version. The latter is $\mathbf{a}$ shifted to the right ($m \geq 0$) or to the left ($m < 0$) and only the overlapping components of **a** and its shifted replica enter the sum in (5.10), all the rest are regarded as being forced to zero. For example, to calculate $\rho_a(1)$ we first write, one above the other, the initial sequence and its copy conjugated and shifted to the right by one position, then compute all of their component-wise products and sum all the products:

$$\begin{array}{cccccc} a_0 & a_1 & a_2 & a_3 & \ldots & a_{N-1} \\ & a_0^* & a_1^* & a_2^* & \ldots & a_{N-2}^* \end{array} \qquad \rho_a(1) = \frac{1}{\|\mathbf{a}\|^2}\sum_{i=1}^{N-1} a_i a_{i-1}^*$$

Let us assume now that the signal is periodic, i.e. $a_{i+N} = a_i, i = \ldots, -1, 0, 1, \ldots$. Then (5.6) defines the *periodic* ACF $\rho_p(m)$, the sum in which always contains N summands, since $a_{-1} = a_{N-1}, a_{-2} = a_{N-2}$, etc:

$$\rho_p(m) = \frac{1}{\|\mathbf{a}\|^2}\sum_{i=0}^{N-1} a_i a_{i-m}^* \tag{5.11}$$

In this case the inner product is computed for the original code sequence and its *cyclically* shifted copy, where under $m \geq 0$ m left 'empty' positions are filled with symbols pushed out rightward. For instance, a scheme of calculating $\rho_p(1)$ looks like this:

$$\begin{array}{cccccc} a_0 & a_1 & a_2 & a_3 & \ldots & a_{N-1} \\ a_{N-1}^* & a_0^* & a_1^* & a_2^* & \ldots & a_{N-2}^* \end{array} \qquad \rho_p(1) = \frac{1}{\|\mathbf{a}\|^2}\sum_{i=0}^{N-1} a_i a_{i-1}^*$$

Since $\rho_p(m)$ is calculated under the assumption of periodicity of a code sequence, it is periodic itself with period N, i.e. $\rho_p(m) = \rho_p(m+N)$, $m = \ldots, -1, 0, 1, \ldots$, which stems directly from (5.11) and, in its turn, reformulates the evenness property as:

$$\rho_p(-m) = \rho_p(N-m) = \rho_p^*(m-N) \tag{5.12}$$

This equation shows that $\rho_p(m)$ is entirely characterized by its values at only shifts

$$m = 1, 2, \ldots, \left\lfloor \frac{N}{2} \right\rfloor,$$

where $\lfloor \cdot \rfloor$ symbolizes rounding towards zero. Another important property of the periodic ACF follows from (5.11) after splitting its sum into two:

$$\rho_p(m) = \frac{1}{\|\mathbf{a}\|^2}\sum_{i=m}^{N-1} a_i a_{i-m}^* + \frac{1}{\|\mathbf{a}\|^2}\sum_{i=0}^{m-1} a_i a_{i-m}^*, \ m \geq 0$$

The first term here is aperiodic ACF $\rho_a(m)$ (see (5.10)), while the second equals $\rho_a(m-N)$, which is again verified immediately from (5.10) by calculating $\rho_a(m-N)$ according to its second row. This produces the equality associating the periodic and aperiodic ACF:

$$\rho_p(m) = \rho_a(m) + \rho_a(m-N), m = 0, 1, \ldots, N \tag{5.13}$$

Equation (5.13) plays quite a crucial role in the synthesis of pulse signals with good autocorrelation properties (see Section 6.10).

Example 5.4.1. Table 5.1 illustrates the technique of computing aperiodic and periodic ACF by the example of a binary sequence of length $N = 8\{+ + + - - + - -\}$. In the table the binary

Table 5.1 Computation of ACF of the sequence of length 8

m	a_0	a_1	a_2	a_3	a_4	a_5	a_6	a_7	$R_a(m)$	$R_p(m)$
0	+	+	+	−	−	+	−	−	+8	+8
1	−	+	+	+	−	−	+	−	+1	0
2	−	−	+	+	+	−	−	+	−2	−4
3	+	−	−	+	+	+	−	−	+1	0
4	−	+	−	−	+	+	+	−	0	0
5	−	−	+	−	−	+	+	+	−1	0
6	+	−	−	+	−	−	+	+	−2	−4
7	+	+	−	−	+	−	−	+	−1	0

code symbols +1 and −1 are designated by just '+' and '−', respectively, as is usually done. Non-normalized ACF is presented and the shading marks the symbols that are ignored in the calculation of the aperiodic ACF. The results after normalization are also used in Figure 5.2, where the autocorrelation functions of the APSK signal with a rectangular chip of duration

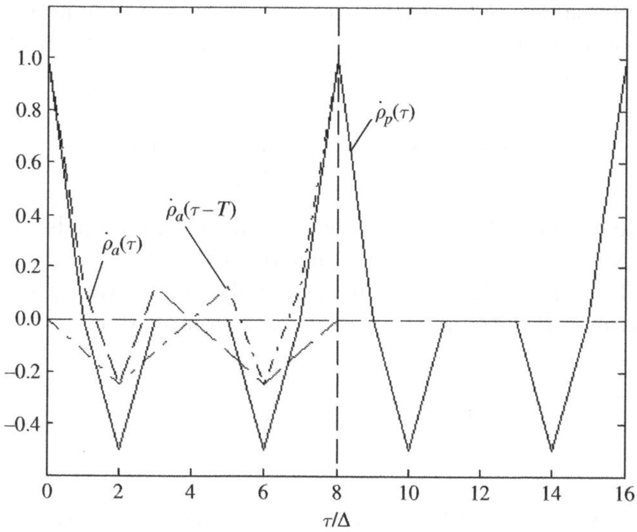

Figure 5.2 Autocorrelation functions of the binary signal of length 8

$\Delta_c = \Delta$ and the considered code sequence are built. The solid, dashed and dotted-dashed lines present the periodic ACF $\dot{\rho}_p(\tau)$ and shifted copies $\dot{\rho}_a(\tau)$, $\dot{\rho}_a(\tau - T)$ of the aperiodic ACF, respectively. Certainly, the plots confirm the validity of (5.12) and (5.13).

When the CCF of two sequences of the same length is calculated we again may discriminate between the aperiodic CCF $\rho_{a,kl}(m)$ and the periodic one $\rho_{p,kl}(m)$ found as:

$$\rho_{a,kl}(m) = \begin{cases} \frac{1}{\|\mathbf{a}_k\|\|\mathbf{a}_l\|} \sum\limits_{i=m}^{N-1} a_{k,i} a^*_{l,i-m}, & m \geq 0 \\ \frac{1}{\|\mathbf{a}_k\|\|\mathbf{a}_l\|} \sum\limits_{i=0}^{N+m-1} a_{k,i} a^*_{l,i-m}, & m < 0 \end{cases} \tag{5.14}$$

and:

$$\rho_{p,kl}(m) = \frac{1}{\|\mathbf{a}_k\|\|\mathbf{a}_l\|} \sum_{i=0}^{N-1} a_{k,i} a^*_{l,i-m} \tag{5.15}$$

Equation (5.13) still stands for CCF:

$$\rho_{p,kl}(m) = \rho_{a,kl}(m) + \rho_{a,kl}(m - N) \tag{5.16}$$

but as for the evenness or unity value at $m = 0$, those, certainly, are no longer inherent features of an arbitrary CCF as they were of any ACF.

5.5 Correlation functions of FSK signals

Let us perform the same work as in the two previous sections but now in application to FSK signals. Following the definition of Section 5.2, a complex envelope of an FSK signal assumes the form:

$$\dot{S}(t) = \sum_{i=-\infty}^{\infty} a_i \dot{S}_0(t - i\Delta) \exp(j2\pi F_i t) \tag{5.17}$$

where in the case of a periodic signal all a_i equal one, while for a pulse signal of length N $a_i = 1, 0 \leq i < N$ and $a_i = 0$ beyond the range $0 \leq i < N$.

Regardless of the periodicity or finiteness of the signal, we again, with the same reasoning as before, may exploit a universal expression of ACF:

$$\dot{\rho}(\tau) = \frac{1}{NE_0} \int_0^T \dot{S}(t)\dot{S}^*(t - \tau)\,dt$$

$$= \frac{1}{NE_0} \sum_{i=0}^{N-1} \sum_{k=-\infty}^{\infty} a_i a_k \exp(j2\pi F_k \tau) \int_0^T \dot{S}_0(t - i\Delta)\dot{S}_0^*(t - k\Delta - \tau) \exp[j2\pi(F_i - F_k)t]\,dt$$

where use is made of the fact that all chips with non-zero amplitudes now have equal energies E_0, i.e. $\|\mathbf{a}\|^2 = N$ and $E = NE_0$.

It is typical of FSK modulation to use a uniform frequency alphabet so that $F_i \in \{0, \pm F, \pm 2F, \ldots\}$, where the frequency step F is no smaller than the chip bandwidth. Thus, the spectra of two chips having frequencies F_i and F_k do not overlap and the chips are orthogonal independently of their time mismatch (see Section 4.4), whenever $F_i \neq F_k$. Allowing for this fact, we arrive at:

$$\dot{\rho}(\tau) = \frac{1}{N}\sum_{k=-\infty}^{\infty}\sum_{i=0}^{N-1} a_i a_k \exp(j2\pi F_k \tau)\delta(F_i - F_k)\dot{\rho}_c[\tau - (i-k)\Delta]$$

$$= \frac{1}{N}\sum_{m'=-\infty}^{\infty}\sum_{i=0}^{N-1} a_i a_{i-m'} \exp(j2\pi F_i \tau)\delta(F_i - F_{i-m'})\dot{\rho}_c(\tau - m'\Delta) \qquad (5.18)$$

where $\dot{\rho}_c(\cdot)$ is, as before, chip ACF and:

$$\delta(x-y) = \begin{cases} 1, & x = y \\ 0, & x \neq y \end{cases}$$

Unlike APSK signals, ACF (5.18) in the general case cannot be further simplified to a form similar to (5.5). It is common practise to analyse the behaviour of the ACF of FSK signals primarily at the delays, which are multiples of the chip duration: $\tau = m\Delta$, where m is integer. Assuming that the integer number of periods l of each frequency fits in the chip duration ($F\Delta = l$) and taking into account that $\dot{\rho}_c(0) = 1, \dot{\rho}_c(\tau) = 0, |\tau| \geq \Delta$, substitution of $\tau = m\Delta$ into (5.18) leaves only one non-zero addend of the sum in m corresponding to $m' = m$, so that:

$$\dot{\rho}(m\Delta) = \frac{1}{N}\sum_{i=0}^{N-1} a_i a_{i-m}\delta(F_i - F_{i-m}) \qquad (5.19)$$

When the signal is finite and $m \geq 0$, all summands possessing indexes beyond the range $m, m+1, \ldots, N-1$ disappear, and the aperiodic ACF of the FSK signal $\dot{\rho}(m\Delta) = \rho_a(m)$ assumes the form:

$$\rho_a(m) = \frac{1}{N}\sum_{i=m}^{N-1} \delta(F_i - F_{i-m}),\ m \geq 0,\ \rho_a(-m) = \rho_a(m) \qquad (5.20)$$

where complex conjugation in the second equation (expressing evenness) is needless, since $\dot{\rho}(m\Delta)$ is always real-valued.

In the case of signal periodicity no zero products $a_i a_{i-m}$ enter the sum of (5.19) and the periodic ACF of the signal in question $\dot{\rho}(m\Delta) = \rho_p(m)$ looks as follows:

$$\rho_p(m) = \frac{1}{N}\sum_{i=0}^{N-1} \delta(F_i - F_{i-m}) \qquad (5.21)$$

The sums in (5.20) and (5.21) just accumulate the number of coincident frequencies in the FSK signal and its replica time-shifted by m chip positions. Therefore, to compute ACF of an FSK signal at the point $m\Delta$, it is quite enough to count the number of pairs $\{F_i, F_{i-m}\}$ with equal F_i and F_{i-m}, where i runs over the ranges $\{m, m+1, \ldots, N-1\}$ (aperiodic ACF, $m \geq 0$) or $\{0, 1, \ldots, N-1\}$ (periodic ACF). Clearly, equation (5.13), linking together periodic and aperiodic ACF, remains valid for FSK signals.

One widely used way to represent an FSK signal is as an $M \times N$ array where the horizontal and vertical directions are assigned to time and frequency, respectively, and M is the size of the frequency alphabet (i.e. the number of frequencies used in modulation). In the ith vertical column of this array only a single entry is labelled (e.g. by a point or shading), which corresponds to a frequency of an ith chip. Then to calculate aperiodic ACF at any specific m we just sum the number of labelled pairs along all rows having distance m and normalize the result if necessary. If periodic ACF is of interest, the sums above obtained for m and $N-m$ should be summed together.

Example 5.5.1. Figure 5.3 shows the modulation law of an FSK signal of length $N = 8$ with $N = 8$, $M = 5$. Its non-normalized aperiodic ACF $R_a(m) = \sum_{i=m}^{7} \delta(F_i - F_{i-m})$ has values 8, 1, 1, 0,

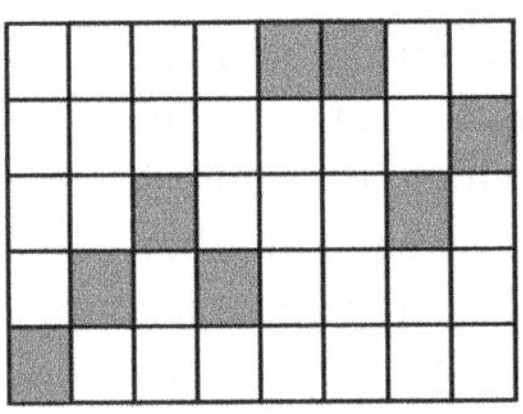

Figure 5.3 FSK signal with $N = 8, M = 5$

1, 0, 0, 0 corresponding to $m = 0, 1, \ldots, 7$, since there is one labelled pair along one line at distance 1, one such pair at distance 2, etc. Directly from (5.13), the values of non-normalized periodic ACF may be found as 8, 1, 1, 0, 2, 0, 1, 1.

Generalization of results (5.19) and (5.20) onto CCF can be done without trouble merely by adjusting the designations:

$$\rho_{a,kl}(m) = \begin{cases} \dfrac{1}{N} \sum\limits_{i=m}^{N-1} \delta(F_{k,i} - F_{l,i-m}), & m \geq 0 \\ \dfrac{1}{N} \sum\limits_{i=0}^{N+m-1} \delta(F_{k,i} - F_{l,i-m}), & m < 0 \end{cases} \tag{5.22}$$

$$\rho_{p,kl}(m) = \frac{1}{N} \sum_{i=0}^{N-1} \delta(F_{k,i} - F_{l,i-m}) \tag{5.23}$$

where $\{F_{k,i}, i = 0, 1, \ldots, N-1\}$ is the frequency code sequence of the kth signal. It is obvious that computation of CCF is again just tallying up a number of coincident frequencies in the pair of signals time-shifted by m chip positions.

5.6 Processing gain of discrete signals

Let us revisit the general model (5.1) to discuss the issue of the processing gain of a discrete signal. Suppose that all F_i belong to the frequency alphabet of size M, in which consecutive frequencies are separated by a chip bandwidth providing orthogonality of chips with different frequencies. Then at each of M available frequencies we have a signal subspace whose dimension is N, since we are fully free in selection of the amplitude-phase code sequence, i.e. N-dimensional vector $\mathbf{a} = (a_0, a_1, \ldots, a_{N-1})$. Orthogonality of these subspaces means that the dimension of the whole signal space covering all M frequencies is MN. In Section 2.5 it was shown that the dimension of bandpass signal space coincides with the total time–frequency resource allocated to signals. Being interested only in spread spectrum signals, each occupying the whole available resource, we may predict that the time–frequency product of a discrete signal, i.e. processing gain, equals MN.

Let us confirm this statement by a straightforward computation, setting $\Delta_c = \Delta$. Estimating chip bandwidth as $1/\Delta$ and taking into account that available bandwidth and time resources are then $W = M/\Delta$ and $T = N\Delta$, we arrive at the result $WT = MN$. Obviously, for APSK signals $M = 1$, and processing gain $WT = N$.

Problems

5.1. A discrete signal of length $N = 5$ has complex amplitudes $a_0 = 1 + j$, $a_1 = -1 + j$, $a_2 = 1 + j$, $a_3 = -1 - j$, $a_4 = 1 - j$ and frequencies $F_i = 0$, $i = 0, 1, 2, 3, 4$. Evaluate the phases and amplitudes of its chips and classify the signal by its modulation mode.

5.2. A discrete signal has amplitude-phase and frequency codes $a_i = \exp[j\pi i(i+1)/2]$, $F_i = 0$, $i = \ldots, -1, 0, 1, \ldots$. Calculate the amplitudes and phases of its chips. Classify the signal by its modulation mode. Is this signal periodic? If so, specify its period.

5.3. Prove the evenness of the periodic and aperiodic autocorrelation functions of code sequences of APSK signals.

5.4. An APSK signal is built of rectangular chips with $\Delta_c = \Delta$ and has the code sequence set by the vector $\mathbf{a} = (1, 1, 0, 1, 0, 0, -1)$. Calculate and sketch its aperiodic and periodic ACF. Do the same for the case $\Delta_c = \Delta/2$.

5.5. What happens to the periodic and aperiodic ACF of an APSK signal under the following transformations of a code sequence:

(a) Cyclic shift of elements?
(b) Changing the signs of all elements?
(c) Changing the signs of only the elements with even numbers?
(d) Multiplying all elements by the same constant?
(e) Mirror-like rearranging (i.e. reading from right to left)?

5.6. Is the combination $|R_a(1)| = 3, |R_p(1)| = 1$ possible for a PSK code? What about the combinations $R_a(1) = -2.1, R_p(1) = 0.8 - 0.6j$; $R_a(1) = 0.6 + 0.8j, R_p(1) = 1.1 + j$? What is the possible value of $|R_p(1) - R_a(1)|$ for a PSK code?

5.7. The spaces between frequencies of an FSK signal are multiples of $F = 1/\Delta$. Prove that time-aligned rectangular chips with different frequencies are orthogonal.

5.8. An FSK signal of length N involves $M < N$ frequencies. Chips of different frequencies are orthogonal. Is it possible that ACF is zero at all non-zero time shifts $\tau = m\Delta$?

Matlab-based problems

5.9. Write a program displaying APSK, BPSK, QPSK and FSK discrete signals. Take $N = 6-10$, carrier frequency $f_0 = (10-20)/\Delta$, chip duration $\Delta_c = \Delta$ and frequency step $F = 1/\Delta$. Run the program for various modulation modes and chip shapes (e.g. rectangular and half-wave sine), adjust the code sequences to provide satisfactory visualization and comment on the waveforms observed. Examples are given in Figure 5.4. Use the program also to demonstrate the finiteness or periodicity of signals.

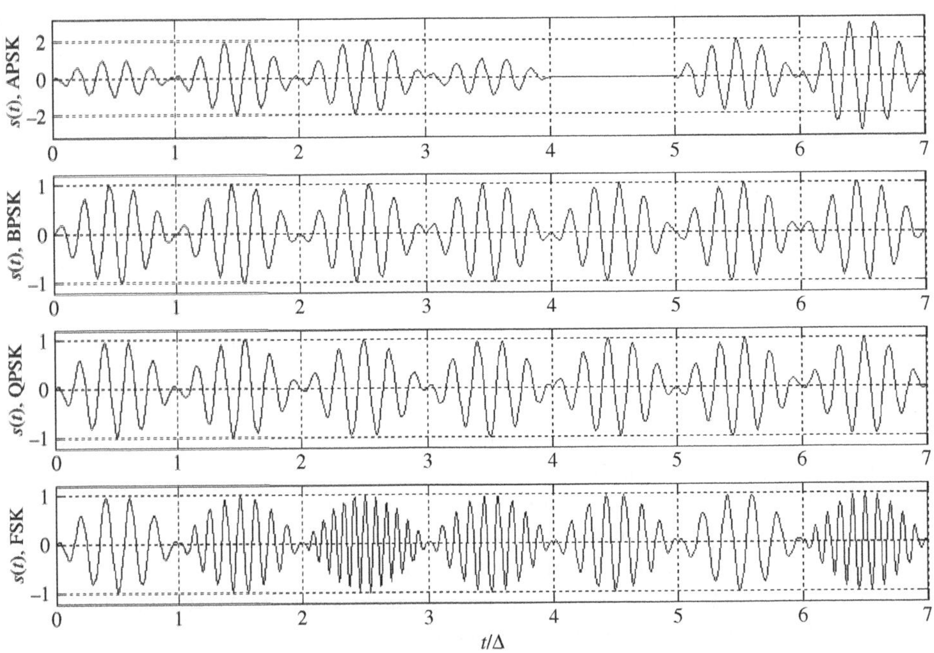

Figure 5.4 Example waveforms of discrete signals

5.10. Write a program demonstrating that the ACF of an APSK signal is the APSK signal itself, its chip being the ACF of the original signal and its code sequence being the ACF of the initial code sequence. Recommended steps:

(a) Form a plain chip (e.g. rectangular or half-wave sine).
(b) Form a real (e.g. binary, ternary etc.) code sequence containing 7–10 elements.
(c) Form and plot the signal with this chip and modulation law.
(d) Calculate and plot its ACF directly using an appropriate Matlab command.
(e) Calculate and plot the ACF of the initial signal chip.
(f) Calculate and plot the initial code ACF.
(g) Verify that the ACF of item (d) reproduces an APSK signal with the chip obtained in item (e) and the code obtained in item (f).
(h) Run the program, varying the chip shape and code of the initial signal, and give your comments. Example plots are shown in Figure 5.5.

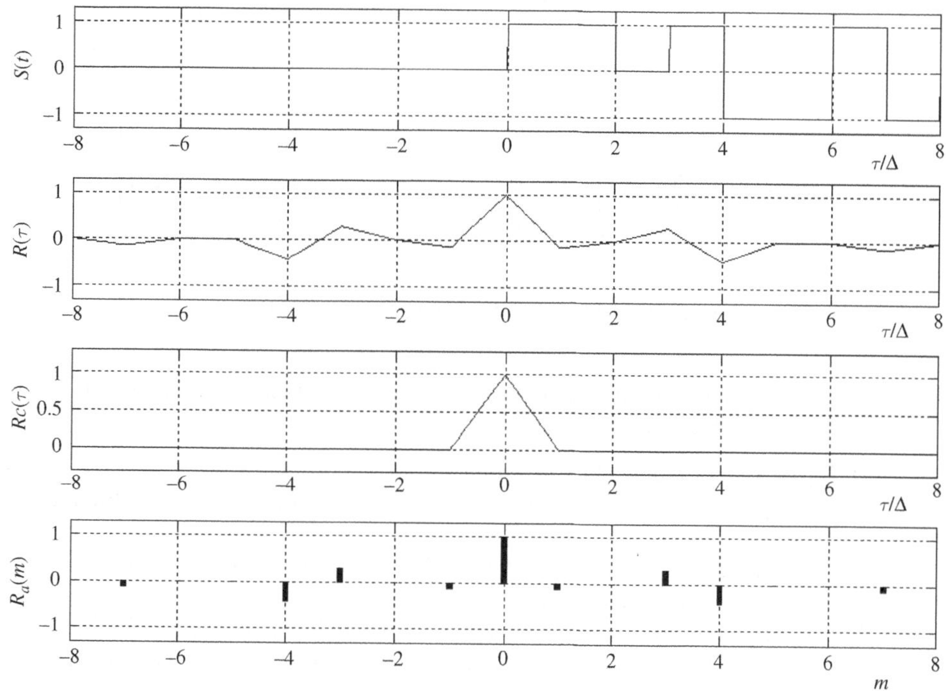

Figure 5.5 Aperiodic ACF of the ternary signal of length $N = 8$

5.11. Write a program verifying the association between periodic and aperiodic ACF of discrete signals (see Figure 5.2). Recommended steps:

(a) Specify some real (more convenient to display versus a complex one) code sequence of length $N = 7–15$.
(b) Calculate its periodic ACF directly from definition.
(c) Plot one period of it for the case when the chip is rectangular.
(d) Calculate the aperiodic ACF.
(e) Plot it and its N-shifted copy.
(f) Run the program for various codes and check the validity of (5.13).

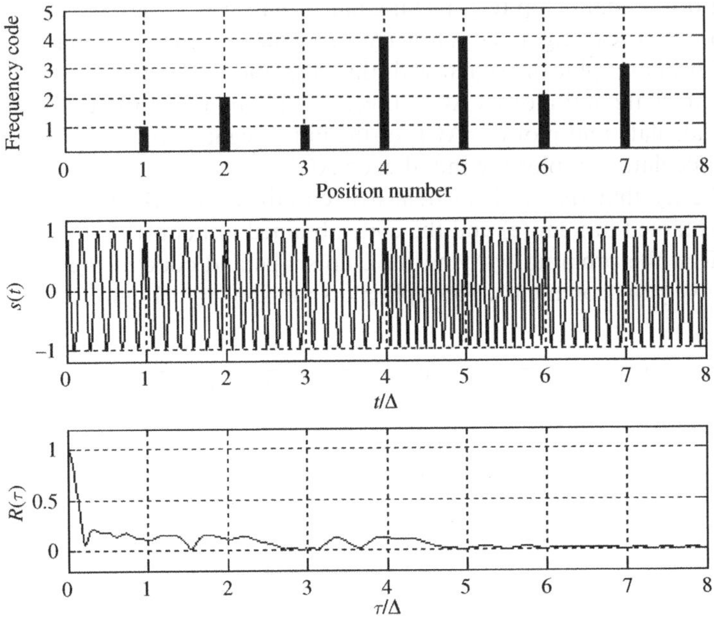

Figure 5.6 ACF of FSK signal of length $N = 8$ (see Example 5.5.1)

5.12. Write and run a program calculating precisely a real envelope of the ACF of an FSK signal. Recommended steps:

(a) Form a rectangular chip envelope.
(b) Specify a frequency code of length $N = 7$–10 and frequency alphabet size $M = 5 \ldots N - 1$;
(c) Form the complex envelope of an FSK signal with a specified frequency code, setting the frequency step $F = 1/\Delta$.
(d) Calculate and plot the real envelope of the ACF of the signal obtained.
(e) Compare the obtained values of ACF at the points $\tau = m\Delta$ to the theoretically predicted ones.
(f) Run the program, varying the frequency code.
(g) Pay attention to the situations where the ACF level between the points $\tau = m\Delta$ is higher than at these points (see Figure 5.6). How would you explain this phenomenon?

6

Spread spectrum signals for time measurement, synchronization and time-resolution

6.1 Demands on ACF: revisited

Let us return to the problems of estimating time delay and time-resolution studied in Sections 2.12 and 2.15, and recollect the demands imposed on a signal if high measuring accuracy and resolution capability are required. The principal condition to be met in both these problems is a response of the matched filter to a signal that is highly concentrated in time, or, equivalently, a 'sharp' signal ACF corresponding in the frequency domain to a wide spectrum. The attractiveness of spread spectrum as opposed to just shortening the signal is that with high processing gain $WT \gg 1$ it is possible to put into a signal the energy dictated by the necessary SNR controlling only the duration of T rather than peak power, which typically has a strict upper limit. Then involvement of an appropriate angle modulation allows widening the signal bandwidth to the extent which provides time compression of the signal by the matched filter, so that the duration of the filter response (correlation spread $\tau_c \approx 1/W$) appears to be many (about WT) times smaller than the duration T of the signal itself.

Let us specify what sort of ACF we may treat as sharp or 'good' concerning the reception problems in question. Actually, the ACF (see definitions (2.66) and (2.67)) of any physically realizable signal cannot equal strict zero at all τ beyond the range $[-\tau_c, \tau_c]$, if the correlation spread τ_c is smaller than the signal duration T. Thus, along

Spread Spectrum and CDMA: Principles and Applications Valery P. Ipatov

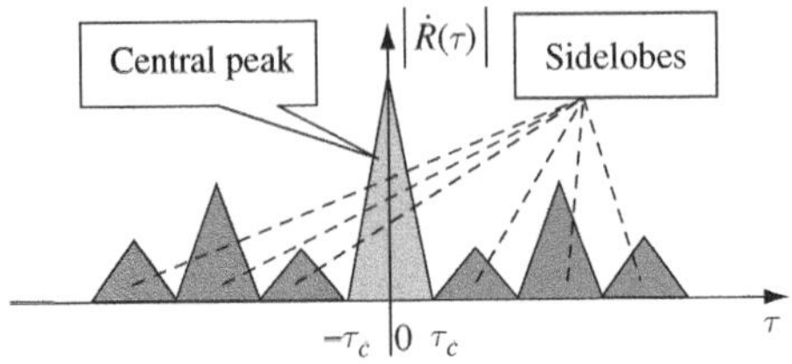

Figure 6.1 Central peak and sidelobes of ACF

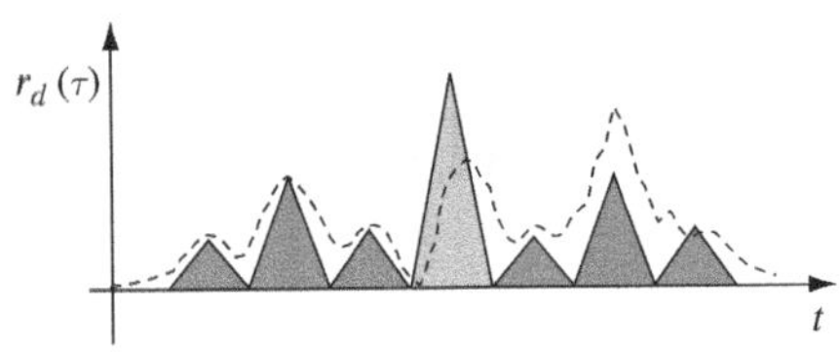

Figure 6.2 ACF sidelobes and abnormal errors

with the so-called *mainlobe* or *central peak* within the interval $[-\tau_c, \tau_c]$, the ACF has also *sidelobes* outside this range (see Figure 6.1). The effect of sidelobes is predominantly obstructive in both delay measuring and time-resolution. Indeed, to measure in an optimal (ML) manner the signal delay one should fix the time position of the maximum of the matched filter output envelope $r_d(t)$ (Section 2.12), and the ACF envelope is exactly the matched filter-detector response to a noiseless signal. In a real situation of noisy observation there is always some risk that somewhere beyond the 'body' of the ACF central peak a false maximum appears higher than a true (i.e. located within the body) one, as the dashed line in Figure 6.2 shows. In this case, an abnormal estimation error occurs, which is a deviation ε of the estimate $\hat{\tau}$ from a genuine value τ exceeding τ_c. It is obvious that confusion of the mainlobe with a false peak emerging in the vicinity of a high sidelobe is more probable than with the false peak located at the 'empty place'. Actually, the closer the levels of mainlobe and sidelobe, the 'easier' it is for the Gaussian noise to raise the second over the first.

To illustrate a harmful effect of sidelobes on time-resolution consider the superposition of two replicas of a bandpass signal, which are time-shifted and scaled as shown in Figure 6.3a. After processing by a matched filter the mainlobe of the weaker replica proves to be fully hidden under the sidelobe of the stronger replica (Figure 6.3b). In these circumstances the observer cannot confidently extract necessary information from both signal copies, or even tell how many copies are received. A situation of this sort is a typical case of non-resolved signals, despite the mainlobe of the ACF being much shorter than the signal duration.

We may now summarize in the most general terms the requirements placed on spread spectrum signals by delay estimation and time-resolution problems: *the ACF of the signal should have a sharp enough central peak and as low as possible level of sidelobes*. In the remainder of this chapter we study ways and instruments of approaching this fundamental objective.

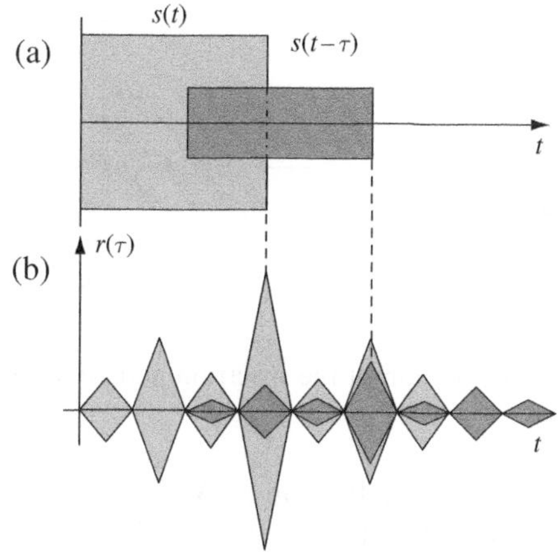

Figure 6.3 Non-resolution due to ACF sidelobes

6.2 Signals with continuous frequency modulation

Historically, among the first discovered signals possessing the matched filter time-compression feature was the linear frequency modulated (LFM) pulse. As the name tells us, the carrier frequency of this signal changes linearly throughout its duration. Consider a bandpass pulse whose instantaneous frequency $f(t)$ grows with time according to the equation:

$$f(t) = f_0 + \frac{W_d t}{T}, |t| \leq \frac{T}{2}$$

where W_d is the frequency deviation, i.e. the entire range of frequency variation, and f_0 is, as usual, the central frequency. A complete phase $\Phi(t)$ of a signal is the integral of a momentary frequency, therefore for an LFM pulse the phase obeys a square law:

$$\Phi(t) = 2\pi \int_0^t f(t)\mathrm{d}t = 2\pi f_0 t + \frac{\pi W_d t^2}{T}, |t| \leq \frac{T}{2}$$

Assuming a rectangular real envelope, the complex envelope of the LFM signal takes the form:

$$\dot{S}(t) = \begin{cases} \exp\left(\frac{j\pi W_d t^2}{T}\right), |t| \leq \frac{T}{2} \\ 0, |t| > \frac{T}{2} \end{cases}$$

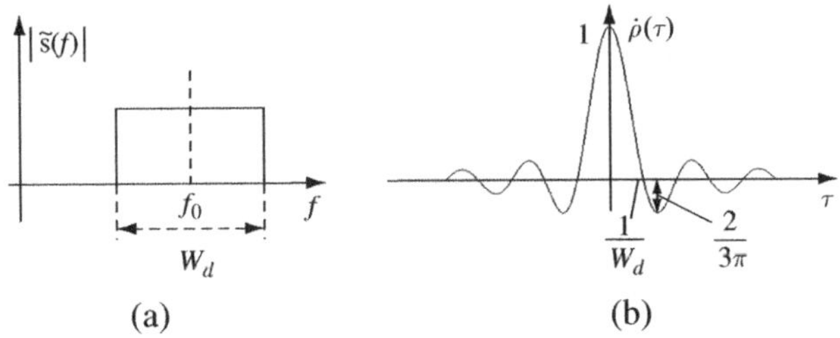

Figure 6.4 Approximation of the spectrum and ACF of an LFM pulse

Substituting this equation into the definition of ACF (2.66), the latter can be calculated formally without any special difficulties. Less formal and more physically transparent logic, however, allows faster achievement of the result. It is well known [1] that when the modulation index $\beta = W_d T$ is sufficiently large ($\beta \gg 1$), the spectrum of a frequency-modulated signal contains components of all momentary frequencies, the shape of the spectrum approaching the signal real envelope. Thus, in our case, the spectrum spans the range $[f_0 - W_d/2, f_0 + W_d/2]$ and has a form close to rectangular (Figure 6.4a). Now, ACF (2.66) may be found from the inverse Fourier transform, as was done in Section 2.12.2. Since the energy spectrum is again rectangular, its inverse Fourier transform is the function of view $\frac{\sin x}{x}$, so that the normalized ACF of the LFM signal:

$$\dot{\rho}(\tau) \approx \frac{\sin(\pi W_d \tau)}{\pi W_d \tau} \tag{6.1}$$

which is shown in Figure 6.4b. To compare this result with the exact one, the reader may turn to Problem 6.41.

As is seen, the complete (i.e. measured between zeros closest to the origin) duration of the ACF mainlobe equals $2\tau_c = 2/W_d$. As a matter of convention, the duration of the mainlobe on some non-zero level may be set equal to $\tau_c = 1/W_d$, so that a matched filter time-compresses the LFM signal $T/\tau_c \approx W_d T \approx WT$ times.

A substantial deficiency of the LFM signal is a high level of ACF sidelobes. The one nearest to the origin has intensity $2/3\pi(-13.5\,\text{dB})$ versus the mainlobe independently of the processing gain WT, i.e. its level cannot be reduced by increasing deviation W_d. To lower the sidelobes, smoothing of the signal envelope is an effective method (Problem 6.41) as well as weighting by a special window or mismatched processing in the receiver. In all of these methods, gain in the sidelobes is obtained in exchange for widening of the mainlobe or/and loss in output SNR.

Example 6.2.1. Take a rectangular LFM pulse with deviation $W_d = 20/T$. Using the program of Problem 2.55 produces waveforms of the signal itself and the matched filter response given in Figure 6.5. Compare the time-compression ratio and level of the first sidelobe to what is expected theoretically.

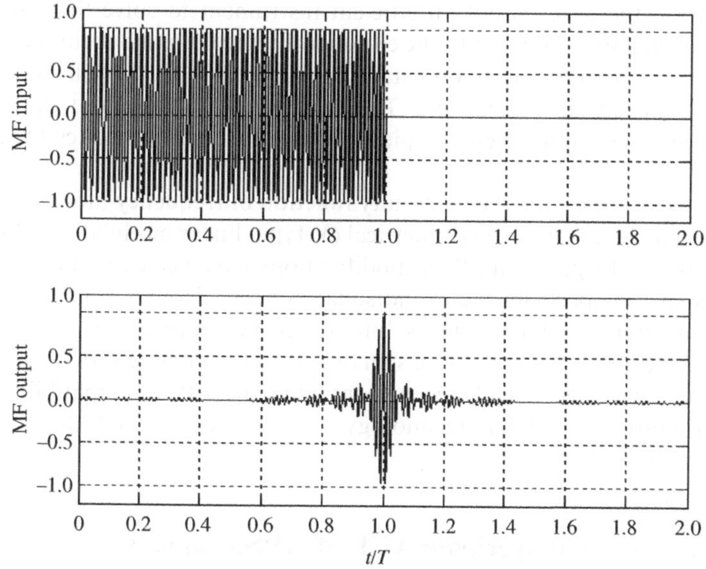

Figure 6.5 Time compression of a rectangular LFM pulse ($W_d = 20$)

The other shortcoming of the LFM signal is its ridge-like ambiguity function $\rho_0(\tau, F)$. From the study of Sections 2.14 and 2.15 we may deduce that for measuring simultaneously time-delay and frequency, as well as for time–frequency resolution, the best sort of ambiguity function is a needle-like one, having one central peak at the origin and falling sharply in all directions of the time–frequency plane. As is seen from Figure 6.6, where plots of the ambiguity function (a) and the ambiguity diagram (b) for an LFM

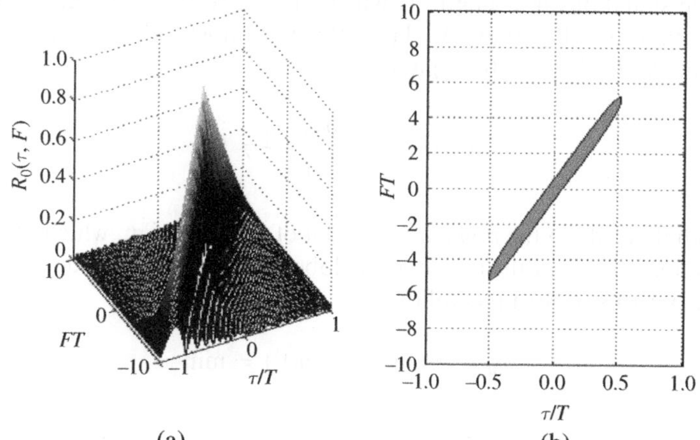

Figure 6.6 Ambiguity function (a) and its horizontal section (b) for an LFM signal ($WT = 10$)

signal are given, the latter is not an efficient instrument to solve the problems under discussion. When pairs τ, F fall into the ellipse of Figure 6.6b, the accuracy of estimating these parameters becomes very poor, since signal replicas with all such pairs are strongly correlated, i.e. hardly distinguishable. The same may be said about the resolution of such time–frequency shifted signal copies: their high similarity makes their separation very problematic.

Several ways of improving the ambiguity function of frequency modulated signals are known; for example, employing symmetrical (V-type) linear modulation (Problem 6.42).

Continuous LFM signals and their modifications have been very popular up to now in numerous spread spectrum radar and sonar systems. However, in modern commercial communication systems or mass long-range navigation they do not find wide application, making room for discrete signals. One of the reasons for such a preference, in addition to what has already been mentioned, is better compatibility of discrete signals with up-to-date digital technology, micro circuitry and the software-based philosophy of radio systems.

6.3 Criterion of good aperiodic ACF of APSK signals

Let us return to equation (5.5) and recollect that the ACF $\dot{\rho}(\tau)$ of an APSK signal is itself built as an APSK signal. Its chip is just the ACF $\dot{\rho}_c(\tau)$ of an original chip, while its code sequence is the ACF $\rho(m)$ of the original code $a_0, a_1, \ldots, a_{N-1}$. Such a construction makes it evident that the profile of the complete ACF, given the chip, is entirely defined by the code ACF $\rho(m)$. In particular, if the chip duration does not exceed the chip repetition period ($\Delta_c \leq \Delta$), the 'height' $|\dot{\rho}(m\Delta)|$ of any sidelobe at point $\tau = m\Delta$ just repeats the magnitude $|\rho(m)|$ of the code ACF at the shift m. It follows from what was stated in Section 6.1 that minimization of the code ACF sidelobe level is the highest priority in signal design, whenever time measurement or time-resolution is among the objectives of a system. Certainly, it would be excellent to have all sidelobes equalling zero, but this is absolutely impossible for finite or aperiodic APSK signals. Indeed, consider some signal of finite length N, which implies that $a_0 \neq 0$ and $a_{N-1} \neq 0$, otherwise the length is smaller than N. Then the rightmost sidelobe of the normalized aperiodic code ACF (5.10) of the signal:

$$\rho_a(N-1) = \frac{a_0 a_{N-1}^*}{\|\mathbf{a}\|^2} \neq 0 \tag{6.2}$$

This immediately leads to the *minimax criterion* of signal design, which requires achieving as small a magnitude of the maximal sidelobe of the aperiodic code ACF as possible. Formally we put this problem as follows:

$$\rho_{a,\max} = \max_{m \neq 0}\{|\rho_a(m)|\} = \min \tag{6.3}$$

In the light of the criterion (6.3) we prefer code sequences with the lowest maximal sidelobe; however, this requirement is always accompanied by a limitation on the modulation mode or, more specifically, on the alphabet to which symbols of the code sequence belong. This constraint reflects technological aspects concerning the complexity of signal forming and processing, and, as will be seen soon, may appear very binding. We summarize our demands towards the best signal as the following optimization task: *on the set of all possible sequences of length N with symbols taken from a pre-assigned alphabet find the sequence(s) with minimum magnitude of maximal sidelobe of the aperiodic* ACF.

6.4 Optimization of aperiodic PSK signals

The optimization task formulated above, like many other discrete optimization problems, does not have any general analytical solution, and exhaustive search is a typical procedure of its handling. Let us limit ourselves to PSK signals, which are usually recognized as most attractive. To explain the reasons for this, just recall that in the problems of time measurement and time-resolution the main advantage of spread spectrum consists in the availability of spreading signal energy over a large time interval, thereby reducing peak power. PSK signals free of amplitude modulation present an extreme version of such spreading, making the signal peak-factor ν (the ratio between peak and average powers) equal to one.

For any PSK signal $|a_i| = 1, i = 0, 1, \ldots, N-1$, so that the product $|a_0 a_{N-1}| = 1$, and the rightmost aperiodic ACF sidelobe (6.2) $|\rho_a(N-1)| = 1/N$. Therefore, the maximal sidelobe of a PSK signal is bordered from below as:

$$\rho_{a,\max} \geq \frac{1}{N} \tag{6.4}$$

Naturally, PSK signals attaining this bound would be optimal. They are called *Barker codes* after the specialist who was a pioneer in their search. Actually, Barker managed to find optimal binary codes lying on the border (6.4). Traditionally, binary sequences of symbols ± 1 are considered most attractive, being especially well fitted to digital circuitry and promising the least complexity of generating and processing. Table 6.1 lists all

Table 6.1 Binary Barker codes

N	Code												
2	+	−											
3	+	+	−										
4	+	+	−	+									
5	+	+	+	−	+								
7	+	+	+	−	−	+	−						
11	+	−	+	+	−	+	+	+	−	−	−		
13	+	+	+	+	+	−	−	+	+	−	+	−	+

binary Barker codes. Some of the codes are not unique in the sense that there are other sequences of the same length meeting the lower bound in (6.4).

Example 6.4.1. Table 6.2 illustrates the calculation of the aperiodic and periodic ACF of the Barker code of length $N = 7$. The structure of the table is identical to that of Table 5.1. As can be seen, not only the aperiodic but also the periodic normalized ACF has maximal sidelobe equal to $1/N$.

Table 6.2 Computation of ACF of the binary Barker code of length 7

m	a_0	a_1	a_2	a_3	a_4	a_5	a_6	$R_a(m)$	$R_p(m)$
0	+	+	+	–	–	+	–	+7	+7
1	–	+	+	+	–	–	+	0	–1
2	+	–	+	+	+	–	–	–1	–1
3	–	+	–	+	+	+	–	0	–1
4	–	–	+	–	+	+	+	–1	–1
5	+	–	–	+	–	+	+	–0	–1
6	+	+	–	–	+	–	+	–1	–1

Let us discuss briefly the matched filtering of a Barker signal. The best way of doing so is again an example.

Example 6.4.2. Figure 6.7 shows the structure of the matched filter for the Barker signal of length $N = 7$. The first unit is a tapped delay line, time separation between taps being equal to chip period Δ. The tap outputs are input to the adder with weights, which are taken in reverse order to the code symbols. The second part of the structure is a matched filter for a single chip. To make sure that the whole scheme is the necessary matched filter, it is enough to test it by applying in one's mind the delta-pulse to its input. The response is a mirror image of the signal, which is exactly what should be the reaction of a matched filter.

Figure 6.8, where the waveforms are numbered correspondingly to the points of Figure 6.7, demonstrates all the details of matched filtering of the Barker baseband signal compounded of rectangular chips. When the last chip of the signal arrives at the filter input, all the previous

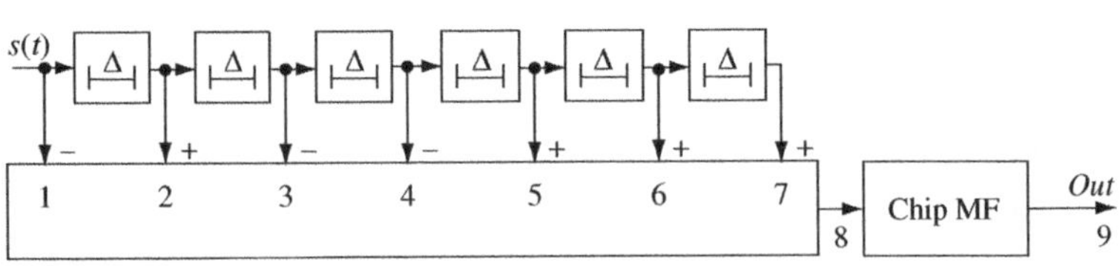

Figure 6.7 Matched filter for the binary Barker signal of length $N = 7$

chips appear at the adder inputs with properly rotated polarities and sum in phase, producing a central ACF peak. Before and after it there are sidelobes having the polarities and levels relative to the mainlobe predicted by Table 6.2.

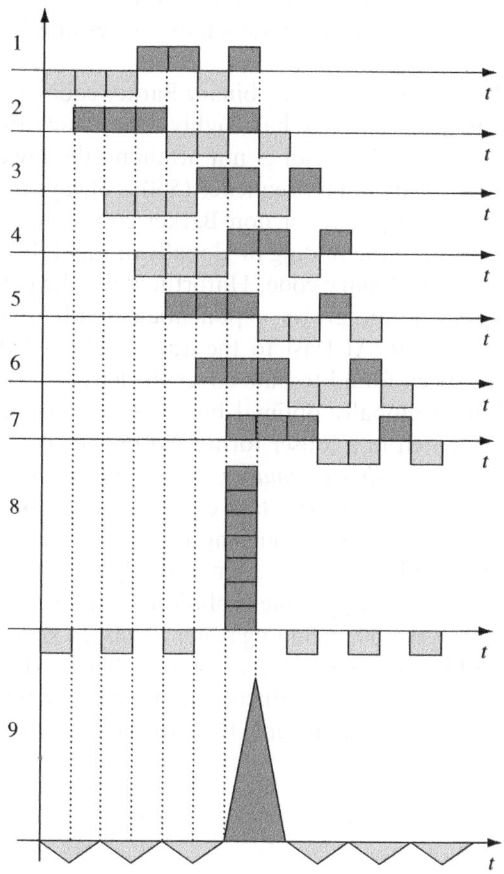

Figure 6.8 Matched filtering of binary Barker signal of length $N = 7$

Regretfully, binary Barker codes exist for only the lengths listed in Table 6.1. As long ago as the early 1960s, Turin and Storer proved their nonexistence for any other odd lengths and for even lengths at least within the range $4 < N < 12\,100$.[1] Extensive efforts have been undertaken to find non-binary PSK Barker codes with an equidistant phase alphabet (polyphase or M-ary PSK codes), but the results achieved so far are not very encouraging. It was found that even a very modest advance towards

[1] According to [22] this range is now extended to 1 898 884 and it is hardly likely that binary Barker codes of even lengths exist beyond it.

longer lengths appears possible only in return for a significant increase of the phase alphabet size M. Probably the longest polyphase Barker codes known to date and found with the aid of the stochastic search have phase alphabets of size either from hundreds to tens of thousands ($N = 32-36$) [23], or 60, 90, 120 ($N = 37-45$) [24]. A large alphabet size unavoidably entails grave hardware complications along with much more severe and difficult to satisfy demands concerning implementation errors, parameter drifts etc.

As can be seen, the lengths of the existing binary Barker codes are rather small to meet the numerous practical needs, which has highly stimulated the search for binary sequences of greater lengths with sidelobes not attaining the lower bound (6.4). Since the non-normalized ACF of any binary sequence (5.9) is always the sum of the plus and minus ones, possible values of $\rho_{a,\max}$ for non-Barker codes are $2/N, 3/N, \ldots$. Only an exhaustive search may result in the finding of globally optimal (having minimal possible $\rho_{a,\max} > 2/N$, length N given) binary code. Unfortunately, the computational resource necessary for such optimization grows exponentially with length N and becomes unrealistic when N exceeds 50. At least to the author's knowledge, globally optimal binary codes which have been found to date cover the length range up to 50 [25,26].

If we accept that longer globally optimal binary sequences are hardly obtainable, problem (6.3) may be restated in a looser form: *find the binary code with satisfactorily small—without guaranteeing global optimality—aperiodic sidelobe* $\rho_{a,\max}$. The general idea of algorithms involved in solving this task consists in a preliminary picking of some limited set of sequences, which seem promising for correlation properties, and subsequent exhaustive search to minimize $\rho_{a,\max}$ among only the sequences entering the set selected. One example of this strategy is the evolutionary algorithm [27], whereby binary codes were found up to length 100, including some which are slightly better than Barker codes: the Barker code of length 13 has $\rho_{a,\max} = 1/13 \approx 0.077$, while the best code of [27] has $\rho_{a,\max} = 6/88 \approx 0.068$. Another productive approach is based on equation (5.13) linking aperiodic ACF with the periodic one. Denoting as $\rho_{p,\max}$ the maximal sidelobe of the periodic ACF:

$$\rho_{p,\max} = \max_{m=1,2,\ldots,N-1} \{|\rho_p(m)|\}$$

and using the inequality $\max\{|x + y|\} \leq \max\{|x| + |y|\} \leq \max\{|x|\} + \max\{|y|\}$ results in the estimate $\rho_{p,\max} \leq 2\rho_{a,\max}$ or:

$$\rho_{a,\max} \geq \frac{1}{2}\rho_{p,\max} \tag{6.5}$$

The implication of this relation is quite remarkable: the necessary condition of 'good' aperiodic ACF is good (having low maximal sidelobe $\rho_{p,\max}$) periodic ACF. To put it differently, sequences with good aperiodic ACF are present only among sequences with good periodic ACF. As will be shown further, there are rather effective analytical tools for constructing sequences with good periodic ACF. Thus, we may prepare some set of sequences with good periodic ACF as the raw material for searching among them for sequences with good aperiodic ACF. This opportunity determines the primary role of periodic ACF in designing sequences with adequate correlation properties and clarifies why, in the next section, we switch to investigating periodic ACF.

6.5 Perfect periodic ACF: minimax binary sequences

Interest in sequences with good periodic ACF is not exhausted by their role as the basic material for designing good aperiodic sequences. Many applications exploit periodic discrete signals (CW-radar, navigation, pilot and synchronization channels of mobile radio etc.), making periodic ACF critically important for system performance. We will address as 'perfect' a periodic ACF which has only zero sidelobes, i.e. values between the periodic mainlobes repeating with the period N. Using normalized notation, we write this condition as:

$$\rho_p(m) = \frac{1}{E}\sum_{i=0}^{N-1} a_i a_{i-m}^* = \begin{cases} 1, m = 0 \bmod N \\ 0, m \neq 0 \bmod N \end{cases} \tag{6.6}$$

where the congruence $m = 0 \bmod N$ is read as m is divisible by N (is multiple of N). It is obvious that for the perfect ACF $\rho_{p,\max} = 0$. Figure 6.9 shows the ACF of a discrete baseband signal with rectangular chips, manipulated by the code with perfect periodic ACF. The practical benefits of perfect ACF are obvious from Figure 6.10, where

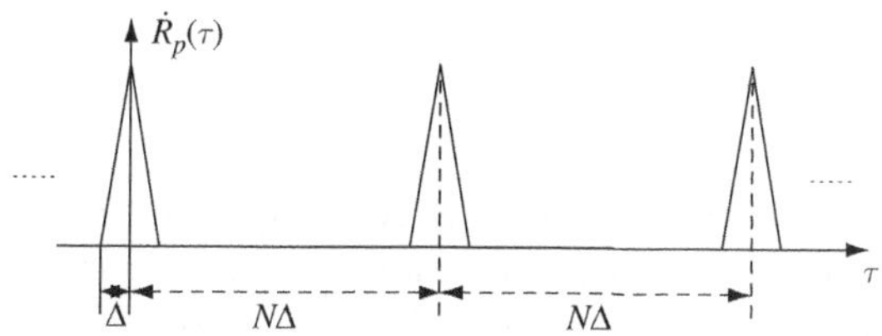

Figure 6.9 Perfect periodic ACF

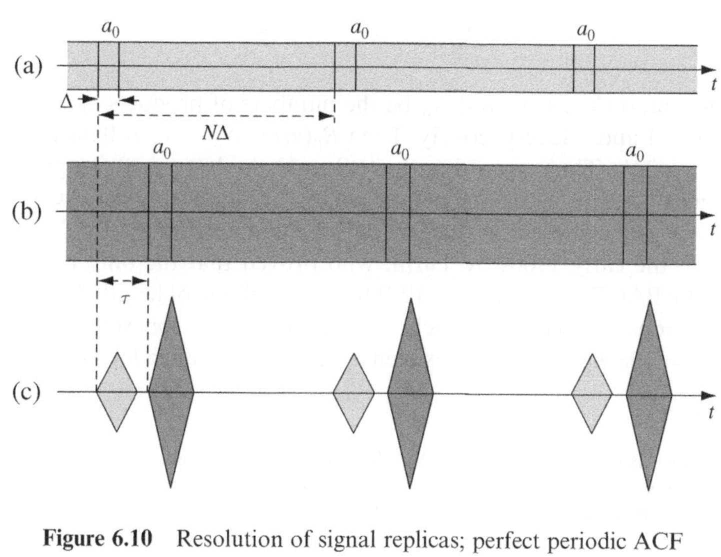

Figure 6.10 Resolution of signal replicas; perfect periodic ACF

waveforms (a) and (b) show two time-shifted copies of the same bandpass periodic signal whose code ACF meets (6.6). When the superposition of these signals arrives at the input of the filter matched to a one-period segment of the signal, two time-shifted replicas of the signal ACF are observed at the output. If the time delay between signal copies is greater than ACF mainlobe duration 2Δ (but smaller than $(N-2)\Delta$), the filter responses to both signals are entirely resolved with no corruption of each other (Figure 6.10c).

Let us investigate non-normalized periodic ACF of a binary sequence composed of elements $\{\pm 1\}$:

$$R_p(m) = \sum_{i=0}^{N-1} a_i a_{i-m} \tag{6.7}$$

where the conjugation asterisk is not required, since all $a_i = \pm 1$. Summing both parts of (6.7) over the range $m = 0, 1, \ldots, N-1$ produces:

$$\sum_{m=0}^{N-1} R_p(m) = \sum_{m=0}^{N-1}\sum_{i=0}^{N-1} a_i a_{i-m} = \sum_{i=0}^{N-1} a_i \sum_{m=0}^{N-1} a_{i-m} = |\tilde{a}_0|^2 \tag{6.8}$$

with:

$$\tilde{a}_0 = \sum_{i=0}^{N-1} a_i$$

being the constant component (*imbalance*) of code sequence $\{a_0, a_1, \ldots, a_{N-1}\}$. Since the constant component of a binary sequence may take on only integer values, the sum in (6.8) is a squared integer. Suppose now that a binary code has perfect periodic ACF. Then $R_p(0) = \sum_{i=0}^{N-1} (a_i)^2 = N$ and $R_p(m) = 0,\ m = 1, 2, \ldots, N-1$, giving:

$$\sum_{m=0}^{N-1} R_p(m) = N = |\tilde{a}_0|^2 \tag{6.9}$$

Assuming $m \neq 0 \bmod N$, let N_e and N_d be the numbers of products $a_i a_{i-m}$ in the sum of (6.7) equalling $+1$ and -1, respectively. Then $R_p(m) = N_e - N_d = 0$ means $N_e = N_d$ and $N = N_e + N_d = 2N_e$. Thus, according to (6.9) and the last result, length N is an even squared integer, i.e. the necessary condition for obtaining perfect ACF for a binary sequence is $N = 4h^2$, where h is integer. All of these lengths (4, 16, 36, 64, ...) were investigated in the early 1960s by Turin, who proved that the only binary code[2] with perfect periodic PACF of length $N \leq 12\,100$ is a trivial one of length 4: $+1+1+1-1$ [28]. Later, the nonexistence of such sequences was proved up to lengths $N < 4 \times 165^2 = 108\,900$ [29]. Their existence beyond this range looks quite improbable.

[2] We do not consider as new (and it is universally adopted) sequences obtained from an initial one by a cyclic shift, mirror imaging or changing signs of all elements. These transforms do not change periodic ACF (Problem 5.5), and sequences obtainable from each other in this way are treated as trivially different or equivalent.

In the light of what has been said, it is important to figure out how low the maximal sidelobe of periodic ACF of binary codes can potentially be. Let N_{++}, N_{+-}, N_{-+} denote numbers of pairs $a_i a_{i-m}$ in (6.7), whose terms are indicated by the subscripts, e.g. N_{+-} is the number of pairs where $a_i = +1, a_{i-m} = -1$. Since both $N_{++} + N_{+-}$ and $N_{++} + N_{-+}$ should produce the same outcome, the total number of positive elements over the code period, $N_{+-} = N_{-+}$. Then the difference:

$$N - R_p(m) = \sum_{i=0}^{N-1} (1 - a_i a_{i-m}) = 2(N_{+-} + N_{-+}) = 4N_{+-}$$

is always divisible by four. It is clear, therefore, that for any binary code the non-normalized periodic ACF always differs from length N by some multiple of four:

$$R_p(m) = N - 4h \tag{6.10}$$

where h is integer.

Clearly, in the absence of binary codes with perfect periodic ACF, next in attractiveness would be binary sequences for which $R_p(m)$ takes on only values ± 1 at $m = 1, 2, \ldots, N-1$, i.e. $\rho_{p,\max} = 1/N$. As is seen from (6.10), value $R_p(m) = +1$ is possible only if length $N = 4h + 1$, while $R_p(m) = -1$ may happen only for length $N = 4h - 1$, h being integer. This shows that binary sequences with $\rho_{p,\max} = 1/N$ may have only two-valued non-normalized periodic ACF, either:

$$R_p(m) = \begin{cases} N, m = 0 \bmod N \\ +1, m \neq 0 \bmod N \end{cases} \tag{6.11}$$

for lengths $N = 4h + 1$, or:

$$R_p(m) = \begin{cases} N, m = 0 \bmod N \\ -1, m \neq 0 \bmod N \end{cases} \tag{6.12}$$

for lengths $N = 4h - 1$.

Sequences meeting (6.11) or (6.12), and therefore having theoretically minimal periodic ACF sidelobe ($\rho_{p,\max} = 1/N$) for a binary code of odd length, are called *minimax*. Only two examples ($N = 5$ and $N = 13$) of sequences complying with (6.11) are known; at the same time, extremely powerful regular rules generating minimax sequences obeying (6.12) do exist! Sections 6.6–6.9 present two of the most popular of them, although at least three more are known.

6.6 Initial knowledge on finite fields and linear sequences

6.6.1 Definition of a finite field

To characterize conceivably constructing binary minimax sequences, we will need some primary idea of finite fields. Our description will be less formal than that found in purely mathematical sources. Let us appropriate the name *field* to a set of elements on which two operations are defined, called *addition* and *multiplication* and denoted by the customary symbols '+' and '·' (or '×', or just writing elements one after another). 'Defined' means that both these operations are closed, i.e. if x, y are elements of the field F, then their sum and product also belong to F: $x + y \in \mathrm{F}, xy \in \mathrm{F}$. Zero '0' and unit '1' elements should be present in any field, which do not change an arbitrary element $x \in \mathrm{F}$ in the addition and multiplication operations, respectively: $x + 0 = x, x \cdot 1 = x$. The tables of addition and multiplication are so built that operations are commutative $(x + y = y + x;\ xy = yx)$, associative $((x + y) + z = x + (y + z);\ x(yz) = (xy)z)$, and invertible, i.e. subtraction and division by non-zero elements are also defined: $x + y = z \Rightarrow x = z - y;\ \ xy = z,\ \ y \neq 0 \Rightarrow x = z/y$. This, in particular, entails the existence of elements *negative* to x, denoted as $-x = 0 - x$, and *inverse* to non-zero x with designation $x^{-1} = 1/x$. Finally, operation tables should follow the distributivity law: $(x + y)z = xz + yz$.

It is obvious that the field is a set, within which we operate with elements, as we do with real numbers in ordinary arithmetic. A field, thereby, is just an abstract generalization of the set of real numbers, or, putting it another way, the set of real numbers is a trivial example of a field. Other examples are sets of rational numbers, complex numbers etc. All of these fields have an infinite *order*, which means the number of elements in a field. In contrast, the constructions below involve *finite* or Galois fields whose orders are finite. In algebra (see, e.g. [30]) it is proved that finite fields exist of any (and only) orders p^m with arbitrary prime p and natural m. The standard designation of a Galois field of order p^m is $GF(p^m)$. For our study *prime* fields $GF(p)$ are sufficient, whose orders are prime numbers $(m = 1)$. The easiest way to treat a prime field $GF(p)$ is to identify all of its elements with p integers $0, 1, \ldots, p - 1$, which are added and multiplied modulo p. Figure 6.11 presents addition and multiplication tables for the three simplest prime fields $GF(2), GF(3), GF(5)$. Note that the negative of any two elements of $GF(2)$ is the same element itself, since $0 + 0 = 1 + 1 = 0$ and the only non-zero element 1 is inverse to itself. Operations in all of the other prime fields are not that degenerated; for instance, in $GF(5)$ $2 + 3 = 0 \Rightarrow -2 = 3$, and $3 \cdot 2 = 1 \Rightarrow 3^{-1} = 2$.

GF(2)

+	0	1
0	0	1
1	1	0

×	0	1
0	0	0
1	0	1

GF(3)

+	0	1	2
0	0	1	2
1	1	2	0
2	2	0	1

×	0	1	2
0	0	0	0
1	0	1	2
2	0	2	1

GF(5)

+	0	1	2	3	4
0	0	1	2	3	4
1	1	2	3	4	0
2	2	3	4	0	1
3	3	4	0	1	2
4	4	0	1	2	3

×	0	1	2	3	4
0	0	0	0	0	0
1	0	1	2	3	4
2	0	2	4	1	3
3	0	3	1	4	2
4	0	4	3	2	1

Figure 6.11 Addition and multiplication tables for the simplest prime fields

6.6.2 Linear sequences over finite fields

Let us introduce a sequence $d_0, d_1, \ldots$ with elements (symbols) of a given finite field $GF(p)$ obeying the linear recurrence:

$$d_i = -f_{n-1}d_{i-1} - f_{n-2}d_{i-2} - \cdots - f_0 d_{i-n}, \ i = n, n+1, \ldots \tag{6.13}$$

where coefficients $f_0, f_1, \ldots f_{n-1}$ are fixed constants belonging to $GF(p)$. Such a sequence is called a *linear recurrent sequence* over $GF(p)$ of *memory n*. Elements of a linear recurrent sequence are calculated one by one, each being determined by the n preceding, so that setting n initial elements $d_0, d_1, \ldots, d_{n-1}$ generates the whole sequence.

Example 6.6.1. Let us build the linear recurrent sequence of memory $n = 3$ over $GF(2)$ (sequences over $GF(2)$ are also attributed as *binary*) starting with initial elements $d_0 = 1$, $d_1 = 0, d_2 = 0$, if coefficients of the recurrence are $f_2 = 0, f_1 = 1, f_0 = 1$. Noting that in the binary field the negative of any element is an element itself, recurrence (6.13) takes the form $d_i = d_{i-2} + d_{i-3}, i \geq 3$, so that the sequence is 1, 0, 0, 1, 0, 1, 1, 1, 0, 0, 1, 0, 1, 1, ... This sequence is periodic of period 7.

Example 6.6.2. Construct the linear sequence over $GF(3)$ (*ternary* sequence) of memory $n = 3$ set by initial symbols $d_0 = 1, d_1 = 0, d_2 = 0$ and coefficients of recurrence (6.13) $f_2 = 0,\ f_1 = 2,\ f_0 = 1$. Since in $GF(3)$ $-2 = 1, -1 = 2$, recursion (6.13) takes the form $d_i = d_{i-2} + 2d_{i-3}, i \geq 3$, so that the desired sequence is 1, 0, 0, 2, 0, 2, 1, 2, 2, 1, 0, 2, 2, 2, 0, 0, 1, 0, 1, 2, 1, 1, 2, 0, 1, 1, ... It is noteworthy that the sequence obtained is again periodic with period 26, each period consisting of two blocks of length 13, and the second block just being a repetition of the first multiplied by 2.

Let us turn to Figure 6.12, which shows a typical generator of a linear recurrent sequence. According to its structure this scheme is called the *linear feedback shift register* (LFSR). The register includes n p-ary delay *stages* or *flip-flops* (shown by squares), each having p possible states and storing some element of $GF(p)$ during a

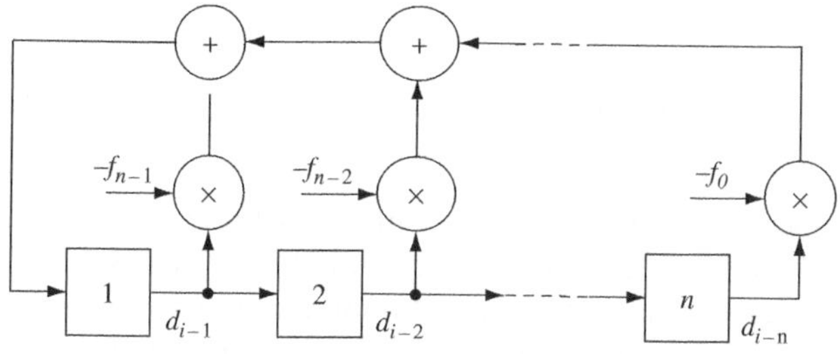

Figure 6.12 LFSR generator of a linear recurrent sequence

clock interval. The clocking circuit (not shown) controls the register in such a manner that every clocking transmits the state of any stage to the next one from left to right. The feedback includes multipliers to multiply elements (states) stored in stages by constants (f_i) and adders. Certainly, both arithmetic operations are made by the rules of the finite field $GF(p)$.

Suppose that initial states (i.e. initial sequence symbols) $d_{n-1}, d_{n-2}, \ldots, d_0$ are fed into the register stages from left to right as shown in Figure 6.12. Then the state of the feedback output will be $-f_{n-1}d_{n-1} - f_{n-2}d_{n-2} - \cdots - f_0 d_0 = d_n$ and after a clocking the contents of the register appears to be $d_n, d_{n-1}, \ldots, d_1$, generating the feedback state $-f_{n-1}d_n - f_{n-2}d_{n-1} - \cdots - f_0 d_1 = d_{n+1}$. After the next clocking the register state is $d_{n+1}, d_n, \ldots, d_2$ and so forth. In general, current register contents $d_{i-1}, d_{i-2}, \ldots, d_{i-n}$ creates the feedback state d_i. Therefore, a complete linear recurrent sequence may be read directly from the rightmost stage, starting with the very first symbol d_0, or from any other stage with an appropriate advance.

Certainly, one or another stage is connected to an adder through a multiplier only if an appropriate feedback coefficient f_i is non-zero, otherwise there is no need for a connection at all.

Example 6.6.3. Figure 6.13a presents an LFSR generator for the binary sequence of Example 6.6.1. Note that in the case of binary sequences multiplication by one is just connection between a stage output and an adder. Figure 6.13b shows successive register contents (stage states) and states of the feedback output (point '*FB*' in the scheme) as clocking happens. The sequence is read out as consecutive states of the rightmost stage. Reading states of other stages results in replicas of the same sequence having a lead of one or two clocks.

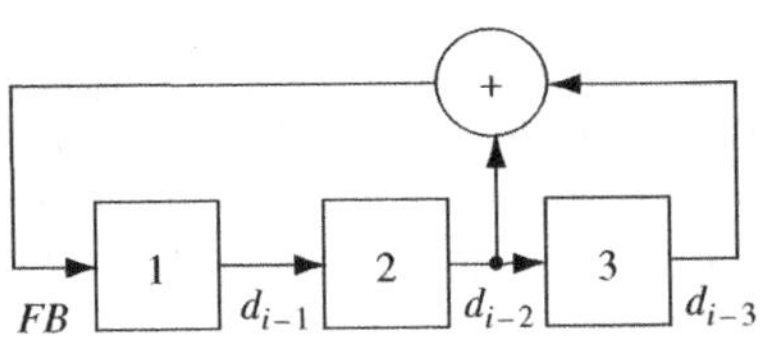

(a)

Clock	*FB*	Register contents		
		1	2	3
1	1	0	0	1
2	0	1	0	0
3	1	0	1	0
4	1	1	0	1
5	1	1	1	0
6	0	1	1	1
7	0	0	1	1

(b)

Figure 6.13 Generator of a binary sequence of length 7 (a) and its table of states (b)

Since the number of different register contents is finite (never greater than p^n) the situation is unavoidable where after some number of clocks the content repeats one that occurred earlier. But starting with some initial loading, i.e. fixed contents, the scheme of Figure 6.12 generates only the unique sequence defined by (6.13).

Therefore, repetition of the register contents leads to a repetition of all subsequent generated symbols, meaning that any linear recurrent sequence is periodic. Furthermore, as is seen directly from (6.13), occurrence of zero contents of the register (zeros in all stages) will always be followed by the infinite degenerated sequence consisting of only zeros. Certainly, such a case is absolutely worthless, so zero contents of the register should be prohibited. This leaves no more than $p^n - 1$ allowed register contents with the implication that the maximal sequence period is never longer than $p^n - 1$.

6.6.3 m-*sequences*

Linear recurrent sequences having the biggest period $L = p^n - 1$ are of particular interest in modern information technology and are called *maximal-length* sequences or *m*-sequences. Being fully deterministic, they possess many properties peculiar to random sequences, e.g. to the sequence of heads and tails in a series of flipping a fair coin. The following features make *m*-sequences extremely valuable for constructing codes with good autocorrelation:

1. *The balance property*. On a single period of a *p*-ary *m*-sequence any non-zero element of $GF(p)$ occurs p^{n-1} times while the zero element occurs $p^{n-1} - 1$ times. To make sure of that it is enough to note that all possible $p^n - 1$ non-zero contents of LFSR should be passed in one or another order during generation of one period of the *m*-sequence, otherwise the period will not be maximal. All those contents are just different *n*-digit *p*-ary numbers from the range $1, 2, \ldots, p^n - 1$ and the *m*-sequence read from the rightmost register stage may be treated as the sequence of rightmost digits of these numbers. Running over the range $0, 1, \ldots, p^n - 1$ would produce any *p*-ary digit at any specific (e.g. rightmost) position of an *n*-digit number exactly p^{n-1} times. Throwing away the all-zero *n*-digit number will reduce by one only the number of occurrences of zeros among the rightmost (or any other) digits. Specifically, the period of a binary *m*-sequence of memory n $L = 2^n - 1$, and there are $L_0 = 2^{n-1} - 1$ zeros and $L_1 = 2^{n-1}$ ones in it. For instance, one may see that the sequence obtained in Example 6.6.1 is a binary *m*-sequence with period $L = 2^3 - 1 = 7$ containing within a period $L_0 = 2^2 - 1 = 3$ zeros and $L_1 = 2^2 = 4$ ones. The sequence of Example 6.6.2 is a ternary *m*-sequence of length $L = 3^3 - 1 = 26$, having in a period $L_0 = 3^2 - 1 = 8$ zeros and $L_1 = L_2 = 3^2 = 9$ repetitions of each of elements 1 and 2.

2. Any two *m*-sequences generated by the same recurrence (6.13) differ from each other by no more than a cyclic shift. Indeed, since with fixed initial *n* elements (6.13) fully determines a sequence, two non-coinciding *m*-sequences generated by a fixed recurrence cannot have absolutely identical *n* initial elements. On the other hand, all contents of the LFSR generator but zero occur at one *m*-sequence period, and after the content reproducing the initial state of the first generator occurs in the second, the second sequence completely repeats the first one, i.e. is just some delayed replica of the first *m*-sequence.

3. *Shift-and-subtract property*. Take some *m*-sequence defined by (6.13) and subtract from it element-wise (certainly, modulo *p*) its own replica cyclically shifted by *m*

(m is an arbitrary integer) positions $d_{i+m} = -f_{n-1}d_{i+m-1} - f_{n-2}d_{i+m-2} - \cdots - f_0 d_{i+m-n}$. The result is:

$$d_i - d_{i+m} = -f_{n-1}(d_{i-1} - d_{i-1+m}) - f_{n-2}(d_{i-2} - d_{i-2+m}) - \cdots - f_0(d_{i-n} - d_{i-n+m})$$

After denoting $d'_i = d_i - d_{i+m}$ we come to the linear recurrent sequence whose elements d'_i are defined by the original recurrence:

$$d'_i = -f_{n-1}d'_{i-1} - f_{n-2}d'_{i-2} - \cdots - f_0 d'_{i-n}. \tag{6.14}$$

Only two possibilities are now expected. Suppose first that the shift m equals the integer number of periods L. Then $d_i = d_{i+m}$ and $d'_i = 0$ for all i, i.e. (6.14) is an all-zero sequence.

Now let m be not a multiple of L. Then d_i and d_{i+m} can not be identical for all $i = 0, 1, \ldots, n-1$, since otherwise we would have the same initial contents of the generator for both the sequence and its shifted replica. This would imply a complete coincidence of the original sequence and its replica, because the initial loading defines a unique sequence generated by the recurrence (6.13). But the shifted replica repeats the original sequence only being shifted by some multiple of the period, which is contrary to the assumption. It is seen then that the recurrence (6.14) completely reproducing (6.13) generates some non-zero sequence. However, according to the previous property, when recurrence (6.13) or, equivalently, a corresponding scheme of Figure 6.12 generates an m-sequence starting with some definite initial loading, it may generate all but another shifted replica of this very m-sequence when starting with different non-zero initial loading. The conclusion we have arrived at is as follows: symbol-wise subtraction of two shifted replicas of the same m-sequence produces either an all-zero sequence, if the shift is a multiple of the period, or some new shifted copy of the same m-sequence otherwise.

For binary sequences subtraction coincides with addition, which explains why in this case the same property is often referred to as *shift-and-add*. Turning again to Example 6.6.1 we may see, for instance, that addition of the sequence obtained there to its two-position left-shifted replica 0, 1, 0, 1, 1, 1, 0, 0, 1, 0, 1, 1, 1, 0, ... results in the sequence 1, 1, 0, 0, 1, 0, 1, 1, 1, 0, 0, 1, 0, 1, ..., being again the original sequence right-shifted cyclically by one position.

The reader is advised to examine the same property for the ternary m-sequence of Example 6.6.2.

Clearly, to generate a p-ary m-sequence, i.e. the sequence with maximal length allowed by a given memory n, rather than a sequence of some shorter length, adequate choice of the coefficients f_i in the recurrence (6.13) (or in the feedback of LFSR) is needed. The necessary and sufficient condition for a linear recurrent sequence to be an m-sequence is that $f_i, i = 0, 1, \ldots, n-1$ are coefficients of the *primitive polynomial* $f(x) = x^n + f_{n-1}x^{n-1} + f_{n-2}x^{n-2} + \cdots + f_0$ of degree n over $GF(p)$. The primitive polynomials are a subclass of *irreducible* polynomials. A polynomial $f(x)$ of degree n over $GF(p)$ (i.e. with coefficients in $GF(p)$) is called irreducible over $GF(p)$ if it cannot be factored into two polynomials of degrees smaller than n. These polynomials play the same role among all polynomials as prime numbers among integers. Not all arbitrary irreducible polynomials need to be primitive, although in a special case $p = 2$ and prime $2^n - 1$ all irreducible polynomials are primitive. Proving the necessity and sufficiency of choosing feedback as indicated above would require some more algebra, which could take us far away from our main purpose. An interested reader may find the details in numerous sources (e.g. [31,32]).

Primitive polynomials are extensively tabulated in books on modern algebra and coding theory or (mostly for $p = 2$) spread spectrum telecommunication [5,6,18,32]. Another option is a computer search, which is not at all a difficult task (e.g. Problem 6.47). In particular, ready-made functions for finding primitive polynomials are present in the Matlab Communications Toolbox.

As a whole, designing an m-sequence generator is pretty straightforward. When p is selected, a necessary length L determines the memory n, and finding an appropriate primitive polynomial exhausts the issue.

Commenting again on Examples 6.6.1 and 6.6.2, note that the binary m-sequence of length 7 is built based on the primitive polynomial $f(x) = x^3 + x + 1$ over $GF(2)$, while the primitive polynomial over $GF(3)$ used to generate the ternary sequence of length 26 is $f(x) = x^3 + 2x + 1$.

6.7 Periodic ACF of *m*-sequences

The results of the previous section lead quickly to the minimax binary sequences with ACF meeting (6.12). Consider a binary m-sequence $\{d_i\}$ of memory n, i.e. length $L = 2^n - 1$. Let us map its symbols 0, 1 onto a binary alphabet ± 1 according to the rule:

$$a_i = (-1)^{d_i} = \begin{cases} +1, d_i = 0 \\ -1, d_i = 1 \end{cases} \tag{6.15}$$

where in raising (-1) to the degree d_i the latter is treated as though it is a real number 0 or 1. The sequence $\{a_i\}$ of real binary symbols ± 1 thus obtained has period $N = L = 2^n - 1$ and is a one-to-one image of the original binary m-sequence $\{d_i\}$. It is natural to keep for it the same name *binary* m-sequence as well. When the confusion is risky, a supplementary label like binary $\{\pm 1\}$ sequence versus binary $\{0, 1\}$ sequence may be used. Let us find the non-normalized periodic ACF (6.7) of $\{a_i\}$:

$$R_p(m) = \sum_{i=0}^{N-1} a_i a_{i-m} = \sum_{i=0}^{L-1} (-1)^{d_i} (-1)^{d_{i-m}} = \sum_{i=0}^{L-1} (-1)^{d_i + d_{i-m}} \tag{6.16}$$

Now the shift-and-add property of binary $\{0, 1\}$ m-sequences may be brought in. Addition in the exponent here may be treated as modulo 2, since it will produce the same result of exponentiation as an ordinary arithmetic summation. But then $\{d'_i\} = \{d_i + d_{i-m}\}$ is a binary $\{0, 1\}$ m-sequence of period L whenever $m \neq 0 \mathrm{mod} L$, or an all-zero sequence otherwise. Due to the balance property, one period of $\{d'_i\} = \{d_i + d_{i-m}\}$ contains $L_0 = 2^{n-1} - 1$ zeros and $L_1 = 2^{n-1}$ ones, therefore the sum in (6.16) contains L_0 plus ones and L_1 minus ones if $m \neq 0 \bmod N$, so that:

$$R_p(m) = L_0 - L_1 = \begin{cases} N, m = 0 \bmod N \\ -1, m \neq 0 \bmod N \end{cases}$$

As is seen this coincides exactly with (6.12), confirming that binary m-sequences are minimax ones.[3]

Example 6.7.1. Consider again the sequence of Example 6.6.1. Its mapping onto the alphabet $\{\pm 1\}$ in accordance with (6.15) produces the $\{\pm 1\}$ m-sequence $-1, +1, +1, -1, +1, -1, -1, \ldots$. Table 6.3, which contains only the minimum necessary number of entries, illustrates calculating the periodic ACF of the sequence. It is interesting also to study the matched filter processing of a discrete signal modulated by this binary sequence. Figure 6.14 presents such a filter matched with one period of a baseband rectangular-chip periodic signal. All units in this structure are absolutely similar to those of Figure 6.7. Waveforms at characteristic points of the filter are shown in Figure 6.15. As is seen, the output waveform has mainlobes repeating with period $N\Delta$ and a uniform sidelobe background of negative polarity seven times smaller than the mainlobe level.

Table 6.3 Calculating periodic ACF of the binary m-sequence $(-++-+--)$

m	a_0	a_1	a_2	a_3	a_4	a_5	a_6	$R_p(m)$
0	−	+	+	−	+	−	−	+7
1	−	−	+	+	−	+	−	−1
2	−	−	−	+	+	−	+	−1
3	+	−	−	−	+	+	−	−1

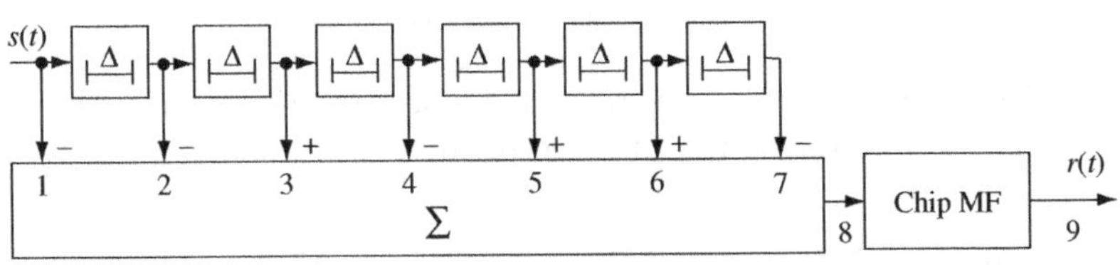

Figure 6.14 Matched filter for the binary m-sequence of length $N = 7$

[3] Generalization of mapping (6.15) onto sequences over $GF(p), p \geq 2$ is $a_i = \exp(j2\pi d_i/p)$, resulting in polyphase (p-phase) code, whose periodic ACF again satisfies (6.12). However, polyphase codes of this sort with $p > 2$ are of less practical interest than minimax binary sequences.

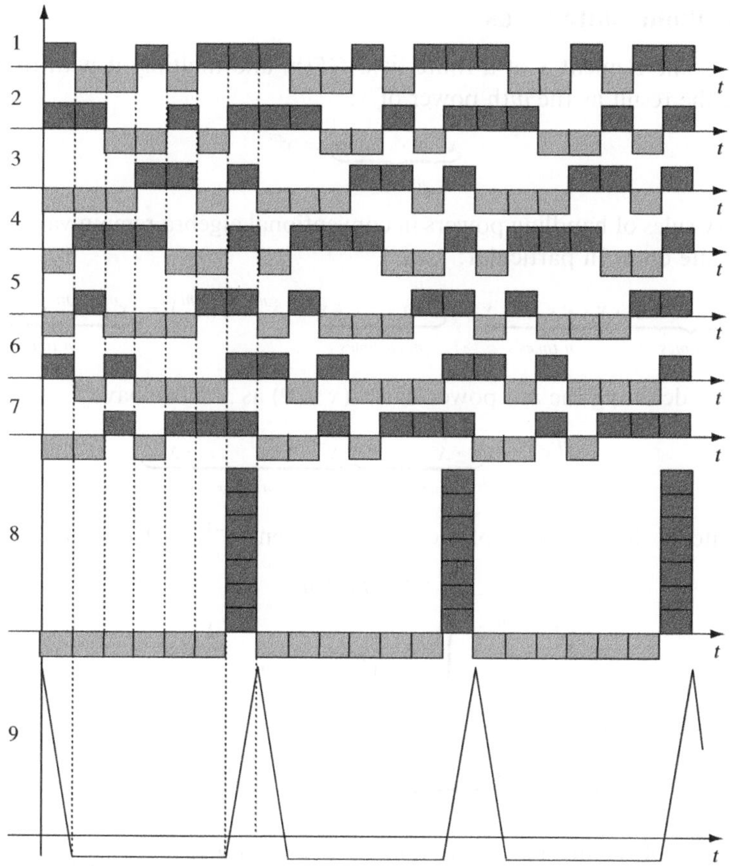

Figure 6.15 Matched filtering of periodic binary m-sequence of length $N = 7$

Binary m-sequences are among the most popular discrete signals in modern information technology due to their optimal periodic correlation properties and very simple generating and processing circuitry. Probably one of the most demonstrative examples of their practical involvement is 2G cdmaOne (IS-95) mobile phone, where m-sequences of various lengths are used as pilot signals for initial synchronization, base station signal multiplexing and data scrambling.

In addition, m-sequences represent the basis for deriving other important signal families (Kasami, Gold and others; see Chapter 7).

At the same time, the set of lengths $N = 2^n - 1 = 3, 7, 15, 31, 63, 127, 255, 511, 1023, \ldots$ where these sequences exist, is rather sparse, which sometimes may appear technologically obstructive. This is a reason for studying one more interesting class of binary minimax sequences, but before attending to that some additional insight into finite fields is required.

6.8 More about finite fields

Let us take some element x of a finite field $GF(p)$ and multiply it with itself m times, designating the result as the mth power of x:

$$\underbrace{x \cdot x \cdot \ldots \cdot x}_{m\ times} = x^m$$

The ordinary rules of handling powers in conventional algebra remain valid in any field, including finite ones. In particular:

$$x^m x^n = \underbrace{x \cdot x \cdot \ldots \cdot x}_{m\ times} \cdot \underbrace{x \cdot x \cdot \ldots \cdot x}_{n\ times} = \underbrace{x \cdot x \cdot \ldots \cdot x}_{m+n\ times} = x^{m+n}, (x^m)^n = \underbrace{x^m \cdot x^m \cdot \ldots \cdot x^m}_{n\ times} = x^{mn}$$

Furthermore, denoting the nth power of $x^{-1} (x \neq 0)$ as x^{-n}, we have:

$$x^m x^{-n} = \underbrace{x \cdot x \cdot \ldots \cdot x}_{m\ times} \cdot \underbrace{x^{-1} \cdot x^{-1} \cdot \ldots \cdot x^{-1}}_{n\ times}$$

Using repeatedly the definition of the inverse element $x^{-1}x = 1$ leads to:

$$x^m x^{-n} = \begin{cases} x^{m-n}, m \geq n \\ \\ (x^{-1})^{n-m}, m < n \end{cases} = x^{m-n}$$

In particular, the equality:

$$\underbrace{x \cdot x \cdot \ldots \cdot x}_{m\ times} \cdot \underbrace{x^{-1} \cdot x^{-1} \cdot \ldots x^{-1}}_{m\ times} = 1 = x^m x^{-m} = x^{m-m}$$

and uniqueness of the inverse of any non-zero element give:

$$x^0 = 1 \text{ and } (x^m)^{-1} = x^{-m}$$

Example 6.8.1. In the field $GF(5)$ (see tables of Figure 6.11):

$2^0 = 1, 2^1 = 2, 2^2 = 2 \cdot 2 = 4, 2^3 = 2^2 \cdot 2 = 4 \cdot 2 = 3, 2^4 = 2^3 \cdot 3 = 3 \cdot 2 = 1$

$2^{-1} = 3, 2^{-2} = (2^{-1})^2 = 3^2 = 4, 2^{-3} = (2^{-1})^3 = 3^3 = 4 \cdot 3 = 2, 2^{-4} = (2^{-1})^4 = 3^4 = 2 \cdot 3 = 1$

Consider now successive degrees of the element $x \neq 0$ of $GF(p)$: $x^0 = 1, x^1, x^2, \ldots$. Since all terms of this series belong to $GF(p)$, i.e. finite field, they cannot all be different, and therefore equality holds $x^i = x^k \Rightarrow x^{i-k} = 1$ for some $i > k$. Suppose that the element α exists whose first $p-1$ powers $\alpha^0 = 1, \alpha^1, \alpha^2, \ldots, \alpha^{p-2}$ are all different. Since $p-1$ is just the number of non-zero elements of $GF(p)$, the powers above are exactly *all* non-zero elements of $GF(p)$. Therefore, the element α, if it really exists, allows constructing the whole field $GF(p)$ but the zero element by just raising α to powers $0, 1, \ldots, p-2$. Such an element is called a *primitive* one.

One of the most important facts about finite fields is that they all contain a primitive element. Proof of this result may be found in many algebraic or coding theory textbooks (e.g. [30,32,33]). A primitive element is not unique: in any finite field whose order exceeds 3, more than one primitive element is present. For instance, as is seen from Example 6.8.1, both 2 and 3 are primitive elements in $GF(5)$.

Since for a primitive element powers $\alpha^0 = 1, \alpha^1, \alpha^2, \ldots, \alpha^{p-2}$ exhaust all non-zero elements of $GF(p)$, α^{p-1} should be equal to one of them. Actually it cannot be equal to anything but 1, because $\alpha^{p-1} = \alpha^l$ with $0 < l \leq p-2$ means that $\alpha^{p-1-l} = 1$. This is not possible, since $1 < p-1-l < p-1$ and among elements $\alpha^1, \alpha^2, \ldots, \alpha^{p-2}$ none may be equal to 1. Hence, $\alpha^{p-1} = 1$. Now it is easy to see that the same is true for any non-zero element of a finite field, not only for a primitive one. Indeed, every non-zero element x of $GF(p)$ is the lth power of a primitive element for a proper integer l: $x = \alpha^l$, so that (small Fermat theorem):

$$x^{p-1} = \left(\alpha^l\right)^{p-1} = \alpha^{l(p-1)} = \left(\alpha^{p-1}\right)^l = 1 \tag{6.17}$$

The next entity bears quite a natural name, fully consistent with the categories of ordinary algebra. The integer exponent l, which after raising α to it produces $x = \alpha^l$, is the *logarithm of x to the base α* with a conventional designation $\log_\alpha x$. Therefore, $\alpha^{\log_\alpha x} = x$.

Now consider only the prime fields of an odd order ($p > 2$) and introduce a new notion of the *binary character* $\psi(x)$ of a non-zero element x defined as follows:

$$\psi(x) = \begin{cases} 1, & \log_\alpha x = 0 \bmod 2 \\ -1, & \log_\alpha x \neq 0 \bmod 2 \end{cases} = (-1)^{\log_\alpha x} \tag{6.18}$$

Clearly, the binary character is simply a mapping of the finite field $GF(p)$ onto a pair of real numbers $\{+1, -1\}$, transforming non-zero element x into $+1$ if its logarithm is even and into -1 otherwise. Note that this mapping does not depend on a specific choice of a primitive element (Problem 6.24). The following properties of a binary character will be used further:

1. The character of the unit element of $GF(p)$ is always one:

$$\psi(1) = 1. \tag{6.19}$$

   This is true because $\alpha^0 = 1 \Rightarrow \log_\alpha 1 = 0$.

2. The character is a multiplicative function, i.e. the character of a product of two non-zero elements is a product of their characters. Indeed, from (6.17) and (6.18):

$$\psi(xy) = (-1)^{\log_\alpha(xy)} = (-1)^{\log_\alpha x + \log_\alpha y} = (-1)^{\log_\alpha x} \cdot (-1)^{\log_\alpha y} = \psi(x)\psi(y) \tag{6.20}$$

3. Balance property: the sum of characters of all non-zero elements of $GF(p)$ is zero:

$$\sum_{x=1}^{p-1} \psi(x) = \sum_{x=1}^{p-1} (-1)^{\log_\alpha x} = 0. \tag{6.21}$$

   To prove this equality note that when x assumes all $p-1$ non-zero values, $\log_\alpha x$ runs in some order over the range of $p-1$ integers $0, 1, \ldots, p-2$. Due to the oddness of p, the number of integers in this range is even, so that $(p-1)/2$ of them are even and $(p-1)/2$ are odd. Therefore, the sum above contains an equal number of plus and minus ones and comes to zero.

4. Consider the character of the element -1, i.e. $\psi(-1)$. Since all elements $\alpha^i, i = 0, 1, \ldots, p-2$ are different, only two of them satisfy the equation $x^2 = 1$, namely 1 and -1. Since the same equation is met for $\alpha^{\frac{p-1}{2}}$, the last element cannot be anything but -1. This entails equalities $\log_\alpha(-1) = \frac{p-1}{2}$ and:

$$\psi(-1) = (-1)^{\frac{p-1}{2}} = \begin{cases} 1, & p = 1 \bmod 4 \\ -1, & p = 3 \bmod 4 \end{cases} \tag{6.22}$$

Example 6.8.2. Continuing Example 6.8.1, note that in $GF(5)$ $\psi(2) = \psi(3) = -1$ and $\psi(1) = \psi(4) = +1$. Therefore, $\psi(2 \cdot 4) = \psi(3) = -1 = \psi(2)\psi(4) = (-1) \cdot (+1)$ in accordance with the multiplicative property and $\psi(1) + \psi(2) + \psi(3) + \psi(4) = 0$ in line with the balance property. Also $\psi(-1) = \psi(4) = 1$, which is in agreement with (6.22), since $5 = 4 \cdot 1 + 1$.

Another name for a binary character especially popular in number theory is Legendre symbol, which explains the name of the sequences studied in the next section.

6.9 Legendre sequences

Let us form a binary sequence of an odd prime length $N = p$ identifying the position number i of its element $a_i = \pm 1$ with an element of the prime field $GF(p)$. Then for every $i \in \{1, 2, \ldots, N-1\}$ the character $\psi(i)$ is determined and the Legendre sequence is just the sequence of characters of numbers i except zero i, for which the sequence element is forced to $+1$.[4] For a periodic version of the Legendre sequence the generation rule is as follows:

$$a_i = \begin{cases} +1, & i = 0 \bmod N \\ \psi(i), & i \neq 0 \bmod N \end{cases} \tag{6.23}$$

The periodicity of the sequence (6.23) with period N stems from the treatment of numbers i in $\psi(i)$ as elements of $GF(p)$, where addition obeys modulo p arithmetic, resulting in $\psi(i+N) = \psi(i+p) = \psi(i)$.

To investigate the periodic ACF of a Legendre sequence, substitute (6.23) into (6.7) and separate from the sum terms containing a_0:

$$\begin{aligned} R_p(m) &= \sum_{i=0}^{N-1} a_i a_{i-m} = a_0 a_{-m} + a_m a_0 + \sum_{\substack{i=1 \\ i \neq m}}^{N-1} a_i a_{i-m} \\ &= \psi(-m) + \psi(m) + \sum_{\substack{i=1 \\ i \neq m}}^{p-1} \psi(i)\psi(i-m) \end{aligned} \tag{6.24}$$

[4] Another option leading to the same final result is forcing this element to -1.

Certainly, we are interested in estimating only the sidelobes, i.e. shifts m, which are not multiples of $N=p$. Exploiting the multiplication property (6.20) gives $\psi(-m)=\psi(-1)\psi(m)$ and $\psi(i-m)=\psi[i(1-mi^{-1})]=\psi(i)\psi(1-mi^{-1})$, where the inverse always makes sense, since zero i is removed from the sum in (6.24). As a result:

$$R_p(m)=\psi(m)[1+\psi(-1)]+\sum_{\substack{i=1\\ i\neq m}}^{p-1}\psi^2(i)\psi(1-mi^{-1}) \tag{6.25}$$

Turn now to (6.22). For any length $N=p=1\,\text{mod}4$ $\psi(-1)=1$ and the first term in the expression above equals $2\psi(m)=\pm 2$ for any $m\neq 0\,\text{mod}N$. On the other hand, for lengths of the type $N=3\,\text{mod}4$ $\psi(-1)=-1$ and the same term of (6.25) vanishes. To cope with the second term in (6.25), note first that for any non-zero i from $GF(p)$ $\psi^2(i)=1$. Second, if i ran over all non-zero elements of $GF(p)$ then i^{-1} as well as $-mi^{-1}$ $(m\neq 0\,\text{mod}p)$ would both run over the same range in some other orders. Therefore, $1-mi^{-1}$ would run over $p-1$ elements of the field including zero but excluding 1, since $-mi^{-1}$ cannot take zero value. Actually, however, the zero element should also be eliminated from the possible values of $1-mi^{-1}$, since i in the second term of (6.24) does not take on the value $i=m$, corresponding to $1-mi^{-1}=0$, and the whole range of values $1-mi^{-1}$ is from 2 to $p-1$. Summarizing all these reasons:

$$\sum_{\substack{i=1\\ i\neq m}}^{p-1}\psi^2(i)\psi(1-mi^{-1})=\sum_{x=2}^{p-1}\psi(x)=\sum_{x=1}^{p-1}\psi(x)-\psi(1)=-1$$

where the final step follows from the character properties (6.19) and (6.21). Now the periodic ACF of a Legendre sequence appears to be one of two types (h below is natural):

1. If length is of the form $N=4h+1$ (i.e. $N=1\,\text{mod}4$), then:

$$R_p(m)=\begin{cases}N, & m=0\,\text{mod}N\\ -3 \text{ or } +1, & m\neq 0\,\text{mod}N\end{cases} \tag{6.26}$$

2. If length is of the form $N=4h+3$ (i.e. $N=3\,\text{mod}4$), then:

$$R_p(m)=\begin{cases}N, & m=0\,\text{mod}N\\ -1, & m\neq 0\,\text{mod}N\end{cases} \tag{6.27}$$

The last result, reproducing (6.12), shows that Legendre sequences of lengths $N=4h+3$ are minimax, i.e. possess optimal periodic correlation properties possible for binary sequences of odd lengths.

Example 6.9.1. Length $N=7$ falls within the set $N=4h+3$. The element 3 is primitive in $GF(7)$, since raising it to the powers 0, 1, ..., 5 gives all different non-zero elements: $3^0=1, 3^1=3, 3^2=2, 3^3=6, 3^4=4, 3^5=5$. As is directly seen from this series, logarithms of 1, 2 and 4 are even, while those of 3, 5 and 6 are odd. Hence, $\psi(1)=\psi(2)=\psi(4)=1$, and $\psi(3)=\psi(5)=\psi(6)=-1$. Now, according to (6.23), placing plus ones in the positions $i=0,1,2,4$ and minus ones in the positions $i=3,5,6$ gives the Legendre sequence of length 7. The computation illustrated by Table 6.4 confirms the optimality of the correlation properties of the binary sequence obtained.

Table 6.4 Calculating periodic ACF of the Legendre sequence {+ + + − + − −}

m	a_0	a_1	a_2	a_3	a_4	a_5	a_6	$R_p(m)$
0	+	+	+	−	+	−	−	+7
1	−	+	+	+	−	+	−	−1
2	−	−	+	+	+	−	+	−1
3	+	−	−	+	+	+	−	−1

Legendre sequences form a very powerful class of binary codes with minimax periodic ACF. The condition of their existence (any prime length of the form $N = 4h + 3$) is significantly looser than that of m-sequences ($N = 2^n - 1$), which means that a lot more Legendre sequences are available as compared to m-sequences. For example, within the range 50–1500 binary m-sequences exist of only 5 lengths, while the number of Legendre sequences is 114.

6.10 Binary codes with good aperiodic ACF: revisited

After collecting necessary knowledge on binary sequences with good periodic ACF, we may return to the idea formulated in Section 6.4 and consisting in utilizing these sequences as a starting point for finding codes with attractive aperiodic ACF. Consider some sequence $a_0, a_1, \ldots, a_{N-1}$ of length N. Any of its cyclic shifts $a_s, a_{s+1}, \ldots, a_{N-1}, a_0, \ldots, a_{s-1}$, where $0 \leq s \leq N - 1$, has the same periodic ACF as the original code, since the periodic ACF is invariant to a cyclic shift (see Problem 5.5). Yet the aperiodic ACF of the cyclically shifted replica may differ from the initial one. Along with the bound (6.5), this fact sets up the basis for a popular algorithm for searching codes with an acceptable aperiodic ACF described below.

At the first step a set of candidate sequences with good periodic ACF, length N pre-assigned, is somehow collected. It may include all known sequences [34–37] of a given N, whose periodic ACF sidelobes—consistent with (6.5)—allow one to hope for a low level of $\rho_{a,\max}$, or be limited according to the designer's technological preferences. For example, if a binary code of length $N = 63$ is sought, the initial set may be limited to only all m-sequences of this length (no Legendre sequence of this length exists, since N is not prime), or include more sequences with promising periodic ACF; with necessary length $N = 127$ it may cover all m-sequences along with Legendre sequences[5] or, again, contain other sequences with sufficiently low periodic sidelobes.

[5] A score of different primitive polynomials of the same degree may exist, each generating a specific m-sequence of the same length. Therefore, a quantity of m-sequences of the fixed length exists and all of them are appropriate for the search for good aperiodic codes. Unlike this, there are only two possible Legendre sequences of the same length differing by the first symbol (+1 in one of them and −1 in the other, see footnote 4 on p. 172).

At the second step the exhaustive search is performed over all one-period segments of candidate sequences by the criterion of the least maximal aperiodic ACF sidelobe. Specifically, one period segment of the first candidate sequence is taken, its aperiodic ACF is computed and its maximal sidelobe is stored in memory along with the numbers of the candidate sequence and its shift. Then the segment is cyclically shifted by one position and all the calculations are repeated. If an updated maximal aperiodic sidelobe is lower than the previous one, its value and new shift number replace the previous data in memory; otherwise registered data are kept unchanged. These iterations are repeated N times, i.e. for all cyclic shifts of the first candidate sequence, after which the next candidate sequence is tried in the same way, etc. The outcome of the search is the sequence with minimal value of $\rho_{a,\max}$ among all sequences picked at the first step. Of course, there is no guarantee that the result will be the best possible among all binary sequences of a given length.

This procedure, which was first proposed in the early 1960s, has been used by many authors, gradually covering wider and wider sets of candidate binary sequences. One of the most detailed lists of binary codes synthesized in this way may be found in [34].

Example 6.10.1. Length $N = 2^3 - 1 = 7$ meets the condition of existence of *m*-sequences. There are two primitive binary polynomials of degree 3 $f(x) = x^3 + x + 1$ and $f(x) = x^3 + x^2 + 1$. An immediate test shows that *m*-sequences generated by them are just mirror images of each other, i.e. one of them is the other read from left to right. Such a transform does not affect either periodic or aperiodic ACF (Problem 5.5). Therefore it is enough to include in the candidate set only a single *m*-sequence: that of Example 6.7.1: $-1, +1, +1, -1, +1, -1, -1$. In addition, $N = 7$ is prime of the sort $N = 4h + 3$, i.e. minimax Legendre sequences of this length exist, too: the one of Example 6.9.1 $(+1, +1, +1, -1, +1, -1, -1)$ and its replica with the first element changed to -1. The latter completely repeats the *m*-sequence selected, while the former—after changing the signs of all elements—coincides with a cyclically shifted skipped *m*-sequence. Since the polarity change again does not affect either periodic or aperiodic ACF (Problem 5.5), only one minimax sequence out of the four analysed is sufficient to enter the candidate set. Let it be the Legendre one starting with $+1$. Calculating its aperiodic ACF gives the following values of $R_a(m)$, $m = 1, 2, \ldots, 6$: $0, +1, 0, -1, -2, -1$, and $\rho_{a,\max} = 2/7$. After one cyclic left shifting the sequence becomes $+1, +1, -1, +1, -1, -1, +1$. For this $R_a(4) = -3$, $\rho_{a,\max} = 3/7$, i.e. the maximal aperiodic sidelobe is worse than for the original one. The next cyclic shift is $+1, -1, +1, -1, -1, +1, +1$ with $R_a(1) = -2$, $\rho_{a,\max} = 2/7$, i.e. no better than for the initial sequence. At the next shift we come to the sequence $-1, +1, -1, -1, +1, +1, +1$, having aperiodic ACF with sidelobes $R_a(m) = 0, -1$; $m \neq 0$, i.e. $\rho_{a,\max} = 1/7$. This sequence is globally optimal among all PSK codes, since no such code can have smaller maximal aperiodic sidelobe (see (6.4)). Actually, the Barker code of length 7 is found which is a mirror replica of that of Table 6.1.

Example 6.10.2. Assume $N = 257 = 64 \times 4 + 1$. Since the number 257 is prime, two Legendre sequences of this length exist differing in only their first symbol. However, N is of the form $4h + 1$, and hence their PACF has maximal sidelobe $\rho_{p,\max} = 3/N = 3/257$ (see (6.26)).

Nevertheless they remain promising from the point of view of $\rho_{a,\max}$, since the lower border (6.4) gives $\rho_{a,\max} \geq 1.5/257$. Applying to them the above procedure results in the sequence with maximal non-normalized aperiodic sidelobe equal to 12, i.e. $\rho_{a,\max} = 12/257$ or -26.6 dB (compare this with the longest binary Barker code, for which $\rho_{a,\max} = 1/13$ or -22.3 dB). The sequence obtained after eliminating the last symbol turns into the code of length $N = 256$ with the same maximal non-normalized sidelobe and $\rho_{a,\max} = 12/256 = 3/64$, i.e. again approximately -26.6 dB. Its aperiodic ACF is shown in Figure 6.16a. Interestingly, in the 3G mobile UMTS standard the primary synchronization code is a binary sequence of this very length, $N = 256$, having aperiodic sidelobes up to 1/4 (Figure 6.16b), i.e. much higher as compared to the sequence just found. On the other hand, the choice of a code for a cell search in UMTS was subject to many other requirements, including implementation issues which might have overpowered the criterion of good autocorrelation.

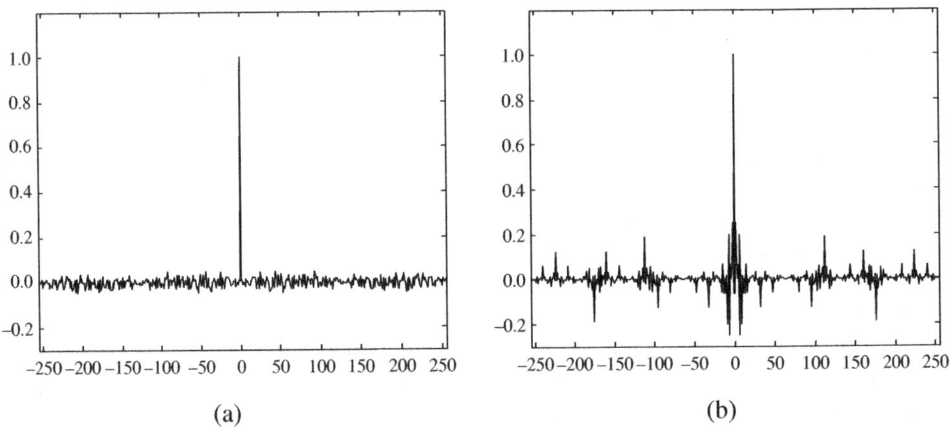

Figure 6.16 Aperiodic ACF of two binary codes of length 256: the code of Example 6.10.2 (a) and primary synchronization code of UMTS (b)

Figure 6.17 presents one more illustration of the optimization of binary codes in the maximal aperiodic ACF sidelobe, showing the dependence of $\rho_{a,\max}$ on length N for presumably the best binary sequences taken from [25–27,34]. The dashed line shows the curve $\rho_{a,\max} \approx 0.77/\sqrt{N}$ approximating the dependence $\rho_{a,\max} = f(N)$ as $a/\sqrt{N}$ with a fitted by the least-squares method. As is seen, the accuracy of this approximation is rather good, especially for $N > 100$.

6.11 Sequences with perfect periodic ACF

As has been indicated time and again, numerous applications exist where the periodicity of the signals makes their periodic correlation properties primarily important. In other words, good periodic ACFs are not only a powerful intermediate tool to design good aperiodic sequences but very valuable in themselves. Examples of this sort include

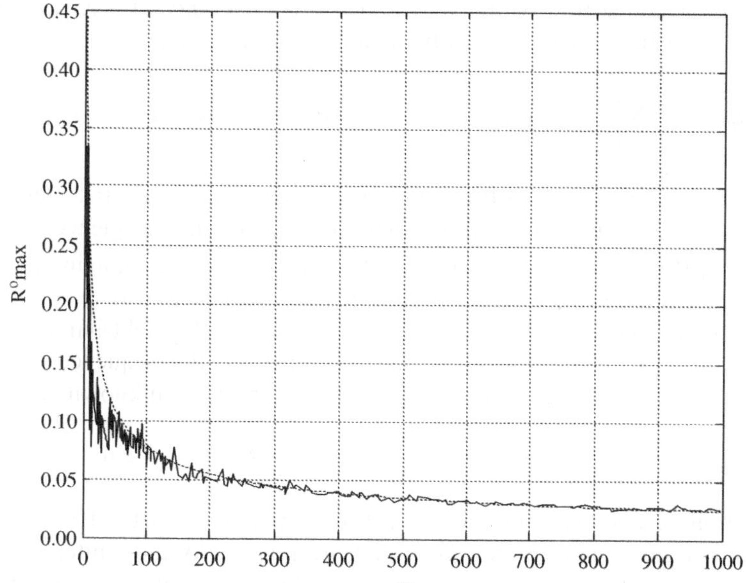

Figure 6.17 Dependence of the minimized maximal aperiodic sidelobe on length

continuous wave ranging systems, in particular in remote space, pilot or synchronization channels for digital data transmission systems (downlink pilot channels of cdmaOne and cdma2000, secondary synchronization channel of UMTS), CW radar and sonar systems etc.

Despite binary $\{\pm 1\}$ minimax sequences look rather practicable, having maximal periodic sidelobe $\rho_{p,\max} = 1/N$ dropping with length, situations are still likely where an acceptable value of $\rho_{p,\max}$ requires an infeasibly long length N. For example, for radar, ranging or sonar systems time-resolution of signals in the dynamic range over 80 dB is not an unusual demand. To meet it with optimal binary sequences, lengths exceeding 10^4 are necessary, which may unreasonably slow down the initial searching procedure (see Section 8.2). Certainly, perfect periodic ACF (6.6) would be the best option for many such scenarios. However, it cannot be realized among binary codes, which are, clearly, the most attractive technologically. In the rest of this chapter we will inspect possible ways of getting perfect periodic ACF when the sequence alphabet is not rigorously limited to just binary symbols $\{\pm 1\}$.

6.11.1 Binary non-antipodal sequences

Replacing the antipodal alphabet $\{+1, -1\}$ by some binary non-antipodal one, it proves to be possible to turn all periodic sidelobes of any binary minimax sequence meeting (6.12) into zero. The simplest way to derive an appropriate alphabet is by adding a constant c (complex in the general case) to the original $\{+1, -1\}$ sequence

$a_0, a_1, \ldots, a_{N-1}$, converting symbols $+1$ and -1 to $1+c$ and $-1+c$, respectively. The periodic ACF of the sequence thus obtained is found directly:

$$R_p(m) = \sum_{i=0}^{N-1} (a_i + c)(a_{i-m} + c^*) = \sum_{i=0}^{N-1} a_i a_{i-m} + 2\mathrm{Re}(c\tilde{a}_0) + N|c|^2 \qquad (6.28)$$

where $\tilde{a}_0 = \sum_{i=0}^{N-1} a_i$ is, as before, a constant component of the initial sequence $a_0, a_1, \ldots, a_{N-1}$. Equation (6.8) shows that for any minimax sequence meeting (6.12) $|\tilde{a}_0|^2 = \sum_{m=0}^{N-1} R_p(m) = N + (N-1)(-1) = 1 \Rightarrow \tilde{a}_0 = \pm 1$. Since changing the signs of all elements does not affect ACF, we may consider only sequences with $\tilde{a}_0 = -1$. Again, for any minimax sequence meeting (6.12), the first sum in the right-hand part of (6.28) equals -1 at any $m \neq 0 \bmod N$. Setting sidelobes of the sequence $a_i + c, i = \ldots, -1, 0, 1, \ldots$ equal to zero leads to the equation in a complex unknown c:

$$|c|^2 + \frac{2}{N}\mathrm{Re}(c\tilde{a}_0) - \frac{1}{N} = |c|^2 - \frac{2}{N}\mathrm{Re}(c) - \frac{1}{N} = 0 \qquad (6.29)$$

This equation in two real unknowns (real and imaginary parts of c) has an infinite number of solutions. Let us find those that are potentially most interesting. If a real alphabet is desired $\mathrm{Re}(c) = c$ and $|c|^2 = c^2$ so (6.29) is a quadratic equation:

$$c^2 - \frac{2}{N}c - \frac{1}{N} = 0$$

with roots $c_{1,2} = \frac{1 \pm \sqrt{N+1}}{N}$. New binary non-antipodal symbols $1+c$ and $-1+c$ may now be divided by $1+c$ to retain $+1$ as one of the symbols in the new alphabet. After this we come to the rule of converting a binary minimax sequence with periodic ACF (6.12) into one with perfect ACF: elements -1 should be changed to:

$$\frac{-N + 1 \pm \sqrt{N+1}}{N + 1 \pm \sqrt{N+1}} = -1 \pm \frac{2}{\sqrt{N+1}}$$

Elements $+1$ remain unchanged.

Example 6.11.1. The m-sequence or Legendre sequence of length $N = 127$ is transformed into a sequence with perfect periodic ACF by replacing all elements -1 by $-1 \pm \frac{1}{4\sqrt{2}}$.

The solution above produces an alphabet with two opposite symbols of unequal magnitude, i.e. results in amplitude modulation (Figure 6.18a). Another possible option is a PSK non-antipodal alphabet. To come to it take a 'pure' imaginary $c = jc_1$. Then (6.29) has the solution $c_1 = \pm j/\sqrt{N}$ and new symbols $1 \pm j/\sqrt{N}$ and $-1 \pm j/\sqrt{N}$ after dividing by $1 \pm j/\sqrt{N}$ become 1 and:

$$\frac{-\sqrt{N} \pm j}{\sqrt{N} \pm j} = -\frac{N-1}{N+1} \pm \frac{2j\sqrt{N}}{N+1} = -\exp(j\Phi)$$

where $\cos\Phi = (N-1)/(N+1)$ (Figure 6.18b).

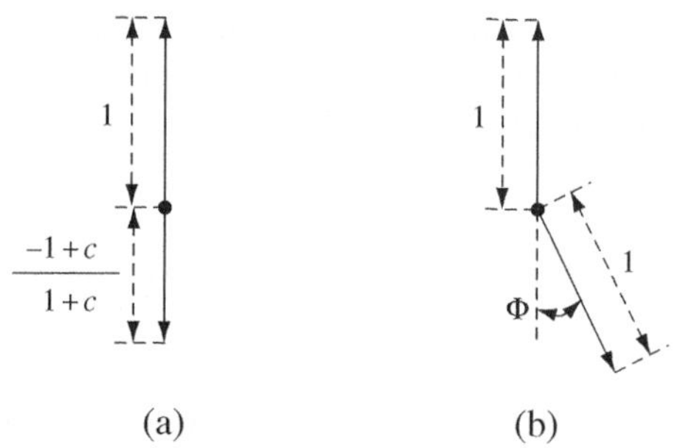

Figure 6.18 Non-antipodal binary alphabets

Example 6.11.2. For $N = 127 \cos\Phi = 63/64$ and $\Phi = \pm \arccos(63/64) \approx \pm 10°8'30''$. Changing all negative elements of the binary *m*-sequence or Legendre sequence of length $N = 127$ to $-\exp(j\Phi)$ produces a sequence with perfect periodic ACF.

The alphabet transformation just discussed, which has been proposed and reopened repeatedly [38,39], can hardly be recognized as very effective practically. As is seen and confirmed by examples, it prescribes rather exotic values of code complex amplitudes, the setting and holding of which with adequate precision may appear technologically infeasible.

6.11.2 Polyphase codes

Involving non-binary PSK modulation with $M > 2$ opens the way to numerous polyphase sequences with perfect periodic ACF. There are various rules for their construction, but more or less all of them originate in two of the most popular algorithms. The first, corresponding to Chu (or quadratic residue) codes, is very straightforward and approximates in a discrete form the law of linear frequency modulation (cf. Section 6.2). The Chu code exists for an arbitrary length N and is generated as:

$$a_i = \begin{cases} \exp\left(\dfrac{j\pi i^2}{N}\right), N \text{ even} \\ \\ \exp\left(\dfrac{j2\pi i^2}{N}\right), N \text{ odd} \end{cases} \tag{6.30}$$

where $i = \ldots, -1, 0, 1, \ldots$.

It is easy to check that $a_i = a_{i+N}$ for all i and therefore, N is at least a multiple of the code period. Calculation of periodic ACF will in passing eventually clarify the issue of a period. For the code of even length non-normalized periodic ACF:

$$R_p(m) = \sum_{i=0}^{N-1} a_i a_{i-m}^* = \exp\left(-\frac{j\pi m^2}{N}\right) \sum_{i=0}^{N-1} \exp\left(\frac{j2\pi im}{N}\right)$$

When $m = 0 \bmod N$ the last sum equals N, while the coefficient in front of it turns into 1. For any other m $\exp(j2\pi im/N)$ depends on i, and the sum above is a sum of the roots of unity of some degree, or, equivalently, a geometric series with the common ratio $\exp(j2\pi m/N)$. Summation of the series gives:

$$R_p(m) = \exp\left(-\frac{j\pi m^2}{N}\right)\frac{1-\exp(j2\pi m)}{1-\exp(j2\pi m/N)}$$

The denominator of the last fraction never turns into zero unless $m = 0 \bmod N$, and therefore $R_p(m) = 0$ at all shifts but multiples of N. Hence, the Chu code defined by the first row in (6.30) has period N and perfect periodic ACF. The solution for an odd N is carried out similarly (Problem 6.29).

Despite Chu codes making a rather convincing academic example of PSK sequences with perfect periodic ACF, their practical feasibility is pretty doubtful, since the size of the phase alphabet grows linearly with length and distances between adjacent phases becomes very small. Because of that excessive demands arise towards the precision of forming code symbols, the fineness of representing a phase, susceptibility to environmental conditions, etc.

The same shortcomings are characteristic (to a slightly lower extent, though) of the second popular family of polyphase sequences: Frank codes. They also realize step-approximation of the linear frequency modulation, but much more roughly, and exist only for lengths that are squares of integers $N = h^2 = 4, 9, 16, 25, 36, 49, \ldots$. Their generation rule is:

$$a_i = \exp\left(\frac{j2\pi i}{h}\left\lfloor\frac{i}{h}\right\rfloor\right), i = \ldots, -1, 0, 1, \ldots \qquad (6.31)$$

where, as usual, $\lfloor x \rfloor$ stands for rounding non-negative x towards zero.

Proof of perfection of periodic correlation properties of Frank codes differs from that above only in minor details and is left for Problem 6.30. As is seen from a comparison of (6.31) and (6.30) the phase step of Frank codes is reduced $\sqrt{N}$ times, so the alphabet size grows with N markedly slower.

Example 6.11.3. Take $N = 4 \Rightarrow h = 2$. Then with reduction phases to the interval $[0, 2\pi)\,\frac{2\pi i}{h}\left\lfloor\frac{i}{h}\right\rfloor = \pi i\left\lfloor\frac{i}{2}\right\rfloor = 0, 0, 0, \pi, i = 0, 1, 2, 3$, and Frank code $+1, +1, +1, -1$ is a unique binary code with perfect ACF.

Example 6.11.4. If $N = 16, h = 4$ and the phase alphabet consists of 4 symbols $\{\pm 1, \pm j\}$, and hence the Frank code of this length exploits QPSK. Since $\frac{2\pi i}{h}\left\lfloor\frac{i}{h}\right\rfloor = \frac{\pi i}{2}\left\lfloor\frac{i}{4}\right\rfloor = 0, 0, 0, 0, 0, \frac{\pi}{2}, \pi, \frac{3\pi}{2}, 0, \pi, 0, \pi, 0, \frac{3\pi}{2}, \pi, \frac{\pi}{2}, i = 0, 1, \ldots, 15$, the code is $+1, +1, +1, +1, +1, +j, -1, -j, +1, -1, +1, -1, +1, -j, -1, +j$. Perfection of its periodic ACF may be tested by a direct computation.

Finishing with polyphase codes, note once more that they are not as attractive technologically as binary antipodal ones. Do codes exist which do not yield to the binary ones in implementation simplicity, but—unlike them—possess perfect periodic ACF? Answering this question is the subject of the next section.

6.11.3 Ternary sequences

Consider sequences whose elements a_i may assume in addition to binary values ± 1 also zero value. In other words, the alphabet is now ternary $\{-1, 0, 1\}$, which technically means combining BPSK with pauses, i.e. time intervals where chips are missing. Clearly, an extension of the binary alphabet $\{\pm 1\}$ into the ternary one $\{-1, 0, 1\}$ does not seriously complicate either the generation or processing circuitry, but, as is shown below, opens the way to obtaining sequences with perfect periodic correlation properties. Remember that one of the main reasons for an interest in spread spectrum in time measuring and resolution is a desire to achieve high performance operating at low peak power, i.e. with signal energy spanning a large time interval. Quite a natural measure of efficiency of energy time-spreading is the peak-factor ν (see Section 2.7.1), i.e. the ratio between peak and time-averaged power. For any PSK, in particular binary, sequence signal energy is uniformly spread over the period so that peak and average powers are the same and $\nu = 1$. Inserting N_p pauses on the sequence period N, as occurs with the ternary alphabet, will make the energy spreading less uniform and increase the peak-factor as $\nu = N/(N - N_p)$. Hence, our interest is to design ternary sequences possessing not only perfect periodic ACF, but also small number of zeros N_p on the period, i.e. peak-factor not much higher than one. Without this constraint the problem would prove degenerated and have a trivial solution: code with only one non-zero symbol on the period N, corresponding to a single chip repeated with period $N\Delta$, certainly has perfect periodic ACF.

A number of rules to generate ternary sequences with the properties just stated are now known. The most powerful of them produces sequences with lengths and values of peak-factor obeying the following equations:

$$N = \frac{q^n - 1}{q - 1}, \quad \nu = \frac{q^n - 1}{q^n - q^{n-1}} < \frac{q}{q - 1} \tag{6.32}$$

where $q = p^w$ is a natural power of a prime p and n is odd. Sequences of this sort exist for any combinations of q, n within stipulated limitations, and therefore, a peak-factor as close to 1 as is wished is achievable by just taking q large enough.

The constructions of the ternary sequences meeting (6.32) are based on some fine features of Galois fields. The simplest of them and at the same time covering the majority of lengths indicated by (6.32) corresponds to the case of odd p $(q = p^w, p > 2)$ [40,41]. In order to present the idea in the most transparent form, let us give a detailed description of the algorithm only for the case $q = p$, i.e. $w = 1$. The easiest way to do it involves p-ary m-sequences.

Let $d_i, i = \ldots, -1, 0, 1, \ldots$, be a p-ary m-sequence, where p is an odd prime. Each of its elements belongs to a prime field $GF(p)$. Let us transform this sequence into a ternary one, mapping its zero elements into a real zero and non-zero elements into their binary

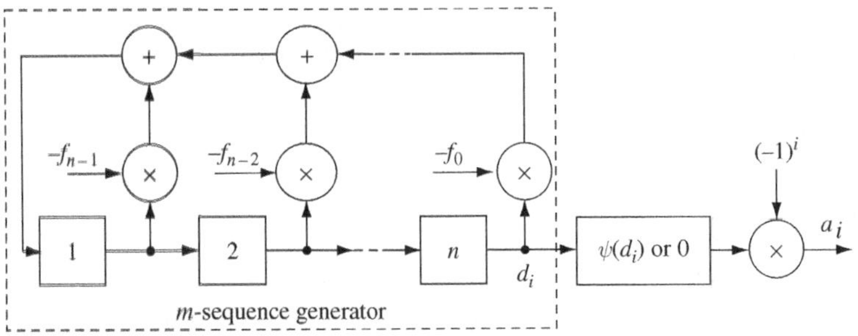

Figure 6.19 Generator of a ternary sequence (6.33)

characters. After this let us change the signs of all elements at the odd positions. The algorithm thus described is formally given by the equation:

$$a_i = \begin{cases} (-1)^i \psi(d_i), d_i \neq 0 \\ 0, d_i = 0 \end{cases} \tag{6.33}$$

where $i = \ldots, -1, 0, 1, \ldots$. Figure 6.19 shows the structure implementing this rule and including the m-sequence generator unit, which maps m-sequence elements into characters or zeros, and the multiplier, providing alternation of polarities.

To compute the peak-factor of a ternary sequence (6.33) it is enough to recollect that the period of m-sequence is $L = p^n - 1$, and the balance property asserts that on this period there are $L_0 = p^{n-1} - 1$ zero symbols. All of them but no others produce zeros in the ternary sequence; therefore, on the periodical segment of L elements of a ternary sequence exactly L_0 elements are zeros and the peak-factor:

$$\nu = \frac{L}{L - L_0} = \frac{p^n - 1}{p^n - p^{n-1}} < \frac{p}{p-1}$$

which coincides with what (6.32) tells us at $q = p$. To prove that a sequence (6.33) has the period obeying (6.32) and perfect periodic ACF, one more pseudorandom property of m-sequences is exploited, proof of which can be found in [42]. To formulate it denote:

$$h = \frac{L}{p-1} = \frac{p^n - 1}{p - 1}$$

and consider all pairs (d_i, d_{i-m}) of elements of p-ary m-sequence separated by m positions if i runs over one period ($i = 0, 1, \ldots, L-1$). Then (*pair property*) if m is not a multiple of h ($m \neq lh$, l being integer), among pairs (d_i, d_{i-m}) the pair $(0, 0)$ occurs $p^{n-2} - 1$ times and any other pair (x, y) of fixed $x, y \in GF(p)$ occurs p^{n-2} times. Otherwise, if $m = lh$, in the pairs (d_i, d_{i-m}) the second element is strictly determined by the first: $d_{i-m} = \alpha^l d_i$, where α is, as usual, a primitive element of the field $GF(p)$.

With understanding that a 'genuine' (i.e. unknown so far) period N of a ternary sequence (6.33) is some divisor of the original m-sequence period L, let us calculate non-normalized periodic ACF of the ternary sequence over the interval L, containing L/N periods:

$$R_p(m) = \frac{N}{L}\sum_{i=0}^{L-1} a_i a_{i-m} = (-1)^m \frac{N}{L}\sum_{\substack{i=0\\ d_i\neq 0\\ d_{i-m}\neq 0}}^{L-1} \psi(d_i)\psi(d_{i-m}) \tag{6.34}$$

where terms in the last sum, for which $d_i d_{i-m} = 0$ resulting in zero contribution, are discarded. Consider first the case when shift m is not a multiple of $h(m \neq lh)$. Then according to the pair property among all pairs (d_i, d_{i-m}) in (6.34), any pair (x, y) of non-zero fixed $x, y \in GF(p)$ occurs exactly p^{n-2} times. This allows computing (6.34) in the following way:

$$\begin{aligned} R_p(m) &= (-1)^m p^{n-2}\frac{N}{L}\sum_{x=1}^{p-1}\sum_{y=1}^{p-1}\psi(x)\psi(y) \\ &= (-1)^m p^{n-2}\frac{N}{L}\sum_{x=1}^{p-1}\psi(x)\sum_{y=1}^{p-1}\psi(y) = 0, m \neq lh \end{aligned} \tag{6.35}$$

due to the character property (6.21). Now turn to the case of shifts divisible by h $(m = lh)$. According to the pair property only pairs $(d_i, d_{i-m}) = (x, \alpha^l x), x \in GF(p)$ enter the sum in (6.34). But due to the balance property each period of a p-ary m-sequence contains exactly p^{n-1} fixed non-zero elements of $GF(p)$. Therefore, making use of the multiplicative property of characters (6.20):

$$R_p(lh) = (-1)^{lh} p^{n-1}\frac{N}{L}\sum_{x=1}^{p-1}\psi(x)\psi(\alpha^l x) = (-1)^{lh} p^{n-1}\psi(\alpha^l)\frac{N}{L}\sum_{x=1}^{p-1}\psi(x^2)$$

and

$$R_p(lh) = (-1)^{l(h+1)} p^{n-1}(p-1)\frac{N}{L}$$

since $\psi(x^2) = 1$ for any non-zero $x \in GF(p)$ and $\psi(\alpha^l) = (-1)^l$ by definition (6.18). Because n is odd, $h = \frac{p^{n-1}-1}{p-1} = p^{n-2} + p^{n-3} + \cdots + 1$ is the sum of an odd number of odd integers and consequently is odd itself. By that $l(h+1)$ is even independently of l and $R_p(lh) = p^{n-1}(p-1)\frac{N}{L}$. As is seen, $R_p(lh)$ is the same for any integer l, while from (6.35) $R_p(m) = 0$ whenever $m \neq lh$. This shows that the $R_p(m)$ as a function of m repeats itself with period h, and hence, the true period of a ternary sequence is $N = h = \frac{L}{p-1} = \frac{p^n-1}{p-1}$ in accordance with the prediction of (6.32). We thus arrive at the final result for the periodic ACF, meaning its perfection:

$$R_p(m) = \begin{cases} p^{n-1}, m = 0 \bmod N \\ 0, m \neq 0 \bmod N \end{cases}$$

with $N = \frac{p^n-1}{p-1}$.

Example 6.11.5. Set $p = 3, n = 3$ which means $N = 26/2 = 13$. To construct a ternary sequence of this period use the ternary m-sequence of Example 6.6.2: 1, 0, 0, 2, 0, 2, 1, 2, 2, 1, 0, 2, 2, 2, 0, 0, 1, 0, 1, 2, 1, 1, 2, 0, 1, 1, ... There are only two non-zero elements in $GF(3)$, of which only 2 is primitive. It is clear that $\psi(1) = 1$, $\psi(2) = -1$ and all non-zero elements of the m-sequence should be replaced as $1 \rightarrow +1, 2 \rightarrow -1$, zeros being mapped onto real zero. This produces the ternary sequence $+1,\ 0,\ 0, -1, 0, -1, +1, -1, -1, +1, 0, -1, -1, -1, 0, 0, +1, 0, +1, -1, +1, +1, -1, 0, +1, +1, \ldots$ of period 26. Changing the signs of the elements with odd numbers (starting with zero) gives the final ternary sequence $+1,\ 0,\ 0, +1,\ 0, +1, +1, +1, -1, -1, 0, +1, -1, +1, 0, 0, +1, 0, +1, +1, +1, -1, -1, 0, +1, -1, \ldots$, having period $N = 13$ and peak-factor $\nu = 13/9 \approx 1.445$. Perfection of its periodic ACF may be verified by a direct computation.

One may eliminate alternation of signs of odd-numbered elements in rule (6.33) and in the generator of Figure 6.19 using instead of m-sequences some special linear sequences of the smaller period. For that coefficients f_i in the recurrence (6.13) and feedback of the LFSR generator should belong to an appropriate non-primitive irreducible polynomial of degree n. The theory behind it may be found in [41]. Examples of such polynomials of the third degree allowing removal of alternating signs in (6.33) are given in Table 6.5 for $p \leq 31$. The last two columns contain non-maximal period L of the linear sequence generated by LFSR and period N of the final ternary sequence. One more advantage of these polynomials is that the negative of at least one of their coefficients is 1, simplifying multiplication down to just connection to the adder.

Example 6.11.6. Form the ternary sequence corresponding to $p = 3$, $n = 3$ starting with polynomial $x^3 + 2x + 2$ of Table 6.5. The recurrence (6.13) then takes the form $d_i = d_{i-2} + d_{i-3}$, generating with initial loading $d_0 = 1, d_1 = d_2 = 0$ the linear sequence over $GF(3)$ 1, 0, 0, 1, 0, 1, 1, 1, 2, 2, 0, 1, 2 of period $L = 13$. After mapping its non-zero elements onto their characters and zeros onto a real zero, the ternary sequence of period $N = 13$ is formed identical to that of the previous example.

The extension of the construction above to the case $q = p^w, p > 2, w > 1$ is rather immediate and rule (6.32) preserves its validity. The only difference is that an m-sequence $\{d_i\}$ in it is now q-ary, i.e. with elements belonging to an *extension* (as opposed to prime) finite field $GF(q)$. The arithmetic of extension fields is a bit trickier than just modulo q operations, and we do not want to dwell on those details here. The reader may consult [40,41].

Table 6.5 Non-primitive polynomials over prime fields

p	$f(x)$	L	N
3	$x^3 + 2x + 2$	13	13
5	$x^3 + 4x^2 + 4$	31	31
7	$x^3 + 6x + 5$	171	57
11	$x^3 + 10x + 7$	665	133
13	$x^3 + 12x + 9$	1098	183
17	$x^3 + 16x + 15$	2456	307
19	$x^3 + 18x + 15$	3429	381
23	$x^3 + 22x + 19$	6083	553
29	$x^3 + 28x + 28$	1742	871
31	$x^3 + 30x + 22$	14895	993

Table 6.6 Parameters of ternary sequences with perfect periodic ACF

N	p	n	q	ν	N	p	n	q	ν
13	3	3	3	1.444	$292 = 4 \times 73$	2	3	8	1.141
21	2	3	4	1.312	307	17	3	17	1.062
31	5	3	5	1.240	341	2	5	4	1.332
$52 = 4 \times 13$	3	3	3	1.444	$364 = 4 \times 91$	3	3	9	1.123
57	7	3	7	1.163	381	19	3	19	1.055
73	2	3	8	1.141	$532 = 4 \times 133$	11	3	11	1.099
$84 = 4 \times 21$	2	3	4	1.312	553	23	3	23	1.045
91	3	3	9	1.123	651	5	3	25	1.042
121	3	5	3	1.494	$732 = 4 \times 183$	13	3	13	1.083
$124 = 4 \times 31$	5	3	5	1.240	757	3	3	27	1.038
133	11	3	11	1.099	781	5	5	5	1.250
183	13	3	13	1.083	871	29	3	29	1.036
$228 = 4 \times 57$	7	3	7	1.163	993	31	3	31	1.033
273	2	3	16	1.066	1057	2	3	32	1.032

In contrast to what has been presented, construction of ternary sequences for $q = 2^w$, and proving the perfection of their periodic ACF, involves much more sophisticated mathematical concepts, such as quadrics in finite fields [43].

If any of the considered ternary sequences is multiplied symbol-wise with a unique binary sequence 1, 1, 1, -1 having perfect periodic ACF, the resulting ternary sequence will have quadrupled length with no effect on the peak factor or ACF perfection. In the same way, a symbol-wise product of two perfect ACF ternary sequences of co-prime lengths N_1, N_2 has again perfect ACF, length $N = N_1N_2$ and peak-factor $\nu = \nu_1\nu_2$, where ν_i stands for the peak-factor of the ith sequence, $i = 1, 2$.

Table 6.6 summarizes the lengths and values of the peak-factor of sequences generated as described along with parameters q, p, n for the range $N \leq 1057$. The rows where lengths are presented as products correspond to symbol-wise products of initial ternary sequences with the binary sequence 1, 1, 1, -1. In this case parameters p, n, q characterize the initial ternary sequence. As is seen, very small to negligible values of ν are characteristic of many of the listed codes, giving the designer rather solid alternatives to the best binary sequences, wherever perfect periodic ACF is desirable.

6.12 Suppression of sidelobes along the delay axis

Suppose that the designer is not inclined to abandon binary $\{\pm 1\}$ sequences and at the same time is dissatisfied with the achievable level ($\rho_{p,\max} \geq 1/N$) of their periodic ACF sidelobes. In such a situation an effective way to settle these contradictory trends is to 'imitate' the perfect periodic ACF by way of rejecting matched filtering in favour of a special mismatched processing, which suppresses sidelobes all over the signal period. Very close ideas underlie the reduction or suppression of aperiodic sidelobes [39,44,45] and combating intersymbol interference with the aid of zero-forcing equalizers [2,5,7], but in the most transparent form they are visible when applied to periodic signals [39,46,47].

6.12.1 Sidelobe suppression filter

Consider some sequence $\ldots, a_{i-1}, a_i, a_{i+1}, \ldots$ of period N, which manipulates chips of duration Δ and a finite impulse response (FIR) filter running the summation of N signal replicas delayed by $i\Delta$ and weighted by coefficients $b_i, i = 0, 1, \ldots, N-1$ as shown in Figure 6.20. In principle, what is presented below may be accomplished for sequences of arbitrary alphabet; however, it seems reasonable to restrict ourselves to only the binary $\{\pm 1\}$ alphabet, since beyond this constraint there exist many sequences with perfect periodic ACF, depriving the sidelobe suppression task of a solid motivation. Accordingly, filter coefficients $b_i, i = 0, 1, \ldots, N-1$ are all assumed real.

When fed by a sequence $a_i, i = \ldots, -1, 0, 1, \ldots$ the filter responds by the sequence $c_i, i = \ldots, -1, 0, 1, \ldots$ whose elements are found as a convolution:

$$c_i = \sum_{l=0}^{N-1} a_{i-l} b_l, \quad i = \ldots, -1, 0, 1, \ldots \tag{6.36}$$

With periodic input sequence $a_i = a_{i+N}, i = \ldots, -1, 0, 1, \ldots$ the output will also be periodic with the same period N: $c_i = c_{i+N}, i = \ldots -1, 0, 1, \ldots$. Then N elements $c_0, c_1, \ldots c_{N-1}$ specify the output sequence exhaustively, and (6.36) becomes a cyclic convolution where subtraction in the index is fulfilled modulo N.

Let us impose on the filter a requirement:

$$c_0 \neq 0, \quad c_i = 0, \quad i = 1, 2, \ldots, N-1 \tag{6.37}$$

meaning physically that the filter output signal has non-zero mainlobes repeated with the period $N\Delta$, while all the sidelobes between them are zero. This filter, the *sidelobe suppression filter* (SLSF), imitates by its response the perfect periodic ACF. Since for a binary code perfect ACF is not achievable (with a single trivial exception), the SLSF is a mismatched filter, and therefore yields in SNR to the matched filter.

The fastest way to come to an explicit expression for the filter coefficients is to use discrete Fourier transform (DFT). A sequence a_i, $i = 0, 1, \ldots, N-1$ and its DFT spectrum $\tilde{a}_k$, $k = 0, 1, \ldots, N-1$ are related to each other by the direct and inverse DFT:

$$\tilde{a}_k = \sum_{i=0}^{N-1} a_i \exp\left(-j\frac{2\pi ik}{N}\right), \quad k = 0, 1, \ldots, N-1$$

$$a_i = \frac{1}{N}\sum_{k=0}^{N-1} \tilde{a}_k \exp\left(j\frac{2\pi ik}{N}\right), \quad i = 0, 1, \ldots, N-1$$

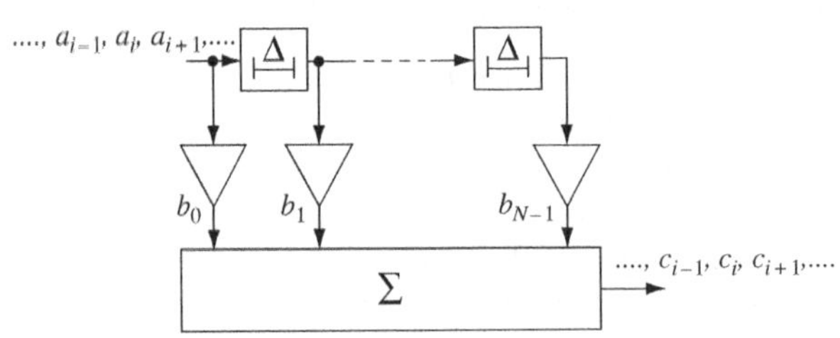

Figure 6.20 FIR filter for a sequence of length N

Our goal is to get to the discrete delta function (6.37) at the filter output having only one non-zero element per period. Its spectrum is uniform: $\tilde{c}_k = c_0, k = 0, 1, \ldots, N-1$. Then on the strength of the convolution theorem [1] the spectrum of the sequence (6.36) at the filter output, $\tilde{c}_k = \tilde{a}_k \tilde{b}_k = c_0, k = 0, 1, \ldots, N-1$, where the spectrum of the sequence of the filter coefficients $\tilde{b}_k$ is nothing but the SLSF transfer function:

$$\tilde{b}_k = \frac{c_0}{\tilde{a}_k}, \quad k = 0, 1, \ldots, N-1 \tag{6.38}$$

As is seen, the SLSF transfer function is inverse to the signal spectrum, which is why filters of this sort are often called inverse filters. An inverse filter just transforms an input spectrum to make it uniform. As the last equation shows, SLSF is physically realizable for any periodic sequence whose DFT spectrum has no zero components. Applying the inverse DFT to (6.38) gives an explicit form of SLSF coefficients:

$$b_i = \frac{c_0}{N} \sum_{k=0}^{N-1} \frac{1}{\tilde{a}_k} \exp\left(j\frac{2\pi ik}{N}\right), \quad i = 0, 1, \ldots, N-1 \tag{6.39}$$

6.12.2 *SNR loss calculation*

The matched filter coefficients (see Figure 6.20) would be (ignoring immaterial common factor) mirror-like to the input sequence:[6] $b_i = a_{N-i}, i = 0, 1, \ldots, N-1$ and output sequence peak $A_{mf} = \sum_{i=0}^{N-1} a_i^2 = N$, since the input sequence is binary. For the input noise having correlation spread within Δ and variance σ^2, the output matched filter variance $\sigma_{mf}^2 = \sigma^2 \sum_{i=0}^{N-1} b_i^2 = \sigma^2 \sum_{i=0}^{N-1} a_{N-i}^2 = N\sigma^2$. Thus, the power SNR q_{mf}^2 at the matched filter output:

$$q_{mf}^2 = \frac{A_{mf}^2}{\sigma_{mf}^2} = \frac{N}{\sigma^2} \tag{6.40}$$

In a similar manner for the SLSF output sequence peak $A_{sl} = c_0$ and noise variance:

$$\sigma_{sl}^2 = \sigma^2 \sum_{i=0}^{N-1} b_i^2$$

Due to the Parseval theorem and the time shift property of DFT, the periodic ACF of an arbitrary sequence $u_0, u_1, \ldots, u_{N-1}$ of period N is linked to the sequence energy spectrum $|\tilde{u}_0|^2, |\tilde{u}_1|^2, \ldots, |\tilde{u}_{N-1}|^2$ by the inverse DFT:

$$R_p(m) = \sum_{i=0}^{N-1} u_i u_{i-m}^* = \frac{1}{N} \sum_{k=0}^{N-1} |\tilde{u}_k|^2 \exp\left(\frac{j2\pi mk}{N}\right), \quad m = 0, 1, \ldots, N-1 \tag{6.41}$$

---

[6] A cyclic shift of coefficients against the case of the aperiodic signal ($b_i = a_{N-1-i}$) serves to make (6.38) more compact by excluding the linear phase exponent. For a periodic signal this shift means an adequate cyclic shift of the periodic output signal, and thereby has no effect on its shape or output SNR.

In particular:

$$R_p(0) = \sum_{i=0}^{N-1} |u_i|^2 = \frac{1}{N} \sum_{k=0}^{N-1} |\tilde{u}_k|^2$$

Using this along with (6.38) in the result for the filter output noise variance gives:

$$\sigma_{sl}^2 = \frac{\sigma^2}{N} \sum_{k=0}^{N-1} |\tilde{b}_k|^2 = \frac{c_0^2 \sigma^2}{N} \sum_{k=0}^{N-1} \frac{1}{|\tilde{a}_k|^2}$$

and output SLSF power SNR (signal peak at the SLSF output $A_{sl} = c_0$):

$$q_{sl}^2 = \frac{A_{sl}^2}{\sigma_{sl}^2} = \frac{N}{\sigma^2} \left(\sum_{k=0}^{N-1} \frac{1}{|\tilde{a}_k|^2} \right)^{-1}$$

Now we may find energy loss of the SLSF versus the matched filter as:

$$\gamma = \frac{q_{mf}^2}{q_{sl}^2} = \sum_{k=0}^{N-1} \frac{1}{|\tilde{a}_k|^2} \tag{6.42}$$

To better comprehend the last result, note that:

$$N = \sum_{i=0}^{N-1} a_i^2 = \frac{1}{N} \sum_{k=0}^{N-1} |\tilde{a}_k|^2 = \overline{|\tilde{a}|^2}$$

and

$$\left(\frac{1}{N} \sum_{k=0}^{N-1} \frac{1}{|\tilde{a}_k|^2} \right)^{-1} = \left(\overline{|\tilde{a}|^{-2}} \right)^{-1}$$

represent, respectively, the arithmetic and harmonic averages of a sequence energy spectrum $|\tilde{a}_k|^2, k = 0, 1, \ldots, N-1$. The harmonic average of any set of non-negative numbers never exceeds the arithmetic one, and they coincide only if the averaged numbers are all equal. Thereby the ratio of these entities may serve as some measure of the range of scattering of the averaged numbers. But in our case this ratio:

$$\frac{\overline{|\tilde{a}|^2}}{\left(\overline{|\tilde{a}|^{-2}} \right)^{-1}} = \sum_{k=0}^{N-1} \frac{1}{|\tilde{a}_k|^2} = \gamma$$

is precisely the energy loss of the SLSF. Therefore, SNR loss γ is determined by the non-uniformity of the sequence energy spectrum $|\tilde{a}_k|^2$, $k = 0, 1, \ldots, N-1$ estimated in terms of the distinction between its harmonic and arithmetic averages.

The possibility of suppressing all periodic sidelobes offers a new criterion for designing binary sequences, which is alternative to the one of minimizing the maximal sidelobe

$\rho_{p,\max}$. Indeed, what is the point of worrying about the sidelobe level, if all the sidelobes may be successfully nullified? It is much more natural to minimize the penalty paid for their elimination—and this penalty is, of course, SNR loss γ. Consider first a binary sequence (hypothetical if $N \neq 4$) whose energy spectrum is uniform: $|\tilde{a}_k|^2 = N$, $k = 0, 1, \ldots, N-1$. In the light of (6.41) this means that the sequence has perfect periodic ACF. It is absolutely predictable that, since there is nothing to suppress, SLSF in this case should coincide with the matched filter, having no SNR loss. Equation (6.42) confirms this, giving $\gamma = 1$ or—measured in decibels—$\gamma_{dB} = 10\lg\gamma = 0\,\mathrm{dB}$. Because binary sequences with perfect periodic ACF do not exist, their sidelobe suppression is done in exchange for SNR loss ($\gamma_{dB} > 0\,\mathrm{dB}$), justifying the introduction of the design criterion $\gamma = \min$.

As in many problems concerning binary sequences (see Section 6.4), a binary globally optimum sequence of fixed length N with minimum loss γ may be found only by an exhaustive search. This work has been run up to length $N = 30$ [48]. Certainly, due to the exponential growth of a computational resource, this search cannot continue far beyond the above mentioned range. However, many regular rules of generating binary sequences of lengths as great as one likes and having very small loss γ (although their global optimality is not guaranteed) are now known.

Let us take a special class of binary sequences having two-level periodic ACF, i.e. constant level R of sidelobes:

$$R_p(m) = \begin{cases} N, m = 0 \bmod N \\ R, m \neq 0 \bmod N \end{cases} \tag{6.43}$$

All minimax binary sequences, among others, are of this type. The energy spectrum of such a sequence, as (6.41) shows, is the direct DFT of ACF:

$$|\tilde{a}_k|^2 = \sum_{m=0}^{N-1} R_p(m)\exp\left(-\frac{j2\pi mk}{N}\right) = N - R + R\sum_{m=0}^{N-1}\exp\left(-\frac{j2\pi mk}{N}\right),\ k = 0, 1, \ldots, N-1$$

The last sum has already appeared in Section 6.11.2 and, as was proved, equals N if $k = 0$ and zero otherwise. Thus:

$$|\tilde{a}_k|^2 = \begin{cases} N + (N-1)R,\ k = 0 \\ N - R,\ k \neq 0 \end{cases} \tag{6.44}$$

Substituting this into (6.42) gives:

$$\gamma = \frac{1}{N + (N-1)R} + \frac{N-1}{N-R} = \frac{1 + (N-2)\rho}{(1-\rho)[1 + (N-1)\rho]} \tag{6.45}$$

where $\rho = R/N$ is the normalized ACF sidelobe.

To come to a SLSF structure, rewrite (6.39) as:

$$b_i = \frac{c_0}{N}\sum_{k=0}^{N-1}\frac{\tilde{a}_k^*}{|\tilde{a}_k|^2}\exp\left(\frac{j2\pi ik}{N}\right),\ i = 0, 1, \ldots, N-1$$

and substitute (6.44) into this equation:

$$b_i = \frac{c_0}{N}\left[\frac{\tilde{a}_0}{N+(N-1)R} - \frac{\tilde{a}_0}{N-R} + \frac{1}{N-R}\sum_{k=0}^{N-1}\tilde{a}_k^* \exp\left(\frac{j2\pi ik}{N}\right)\right], i = 0, 1, \ldots, N-1$$

The sum in k here produces coefficients of the matched filter, since it may be presented as:

$$\frac{1}{N}\sum_{k=0}^{N-1}\tilde{a}_k^* \exp\left(-\frac{j2\pi(N-i)k}{N}\right)$$
$$= \left[\frac{1}{N}\sum_{k=0}^{N-1}\tilde{a}_k \exp\left(\frac{j2\pi(N-i)k}{N}\right)\right]^* = a_{N-i}, i = 0, 1, \ldots, N-1$$

where use is made of the fact that binary sequence elements are real. Since c_0 is an arbitrary scaling factor, let us choose it equal to $N - R$. Then:

$$b_i = a_{N-i} - \frac{\rho\tilde{a}_0}{1+(N-1)\rho}, i = 0, 1, \ldots, N-1 \tag{6.46}$$

The first term here corresponds to the sequence $\{a_i\}$ read from right to left, i.e. coefficients of the matched filter. Therefore, for the sequences possessing two-valued ACF (6.43), SLSF is obtained by a slight modification of the matched filter: subtracting a constant from all coefficients. Moreover, for binary sequences of this sort coefficients of SLSF take on only two possible values:

$$\pm 1 - \frac{\rho\tilde{a}_0}{1+(N-1)\rho}$$

where $\tilde{a}_0$ is a sequence constant component, i.e. the difference between the numbers of plus and minus ones over the period $\tilde{a}_0 = N_+ - N_-$.

As is known from the previous material, lots of binary sequences exist with ACF (6.43) where $R = -1$ (m-sequences, Legendre sequences and other minimax ones with ACF (6.12)). Evaluating their SLSF loss by (6.45) results in $\gamma = 2N/(N+1)$, i.e. $\gamma \approx 2$ (3 dB) for lengths that are of practical interest. It is seen then that the most popular minimax binary sequences are of no great value in the light of the criterion of SLSF loss: half of their energy is lost under SLSF processing.

On the other hand, when R is positive and low enough as compared to N $\gamma < 1/(1-\rho)$, i.e. may appear rather small. So-called Singer codes [34,48] are a good example of such binary sequences. A Singer code exists for any length of the form $N = (q^n - 1)/(q - 1)$ and has two-valued ACF (6.43) with $R = N - 4q^{n-2}$, where $q = p^w$ is a natural power of prime p and n is natural. The most interesting modification of Singer codes in our context corresponds to $q = 3$, in which case $\rho = (3^{n-2} - 1)/(3^n - 1)$ and $\gamma < (3^n - 1)/(8 \cdot 3^{n-2}) < 9/8 = 1.125$, i.e. $\gamma_{dB} < 0.51$ dB. As is seen, these codes, being processed by SLSF, are rather attractive, since SNR loss accompanying a total sidelobe suppression for them is small.

Example 6.12.1. Let us take a periodic version of the binary Barker code of length $N = 5$ from Table 6.1: $+1, +1, +1, -1, +1$, for which $N_+ = 4$, $N_- = 1$ and constant component $\tilde{a}_0 = 3$. Its periodic ACF, as may be checked by direct evaluation, obeys (6.43) with $R = 1$ ($\rho = 1/5$). Actually, this sequence is a Singer one with $q = 4$, $n = 2$. Figure 6.21b shows the ACF (i.e. the matched filter response) of the periodic signal modulated by this sequence (Figure 6.21a). As is seen from (6.46), a matched filter for this sequence is easily transformed into an SLSF by changing all coefficients +1 to +2/3 and −1 to −4/3, which after appropriate scaling means changing −1 to −2 with no change of +1 (see Figure 6.22). The SLSF response to the same signal is constructed in Figure 6.23, where the waveform labels correspond to the points in Figure 6.22. The filter output has the desired form, i.e. with zero level of sidelobes. Using (6.45), it is easy to find the energy loss of the SLSF as $\gamma = 10/9 = 1.1111\ldots$ (0.46 dB).

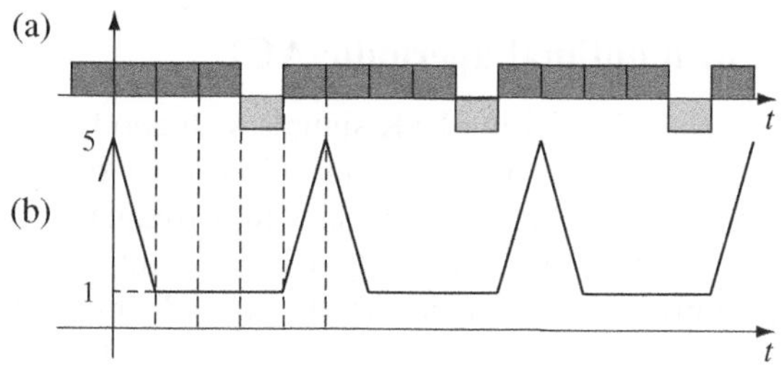

Figure 6.21 Binary signal of length $N = 5$ and its periodic ACF

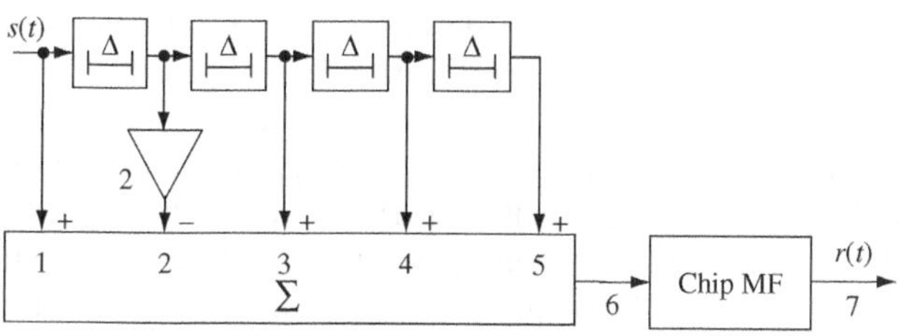

Figure 6.22 SLSF for a sequence of length $N = 5$

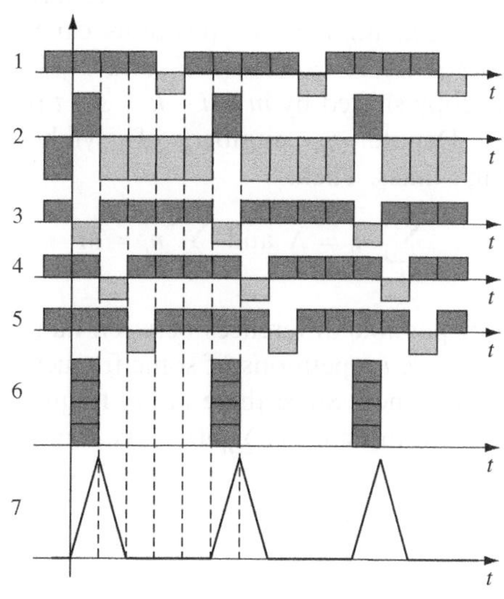

Figure 6.23 SLSF response for the binary sequence of length $N = 5$

A multitude of even more effective binary codes were found, based on converting linear recurrent sequences over finite fields to the binary $\{\pm 1\}$ alphabet, as well as on Singer codes. All of them possess a similar appealing feature: a simple SLSF structure whose coefficients take on no more than three different values. Without going into the details, which are sophisticated enough, and referring the curious reader to [49, 50], note that among those sequences families are present having asymptotically vanishing SLSF loss: $\gamma_{\mathrm{dB}} \to 0\,\mathrm{dB}$ with $N \to \infty$.

6.13 FSK signals with optimal aperiodic ACF

To discuss briefly the issue of designing FSK signals with good correlation properties let us come back to (5.20), recalling that having a low level of $R_p(m)$ is equivalent to minimizing the number of coincident frequencies in a frequency code $F_0, F_1, \ldots, F_{N-1}$ and its replica shifted by m positions. Certainly, when the number of chips N (i.e. length) does not exceed the number of frequencies M, it is trivially easy to obtain zero level of sidelobes $\rho_a(m) = 0, m \neq 0$ by using a frequency code whose elements F_i are all different. Practically, however, the case $N > M$ is much more interesting, entailing repetitions among elements $F_i, i = 0, 1, \ldots, N-1$ and, thus, at least one coincidence in the shifted replicas of the frequency code, i.e. $\rho_{a,\max} \geq 1/N$.

Since a frequency sequence may be described as an $M \times N$ array (Section 5.5), minimizing $\rho_{a,\max}$ means inventing an array with minimal possible number of coincident labels (dots) in the array itself and its replica, which is shifted horizontally by m positions. One of the topical problems is constructing the so-called *radar arrays* defined as $M \times N$ arrays having only one labelled entry in every column and $\rho_{a,\max} = 1/N$, i.e. the number of abovementioned coincidences within one. The desire to find a radar array as long as possible is understandable, given M, because it would mean minimizing $\rho_{a,\max}$ under limitations imposed on the frequency resource. Following [51] let us proof a simplest upper bound on the length of a radar array.

Consider a sequence $F_0, F_1, \ldots, F_{N-1}$ and note that to have no more than one coincidence all differences between numbers of positions carrying the same frequencies should be different. Indeed, let $F_i = F_k, F_s = F_t$ and $i - k = s - t > 0$. Then in the original sequence and its copy shifted by $m = i - k = s - t$ positions at least two coincidences will happen. Denote n_i number of symbols (frequencies) among $F_0, F_1, \ldots, F_{N-1}$ occurring i times. Then:

$$\sum_i i n_i = N \text{ and } \sum_i n_i = M \tag{6.47}$$

Now count the number of possible differences between numbers of positions carrying identical frequencies. There are i repetitions of some frequency and hence $i(i-1)$ such differences for this very frequency. Since there are n_i frequencies repeated i times, the total number of differences in question is $\sum_i i(i-1)n_i$, and because among the differences no repetition is allowed:

$$\sum_i i(i-1)n_i \leq N - 1 \tag{6.48}$$

where the right-hand side gives the maximal number of unequal positive differences among the numbers $\{0, 1, \ldots, N-1\}$. The trinomial $i(i-1)+3-2i = i^2-3i+3$ has no real roots, and, hence, is positive at any i. Therefore the sum:

$$\sum_i [i(i-1)+3-2i]n_i = \sum_i i(i-1)n_i + 3\sum_i n_i - 2\sum_i in_i \geq 0$$

which, being combined with (6.47), (6.48), gives $N-1+3M-2N \geq 0$ or:

$$N \leq 3M-1 \tag{6.49}$$

In fact this bound is not the tightest one. More accurate bounds are known, e.g. in [52] the asymptotic result is derived:

$$N \leq \frac{20+\sqrt{6}}{8}M, M \gg 1, \tag{6.50}$$

lowering the right-hand side of (6.49) by approximately $0.194M$.

Absolutely tight, i.e. really achievable, upper bounds on the length N are now known up to $M = 16$. The table given in [52] allows the maximal length N_{max} of a radar array in this range of M to be expressed as

$$N_{max} = \begin{cases} 3M-2, & 2 \leq M \leq 4 \\ 3M-3, & 5 \leq M \leq 9 \\ 3M-4, & 10 \leq M \leq 13 \\ 3M-5, & 14 \leq M \leq 16 \end{cases} \tag{6.51}$$

Example 6.13.1. The frequency code 1, 2, 3, 4, 5, 6, 7, 8, 7, 4, 3, 9, 9, 5, 8, 2, 6, 5, 1, 4, 2, 1, 3, 7, where the numbers of frequencies in an alphabet containing $M = 9$ frequencies or, equivalently, the numbers of dotted rows in every column of the array are given, has maximal possible length $N = 24$. Its radar array property, i.e. possessing only one frequency coincidence at all non-zero shifts, is verified by a direct test (Problem 6.54).

In addition, a regular rule for constructing radar arrays of length $N = 2.5M$ exists (see details in [51]) whenever M is even and $M/2$ is a product of primes having remainder one of division by 4, i.e. $M = 10, 26, 34, 58, \ldots$.

A *sonar* array is a further generalization of a radar array, preserving the 'no more than one coincidence' property for arbitrary non-zero combinations of horizontal and vertical shifts [53]. Physically, this requirement reflects the desire to have a small correlation of signal replicas detuned in both time and frequency. Considering an approach to the choice of frequency space for FSK signals (Section 5.5), frequency shifts turning a current frequency into the adjacent one are more typical of sonar than of radar systems, where the terminology stems from. For instance, a great number of regular algorithms for constructing Costas arrays [54], i.e. square ($M = N$) sonar arrays or FSK sequences having equal length and number of frequencies, are known.

Problems

6.1. The frequency of a rectangular pulse drops linearly throughout its duration $T = 10\,\mu s$ from 110 to 90 MHz. Calculate the processing gain of the signal. What is the approximate duration of the signal at the matched filter output? Sketch the ambiguity function and ambiguity diagram.

6.2. The frequency of a rectangular pulse drops linearly over the first half of its duration $T = 10\,\mu s$ from 110 to 90 MHz and then grows linearly from 90 to 110 MHz over the second half. Calculate the processing gain of the signal. Sketch the ambiguity function and its low-level and high-level horizontal sections.

6.3. Calculate aperiodic and periodic ACF for a binary Barker code of length $N = 11$. Try to do it the most economical way.

6.4. Take a periodic sequence $\{a_i\}$ of period N and form a new sequence $\{b_i\}$ picking each dth element of $\{a_i\}$: $b_i = a_{di}$, where multiplication in the index is modulo N. Such a transform is called the *decimation* of $\{a_i\}$ with *index d*. Prove that if $\{a_i\}$ has perfect periodic ACF and d is co-prime to N, $\{b_i\}$ also has perfect periodic ACF.

6.5. The binary $\{\pm 1\}$ code of length $N = 5$ has periodic ACF $R_p(m) = 1, m \neq 0 \bmod 5$. Values of its aperiodic ACF are $R_a(1) = 0$, $R_a(2) = -1$. Find $R_a(3)$ and $R_a(4)$.

6.6. The binary $\{\pm 1\}$ code of length $N = 5$ has constant component $\tilde{a}_0 = +3$ and $R_a(4) = -1$. Find $R_p(m)$ and the rest of the values of $R_a(m)$.

6.7. Can a binary $\{\pm 1\}$ code of an odd length $N > 5$ have $R_a(5) = -1$? Can a binary code of an even length $N > 6$ have $R_a(6) = -1$? For an arbitrary binary code formulate and prove the relation between parities of the three values: length N, shift m and level of $R_a(m)$.

6.8. Is it possible for a binary $\{\pm 1\}$ sequence that $R_a(m) = 1$, $R_a(m+1) = 3$ for some m? What about parities of $R_a(m)$ and $R_a(m+1)$?

6.9. Is it possible for a binary $\{\pm 1\}$ sequence that $R_a(2) = 2$, $R_p(2) = -1$?

6.10. Suppose someone has found that each of the PSK sequences of length $N = 100$ at his disposal has non-normalized periodic ACF taking on values ± 12 at some shifts $m \in \{1, 2, \ldots, N-1\}$. Can a code with $\rho_{a,\max} < 0.05$ be present among them?

6.11. Construct a matched filter for a BPSK signal manipulated by a sequence $\{+-++\}$ and show waveforms at its characteristic points, when the filter is input by aperiodic and periodic versions of the signal.

6.12. Construct a matched filter for a BPSK signal manipulated by a sequence $\{+-++---\}$ and show waveforms at its characteristic points when aperiodic and periodic versions of the signal are applied to its input.

6.13. A student has calculated periodic ACF of the binary $\{\pm 1\}$ sequence of length $N = 21$ and obtained the following values: $R_p(m_1) = 9$, $R_p(m_2) = -3$, $R_p(m_4) = -5$, $R_p(m_5) = -7$, $R_p(m_6) = 7$. Can all of these results be correct? Which (if any) are definitely incorrect?

6.14. Prove the non-existence of minimax binary sequences ($R_p(m) = \pm 1,\ m = 1, 2, \ldots, N - 1$) for lengths $N = 17, 21, 29, 33, 37, 45$. (Hint: use the same technique as in deriving the necessary conditions for perfect ACF of binary sequences.)
6.15. Prove that decimation of a minimax binary sequence again produces a minimax binary sequence whenever the decimation index is co-prime to the period N.
6.16. Prove that decimation of any periodic sequence does not change the maximal periodic sidelobe, whenever the decimation index is co-prime to the period N.
6.17. Calculate $(3 + 7) - 5(6 \cdot 7 + 4) + 1$ in $GF(11)$.
6.18. Solve the equation $6x + 7(5 + 4 \cdot 2)^{-1} = 1$ in $GF(11)$.
6.19. Prove that $GF(4)$ cannot be built on the basis of modulo 4 operations.
6.20. Is the sequence of length $L = 7$ $\{0100110\}$ a binary m-sequence? What is the answer if all zeros are replaced by ones and vice versa?
6.21. Construct a binary m-sequence of length $L = 15$ with an initial loading $d_0 = 1, d_1 = d_2 = d_3 = 0$, draw the structure of its generator and compile a table exhibiting its state changes.
6.22. Prove that in one period of a binary m-sequence of memory n the number of series of successive symbols (01), (10), (11) is 2^{n-2} while the number of series (00) is smaller by one.
6.23. Someone observes an m-sequence knowing its alphabet and memory but not its coefficients of recurrence (6.13). What is the minimal necessary and sufficient number of observed symbols to recover the coefficients?
6.24. Prove the independence of binary character of a specific choice of primitive element.
6.25. Find a primitive element of $GF(13)$. Build a table of logarithms and binary characters of all non-zero elements of $GF(13)$.
6.26. Build the Legendre sequence of length $N = 11$, calculate its periodic ACF and compare it with the theoretically predicted one.
6.27. Build the Legendre sequence of length $N = 13$, calculate its periodic ACF and compare it with the theoretically predicted one.
6.28. Find the cyclic shift of the sequence of Problem 6.26 with minimal aperiodic ACF sidelobe.
6.29. Prove the perfection of periodic ACF for Chu codes of odd lengths.
6.30. Prove the perfection of periodic ACF for Frank codes. (Hint: use representation $i = i_1 h + i_2,\ m = m_1 h + m_2;\ 0 \le i_1, i_2, m_1,\ m_2 \le h - 1$ and summation over i_1, i_2 in (5.9).)
6.31. Does an 8-PSK sequence of length $N = 64$ with perfect periodic ACF exist? If so, construct it.
6.32. Prove the non-existence of QPSK sequences with perfect periodic ACF for odd lengths N.
6.33. Prove the non-existence of QPSK sequence of length $N = 30$ with perfect periodic ACF.
6.34. Prove that for a ternary $\{0, \pm 1\}$ sequence with perfect periodic ACF, the number of non-zero elements per period is always a square of an integer.
6.35. Prove the non-existence of ternary $\{0, \pm 1\}$ sequences having perfect periodic PACF and a single zero per period for any odd length.
6.36. Construct an SLSF, show the effect of sidelobe suppression and find SNR loss for a binary sequence $\{+ + + + + -\}$.

6.37. Find the SLSF energy loss for a sequence $\{++---++-\}$.

6.38. Find the maximal length FSK sequence with alphabet size $M = 4$ and no more than one frequency coincidence.

Matlab-based problems

6.39. Using binary codes $\{++-+++-+++-\}$ and $\{+-++-+++---\}$ demonstrate the destructive role of high sidelobes of ACF in time delay measurement and synchronization (Figure 6.24). Recommended steps:

(a) Form the codes given above and oversample them 10 times to simulate baseband signals (complex envelopes) with a rectangular chip.
(b) Plot the signals in two separate subplots.
(c) Calculate the ACFs of the signals above and plot them in two separate subplots.
(d) Extend the signals by appending 110 initial and 110 tail zeros to imitate the stationary regime of a receiver.
(e) Form the observation vector for each signal, adding to it complex Gaussian noise having standard deviations of real and imaginary components three times higher than the signal amplitude.
(f) Calculate the real envelope at the matched filter output for each of the two signals and fix the time moment of its maximum.

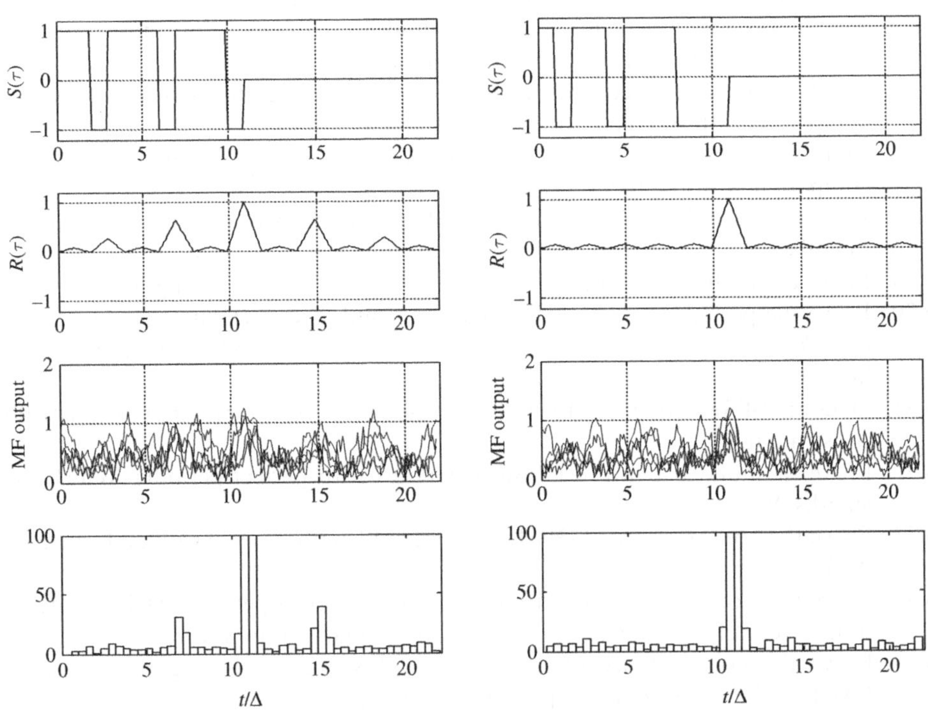

Figure 6.24 ACF sidelobes and time-delay measurement accuracy

(g) Do item (e) several times and plot the superposition of waveforms for the two signals separately.
(h) Repeat items (e)–(g) 1000 times, display histograms of time estimations for the two signals and interpret the differences between them.

6.40. Using binary codes $\{+++-+-+-+-+\}$ and $\{+-++-+++---\}$ demonstrate the destructive role of high sidelobes of ACF in time resolution of signals with different intensities (Figure 6.25). Recommended steps:

(a) Form a plain bandpass chip with 10–15 periods of carrier frequency per chip duration and find its ACF.
(b) Find the ACF of the code vectors.
(c) Find and plot matched filter responses to 'pure' signals for the two codes.
(d) Display the delayed and attenuated copies of the matched filter responses.
(e) For each of the two signals display the full matched filter responses to the superposition of direct and delayed signals.
(f) Varying the delay and attenuation, compare the masking effect of the sidelobes of the stronger signal on the visibility of the weaker one for the two codes; explain the results.

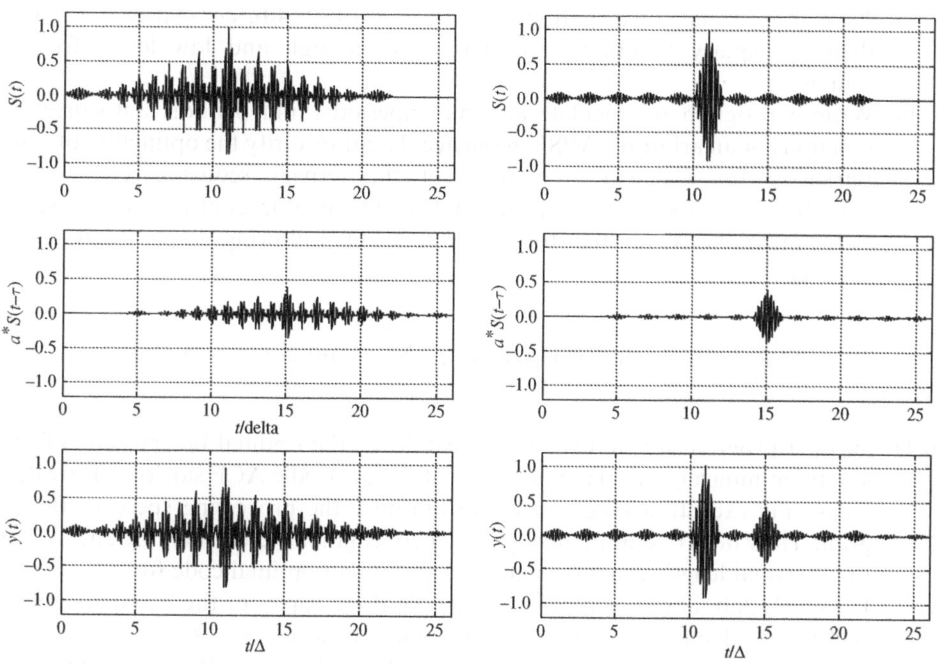

Figure 6.25 ACF sidelobes and signal time resolution

6.41. Write a program demonstrating the behaviour of the ACF and spectrum of an LFM pulse with changing deviation. Take three values of deviation (e.g. $W_d T = 10, 25, 40$). Compare the exact form of the ACF with the approximation

(6.1). Run the program for rectangular and bell-shaped pulses and explain the results.

6.42. Write a program calculating and plotting the ambiguity function and its horizontal sections (ambiguity diagrams) at different levels for the LFM (see Figure 6.6) and V-LFM rectangular pulses, whose complex envelopes are:

$$\dot{S}(t) = \begin{cases} \exp\left(j\dfrac{\pi W_d t^2}{T}\right), & |t| \le T/2 \\ 0, & |t| > T/2 \end{cases}$$

$$\dot{S}(t) = \begin{cases} \exp\left(-j2\pi\dfrac{W_d t^2}{T}\right), & \dfrac{-T}{2} \le t < 0 \\ \exp\left(j2\pi\dfrac{W_d t^2}{T}\right), & 0 \le t \le \dfrac{T}{2} \\ 0, & |t| > T/2. \end{cases}$$

Take three values of deviation (e.g. $W_d T = 10, 25, 40$), comparing for each of them ambiguity functions and diagrams on high and low levels for both signals.

6.43. Write a program to calculate and plot aperiodic and periodic autocorrelation functions of an arbitrary APSK sequence. Use it to verify the optimality of binary Barker codes. Calculate the ACF of the ternary sequence $\{+++++-+0+0-++-00+-0--\}$ and find the maximal level of its sidelobe relative to the mainlobe. Verify that for binary codes the following properties of periodic ACF are true:

$$R_p(m) - R_p(l) = 0 \bmod 4, \forall m, l; \sum_{m=0}^{N-1} R_p(m) = (N_+ - N_-)^2$$

6.44. Write a program for an exhaustive search for the optimal binary code of given length, minimizing the maximum level of the aperiodic ACF sidelobe. Provide for measuring execution time and try to optimize the processing speed of the program. How does operating time grow when length is incremented by one? What is the maximal length you have managed to find an optimal code for?

6.45. Using Matlab, illustrate time-resolution of three copies of the bandpass periodic APSK signal manipulated with the ternary code $(+++++-+0+0-++-00+-0--)$ at the matched filter output (Figure 6.26). Take time shifts between consecutive copies of 2–3 and 5.5–6.5 chip durations Δ correspondingly. Recommended steps are:

(a) Form several (3–4) periods of ternary code and oversample them 100 times to simulate rectangular baseband chips manipulated by the given code: the signal complex envelope.

(b) Form the sum of three overlapped delayed copies of the complex envelope having remarkably different amplitudes, say in the proportions 1 : 4 : 8.
(c) When summing copies, be mindful of the proper (consistent with delay) phase of a copy, specifying the carrier frequency to be $5/\Delta$.
(d) Calculate and plot the resulting bandpass signal at the matched filter input.
(e) Filter the resulting complex envelope with a filter matched to one period of the signal.
(f) Calculate and plot the bandpass filter output.
(g) Run the program for other values of delays and amplitudes, and comment on the results.

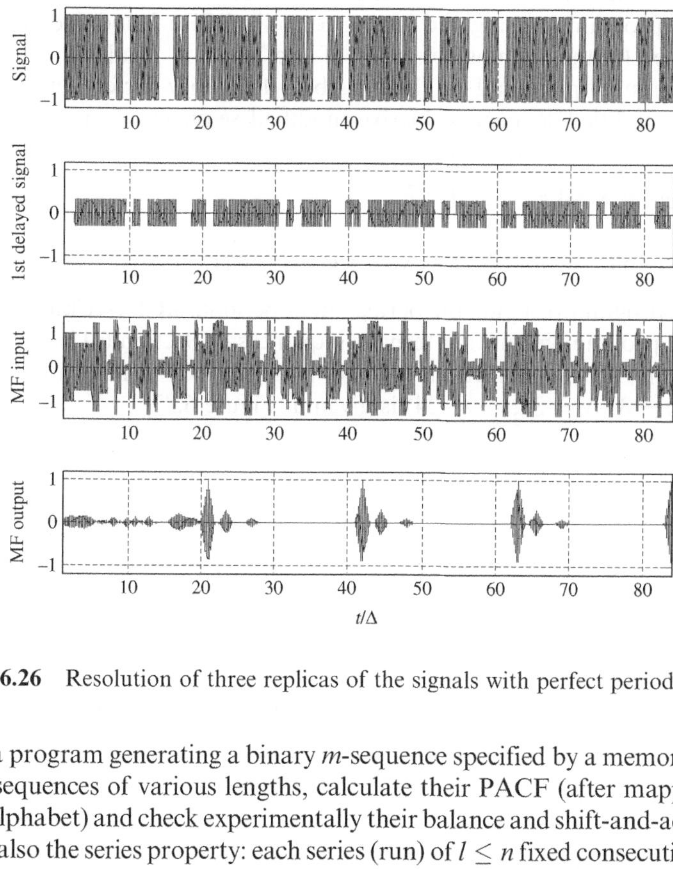

Figure 6.26 Resolution of three replicas of the signals with perfect periodic ACF

6.46. Write a program generating a binary m-sequence specified by a memory n. Generating m-sequences of various lengths, calculate their PACF (after mapping onto the $\{\pm 1\}$ alphabet) and check experimentally their balance and shift-and-add properties. Check also the series property: each series (run) of $l \leq n$ fixed consecutive bits occurs 2^{n-l} times during the period except for the all-zero run, which occurs $2^{n-l} - 1$ times.

6.47. Testing the binary polynomial of degree n in the LFSR generator may answer the question of whether it is primitive. If the LFSR does generate an m-sequence of length $N = 2^n - 1$ then the polynomial is primitive. Write a program performing such a test and tabulate all primitive polynomials of degrees from 4 to 12. Is polynomial $f(x) = x^6 + x^4 + x^2 + x + 1$ primitive? What sequence is generated by LFSR with a feedback defined by this polynomial?

6.48. Write a program to build a Legendre sequence. First find a primitive element of a prime field $GF(p)$. Then plus ones should be put in the positions whose numbers are even powers of the primitive element in $GF(p)$ and in position number zero (the latter may alternatively be assigned to -1, too), while the rest of the elements should be minus ones. Run the program for lengths $N = 11, 23, 31, 43$ and those of your own choice, checking in each case the PACF of the sequence obtained.

6.49. Write a program finding the sequence with minimax aperiodic ACF sidelobes among one-period segments of a given sequence with promising periodic ACF. Running this program along with that of Problem 6.48, find the best (in aperiodic sidelobe level) one-period segments of Legendre sequences having lengths $N = 23, 29, 31, 37, 43, 47, 53, 59, 61, 67$. Have you found anything better in $\rho_{a,\max}$ than the binary Barker code of length $N = 13$?

6.50. Write a program forming and testing periodic ACF of modified m- and Legendre sequences (if the leftmost element of the latter is 1 take the negative of the sequence, multiplying all its elements by -1). Verify perfection of periodic ACF for the modified sequences. Derive modified sequences from the initial ones by replacing the -1 symbol by:

(a) a real symbol $b = -1 \pm \frac{2}{\sqrt{N+1}}$

(b) a complex symbol $b = -\exp\left[j \arccos\left(\frac{N-1}{N+1}\right)\right]$.

6.51. Write a program verifying the perfection of periodic ACF of Chu and Frank codes.

6.52. Write a program generating ternary sequences of Section 6.11.3 over prime fields and verifying the perfection of their periodic ACF. An example for the case $p = 5, n = 3$ ($N = 31, \nu = 1.24$) is given in Figure 6.27.

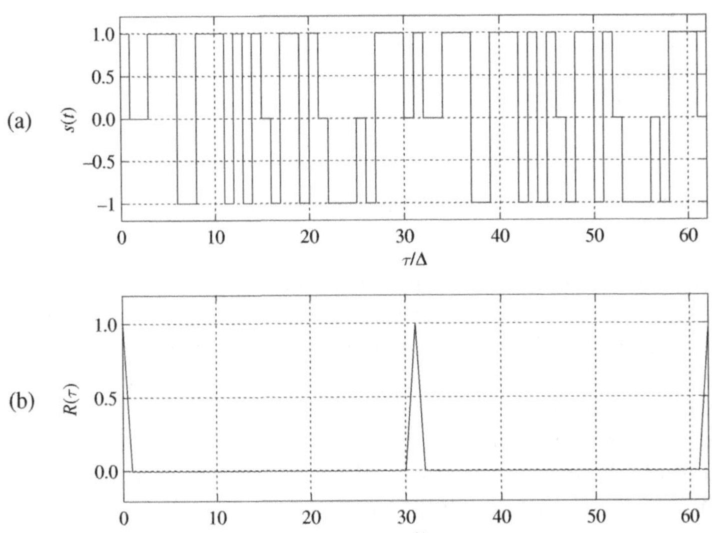

Figure 6.27 Rectangular-chip signal modulated by the ternary sequence (a) and its periodic ACF (b)

6.53. Write a program finding SLSF and calculating its SNR loss for a given binary sequence. Plot the signal manipulated by this sequence, its periodic ACF and SLSF response. An example for a binary sequence $\{-++----+-++-+-+++-++--+++++\}$ of length $N = 27(\gamma_{\text{dB}} = 0.4\,\text{dB})$ is given in Figure 6.28.

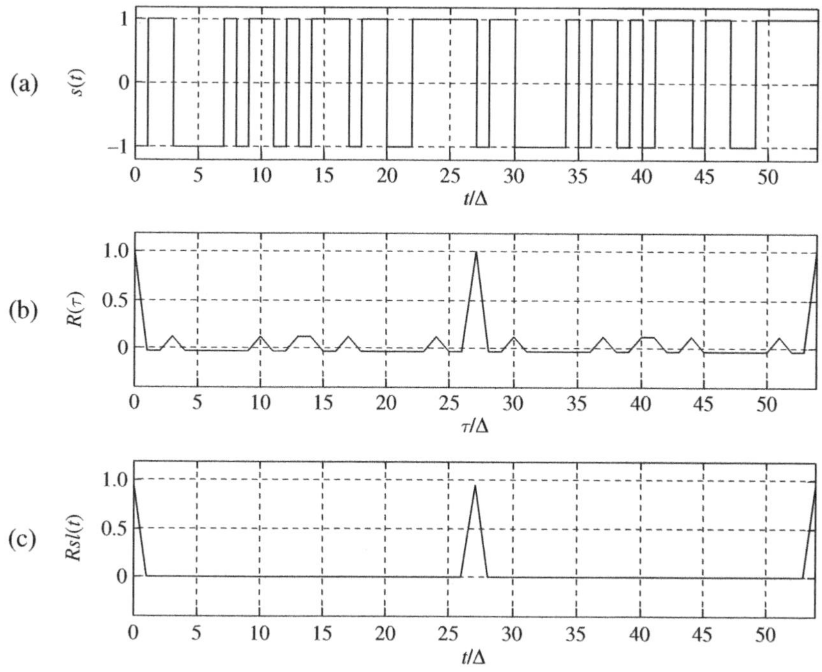

Figure 6.28 Rectangular-chip signal manipulated by a binary sequence (a), its periodic ACF (b) and SLSF response (c)

6.54. Write a program to calculate aperiodic ACF of a radar array. Run the program to verify the properties of the FSK code in Example 6.13.1.

7

Spread spectrum signature ensembles for CDMA applications

7.1 Data transmission via spread spectrum

It is clear from the discussion of Sections 4.3–4.6 that in a CDMA network each of K users transmits or receives its individual data employing some user-specific signature, ensemble of K signatures being carefully designed to provide the best possible compatibility. In order to make the kth signature transport the datastream some sort of modulation is necessary, which—due to the spread spectrum nature of CDMA signatures—is often called spread spectrum modulation. There are two classical versions: *direct sequence* (DS) and *frequency hopping* (FH) modulation. The first is more typical of modern commercial wireless multiuser applications, and so the second will be considered below only briefly.

7.1.1 Direct sequence spreading: BPSK data modulation and binary signatures

The general idea of direct spread spectrum is APSK modulation of the APSK signature by a datastream. To make the concept easier to grasp, let us start with the simplest case of BPSK non-spread spectrum data transmission. Let $B_k(t)$ be the data waveform of the kth user (Figure 7.1) where positive and negative polarities during one bit interval T_b correspond to transmitting a bit equal to 0 and 1, respectively. If $\mathbf{b}_k = (\ldots, b_{k,-1}, b_{k,0}, b_{k,1}, \ldots), b_{k,i} = \pm 1$, is, as it was in Chapter 4, the kth user bit (or binary symbol) stream, then $B_k(t) = b_{k,i} = \pm 1, (i-1)T_b < t \leq iT_b$. Transmitting $B_k(t)$ by

Spread Spectrum and CDMA: Principles and Applications Valery P. Ipatov

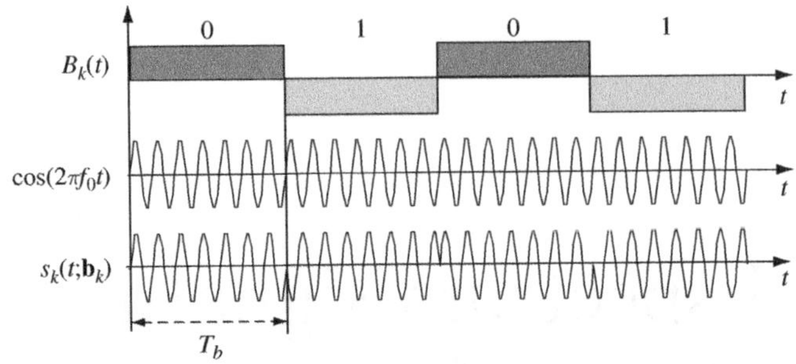

Figure 7.1 Illustration of BPSK data transmission

BPSK just means multiplying it with a CW carrier of frequency f_0 to come to a sent modulated signal (see Figure 7.1):

$$s_k(t; \mathbf{b}_k) = B_k(t)\cos(2\pi f_0 t) \tag{7.1}$$

Therefore, technically a BPSK modulator is just the multiplier shown in Figure 7.2a. After passing through the channel the signal assumes time delay τ_k and initial phase ϕ_k as well as attenuation, the latter being ignored as immaterial in our current study. Then the received useful signal:

$$s_{kr}(t; \mathbf{b}_k) = B_k(t - \tau_k)\cos(2\pi f_0 t + \phi_k) \tag{7.2}$$

A typical receiver of BPSK data contains timing and carrier phase recovery loops, which estimate running values of delay τ_k and initial phase ϕ_k. At the moment the issue of estimation precision may be left aside, and we assume that the receiver knows 'genuine' τ_k and ϕ_k. If the signal above is corrupted by AWGN, the optimal (ML) procedure (see Section 2.2) to retrieve the ith transmitted bit is to calculate the correlation of the observation $y(t) = s_{kr}(t; \mathbf{b}_k) + n(t)$ with the difference of signals carrying bit contents 0 and 1, respectively, which in the considered case is just $2\cos(2\pi f_0 t + \phi_k)$. Since only polarity of the correlation is used for the decision on the received bit, and since the ith

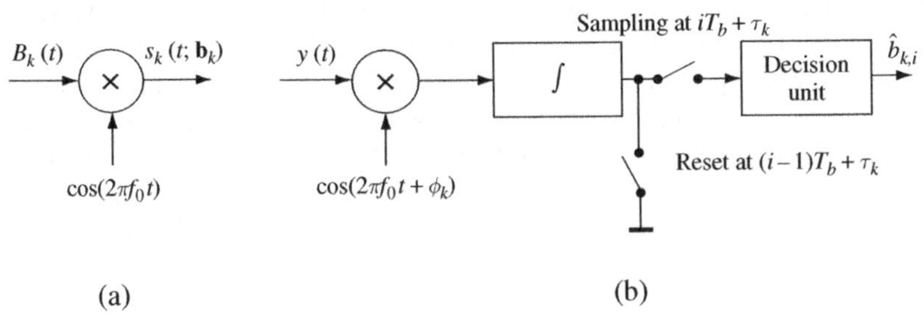

Figure 7.2 BPSK modulator (a) and demodulator (b)

bit at the channel output occupies time interval $((i-1)T_b+\tau_k, iT_b+\tau_k]$, the correlation discussed is:

$$z_k = \int_{(i-1)T_b+\tau_k}^{iT_b+\tau_k} y(t)\cos(2\pi f_0 t+\phi_k)\,\mathrm{d}t$$

and the decision $\hat{b}_{k,i}=0$ or $\hat{b}_{k,i}=1$ is taken depending on the positive or negative sign of z_k. A possible and very popular structure of a *demodulator* implementing this rule is given in Figure 7.2b. It contains the correlator realized as a multiplier multiplying the observation with a locally generated CW reference $\cos(2\pi f_0 t+\phi_k)$ and an integration-and-reset unit. At the end of every consecutive bit interval a sample is taken from the integrator output, a decision on the current bit is made according to its polarity, and the integrator is zeroed in preparation for operation over the next bit interval.

Consider now the changes that need to be done for transmitting BPSK data with BPSK DS *spreading*. Let $s_k(t)$ be the kth user signature, i.e. a discrete signal consisting of chips of duration Δ, manipulated by some user-specific binary sequence. Let there be N signature chips per one data bit. Then DS spreading of the BPSK signal just involves inserting one more multiplication in (7.1)—by a signature $s_k(t)$:

$$s_k(t;\mathbf{b}_k) = s_k(t)B_k(t)\cos(2\pi f_0 t) \tag{7.3}$$

Since the bandwidths of signals (7.1) and (7.3) are inverse to bit duration $T_b = 1/R$ and chip duration $\Delta = T_b/N = 1/RN$, respectively, the DS spreading widens the spectrum N times. This explains one more name: the *spreading factor* for the time–frequency product or processing gain $WT = N$. In practice, multiplications in (7.3) may be fulfilled in an arbitrary order, e.g. as Figures 7.3 (spreading by a binary m-sequence of length $N=7, T_b=N\Delta$) and 7.5a show, the bit stream $B_k(t)$ may first be multiplied with a signature $s_k(t)$ to further modulate the CW carrier by the product $s_k(t)B_k(t)$. We may say in this case that the bit stream first modulates the baseband signature and then the result BPSK-modulates the carrier.

After passing the channel and acquiring delay τ_k and phase ϕ_k, the signal takes the form:

$$s_{kr}(t;\mathbf{b}_k) = s_k(t-\tau_k)B_k(t-\tau_k)\cos(2\pi f_0 t+\phi_k) \tag{7.4}$$

Assuming again a perfect knowledge of parameters τ_k, ϕ_k, the receiver for retrieving the current (ith) bit just needs to distinguish between the signal $s_k(t-\tau_k)\cos(2\pi f_0 t+\phi_k)$ and its antipodal copy. To perform it optimally a correlation:

$$z_k = \int_{(i-1)T_b+\tau_k}^{iT_b+\tau_k} y(t)s_k(t-\tau_k)\cos(2\pi f_0 t+\phi_k)\,\mathrm{d}t$$

of the observed waveform $y(t)$ with a local reference bandpass signature replica $s_k(t-\tau_k)\cos(2\pi f_0 t+\phi_k)$ may be found and its polarity used for the decision. Interestingly, however, the same optimal operation may be realized in two stages, first removing

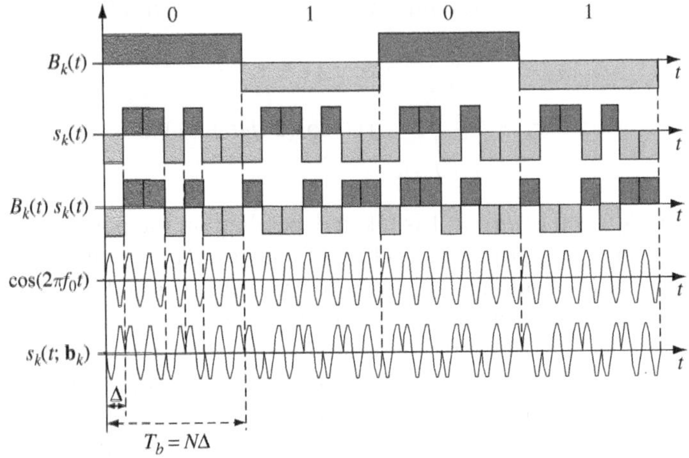

Figure 7.3 DS spreading of BPSK data with binary signature

the spreading and then demodulating the data as though they had been transmitted directly with no spreading. Let the observation $y(t)$ be multiplied by the local replica $s_k(t-\tau_k)$ of the baseband signature synchronized accurately with the arriving signal. The useful component (7.4) of the observation after this operation changes as:

$$s_{kr}(t;\mathbf{b}_k)s_k(t-\tau_k) = s_k^2(t-\tau_k)B_k(t-\tau_k)\cos(2\pi f_0 t+\phi_k)$$
$$= B_k(t-\tau_k)\cos(2\pi f_0 t+\phi_k)$$

where the binary nature of the signature ($s_k(t) = \pm 1$) is used, on the strength of which $s_k^2(t) = 1$. As is seen, after this step the received signal has no more features of spread spectrum, coinciding entirely with the plain signal (7.2) BPSK-modulated by the datastream. Due to this, multiplying of the observation by a signature replica is called *despreading*. Figure 7.4 shows the procedure of transforming a DS-spread signal into a conventional BPSK data-modulated signal.

Since a despread signal is a conventional BPSK data-modulated CW carrier, further data recovery is fulfilled by an ordinary BPSK demodulator, e.g. by the one of Figure 7.2b. The entire spreading–despreading cycle is illustrated in Figure 7.5.

To support the discussion in terms of the frequency domain, consider Figure 7.6, which shows the power spectra densities $\tilde{S}_b(f)$, $\tilde{S}_{bs}(f)$ of the initial datastream $B_k(t)$ and its spread version $s_k(t)B_k(t)$, respectively. For a sequence $B_k(t)$ of bit-pulses of duration T_b whose polarities are random and independent, the power spectrum $\tilde{S}_b(f) = T_b \mathrm{sinc}^2(\pi f T_b)$. Treating the spread datastream again as a random sequence of pulses with independent polarities—this time of duration Δ—leads to a power spectrum of the same shape, but occupying N times wider bandwidth: $\tilde{S}_{bs}(f) = \Delta \mathrm{sinc}^2(\pi f \Delta) = (T_b/N)\mathrm{sinc}^2(\pi f T_b/N)$. Transmitting a wide-spectrum signal on the air utilizes all the benefits of spread spectrum (see Chapters 3 and 4) but at the receiving end despreading returns the spectrum into its original bandwidth, converting the signal into narrowband and allowing use of the simplest technologies of data demodulation.

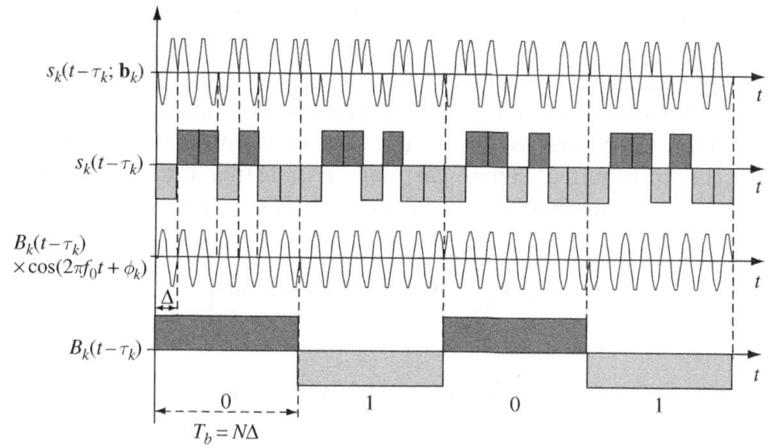

Figure 7.4 Despreading of BPSK data-modulated signal

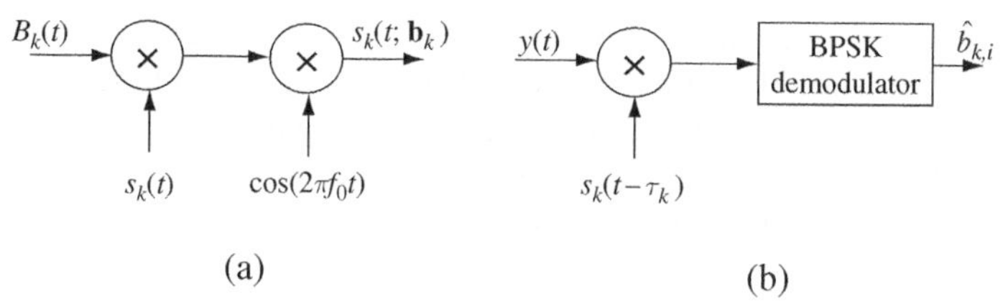

Figure 7.5 Spreading (a) and dispreading (b) a BPSK data-modulated signal

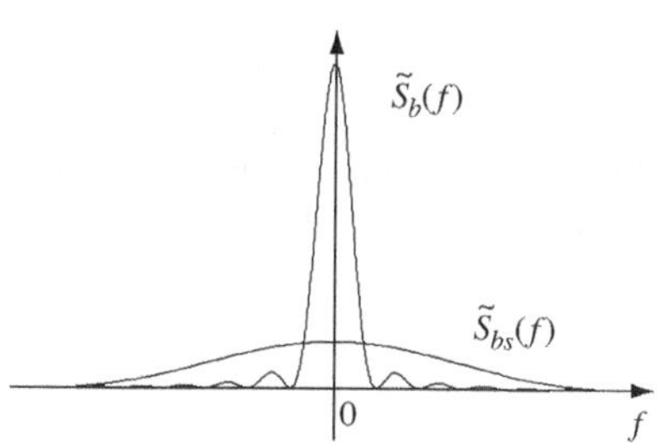

Figure 7.6 Power spectra of original and spread datastreams

7.1.2 DS spreading: general case

The idea of direct spreading considered above for the BPSK data transmission and a binary signature may be readily generalized to comprise a much wider range of data modulation modes and signatures. Let $\dot{B}_k(t)$ denote a complex waveform corresponding to a datastream $\mathbf{b}_k$ of the kth user transmitted in some digital data modulation format (ASK, M-ary PSK, QAM etc.). With M-ary digital data transmission $\dot{B}_k(t)$ consists of contiguous rectangles having duration $T = (\log M)T_b$ and manipulated by complex symbols belonging to the specific M-ary modulation alphabet. For example, in the case of 8-PSK, rectangles of duration $T = 3T_b$ are manipulated by complex amplitudes from the alphabet $\{\exp(jl\pi/4): l = 0, 1, \ldots, 7\}$ shown in Figure 2.6c; if 16-QAM is preferred, then the rectangles have duration $T = 4T_b$, their complex amplitudes taking the values defined by Figure 2.6b and so on. In the case of an ordinary (non-spread) M-ary modulation the transmitted signal carrying a datastream $\mathbf{b}_k$ (it is convenient to assume now that components $b_{k,i}$ of $\mathbf{b}_k$ are M-ary complex symbols: $\dot{B}_k(t) = b_{k,i}, (i-1)T < t \leq iT$):

$$s_k(t; \mathbf{b}_k) = \mathrm{Re}\left[\dot{B}_k(t)\exp(j2\pi f_0 t)\right]$$

The received signal:

$$s_{kr}(t; \mathbf{b}_k) = \mathrm{Re}\left\{\dot{B}_k(t-\tau_k)\exp[j(2\pi f_0 t + \phi_k)]\right\}$$

has a complex envelope:

$$\dot{S}_{kr}(t; \mathbf{b}_k) = \dot{B}_k(t-\tau_k)\exp(j\phi_k) \tag{7.5}$$

The transmitted ith data symbol $b_{k,i}$ is just the complex amplitude of a CW carrier, which is constant within the interval $((i-1)T, iT]$. Consequently, to recover this symbol the receiver should decide between M replicas of the rectangular pulse existing at $((i-1)T + \tau_k, iT + \tau_k]$ and having different competitive values of complex amplitude. To fulfil this, the observed complex envelope $\dot{Y}(t)$ retrieved from the noisy observation $y(t)$ should be correlated with $\exp(j\phi_k)$, resulting in:

$$\dot{z}_k = \int_{(i-1)T+\tau_k}^{iT+\tau_k} \dot{Y}(t)\exp(-j\phi_k)\,\mathrm{d}t$$

used (after normalization to the reference symbol energy) to get the necessary estimate $\hat{b}_{k,i}$ of $b_{k,i}$.

The demodulator of Figure 7.7, where the double circle symbolizes complex multiplication, implements the procedure. This scheme just generalizes the correlator

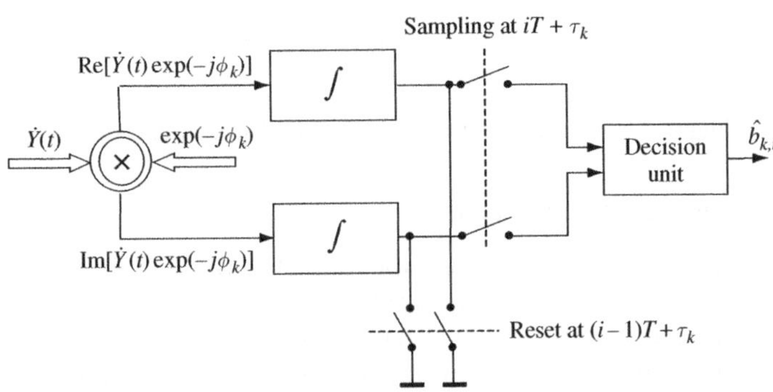

Figure 7.7 M-ary demodulator

structure of Figure 7.2b to the case of arbitrary digital data modulation. The observed complex envelope $\dot{Y}(t)$ is first multiplied by $\exp(-j\phi_k)$ to compensate for the channel phase shift ϕ_k. The useful component (7.5) of $\dot{Y}(t)$ after this operation becomes $\dot{B}_k(t-\tau_k)$. Any complex waveform is equivalent to two real ones (real and imaginary parts), so the multiplier output in Figure 7.7 is treated in terms of real and imaginary parts of the product $\dot{Y}(t)\exp(-j\phi_k)$. The useful components of them are the real and imaginary parts of $\dot{B}_k(t-\tau_k)$ and further integration, as always, serves to clear them off noise. Samples at the integrators' outputs at the moment $iT+\tau_k$ are estimations of the real and imaginary parts of the received M-ary data symbol used by a decision unit to give out the demodulated symbol.

Consider now how direct spreading may be incorporated into this modulation–demodulation scheme. Let $\dot{S}_k(t)$ be the complex envelope of the kth user CDMA signature. Its alphabet may be chosen independently of the data modulation alphabet, e.g. may be binary, quaternary, APSK etc. Then a direct spreading means multiplication of the data modulation $\dot{B}_k(t)$ and signature $\dot{S}_k(t)$ waveforms to use their product $\dot{S}_k(t)\dot{B}_k(t)$ as a complex envelope of a transmitted signal:

$$s_k(t;\mathbf{b}_k) = \mathrm{Re}\left[\dot{S}_k(t)\dot{B}_k(t)\exp(j2\pi f_0 t)\right] \tag{7.6}$$

The received useful signal is a delayed and phase-shifted copy of (7.6):

$$s_{kr}(t;\mathbf{b}_k) = \mathrm{Re}\left\{\dot{S}_k(t-\tau_k)\dot{B}_k(t-\tau_k)\exp[j(2\pi f_0 t+\phi_k)]\right\}$$

with complex envelope:

$$\dot{S}_{kr}(t;\mathbf{b}_k) = \dot{S}_k(t-\tau_k)\dot{B}_k(t-\tau_k)\exp(j\phi_k) \tag{7.7}$$

The constancy of $\dot{B}(t-\tau_k)$ at the interval $((i-1)T+\tau_k, \tau_k]$ means again that to retrieve the ith symbol the receiver should decide between M competitive replicas of the same signature $\dot{S}_k(t-\tau_k)\exp(j\phi_k)$ multiplied by different data symbols $b_{k,i}$. Correlation:

$$\dot{z}_k = \int\limits_{(i-1)T+\tau_k}^{iT+\tau_k} \dot{Y}(t)\dot{S}_k^*(t-\tau_k)\exp(-j\phi_k)\,\mathrm{d}t \tag{7.8}$$

then serves (after proper normalization) to get the necessary estimate of $b_{k,i}$ and may again be treated as a couple of samples at the integrator outputs of the demodulator in Figure 7.7, provided the reference signal in the complex multiplier is changed from $\exp(-j\phi_k)$ to $\dot{S}_k^*(t-\tau_k)\exp(-j\phi_k)$. After multiplying with such a reference the useful component of the observed complex envelope:

$$\dot{S}_{kr}(t;\mathbf{b}_k)\dot{S}_k^*(t-\tau_k)\exp(-j\phi_k) = \left[\left|\dot{S}_k(t-\tau_k)\right|^2\dot{B}_k(t-\tau_k)\exp(j\phi_k)\right]\exp(-j\phi_k)$$

at the ith data symbol interval becomes just one of M possible replicas of the baseband signal $\left|\dot{S}_k(t-\tau_k)\right|^2$ multiplied by different complex coefficients $b_{k,i}$. If signature chips are not amplitude-manipulated, i.e. $\dot{S}_k(t)$ is of PSK type, $\left|\dot{S}_k(t-\tau_k)\right|^2=1$ and, as the previous equation shows, multiplication by $S_k^*(t-\tau_k)$ converts the received signal complex envelope into the one characteristic of an ordinary (non-spread) M-ary data modulation, i.e. performs the despreading. Thanks to this the receiver may again be thought of as a two-stage one: fulfilling first despreading and then a conventional M-ary demodulation, using, e.g., the scheme of Figure 7.7.

Let us dwell now on implementing complex multiplication and extracting the complex envelope $\dot{Y}(t)$ from a truly observed real waveform $y(t)$. Recalling the basic rules of complex arithmetic:

$$\mathrm{Re}(xy) = \mathrm{Re}(x)\mathrm{Re}(y) - \mathrm{Im}(x)\mathrm{Im}(y), \quad \mathrm{Im}(xy) = \mathrm{Re}(x)\mathrm{Im}(y) + \mathrm{Im}(x)\mathrm{Re}(y)$$

the multiplier of two complex entities x and y contains four conventional multipliers and two adders (Figure 7.8). Input complex operands x, y specified by their real and imaginary parts produce two outcomes, being the real and imaginary parts of the product xy.

Obtaining the observed complex envelope is based on the definition of $\dot{Y}(t)$: $y(t) = \mathrm{Re}[\dot{Y}(t)\exp(j2\pi f_0 t)]$. Applying the complex multiplication rule above and the Euler formula we have $y(t) = [\mathrm{Re}\,\dot{Y}(t)]\cos(2\pi f_0 t) - [\mathrm{Im}\,\dot{Y}(t)]\sin(2\pi f_0 t)$. Multiplying both parts of this equation by $2\cos(2\pi f_0 t)$ and $-2\sin(2\pi f_0 t)$, along with using trigonometric identities, leads to:

$$2y(t)\cos(2\pi f_0 t) = \mathrm{Re}\,\dot{Y}(t) + [\mathrm{Re}\,\dot{Y}(t)]\cos(4\pi f_0 t) - [\mathrm{Im}\,\dot{Y}(t)]\sin(4\pi f_0 t)$$

$$-2y(t)\sin(2\pi f_0 t) = \mathrm{Im}\,\dot{Y}(t) - [\mathrm{Im}\,\dot{Y}(t)]\cos(4\pi f_0 t) - [\mathrm{Re}\,\dot{Y}(t)]\sin(4\pi f_0 t) \qquad (7.9)$$

The first terms of the right-hand sides of equations (7.9) are baseband waveforms (since the complex envelope is a modulation law, i.e. baseband), while the rest are bandpass waveforms, having a central frequency $2f_0$. The bandwidth of the modulation law is smaller than f_0 (see Figure 7.9a), therefore the lowpass filters may easily filter out high-frequency components in (7.9), preserving only the real and imaginary parts of a desired complex envelope $\dot{Y}(t)$. This principle of restoring the complex envelope from the real observation $y(t)$ is realized in the scheme shown in Figure 7.9b.

To conclude this discussion, Figure 7.10 illustrates the operations run by the transmitting and receiving sides in a generic DS spread spectrum system. The modulator (Figure 7.10a) implements (7.6) using only the real part of the complex product. In the demodulator (Figure 7.10b) the observed complex envelope retrieved according to the scheme of Figure 7.9b is first despread by multiplication with the reference $\dot{S}_k^*(t - \tau_k)$

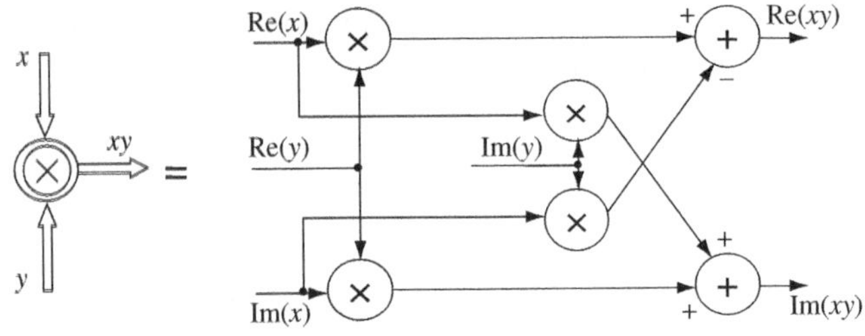

Figure 7.8 Complex multiplication

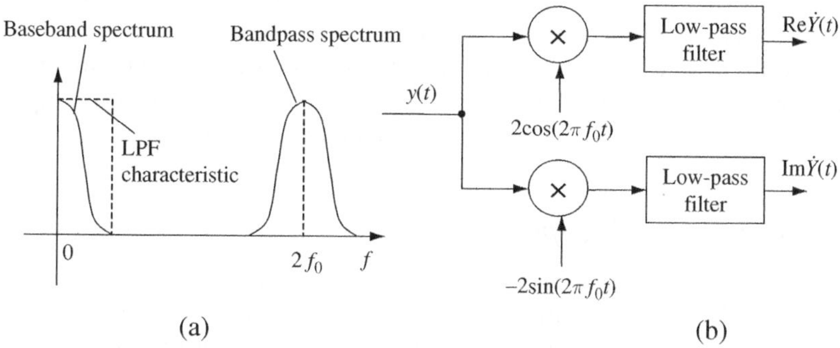

Figure 7.9 Retrieving the complex envelope

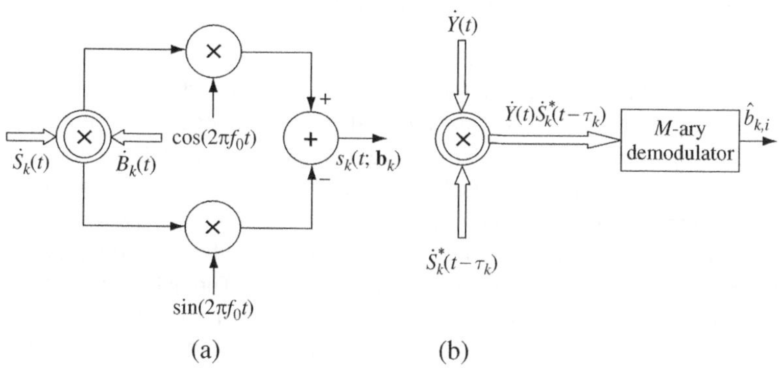

Figure 7.10 Modulation (a) and demodulation (b) in a generic DS spreading

and is then input to a conventional M-ary demodulator (see Figure 7.7) to produce decisions on the received symbols.

Note in passing that the method of spreading–despreading just described is not unique and a variety of concrete circuitry solutions exists to implement those operations. For instance, multiplying complex envelopes may be done indirectly in the course of heterodyning. Specifically, if $u_i(t) = U_i(t)\cos[2\pi f_i t + \gamma_i(t)], i = 1, 2$ are two bandpass signals with carrier frequencies f_i and complex envelopes $\dot{U}_i(t) = U_i(t)\exp[j\gamma_i(t)]$, their product:

$$u_1(t)u_2(t) = \frac{1}{2}U_1(t)U_2(t)\cos[2\pi(f_1 - f_2)t + \gamma_1(t) - \gamma_2(t)] + \frac{1}{2}U_1(t)U_2(t)\cos[2\pi(f_1 + f_2)t + \gamma_1(t) + \gamma_2(t)]$$

The two terms here are bandpass signals of carrier frequencies $f_1 - f_2$ and $f_1 + f_2$. If the lower carrier frequency $f_1 - f_2$ exceeds the bandwidth of the product $\dot{U}_1(t)\dot{U}_2^*(t)$, then after filtering out the higher-frequency term the remaining lower-frequency bandpass signal will have complex envelope $\dot{U}_1(t)\dot{U}_2^*(t)$, i.e. exactly the product necessary in despreading. In the same way, the higher-frequency term is a bandpass signal, whose complex envelope is a similar product without conjugation.

DS spreading is used in all 2G and 3G CDMA standards: IS-95 (cdmaOne), UMTS and cdma2000. Various combinations of data modulation and signature alphabets are involved there, which will be discussed in more detail in Section 11.3.

7.1.3 Frequency hopping spreading

In FH spreading FSK signatures are used and data modulation is also typically FSK. Two sorts of FH are traditionally distinguished—*fast* and *slow*—the relation between the chip Δ and data symbol T durations being the criterion of this classification. For a fast FH $\Delta = T/l$, where $l > 1$ is natural, while for a slow FH $\Delta = lT$ with a natural $l \geq 1$. In other words, for fast FH there are several frequency hops per data symbol, while with slow FH several data symbols may be transmitted during one frequency hop of a signature. To better imagine how it all works let us turn to an example.

Example 7.1.1. Let us take the FSK signature of Example 5.5.1 and use it for fast FH spreading in combination with a binary FSK data modulation. In this case the number of different frequencies in a signature $M = 5$, signature length $N = 8$ and one data symbol transmits one bit, so that $T = T_b$. Suppose that in a fast FH scheme $l = N = 8$, i.e. there are 8 frequency hops per data bit. Then the whole sequence of FSK chips shown in Figure 5.3 is transmitted during one bit. If the data bit is zero, this frequency pattern is transmitted on the carrier frequency f_0, while for the bit equal to one the carrier frequency jumps to f_1. The difference between f_1 and f_0 should certainly be no smaller than the bandwidth occupied by the signature, i.e. MF. The transmitted frequency pattern corresponding to data bit stream 01011 is shown in Figure 7.11a.

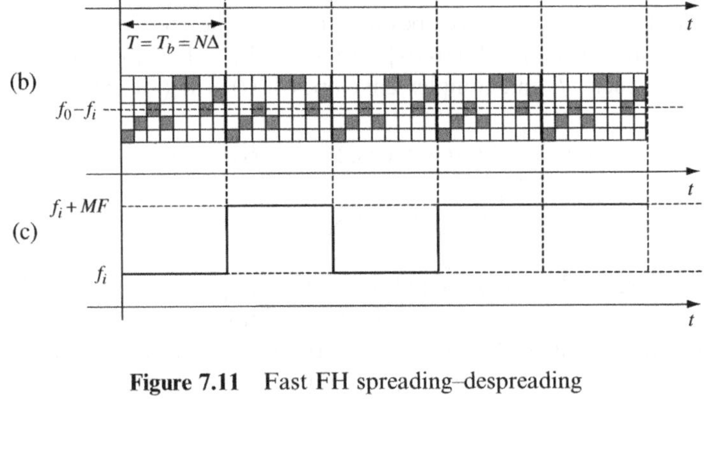

Figure 7.11 Fast FH spreading–despreading

As may be seen, the spectrum of a single data bit whose width before spreading was about $1/T_b$ is spread to span a bandwidth around $MF \approx M/\Delta = MN/T_b$, i.e. MN times wider (see Section 5.6). At the receiving end despreading consists in down-converting the observed waveform to the intermediate frequency f_i. For that the reference waveform of the carrier $f_0 - f_i$ is used, modulated according to a signature FSK pattern, properly delayed in time (Figure 7.11b). As a result a despread signal is an ordinary narrowband FSK data-modulated waveform, where the zero data bit is transmitted by lower frequency than the bit one (Figure 7.11c). The spectrum of an isolated data symbol is now returned to the bandwidth $1/T_b$ and a conventional binary FSK demodulator may be used to recover the received data.

Let us illustrate in the next example how to run slow FH spreading.

Example 7.1.2. Take the same signature in combination with, again, binary FSK data modulation and set equality between chip and symbol durations: $T = T_b = \Delta$. This means that the current frequency remains constant over all the data bit duration and frequency hops happen only from one bit to the next. The frequency pattern of the signature is stretched in time and its length now covers N data bits (Figure 7.12a). Suppose that during a data bit number i signature frequency equals F_i. Then the transmitted frequency becomes $f_0 + F_i$ for a zero data

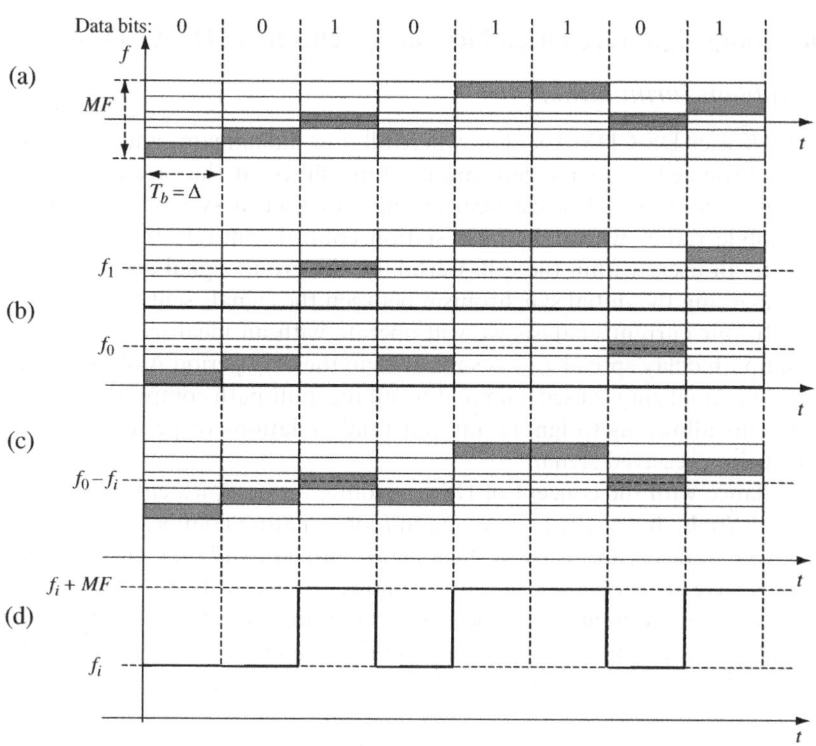

Figure 7.12 Slow FH spreading–despreading

bit and $f_1 + F_i$ for a data bit equal to one. Figure 7.12b shows this for the bit stream 00101101. The principal difference between fast and slow FH is now seen: the latter does not spread the spectrum of an individual data symbol, widening just the total bandwidth occupied by the system. It looks like a system that merely switches from one operational frequency to another from time to time but within a fixed group of data symbols no switching happens. At the receiving end down-conversion to an intermediate frequency f_i is accomplished with the aid of a reference signal repeating the signature frequency pattern (with an appropriate delay) on the carrier $f_0 - f_i$ (Figure 7.12c). This returns the waveform to the bandwidth inherent to a plain (non-frequency-hopping) FSK data modulation (Figure 7.12d), so that an ordinary FSK demodulator may restore the transmitted data (Figure 7.12c).

The techniques illustrated in the examples above for binary data transmission are easily extended to a general FSK data modulation (see Problems 7.5–7.7).

FH spreading has some features that make it especially attractive for military applications, in particular in various antagonistic scenarios of games against jamming systems [3,6]. At the same time, its commercial use had not until recently been significant, at least as regards fast FH. However, the advent of Bluetooth technology [55] indicates that this kind of spread spectrum may also possess good commercial prospects.

7.2 Designing signature ensembles for synchronous DS CDMA

7.2.1 Problem formulation

Consider a K-user DS CDMA system where all user datastreams and all signatures are strictly synchronized, i.e. have zero mutual time shifts, at the receiver input. As was pointed out in Section 4.4, a classical example of such a system is the downlink of CDMA mobile radio, where the base station controls entirely the timing of signals addressed to all users within the cell. Certainly, the group signal arrives at the mobile receiver preserving the initial synchronism between the signals sent to different individual users. In our current analysis we will operate with an idealized channel model, in which multipath delay spread $\tau_{\max}$ is smaller than the chip period Δ of users' signatures, or an efficient equalizing is used, eliminating all the multipath components whose delays exceed Δ. This allows us to ignore any potential violations of perfect synchronism of components in a received signal.

In accordance with the concept of DS spreading, the complex envelope of a received group signal $\dot{S}(t; \mathbf{b}_1, \mathbf{b}_2, \ldots, \mathbf{b}_K)$ is the sum in k of signature complex envelopes manipulated by users' datastreams, each of them being defined by (7.6). Generally, each user's signals may have its individual amplitude; however, we will restrict ourselves to the simplest case of equal intensities. Since the assumption of perfect synchronism permits us to set all delays τ_k and initial phases ϕ_k in (7.7) equal to zero, we then arrive at the expression of the received complex envelope (subscript r discarded as needless now):

$$\dot{S}(t; \mathbf{b}_1, \mathbf{b}_2, \ldots, \mathbf{b}_K) = \sum_{k=1}^{K} \dot{B}_k(t)\dot{S}_k(t) \tag{7.10}$$

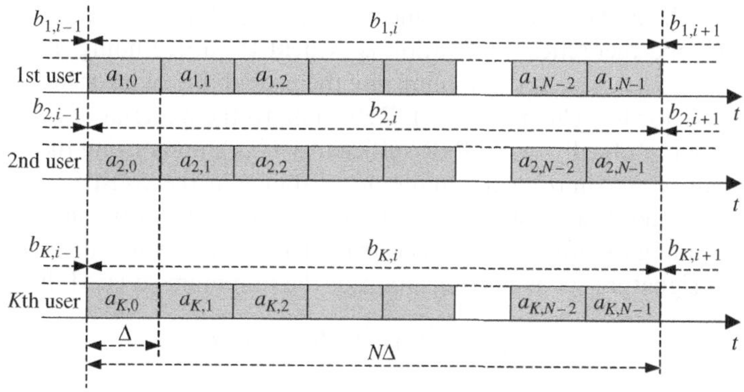

Figure 7.13 Data symbol and chip alignment in synchronous CDMA

Let us focus on a single data symbol interval of duration T. Again, due to the complete synchronism the current data symbols of all users start and end strictly simultaneously. With b_k being the kth user's current data symbol, (7.10) during a single symbol interval may be written as follows:

$$\dot{S}(t; \mathbf{b}) = \sum_{k=1}^{K} b_k \dot{S}_k(t) \tag{7.11}$$

where $\mathbf{b} = (b_1, b_2, \ldots, b_K)$ is the K-dimensional vector of current data symbols of all users.

Remember now that every signature in the DS CDMA system is an APSK signal described by the model (5.2):

$$\dot{S}(t) = \sum_{i=0}^{N-1} a_{k,i} \dot{S}_0(t - i\Delta) \tag{7.12}$$

where $\{a_{k,0}, a_{k,1}, \ldots a_{k,N-1}\}$ is a code sequence, manipulating chips of the kth signature, and N is a spreading factor, i.e. the number of chips per one data symbol. Figure 7.13 emphasizes the strong mutual alignment between signature chips and boundaries of transmitted data symbols characteristic of synchronous CDMA.

Based on (7.11) and (7.12), several approaches to designing signature ensembles for synchronous DS CDMA networks may be formulated. Among the main factors influencing the procedure and results of the signature set optimization is the relation between the number of users K and spreading factor N, as well as the receiver algorithm (multiuser or conventional).

7.2.2 Optimizing signature sets in minimum distance

Suppose that the receiver of any complexity is admissible and, therefore, we are allowed to use the optimal (multiuser) algorithm of estimating the data vector $\mathbf{b}$ based on the search for the value of $\mathbf{b}$ minimizing the distance between observation $y(t)$ and the candidate group signal $s(t; \mathbf{b})$ (see Section 4.1). In terms of the complex envelope this means minimization in $\mathbf{b}$ of the squared distance $\left\|\dot{Y}(t) - \dot{S}(t;\mathbf{b})\right\|^2$, where $\dot{S}(t;\mathbf{b})$ is given by (7.11). Then a solid theoretical motivation is evident towards finding the ensemble of K signatures $\{\dot{S}_1(t), \dot{S}_2(t), \ldots, \dot{S}_K(t)\}$ minimizing the probability of error in the estimate $\hat{\mathbf{b}}$ of a K-user data vector $\mathbf{b} = (b_1, b_2, \ldots, b_K)$. Returning to the material of Section 2.3, let us recollect that asymptotically (with SNR sufficiently high), minimizing the error probability is equivalent to maximizing the minimum distance in the constellation of M transmitted signals. In the studied case the alternative signals to be distinguished are copies of (7.11) corresponding to different data vectors $\mathbf{b}$. Hence, we may formulate the problem of optimizing the signature set as maximization of the minimum squared distance:

$$d^2_{\min} = \min_{\substack{\mathbf{b},\mathbf{b}' \\ \mathbf{b}\neq\mathbf{b}'}} d^2(\mathbf{b},\mathbf{b}') = \max \tag{7.13}$$

where the minimum distance $d_{\min}$ is found over all different pairs of data vectors $\mathbf{b} = (b_1, b_2, \ldots, b_K), \mathbf{b}' = (b'_1, b'_2, \ldots, b'_K), \mathbf{b} \neq \mathbf{b}'$ and (see (2.43)):

$$d^2(\mathbf{b},\mathbf{b}') = \|s(t;\mathbf{b}) - s(t;\mathbf{b}')\|^2 = \frac{1}{2}\left\|\dot{S}(t;\mathbf{b}) - \dot{S}(t;\mathbf{b}')\right\|^2 \tag{7.14}$$

Let us study in more detail binary data transmission when the data symbols are bits transmitted directly by BPSK so that $b_k, b'_k = \pm 1, k = 1, 2, \ldots, K$. This narrowing of the scope makes the analysis a bit easier, subsequent spreading to a general PSK being straightforward. Then using (7.11), (2.41) and (2.42) in (7.14) results in:

$$d^2(\mathbf{b},\mathbf{b}') = \frac{1}{2}\left\|\sum_{k=1}^{K} \varepsilon_k \dot{S}_k(t)\right\|^2 = \frac{1}{2}\int_0^T \left|\sum_{k=1}^{K} \varepsilon_k \dot{S}_k(t)\right|^2 \mathrm{d}t = E_b \sum_{k=1}^{K}\sum_{l=1}^{K} \varepsilon_k \varepsilon_l \dot{\rho}_{kl} \tag{7.15}$$

where $\varepsilon_k = b_k - b'_k$ takes on one of three possible values: 0 or ± 2; $E_b = \frac{1}{2}\int_0^T \left|\dot{S}_k(t)\right|^2 \mathrm{d}t$ is the energy of the kth signature used for transmitting one bit, these energies for all K users assumed the same, and:

$$\dot{\rho}_{kl} = \frac{1}{2E_b}\int_0^T \dot{S}_k(t)\dot{S}^*_k(t)\,\mathrm{d}t$$

is the correlation coefficient between the complex envelopes of the kth and lth signatures. Using properties of the correlation coefficients seen from its definition, $\dot{\rho}_{kk} = 1, \dot{\rho}_{kl} = \dot{\rho}^*_{lk}$ (7.15) takes the form explicitly showing that distance is always real:

$$d^2(\mathbf{b},\mathbf{b}') = E_b \sum_{k=1}^{K} \varepsilon_k^2 + 2E_b \sum_{k=1}^{K-1}\sum_{l=k+1}^{K} \varepsilon_k \varepsilon_l \mathrm{Re}(\dot{\rho}_{kl}) \tag{7.16}$$

Take two data vectors (bit patterns) $\mathbf{b}, \mathbf{b}'$ differing in only one, for instance the first, component. Then $\varepsilon_k = 0, k = 2, 3, \ldots, K, \varepsilon_1 = \pm 2$ and from (7.16) $d^2(\mathbf{b},\mathbf{b}') = 4E_b$. Since $d^2_{\min}$ is never greater than the squared distance for any specific pair of $\mathbf{b}, \mathbf{b}'$:

$$d^2_{\min} \leq 4E_b \tag{7.17}$$

This upper bound tells us that a signature ensemble, for which $d^2_{\min} = 4E_b$, should be treated as optimal according to the criterion of maximum minimum distance (7.13). One of the sufficient conditions of achieving the bound (7.17) is the *weak* orthogonality of complex envelopes of signatures:

$$\mathrm{Re}(\dot{\rho}_{kl}) = \begin{cases} 1, k = l \\ 0, k \neq l \end{cases} = \delta_{kl} \tag{7.18}$$

The reason why complex envelopes satisfying (7.18) are called weakly orthogonal becomes clear after comparison of (7.18) and (2.46) ($\dot{\rho}_{kl} = \delta_{kl}$). The latter is much more demanding and forces the signals $s_k(t), s_l(t)$ with the complex envelopes $\dot{S}_k(t), \dot{S}_l(t)$ to preserve orthogonality under any mutual phase shifts. At the same time, two signals modulated by $\dot{S}(t)$ and $\dot{S}(t)\exp[j(\pi/2)] = j\dot{S}(t)$, i.e. just quadrature (phase shifted by $\pi/2$) replicas of the same signal, are orthogonal, but lose orthogonality if their mutual phase shift differs from $\pm\pi/2$. Hence, $\dot{S}(t)$ and $j\dot{S}(t)$ are only weakly orthogonal. Of course, any orthogonal (in terms of (2.46)) signatures are weakly orthogonal, but not vice versa.

For the signatures meeting (7.18), equation (7.16) becomes $d^2(\mathbf{b}, \mathbf{b}') = E_b \sum_{k=1}^{K} \varepsilon_k^2$. At least one of the summands in this sum is non-zero, so that $d^2_{\min} \geq 4E_b$, whereby along with (7.17) $d^2_{\min} = 4E_b$. The implication of this is that the ensemble of K weakly orthogonal signatures is optimal in minimum distance, and hence (asymptotically) in probability of confusion between different users' bit patterns.

A lot of techniques exist for generating orthogonal (meeting (2.46)) spread spectrum signals for various lengths (spreading factors) N. One example is Walsh functions or, more generally, Hadamard matrices, discussed in Section 2.7.3 and providing binary orthogonal codes. Another possible option is cyclically shifted replicas of any sequence with perfect periodic ACF, e.g. ternary, polyphase etc. (see Section 6.11). Any ensemble of K' orthogonal signatures is trivially transformed into the set of weakly orthogonal signatures containing $2K'$ signals by adding quadrature replicas of any signal—a fact repeatedly referred to before (see Sections 2.5 and 4.1).

Under any specific choice of orthogonal signatures the signal space dimension strictly limits their number (and hence, number of users K) (see Section 2.5). According to (7.12), given the chip, N-dimensional vector $\mathbf{a}_k = (a_{k,0}, a_{k,1}, \ldots, a_{k,N-1})$ of the kth code sequence exhaustively determines the kth signature, and the orthogonality of the kth and lth signatures is equivalent to the orthogonality of vectors $\mathbf{a}_k, \mathbf{a}_l$. Indeed, repeating the derivation of (2.52) for complex envelopes (alternatively using (5.7)) allows the inner product of $\dot{S}_k(t), \dot{S}_l(t)$ to be found, as:

$$(\dot{\mathbf{S}}_k, \dot{\mathbf{S}}_l) = 2E_0 \sum_{i=0}^{N-1} a_{k,i} a^*_{l,i} = 2E_0(\mathbf{a}_k, \mathbf{a}_l) \tag{7.19}$$

confirming that the orthogonality of $\mathbf{a}_k, \mathbf{a}_l$ is necessary and sufficient for the orthogonality of $\dot{S}_k(t), \dot{S}_l(t)$. Dimension N of the space of code sequence vectors $\mathbf{a}_k$ is evidently a maximal number K' of orthogonal signatures $\dot{S}_k(t)$. Let us stress again that when quadrature splitting of every signature is allowed, the maximal number of users accommodated within the signature ensemble is defined as $K = 2K' = 2N$. If, however, for some reason the accurate phase shift $\pm\pi/2$ between the quadrature copies of the same signature cannot be maintained, weak orthogonality is insufficient, and the maximal number of users is two times smaller: $K = K' = N$.

Note that weak orthogonality is only a sufficient but not a necessary condition of equality in (7.17) and, in particular, it is quite an interesting issue whether it is possible to achieve the upper bound in (7.17) with the number of signatures exceeding the dimension of the signal space n_s. As follows from the previous reasoning, n_s is either $2N$ or N depending on whether or not a quadrature splitting of signatures is allowed. A synchronous CDMA system in which $K > n_s$ is called *oversaturated*, emphasizing that the excessive number of code vectors involved excludes the chance of their orthogonality (possibly a weak one).

The opportunity and algorithm for obtaining the minimum distance equal to the upper limit (7.17) in an oversaturated system was proved in [56]. To discuss the idea more transparently and simplify the notation, let us first ignore the opportunity for doubling the signal space dimension due to a quadrature splitting, putting $n_s = N$. Let us take N orthonormal N-dimensional vectors $\mathbf{a}_k, k = 1, 2, \ldots, N$; $(\mathbf{a}_k, \mathbf{a}_l) = \delta_{kl}$, and add to them one more vector built as:

$$\mathbf{a}_{N+1} = \frac{1}{\sqrt{N}} \sum_{k=1}^{N} \mathbf{a}_k \tag{7.20}$$

Using $N + 1$ vectors $\mathbf{a}_k, k = 1, 2, \ldots, N + 1$ thus obtained to form $K = N + 1$ signatures according to (7.12), we have the $N + 1$th signature:

$$S_{N+1}(t) = \frac{1}{\sqrt{N}} \sum_{k=1}^{N} \dot{S}_k(t)$$

and, modulating all the signatures by binary data symbols $b_k = \pm 1$, a group signal:

$$\begin{aligned} \dot{S}(t; \mathbf{b}) &= \sum_{k=1}^{N+1} b_k \dot{S}_k(t) = \sum_{k=1}^{N} b_k \dot{S}_k(t) + \frac{1}{\sqrt{N}} b_{N+1} \sum_{k=1}^{N} \dot{S}_k(t) \\ &= \sum_{k=1}^{N} (b_k + \frac{1}{\sqrt{N}} b_{N+1}) \dot{S}_k(t). \end{aligned} \tag{7.21}$$

The difference between the two versions of the group signal corresponding to two bit patterns $\mathbf{b} = (b_1, b_2, \ldots, b_{N+1}), \mathbf{b}' = (b'_1, b'_2, \ldots, b'_{N+1})$ is:

$$\dot{S}(t; \mathbf{b}) - \dot{S}(t; \mathbf{b}') = \sum_{k=1}^{N} \left(\varepsilon_k + \frac{1}{\sqrt{N}} \varepsilon_{N+1} \right) \dot{S}_k(t), \varepsilon_k = b_k - b'_k = 0, \pm 2$$

Using the same technique as in (7.15) and the orthogonality of the first N signatures, we arrive at:

$$d^2(\mathbf{b}, \mathbf{b}') = \left\| \dot{S}(t; \mathbf{b}) - \dot{S}(t; \mathbf{b}') \right\|^2 = E_b \sum_{k=1}^{N} \left(\varepsilon_k + \frac{1}{\sqrt{N}} \varepsilon_{N+1} \right)^2 \tag{7.22}$$

Since the bit patterns $\mathbf{b}, \mathbf{b}'$ are different, at least one of $\varepsilon_k, k = 1, 2, \ldots, N+1$ is non-zero, i.e. equals ± 2. If $\varepsilon_{N+1} = 0$ then such an ε_k is present among $\varepsilon_1, \varepsilon_2, \ldots, \varepsilon_N$ and $d^2(\mathbf{b}, \mathbf{b}') \geq 4E_b$. If $\varepsilon_{N+1} = \pm 2$, then the summands in (7.22) with $\varepsilon_k = 0$ equal $4/N$, while all the rest are $4(\sqrt{N} \pm 1)^2/N$, resulting in $d^2(\mathbf{b}, \mathbf{b}') \geq 4E_b \min\{1, (\sqrt{N}-1)^2\}$. Combining these results we come to the estimate of the minimum squared distance from below:

$$d_{\min}^2 \geq \min\left\{4E_b, 4\left(\sqrt{N}-1\right)^2 E_b\right\} = \begin{cases} 4\left(\sqrt{N}-1\right)^2 E_b, N < 4 \\ 4E_b, N \geq 4 \end{cases}$$

Comparing this with (7.17) shows the possibility of adding one extra signature to the N orthogonal ones without sacrificing the minimum distance, whenever $N \geq 4$. Generalization of this idea underlies the following procedure of building an optimal oversaturated signature ensemble [56,57]. Let vectors $\mathbf{a}_0^0, \mathbf{a}_1^0, \ldots, \mathbf{a}_{N-1}^0$ be an orthonormal basis of N-dimensional space where $N = 4^l$, l is natural. Let us use them as codes of N primary signatures. We arrange oversaturating supplementary signatures as an l-layer procedure. Supplementary signature code sequences of the sth layer are:

$$\mathbf{a}_k^s = \frac{1}{2}\sum_{m=0}^{3} \mathbf{a}_{4k+m}^{s-1} = \frac{1}{2^s}\sum_{m=0}^{4^s-1} \mathbf{a}_{4^s k+m}^0, \quad k = 0, 1, \ldots, \frac{N}{4^s} - 1, \quad s = 1, 2, \ldots, l \tag{7.23}$$

In other words, in the first layer of supplementary signatures we perform splitting of the basic set $\{\mathbf{a}_0^0, \mathbf{a}_1^0, \ldots, \mathbf{a}_{N-1}^0\}$ into 4^{l-1} groups each containing four primary signatures. The linear combination (7.20) (where $N = 4$) of these four basic signatures is added to their group, producing the total of $N + N/4$ signatures. At the second layer we split all supplementary signatures of the first layer the same way into groups of four and introduce again in each group linear combinations (7.20), and so forth. The tree in Figure 7.14 illustrates the whole procedure. Then there are 4^{l-1} supplementary signatures at the first layer, 4^{l-2} at the second and generally 4^{l-s} at the sth layer totalling:

$$4^{l-1} + 4^{l-2} + \cdots + 4 + 1 = \frac{4^l - 1}{3} = \frac{N-1}{3}$$

supplementary signatures, or—together with the primary ones identified with a layer zero—an overall number of signatures:

$$K = \frac{4N-1}{3} = \left\lfloor \frac{4N}{3} \right\rfloor$$

Since norms of all vectors (7.23) remain equal to one, supplementary signatures preserve the same energy per bit E_b as the primary ones.

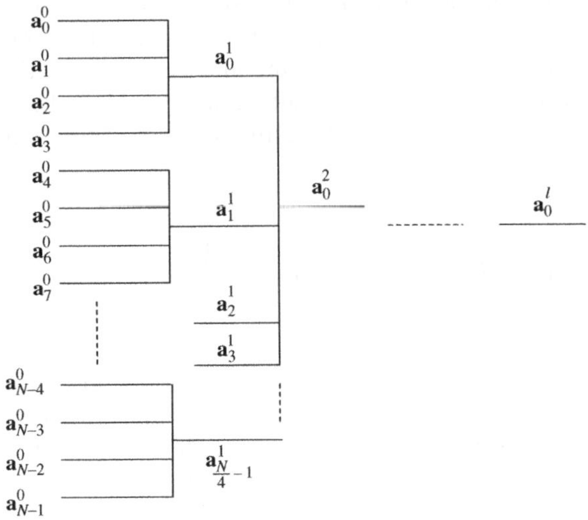

Figure 7.14 Constructing an oversaturated signature set

Let $\dot{S}_k^s(t)$ and b_k^s be the complex envelope of the kth signature on the sth layer and the user's bit transmitted by this signature, respectively. Then composing a group signal similarly to (7.21) leads to:

$$\dot{S}(t;\mathbf{b}) = \sum_{k=0}^{N-1} b_k^0 \dot{S}_k^0(t) + \sum_{k=0}^{\frac{N}{4}-1} b_k^1 \dot{S}_k^1(t) + \cdots + \sum_{k=0}^{\frac{N}{4^{l-1}}-1} b_k^{l-1} \dot{S}_k^{l-1}(t) + b_0^l \dot{S}_0^l(t) = \sum_{s=0}^{l} \sum_{k=0}^{\frac{N}{4^s}-1} b_k^s \dot{S}_k^s(t)$$

which after substituting (7.23) turns into:

$$\dot{S}(t;\mathbf{b}) = \sum_{s=0}^{l} \frac{1}{2^s} \sum_{k=0}^{\frac{N}{4^s}-1} \sum_{m=0}^{4^s-1} b_k^s \dot{S}_{4^s k+m}^0(t)$$

The double sum in k, m here contains N summands independently of s. It can be rearranged into a single sum after changing the summation index as:

$$4^s k + m = n \Rightarrow k = \left\lfloor \frac{n}{4^s} \right\rfloor$$

leading to (with redesignation $n \to k$):

$$\dot{S}(t;\mathbf{b}) = \sum_{s=0}^{l} \frac{1}{2^s} \sum_{k=0}^{N-1} b^s_{\left\lfloor \frac{k}{4^s} \right\rfloor} \dot{S}_k^0(t) = \sum_{k=0}^{N-1} \left(b_k^0 + \frac{1}{2} b^1_{\left\lfloor \frac{k}{4} \right\rfloor} + \cdots + \frac{1}{2^{l-1}} b^{l-1}_{\left\lfloor \frac{k}{4^{l-1}} \right\rfloor} + \frac{1}{2^l} b_0^l \right) \dot{S}_k^0(t)$$

Then the squared distance between the group signals corresponding to different bit patterns generalizes (7.22) as:

$$d^2(\mathbf{b}, \mathbf{b}') = E_b \sum_{k=0}^{N-1} \left(\varepsilon_k^0 + \frac{1}{2} \varepsilon^1_{\lfloor \frac{k}{4} \rfloor} + \cdots + \frac{1}{2^{l-1}} \varepsilon^{l-1}_{\lfloor \frac{k}{4^{l-1}} \rfloor} + \frac{1}{2^l} \varepsilon_0^l \right)^2 \tag{7.24}$$

where, as before, $\varepsilon_n^s = 0, \pm 2$ is the difference of bits transmitted on the signature $\dot{S}_k^s(t)$ in the user's bit patterns $\mathbf{b}, \mathbf{b}'$. If all $\varepsilon_m^s, s > 0$ are zeros (bits of $\mathbf{b}, \mathbf{b}'$ on all the supplementary signatures are identical), then at least one of ε_k^0 equals ± 2 and $d^2(\mathbf{b}, \mathbf{b}') \geq 4E_b$. If u is a maximal layer number for which $\varepsilon_m^u = \pm 2, u > 0$, then a summand of (7.24) containing ε_m^u may be presented as:

$$\frac{1}{4^{u-1}} \left| (2^u x_0 \pm 2^{u-1} x_1 \pm \cdots \pm 2 x_{u-1}) \pm 1 \right|^2$$

where $x_s = \varepsilon_n^s / 2 = 0, \pm 1, s = 0, 1, \ldots, u-1$. The number in the round brackets of the modulus above is always even so the squared modulus is never smaller than one. Since there are exactly 4^u terms in (7.24) entered by ε_m^u with any fixed m, we come to an estimation $d^2(\mathbf{b}, \mathbf{b}') \geq 4^u E_b / 4^{u-1} = 4E_b$, proving that the oversaturated ensemble of this sort does not reduce the minimum distance of the primary orthogonal set.

In its general form the procedure described does not guarantee that the supplementary code sequences (7.23) obtained from the binary primary sequences will also be binary. To meet this latter demand, a version of the procedure may be used [57] in which primary sequences are generated as rows of the lth Kronecker power of the 4th order Hadamard matrix having an odd number of plus ones in any column.

Example 7.2.1. Let us build up an oversaturated ensemble of binary signatures of length $N = 16 = 4^2$. According to the scheme just discussed, 5 supplementary orthogonal signatures may be added to the $N = 16$ primary ones (four at layer $s = 1$ and one at layer $s = 2$) providing total number of users $K = 21$. In order to have all signatures binary take the Hadamard matrix:

$$\mathbf{H}_4 = \begin{bmatrix} + & + & + & + \\ - & - & + & + \\ + & - & + & - \\ + & - & - & + \end{bmatrix}$$

having one or three plus ones in its columns and form its Kronecker square:

$$\mathbf{H}_{16} = \mathbf{H}_4 \otimes \mathbf{H}_4 = \begin{bmatrix} \mathbf{H}_4 & \mathbf{H}_4 & \mathbf{H}_4 & \mathbf{H}_4 \\ -\mathbf{H}_4 & -\mathbf{H}_4 & \mathbf{H}_4 & \mathbf{H}_4 \\ \mathbf{H}_4 & -\mathbf{H}_4 & \mathbf{H}_4 & -\mathbf{H}_4 \\ \mathbf{H}_4 & -\mathbf{H}_4 & -\mathbf{H}_4 & \mathbf{H}_4 \end{bmatrix}$$

Primary signatures are just rows of this matrix, i.e. in the normalized form:

$$\left(\mathbf{a}_0^0, \mathbf{a}_1^0, \mathbf{a}_2^0, \mathbf{a}_3^0, \mathbf{a}_4^0, \mathbf{a}_5^0, \mathbf{a}_6^0, \mathbf{a}_7^0, \mathbf{a}_8^0, \mathbf{a}_9^0, \mathbf{a}_{10}^0, \mathbf{a}_{11}^0, \mathbf{a}_{12}^0, \mathbf{a}_{13}^0, \mathbf{a}_{14}^0, \mathbf{a}_{15}^0\right)^{\mathrm{T}} =$$

$$= \frac{1}{4}\begin{pmatrix}
+ & + & + & + & + & + & + & + & + & + & + & + & + & + & + & + \\
- & - & + & + & - & - & + & + & - & - & + & + & - & - & + & + \\
+ & - & + & - & + & - & + & - & + & - & + & - & + & - & + & - \\
+ & - & - & + & + & - & - & + & + & - & - & + & + & - & - & + \\
- & - & - & - & - & - & - & - & + & + & + & + & + & + & + & + \\
+ & + & - & - & + & + & - & - & - & - & + & + & - & - & + & + \\
- & + & - & + & - & + & - & + & + & - & + & - & + & - & + & - \\
- & + & + & - & - & + & + & - & + & - & - & + & + & - & - & + \\
+ & + & + & + & - & - & - & - & + & + & + & + & - & - & - & - \\
- & - & + & + & + & + & - & - & - & - & + & + & + & + & - & - \\
+ & - & + & - & - & + & - & + & + & - & + & - & - & + & - & + \\
+ & - & - & + & - & + & + & - & + & - & - & + & - & + & + & - \\
+ & + & + & + & - & - & - & - & - & - & - & - & + & + & + & + \\
- & - & + & + & + & + & - & - & + & + & - & - & - & - & + & + \\
+ & - & + & - & - & + & - & + & - & + & - & + & + & - & + & - \\
+ & - & - & + & - & + & + & - & - & + & + & - & + & - & - & +
\end{pmatrix}$$

Applying (7.23) to the rows of this matrix gives five supplementary binary signatures:

$$\begin{pmatrix} \mathbf{a}_0^1 \\ \mathbf{a}_1^1 \\ \mathbf{a}_2^1 \\ \mathbf{a}_3^1 \\ \mathbf{a}_0^2 \end{pmatrix} = \frac{1}{4}\begin{pmatrix}
+ & - & + & + & + & - & + & + & + & - & + & + & + & - & + & + \\
- & + & - & - & - & + & - & - & + & - & + & + & + & - & + & + \\
+ & - & + & + & - & + & - & - & + & - & + & + & - & + & - & - \\
+ & - & + & + & - & + & - & - & - & + & - & - & + & - & + & + \\
+ & - & + & + & - & + & - & - & + & - & + & + & + & - & + & +
\end{pmatrix}$$

It may be quite a challenge for the reader to check the minimum distance property of this oversaturated ensemble.

Let us remind ourselves that the criterion of minimum distance is adequate (at least asymptotically) whenever multiuser reception is affordable. Up to this point we have not worried about the multiuser receiver complexity. For the case of non-oversaturated systems ($K \leq N$) this is not a critical matter, since—the orthogonal ensemble being optimal—multiuser reception degenerates in this case to a single-user one (see Section 4.1). On the other hand, when an oversaturated system is analysed, the opportunities for simplifying the multiuser algorithm at the cost of proper design of signatures are very important. One way to realize this approach again exploits the idea of splitting the overall N-dimensional signal space into orthogonal subspaces of smaller dimension n. However, in contrast with what was discussed above, every subspace is further oversaturated autonomously, providing $n+n_{ov}$ signatures, so that all signatures from different subspaces remain orthogonal. The reason for so doing is to split an overall multiuser algorithm into N/n parallel ones, each operating in the n-dimensional subspace independently of the others. With a moderate n these partial algorithms are simple, making the whole receiver structure technologically feasible. The total number of users achievable in such a system is:

$$\frac{N}{n}(n+n_{ov}) = N\left(1+\frac{n_{ov}}{n}\right)$$

The problem of optimizing an ensemble of this sort is in a sense non-trivial, and just adding n_{ov} supplementary signatures to the n primary orthogonal ones does not solve it. We refer the curious reader to [58,59] for details. Another alternative is the design of signature ensembles allowing implementation of multiuser algorithms in various computational-effective iterative forms [60,61].

7.2.3 Welch-bound sequences

Let us get down to another scenario, where a priori tough limitation on the receiver complexity makes acceptable only the simplest, i.e. single-user or conventional, reception algorithm. In this case a decision $\hat{b}_k$ on the current data symbol b_k of the kth user is defined only by correlation (7.8), as though no interference except AWGN is present at the receiver input. With no loss of generality, we may admit that a current symbol is received at the interval $(0,T]$ and put delay τ_k and phase ϕ_k in (7.8) equal to zero:

$$\dot{z}_k = \int_0^T \dot{Y}(t)\dot{S}_k^*(t)\mathrm{d}t \qquad (7.25)$$

When all signatures are perfectly synchronized and their number K does not exceed N, orthogonal signatures are again the best choice, since they make a single-user algorithm identical to a multiuser (ML) one. Certainly, no MAI arises in this case so ignoring all the signals of the other users does not undermine the receiver optimality. In contrast to this, the case of an oversaturated ($K > N$) system is of separate interest, because all signatures then cannot be orthogonal and MAI is unavoidable. Returning to (7.11), let us present the observed complex envelope as:

$$\dot{Y}(t) = \dot{S}(t;\mathbf{b}') + \dot{N}(t) = \sum_{l=1}^{K} b'_l \dot{S}_l(t) + \dot{N}(t)$$

where $\dot{N}(t)$ is the noise complex envelope and designation $\mathbf{b}' = (b'_1, b'_2, \ldots, b'_K)$ symbolizes again (as in (4.8)) the genuine (i.e. unknown at the receiver) data pattern transmitted by K users to distinguish it from the one $\mathbf{b} = (b_1, b_2, \ldots, b_K)$ hypothesized in the course of the decision. After substituting this into (7.25) we obtain:

$$\dot{z}_k = 2b'_k E + 2E\sum_{\substack{l=1\\ l\neq k}}^{K} b'_l \dot{\rho}_{lk} + \int_0^T \dot{N}(t)\dot{S}_k^*(t)\mathrm{d}t \qquad (7.26)$$

where $E = \frac{1}{2}\int_0^T |\dot{S}_k(t)|^2 \mathrm{d}t$ is (assumed the same for all users) the signature energy per one transmitted symbol, and $\dot{\rho}_{lk} = \dot{\rho}^*_{kl}$ is, as always, the correlation coefficient of the complex envelopes of the lth and kth signatures. The second term of (7.26) presents MAI, i.e. mutual interference created by the alien signals at the output of the receiver 'tuned' to the kth user signal. Each summand $b'_l\dot{\rho}_{lk}$ of the sum in l (i.e. the contribution to total MAI of the lth user signal) is random due to the randomness of users' data symbols b'_l. For any PSK, data modulation $|b'_l| = 1$ and average power (variance) of each contribution to MAI is $4E^2|\dot{\rho}_{lk}|^2$. Naturally, all the users transmit their data independently, so that the total average power (variance) of MAI P_{Ik} at the kth receiver output is a sum in l of the powers of individual contributors:

$$P_{Ik} = 4E^2 \sum_{\substack{l=1 \\ l \neq k}}^{K} |\dot{\rho}_{kl}|^2$$

Since this quantity evaluates the power of MAI for only the kth user receiver, to cover the whole system we may sum it in k, coming to the result:

$$P_I = \sum_{k=1}^{K} P_{Ik} = 4E^2 \sum_{k=1}^{K} \sum_{\substack{l=1 \\ l \neq k}}^{K} |\dot{\rho}_{kl}|^2 \tag{7.27}$$

Now we may see that an adequate criterion of optimizing synchronous signatures, single-user reception postulated, is the minimum of the total MAI power or, equivalently, the sum of squared correlations in the expression above. Of course, again for the case $K \leq N$, the orthogonal signature set creates no MAI, i.e. turns this sum into zero so that only oversaturated ensembles are of a special interest.

The criterion just introduced typically emerges in the literature as the minimum of the *total squared correlation* (TSC):

$$TSC = \sum_{k=1}^{K} \sum_{l=1}^{K} |\dot{\rho}_{kl}|^2 = \min \tag{7.28}$$

which does not differ from the original one, since the sum in it is greater than the one in (7.27) by a constant K ($\dot{\rho}_{kk} = 1$).

There is a fundamental lower limit on the TSC known as the Welch bound [62]. Let us derive it, expressing first the correlation coefficients in terms of elements $a_{k,i}$ of the signature code sequences. Assuming all code sequence vectors $\mathbf{a}_k = (a_{k,0}, a_{k,1}, \ldots, a_{k,N-1})$ normalized so that $\|\mathbf{a}_k\|^2 = N$, (7.19) gives:

$$\dot{\rho}_{kl} = \frac{(\dot{\mathbf{S}}_k, \dot{\mathbf{S}}_l)}{2E} = \frac{(\dot{\mathbf{S}}_k, \dot{\mathbf{S}}_l)}{2\|\mathbf{a}_k\|\|\mathbf{a}_l\|E_0} = \frac{(\mathbf{a}_k, \mathbf{a}_l)}{N} = \frac{1}{N}\sum_{i=0}^{N-1} a_{k,i}a^*_{l,i}$$

Substituting this into the definition of TSC in (7.28) results in:

$$TSC = \frac{1}{N^2}\sum_{k=1}^{K}\sum_{l=1}^{K}\sum_{i=0}^{N-1}\sum_{j=0}^{N-1} a_{k,i}a_{l,i}^{*}a_{k,j}^{*}a_{l,j}$$

$$= \frac{1}{N^2}\sum_{i=0}^{N-1}\sum_{j=0}^{N-1}\sum_{k=1}^{K} a_{k,i}a_{k,j}^{*}\sum_{l=1}^{K} a_{l,i}a_{l,j}^{*} = \frac{1}{N^2}\sum_{i=0}^{N-1}\sum_{j=0}^{N-1}\left|\sum_{k=1}^{K} a_{k,i}a_{k,j}^{*}\right|^2$$

Since summands in i, j are all non-negative, omitting those with different i, j never increases the sum, so that:

$$TSC \geq \frac{1}{N^2}\sum_{i=0}^{N-1}\left(\sum_{k=1}^{K}|a_{k,i}|^2\right)^2 \tag{7.29}$$

To come to the final result one may further use the Schwarz inequality, but this step becomes unnecessary in the most interesting case of PSK signatures. For any PSK alphabet $|a_{k,i}| = 1$, which concludes the derivation of the Welch bound:

$$TSC \geq \frac{1}{N^2}\sum_{i=0}^{N-1}\left(\sum_{k=1}^{K}1\right)^2 = \frac{K^2}{N}$$

In the absence of oversaturation ($K \leq N$) the straightforward corollary of definition (7.28) is a tighter bound, based on the fact that with orthogonal signatures all summands in (7.28) with unequal k, l vanish and TSC achieves its minimum equal to K. Combining the results brings about the following general form of the Welch bound:

$$TSC \geq \begin{cases} K, K \leq N \\ \frac{K^2}{N}, K > N \end{cases} \tag{7.30}$$

Certainly, the set of sequences achieving (7.30) (*Welch-bound sequences*) is the best possible in total MAI criterion for a single-user receiver. But in fact, the significance of these sets goes far beyond only this feature, since Welch-bound sequences maximize the Shannon capacity of CDMA channels with AWGN and Gaussian input, the latter constraint losing its importance whenever a receive symbol SNR becomes small enough. Details of the proof of this remarkable property can be found in [63].

Since TSC includes K squared correlations of vectors with themselves, each equalling one, the difference $TSC - K$ covers only unwanted correlations between non-coinciding vectors, which we are interested to have as small as possible. There are $K(K-1)$ such vector pairs entering TSC, so that average squared correlation $\overline{\rho^2}$ per pair is:

$$\overline{\rho^2} = \frac{TSC - K}{K(K-1)}$$

giving, together with (7.30), the lower bound on this parameter:

$$\overline{\rho^2} \geq \begin{cases} 0, & K \leq N \\ \dfrac{K-N}{N(K-1)}, & K \geq N \end{cases} \tag{7.31}$$

From the way of obtaining (7.30), we may deduce how to come to the Welch bound-ensemble. Of course, only a non-trivial case of oversaturation should be discussed, since ways of generating orthogonal sequences have been considered previously. First of all, equality in (7.29) is a sufficient (and, of course, necessary) condition of equality in (7.30), or, considering the equation preceding (7.29), sequences for which:

$$\sum_{k=1}^{K} a_{k,i} a_{k,j}^{*} = 0, \ i \neq j \tag{7.32}$$

are Welch-bound sequences. Suppose all vectors $\mathbf{a}_1, \mathbf{a}_2, \ldots, \mathbf{a}_K$ of signature code sequences are written as columns of an $N \times K$ signature matrix $\mathbf{A}$:

$$\mathbf{A} = [\mathbf{a}_1 \mathbf{a}_2 \ldots \mathbf{a}_K] = \begin{bmatrix} a_{1,0} & a_{2,0} & \cdots & a_{K,0} \\ a_{1,1} & a_{2,1} & \cdots & a_{K,1} \\ \cdots & \cdots & \cdots & \cdots \\ a_{1,N-1} & a_{2,N-1} & \cdots & a_{K,N-1} \end{bmatrix}$$

then (7.32) means nothing but orthogonality of the rows of $\mathbf{A}$. Therefore, to build up an oversaturated ($K > N$) ensemble of Welch-bound sequences, one should just construct an $N \times K$ matrix $\mathbf{A}$ with orthogonal rows. Since the dimension of rows of such a matrix is greater than their number, there is no principal prohibition on its existence. Then the desired sequences are simply columns of $\mathbf{A}$.

We may now estimate the floor (i.e. noise-neglected) SIR for an oversaturated Welch-bound ensemble. The total MAI power P_I may be found from (7.27) and (7.28) as $P_I = 4E^2(TSC - K)$. Since this quantity is MAI summed over K single-user receivers, an average output MAI power per receiver will be $\bar{P}_{Ik} = P_I/K$. The useful (i.e. caused by the kth signature) effect at the kth receiver output expressed by the first term of (7.26) has power $4E^2$ (PSK modulation assumed), so that the floor power SIR with respect to an average MAI power according to (7.30) is:

$$q_I^2 = \frac{4E^2}{\bar{P}_{Ik}} = \frac{K}{TSC - K} = \frac{N}{K - N} \tag{7.33}$$

Example 7.2.2. Let us construct the binary Welch-bound ensemble of $K = 16$ sequences of length $N = 14$. For this we may make use of the matrix $\mathbf{H}_{16}$ of Example 7.2.1 and discard two arbitrary (e.g. the last two) rows. The matrix **A** obtained this way is exactly what is needed, and its 16 columns are Welch-bound signatures of length 14. TSC for the ensemble thus found equates accurately to its minimum value determined by (7.30):

$$TSC = \frac{K^2}{N} = \frac{256}{14}$$

The floor SIR estimated with respect to average MAI power per receiver is according to (7.33) $q_l^2 = N/(K - N) = 7$.

If belonging to a PSK alphabet is the only restriction on the signature code sequences, then the algorithm for constructing Welch-bound sets described above works universally. For example, rows of the matrix **A** may always be taken as K cyclically shifted replicas of the Chu sequence of length K. As was shown in Section 6.11.2, Chu codes exist for any length and all their different cyclic replicas are orthogonal. On the other hand, when all signatures should be binary ($a_{k,i} = \pm 1$) orthogonality of all N rows of the matrix **A** with $N > 2$ is possible only for K divisible by four (see Problem 7.14). This implies that for $K \neq 0 \bmod 4$ binary signatures the Welch bound (7.30) is not tight, and more precise lower borders should exist. Derivations of them may be found in [64,65] (see also Problem 7.17).

7.3 Approaches to designing signature ensembles for asynchronous DS CDMA

Let us extend the issue of signature design to the case of asynchronous DS CDMA, where time and phase shifts between individual user signals are random. On the assumption of employing a single-user receiver, the decision on the current symbol of the kth user is again done on the basis of correlation (7.25). Now, however, a tough alignment between the boundaries of data symbols and the chips of different users is not maintained due to arbitrary mutual time shifts of users' signals. Suppose that the kth user data receiver is explored and τ_l is the delay of the lth signal against the kth signal. In order to concentrate only on the issue of designing signature codes let us assume that chip boundaries of all K signatures are synchronized, i.e. mutual delays are multiples of Δ: $\tau_l = n_l\Delta$, where n_l is integer, $0 \leq n_l < N$. Then the situation is well explained by Figure 7.15 (for $k = 1$), stressing that in asynchronous CDMA, unlike synchronous (see Figure 7.13), data symbols of other users may change during the reception of the kth user's current data symbol. Still the main factor making design of asynchronous signature sets harder is the necessity to distinguish every signature from all possible shifted replicas of the others, which is not necessary in synchronous CDMA.

Suppose first that no change of data symbols of all users happens during the received symbol of the kth user, i.e. $b_{l,i-1} = b_l, l = 2, 3, \ldots, K$. Then the situation is different from the synchronous one only in the mutual time-mismatch of signatures. Start with the assumption that the signature period L coincides with the processing gain N, which

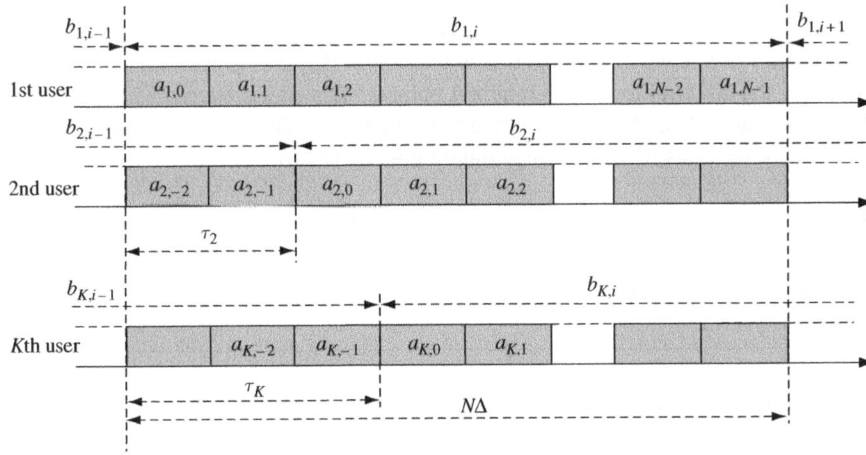

Figure 7.15 Datastreams and signatures in asynchronous CDMA

is the number of chips per data symbol duration or, equivalently, of chips integrated in the correlator. If no restriction is imposed on the possible range of mutual delays, the lth signature may be presented by any of its N cyclically shifted replicas, so that there are $N(K-1)$ different N-dimensional vectors, each of which is a potential source of MAI in the kth receiver. When a channel is subject to multipath effects every cyclic replica of the kth signal may also create interference at the kth receiver. Let us admit that up to $N-1$ such replicas may exist, i.e. the multipath delay spread ranges up to the period of the signature. Another reason to include the cyclic replicas in the explored vector set is a desire to have low autocorrelation sidelobes, which is important in the search problem (see Section 8.2). With such an extension we have a total of KN vectors, whose correlations should be as small as possible.

The Welch bound is again a good instrument to estimate the lower limit of the average squared correlation $\overline{\rho^2}$ of those KN vectors. For that it is enough to replace K by KN in (7.31). Since $KN > N$ for any $K \geq 2$ this gives:

$$\overline{\rho^2} \geq \frac{K-1}{KN-1} \tag{7.34}$$

This inequality shows the fundamental lower limit, which an average squared correlation between all cyclic replicas of all K signatures of length N (own replicas of each one included) can never fall below. When the number of users is about ten or more this version of the Welch bound becomes especially simple:

$$\overline{\rho^2} \geq \frac{1}{N}, K \gg 1 \tag{7.35}$$

Suppose now that the signature period in the number of L chips covers several data symbols $L > N$ and data of no user changes during the kth current data symbol, as before.[1] Again, let the delays range up to the signature period. Since the number of chips per data symbol (integration interval) remains N, we, as previously, deal with N-dimensional vectors, but the number of vectors whose correlations are controlled is now KL instead of KN, so that the bound stems from (7.31):

$$\overline{\rho^2} \geq \frac{KL - N}{N(KL - 1)} \tag{7.36}$$

which again turns into (7.35) with $K \gg 1$. The last result makes it possible to demonstrate that data modulation can in no way lower the bounds obtained. Indeed, every data-modulated signature may be considered as a new sequence of some (possibly very big) period L_k. Then all modulated signatures will have a common period L, being a least common multiple of all L_k, and the average squared correlation will be bordered from below by (7.36), again meaning that (7.35) is valid for the case of many users.

The derivations just undertaken establish a criterion of asynchronous signature set design: the ensemble of many signatures may be considered appropriate if its average squared correlation is close to the bound (7.35). Let us demonstrate that ensembles of random signatures attain this bound. Let all signatures be composed independently of each other by a random independent choice of elements of each of them. The whole procedure is similar to drawing balls out of an urn. Set an M-ary PSK alphabet and treat it as an urn with different M balls (code symbols). Pick out one ball K times, each time noting the result and returning the ball to the urn. This gives the first symbols of K signatures. The next symbols of all the signatures are generated the same way. Since all M-ary symbols in this scheme are equiprobable, uniformly spaced on the plane (see Figure 2.6c) and independent of each other, we have the following expectations:

$$\overline{a}_{k,i} = 0, \quad \overline{a_{k,i}a_{l,j}^*} = \begin{cases} 1, k = l \text{ and } i = j \\ \\ 0, \text{otherwise} \end{cases} = \delta_{kl}\delta_{ij} \tag{7.37}$$

the second equation stemming from the fact that the expectation of the product of independent entities equals the product of their expectations. Let us use this in an estimation of the average squared correlation of signatures at the integration interval of N chips:

$$\overline{|\dot{z}_{kl}(m)|^2} = \overline{\left|\sum_{i=0}^{N-1} a_{k,i}a_{l,i-m}^*\right|^2} \tag{7.38}$$

Physically (7.38) is nothing but the expectation of MAI power ($k \neq l$), or of multipath interference power ($k = l$) created by the lth signature shifted by m chips at the kth

[1] Saving symbol N for the processing gain, i.e. the number of chips per data symbol, we will from now on denote a signature period by L, whenever they may be different.

correlator output. Squaring the modulus and interchanging the summation and averaging (expectation of sum equals sum of expectations) gives:

$$\overline{|\dot{z}_{kl}(m)|^2} = \sum_{i=0}^{N-1}\sum_{j=0}^{N-1} \overline{a_{k,i}a^*_{l,i-m}a^*_{k,j}a_{l,j-m}} \tag{7.39}$$

To estimate the MAI effect set $k \neq l$ and split the summands into products of independent random variables:

$$\overline{|\dot{z}_{kl}(m)|^2} = \sum_{i=0}^{N-1}\sum_{j=0}^{N-1} \overline{a_{k,i}a^*_{k,j}} \cdot \overline{a^*_{l,i-m}a_{l,j-m}}$$

Now applying (7.37) to this equation leaves only terms with $i = j$ in the right-hand side sum, resulting in $\overline{|\dot{z}_{kl}(m)|^2} = N$. Estimating the effect of multipath interference let us put $k = l,\ m \neq 0$, which brings (7.39) to the form:

$$\overline{|\dot{z}_{kk}(m)|^2} = \sum_{i=0}^{N-1}\sum_{j=0}^{N-1} \overline{a_{k,i}a^*_{k,i-m}a^*_{k,j}a_{k,j-m}}$$

In terms here having different i and j $a_{k,i}$ is independent of both $a_{k,i-m}$ (since $m \neq 0$) and $a_{k,j}$ (since $i \neq j$). By the same reasoning, $a_{k,j-m}$ is independent of both $a_{k,j}$ and $a_{k,i-m}$. Therefore:

$$\overline{a_{k,i}a_{k,j-m} \cdot a^*_{k,i-m}a^*_{k,j}} = \overline{a_{k,i}a_{k,j-m}} \cdot \overline{a^*_{k,i-m}a^*_{k,j}},\ \ i \neq j$$

Due to the independence of different symbols of the same sequence $\overline{a_{k,i}a_{k,j-m}} = \overline{a_{k,i}} \cdot \overline{a_{k,j-m}} = 0$, whenever $i \neq j - m$, and $\overline{a^*_{k,i-m}a^*_{k,j}} = \overline{a^*_{k,i-m}} \cdot \overline{a^*_{k,j}} = 0$, whenever $i - m \neq j$. Hence, addends of the double sum above with different i, j may appear non-zero only if both equations are true: $i = j - m$ and $i - m = j$, which is impossible for any non-zero m. Consequently, only summands with equal i and j produce non-zero contribution, and:

$$\overline{|\dot{z}_{kk}(m \neq 0)|^2} = \sum_{i=0}^{N-1} |a_{k,i}|^2 |a_{k,i-m}|^2 = N$$

The useful effect, i.e. the power, created by the non-shifted kth signature at the kth receiver output:

$$|\dot{z}_{kk}(0)|^2 = \sum_{i=0}^{N-1} |a_{k,i}|^2 = N^2$$

Then the normalized unwanted effects created by either MAI or multipath interference (unwanted squared correlations):

$$\frac{\overline{|\dot{z}_{kl}(m)|^2}}{|\dot{z}_{kk}(0)|^2} = \frac{1}{N},\ \ \frac{\overline{|\dot{z}_{kk}(m \neq 0)|^2}}{|\dot{z}_{kk}(0)|^2} = \frac{1}{N} \tag{7.40}$$

It is clearly seen now that all unwanted squared correlations in the ensemble under consideration attain the lower bound (7.35), i.e. sets of random signatures are optimal when the number of users is around ten or more. It is extremely important to emphasize that the data modulation of random sequences meeting (7.37) (multiplying them by data symbols independent of them) does not destroy (7.37) (see Problem 7.20). Therefore, the presence or absence of modulation does not affect all of the derivations above, as well as the final result (7.40) and conclusion on the set optimality.

Equations (7.37) seem to give an unequivocal instruction for designing signature ensembles. In practice, however, signatures cannot be random, since the receiver should be a priori aware of the signature modulation law in order to generate the necessary correlator reference. To realize the randomness properties (7.37) by the deterministic coding rules, so-called *pseudorandom* sequences are necessary.

Take the deterministic PSK signature of period L and treat it as though it is one of several equiprobable realizations of a stationary ergodic random sequence $\{a_{k,i}\}$ (random discrete-time process) [14,66]. The other realizations may be all cyclic shifts of the initial sequence. Then, due to the ergodicity property, each realization presents the whole random process exhaustively, and statistical averaging $\{a_{k,i}\}$ over all realizations is equivalent to time averaging, i.e. evaluating the expectation $\overline{a_{k,i}}$ and correlation moment $\overline{a_{k,i}a^*_{k,i-m}}$ via constant component and periodic ACF of the deterministic signature, respectively:

$$\overline{a_{k,i}} = \frac{1}{L}\sum_{i=0}^{L-1} a_{k,i} = \frac{\tilde{a}_{k,0}}{L}, \quad \overline{a_{k,i}a^*_{k,i-m}} = \frac{1}{L}\sum_{i=0}^{L-1} a_{k,i}a^*_{k,i-m} = \rho_{kk}(m) \tag{7.41}$$

In the same way, treating two deterministic signatures as realizations of two jointly ergodic random sequences $\{a_{k,i}\}$ and $\{a_{l,i}\}$, we have equality between the correlation moments of two random sequences and CCF of two deterministic signatures:

$$\overline{a_{k,i}a^*_{l,i-m}} = \frac{1}{L}\sum_{i=0}^{L-1} a_{k,i}a^*_{l,i-m} = \rho_{kl}(m) \tag{7.42}$$

The comparison of (7.41) and (7.42) with (7.37) sets a criterion of pseudorandomness: to serve as signatures in asynchronous DS CDMA all the deterministic sequences of the ensemble should ideally have zero constant component, perfect periodic ACF and zero periodic CCF:

$$\tilde{a}_{k,0} = 0; \rho_{kk}(m) = 0, m \neq 0 \bmod L; \rho_{kl}(m) = 0, k, l = 1, 2, \ldots, K \tag{7.43}$$

In the case of unconstrained mutual time shifts (any m in the range $0, 1, \ldots, L-1$ are probable) the last demands obviously contradict each other, making ensembles of this sort hypothetical for any finite L. Indeed (see also Problem 7.21), the requirements for perfect ACF and zero CCF mean nothing but zero level of correlations between all cyclic shifts of K sequences of period L, i.e. zero value of the average of unwanted squared correlations $\overline{\rho^2}$. As (7.34) and (7.35) show, this is impossible with $K \geq 2$, and in particular with many users $\overline{\rho^2}$ cannot fall smaller than $1/L$.

The conclusion we just arrived at explains why so many efforts have been dedicated to searching for ensembles whose characteristics approach those of the hypothetical ensembles mentioned above when length L grows. Quite a popular criterion of this approximation is the minimax one, orienting the ensemble design towards minimizing maximum value among all unwanted correlations. Define the *correlation peak* $\rho_{\max}$ as the greater of two entities: the maximal autocorrelation sidelobe $\rho^a_{\max}$ among all sequences and the maximal cross-correlation peak $\rho^c_{\max}$ among all pairs of sequences:

$$\rho_{\max} = \max\{\rho^a_{\max}, \rho^c_{\max}\}, \rho^a_{\max} = \max_{\substack{k,m \\ m\neq 0}} |\rho_{p,kk}(m)|, \rho^c_{\max} = \max_{\substack{k,l,m \\ k\neq l}} |\rho_{p,kl}(m)| \tag{7.44}$$

Naturally, for the hypothetical perfect ensemble, $\rho_{\max}$ along with $\overline{\rho^2}$ is zero, and for any real ensemble $\rho_{\max}$ may serve as an adequate measure of its proximity to the perfect one.

Since the maximal value of any variable can never be smaller than its average, $\rho^2_{\max} \geq \overline{\rho^2}$, which spreads the Welch bounds (7.34) and (7.35) on the correlation peak:

$$\rho^2_{\max} \geq \frac{K-1}{KL-1} \approx \frac{1}{L} \tag{7.45}$$

where, again, the last approximation corresponds to the case $K \gg 1$. With additional limitations on the PSK alphabet, the bound above may appear rather loose, especially when the number of sequences approaches L. In particular, for sufficiently large ensembles of binary $\{\pm 1\}$ sequences the Sidelnikov bound holds [67,68]:

$$\rho^2_{\max} \geq \frac{2}{L}, K \geq \frac{L}{2} \tag{7.46}$$

Ensembles having $\rho_{\max}$ attaining the limit predicted by the lower bounds are certainly optimal in the correlation peak criterion and are sometimes called *minimax*. Some of them are discussed in Section 7.5.

7.4 Time-offset signatures for asynchronous CDMA

In many real situations mutual time shifts of asynchronous signatures may vary only within a restricted range. The finiteness of a channel delay spread on the one hand, and system geometry on the other are the most typical factors setting such limitations. To be specific, let us turn to the uplink of a cellular mobile radio. The local clock of an active MS is synchronized with the received BS signal and has a delay τ_1 versus the BS clock determined by the distance D from BS to MS as $\tau_1 = D/c$, where c is the speed of light. Since the signal transmitted by a specific MS reaches the BS receiver with the same delay, the total delay of the signal arriving at BS versus the BS clock is $\tau_2 = 2\tau_1 = 2D/c$. Let $D_{\max}$ be the maximal distance, the signal from which has intensity perceptible by the BS receiver. Strong path attenuation (see Section 4.6) permits signals arriving from distances markedly exceeding the cell radius D_c to be ignored, which gives a rough estimation $D_{\max} \approx D_c$. Then the maximal value of τ_2 is $2D_c/c$ and signals from mobiles at distances from BS ranging between zero and D_c arrive at BS within the time window

$[0,2D_c/c]$. Besides, multipath replicas of signals are also present, so that a complete extension τ_{max} of the window spanning the delays of all multipath signals increases by the channel delay spread τ_{ds}: $\tau_{max} = 2D_c/c + \tau_{ds}$, where τ_{ds} may be maximized over all possible locations of MS. Figure 7.16 helps to show the details of these deliberations. The signal of some specific MS may have an advance as well as a delay compared to some other, and all multipath replicas of any MS signal are potentially usable by a BS receiver (RAKE processing; see Section 3.7). Therefore, the entire range of possible mutual time shifts between any multipath replicas of any signatures proves to be $[-\tau_{max}, \tau_{max}]$, $\tau_{max} = 2D_c/c + \tau_{ds}$.

Certainly, in circumstances like these, one should take care to observe the second and third conditions (7.43) within the range of only really likely values of m. Let us denote the contents of τ_{max} in the number of chips rounded upwards as m_{max}: $m_{max} = \lceil \frac{\tau_{max}}{\Delta} \rceil$. Then the range of m, where (7.43) should be obeyed, is $[-m_{max}, m_{max}]$. Now take a sequence $\{a_{1,i}\}$ of the period $L \geq K(m_{max} + 1)$ and use as K signatures its cyclic replicas offset from each other by $m_{max} + 1$ positions:

$$a_{k,i} = a_{1,i-(k-1)(m_{max}+1)}, k = 1, 2, \ldots, K; i = \ldots, -1, 0, 1, \ldots,$$

as is shown in Figure 7.17. Evidently all correlations between signatures thus arranged will be expressed in terms of ACF $\rho_{11}(m)$ of the initial sequence $\{a_{1,i}\}$. Evaluating the CCF of the kth and lth signatures results in:

$$\rho_{kl}(m) = \frac{1}{L}\sum_{i=0}^{L-1} a_{k,i}a^*_{l,i-m} = \frac{1}{L}\sum_{i=0}^{L-1} a_{1,i-(k-1)(m_{max}+1)}a^*_{1,i-(l-1)(m_{max}+1)-m}$$

or:

$$\rho_{kl}(m) = \rho_{11}[(k-l)(m_{max}+1) + m] \tag{7.47}$$

Suppose now that the initial sequence $\{a_{1,i}\}$ has either perfect or good enough periodic ACF $\rho_{11}(m)$. The former is possible, e.g. for ternary or polyphase sequences (see Section 6.11), while any minimax binary sequence (Sections 6.7 and 6.9) may serve as an example of the latter. The idea is that all sidelobes of $\rho_{11}(m)$ are negligible. Then with $|m| \leq m_{max}$ the argument in the square brackets of (7.47) turns into zero modulo L only for the case $k = l$ and $m = 0 \bmod L$, corresponding to the mainlobe of the ACF of the kth signature. For any other combination of k,l,m, the right-hand side of (7.47) gives

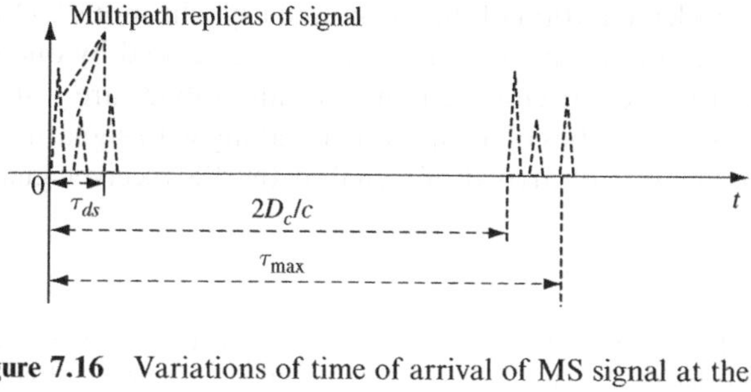

Figure 7.16 Variations of time of arrival of MS signal at the BS

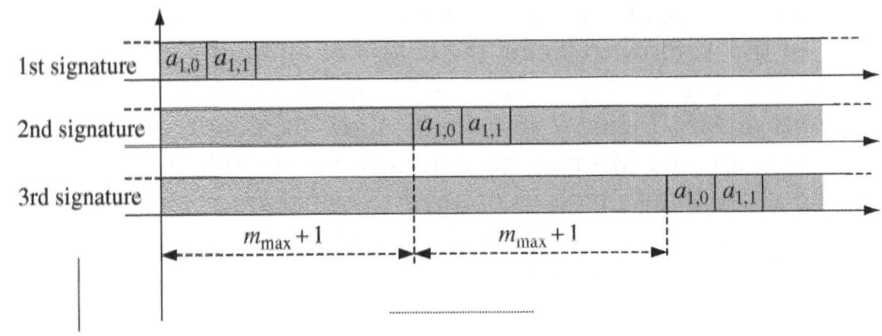

Figure 7.17 Signatures formed as time-shifted copies of the initial one

a sidelobe of $\rho_{11}(m)$, whose level was assumed to be negligible. We have thus proved that properly offset copies of the initial sequence with good periodic ACF produce the ensemble where conditions of pseudorandomness (7.43) hold within the full range of possible mutual signature shifts $|m| \leq m_{\max}$. It follows immediately that this ensemble achieves (when $\rho_{11}(m)$ is perfect) or approaches very closely the lowest level (7.40) of average unwanted effects due to MAI and multipath propagation or, equivalently, the Welch bounds (7.36) or (7.35). We again emphasize strongly the validity of this statement in the presence of DS data modulation of signatures, since conditions (7.43) are sufficient to minimize unwanted MAI/multipath effects in this case (see the remark following (7.40)).[2]

Example 7.4.1. Consider the system with chip duration $\Delta = 1\,\mu\text{s}$, number of users $K = 60$, channel delay spread $\tau_{ds} = 20\,\mu\text{s}$ and cell radius $D_c = 15\,\text{km}$. In this case $\tau_{\max} = 2D_c/c + \tau_{ds} = 120\,\mu\text{s}$ and $m_{\max} = 120$. The signature ensemble may be arranged starting with the initial sequence $\{a_{1,i}\}$ whose period $L \geq K(m_{\max}+1) = 60 \times 121 = 7260$. Since $\{a_{1,i}\}$ should have a good periodic ACF the relevant candidates may be the ternary perfect ACF sequence of length $L = 8011$, a binary m-sequence ($L = 2^{13} - 1 = 8191$) or a Legendre sequence ($L = 7283$). The 60 signatures are then just 60 cyclic replicas of the $\{a_{1,i}\}$ offset from each other by 121 chips. Clearly, there is no upper limit on the length of the sequence and it may be advisable to take it longer with an appropriate increase in signature offset to secure some safety margin.

The uplinks of the 2G cdmaOne (IS-95) and 3G cdma2000 standards present very good examples of implementation of this version of asynchronous CDMA [69]. A binary m-sequence of an extremely long length $L = 2^{42} - 1$ extended by one symbol is used as the initial one and the user-specific signatures of all mobiles are just its relevant cyclic replicas. Pseudonoise properties of an m-sequence along with signature offsets exceeding possible variations of time of arrival of signal at the BS receiver guarantee a minimal

[2] Without DS data modulation, perfection of periodic ACF of the initial sequence secures zero level of both MAI and multipath interference for any $m \leq m_{max}$ in the described signature construction.

(see (7.40)) level of average power of MAI and multipath interference at the correlator output.

7.5 Examples of minimax signature ensembles

The signature ensembles considered in the previous section may be regarded as adequate only in situations where mutual time shifts of users' signals are entirely controllable by the system and may be kept within the predicted range. If this is not the case, asynchronous CDMA based on shifted replicas of the same sequence risks collisions: the signal of one user may acquire delay, making it indistinguishable from the signal of some other user. This may be the reason for employing minimax signature ensembles, i.e. those whose correlation peaks achieve or approach bounds (7.45) or (7.46). Since the correlation peak of a minimax ensemble is maximized over the whole period, its small value (achievable at the cost of long enough length L) secures the proximity of ensemble correlation properties to the perfect ones (7.43), guaranteeing pseudorandomness of signatures.

A survey of all known minimax ensembles would take a lot of space, so we will confine ourselves to a brief discussion of those that either enjoy wider practical application or seem more indicative among others. Readers interested in learning more about them may consult [9,67,70].

7.5.1 Frequency-offset binary m*-sequences*

Take a binary $\{\pm 1\}$ m-sequence $\{a_{1,i}\}$ of period $L = 2^n - 1$ and use it as a signature for the first user. The rest of the $K - 1$ signatures are generated by a symbol-wise multiplication of $\{a_{1,i}\}$ with discrete harmonics of frequencies $(k-1)/L, k = 2, 3, \ldots, K$:

$$a_{k,i} = a_{1,i} \exp\left(j\frac{2\pi(k-1)i}{L}\right), i = \ldots, -1, 0, 1, \ldots, k = 1, 2, \ldots, K. \tag{7.48}$$

Thus the squared modulus of the periodic CCF of the kth and lth sequences is:

$$|R_{p,kl}(m)|^2 = \left|\sum_{i=0}^{L-1} a_{k,i} a_{l,i-m}^*\right|^2 = \left|\sum_{i=0}^{L-1} a_{1,i} a_{1,i-m} \exp\left(j\frac{2\pi(k-l)i}{L}\right)\right|^2 \tag{7.49}$$

Consider first the case $m = 0 \bmod L$, i.e. $a_{1,i}a_{1,i-m} = |a_{1,i}|^2 = 1$. Then, if $k = l$ (7.49) gives the mainlobe of the ACF of the kth signature, i.e. $|R_{p,kk}(0)|^2 = L^2$. If $k \neq l$, the sum in (7.49) is the sum of all roots of unity of degree L and equals zero (see Section 6.11.2). Now let $m \neq 0 \bmod L$. Then according to the shift-and-add property (Section 6.11) of the m-sequence, $a_{1,i}a_{1,i-m} = a_{1,i-t}$ for some t, and the squared CCF modulus:

$$|R_{p,kl}(m)|^2 = \left|\sum_{i=0}^{L-1} a_{1,i-t} \exp\left(j\frac{2\pi(k-l)i}{L}\right)\right|^2 = \left|\sum_{i=0}^{L-1} a_{1,i} \exp\left(j\frac{2\pi(k-l)i}{L}\right)\right|^2$$

which is the $(k-l)$th component of the DFT energy spectrum of the sequence $\{a_{1,i}\}$. Since the energy spectrum of $\{a_{1,i}\}$ is the DFT of its periodic ACF, and the ACF equals -1 everywhere except at the zero point, where it is equal to L, we have:

$$|R_{p,kl}(m)|^2 = \sum_{m=0}^{L-1} R_{p,11}(m) \exp\left(j\frac{2\pi(k-l)m}{L}\right) = L+1-\sum_{m=0}^{L-1} \exp\left(j\frac{2\pi(k-l)m}{L}\right)$$

The last sum differs from zero and equals L only for $k=l$, so that collecting all the results together and passing over to normalized correlations gives:

$$|\rho_{p,kl}(m)|^2 = \begin{cases} 1, k=l, m=0 \bmod L \\ \frac{1}{L^2}, k=l, m \neq 0 \bmod L \\ 0, k \neq l, m=0 \bmod L \\ \frac{L+1}{L^2}, k \neq l, m \neq 0 \bmod L \end{cases}$$

It is seen now that the squared correlation peak of the ensemble (7.48):

$$\rho_{\max}^2 = \frac{L+1}{L^2} \approx \frac{1}{L}$$

i.e. practically coincides with the Welch bound (7.45). Thus, the ensemble under scrutiny is a minimax one, realizing the optimal asynchronous CDMA mode.

The description of the above ensemble one may find, e.g. in [71], yet earlier and independently it was used in the global satellite-based navigation system GLONASS (see Section 11.2). One of the advantages of this signature set versus other polyphase ones is the possibility of generating signatures by a simple offset of carrier frequency. Indeed, incrementing the carrier frequency f_0 by $(k-1)/L\Delta$ is equivalent to a linear phase progression between adjacent chips equalling $2\pi(k-1)/L$, which is exactly what is prescribed by the rule (7.48).

Despite many other minimax polyphase ensembles being known, the binary $\{\pm 1\}$ ones are traditionally considered more attractive from a hardware point of view, and the rest of this section is dedicated to some important examples of binary signature sets.

7.5.2 Gold sets

The following properties of binary $\{\pm 1\}$ m-sequences may serve to explaining the set construction found by Gold:

1. If a binary $\{\pm 1\}$ m-sequence $\{u_i\}$ of period $L = 2^n - 1$ is *decimated* with the *decimation index d*, where d is co-prime to L, the resulting sequence $\{v_i\}$ is again a binary m-sequence of the same period. To decimate means to pick out every dth symbol of $\{u_i\}$ and write symbols thus obtained one by one, so that $v_i = u_{di}$. We call the sequence $\{v_i\}$ produced this way a decimation of $\{u_i\}$.

2. Let the memory n of a binary m-sequence $\{u_i\}$ be odd and in the decimation index $d = 2^s + 1$ s be co-prime to n. Then d is co-prime to the length $L = 2^n - 1$ of $\{u_i\}$, the decimation $\{v_i\}$ is an m-sequence of the same period L, and the non-normalized periodic CCF $R_{p,uv}(m)$ of $\{u_i\}$, $\{v_i\}$ takes only three values:

$$R_{p,uv}(m) \in \{\pm\sqrt{2(L+1)} - 1, -1\} = \{\pm 2^{\frac{n+1}{2}} - 1, -1\}, m = 0, 1, \ldots, L-1 \qquad (7.50)$$

3. Let the memory n of a binary m-sequence $\{u_i\}$ be even, but not a multiple of four, and in the decimation index $d = 2^s + 1$ s be even and co-prime to $n/2$. Then d is co-prime to the length $L = 2^n - 1$ of $\{u_i\}$, the decimation $\{v_i\}$ is an m-sequence of the same period L, and the non-normalized periodic CCF $R_{p,uv}(m)$ of $\{u_i\}$, $\{v_i\}$ takes only three values:

$$R_{p,uv}(m) \in \{\pm 2\sqrt{L+1} - 1, -1\} = \{\pm 2^{\frac{n+2}{2}} - 1, -1\}, m = 0, 1, \ldots, L-1 \qquad (7.51)$$

Proof of these propositions is rather sophisticated and demands more insight into the algebra of extension finite fields. We leave it aside and refer the interested reader to the original paper by Gold [72] or other sources (e.g. [9,70]).

Now take a pair of m-sequences, $\{u_i\}$ and its decimation $\{v_i\}$, satisfying the conditions of item 2 or 3 above and form the ensemble of K signatures by the rule:

$$\begin{aligned} a_{k,i} &= u_i v_{i-k}, k = 1, 2, \ldots, L \\ a_{L+1,i} &= u_i \\ a_{L+2,i} &= v_i \end{aligned} \qquad (7.52)$$

where $i = \ldots, -1, 0, 1, \ldots$. Expressing this in words, we build up L signatures multiplying symbol-wise $\{u_i\}$ with cyclic replicas of $\{v_i\}$, and two more signatures are initial m-sequences themselves. In total, therefore, we may have up to $K = L + 2 = 2^n + 1$ signatures. In practice, the $\{\pm 1\}$ m-sequence is traditionally generated as a binary $\{0,1\}$ sequence, i.e. over $GF(2)$ using an LFSR generator, with a subsequent mapping of elements of $GF(2)$ onto the real pair $\{\pm 1\}$ (see Sections 6.6 and 6.7). Thus, to implement (7.52) two n-cell LFSRs may be used, generating $\{0,1\}$ predecessors $\{u'_i\}$ and $\{v'_i\}$ of $\{u_i\}$ and $\{v_i\}$. Instead of multiplication of $\{u_i\}$ with $\{v_{i-k}\}$ their predecessors may be added modulo 2 with a subsequent mapping of the result onto $\{\pm 1\}$: $u_i v_{i-k} = (-1)^{u'_i + v'_{i-k}}$. Figure 7.18 illustrates the implementation of the Gold construction according to the above description.

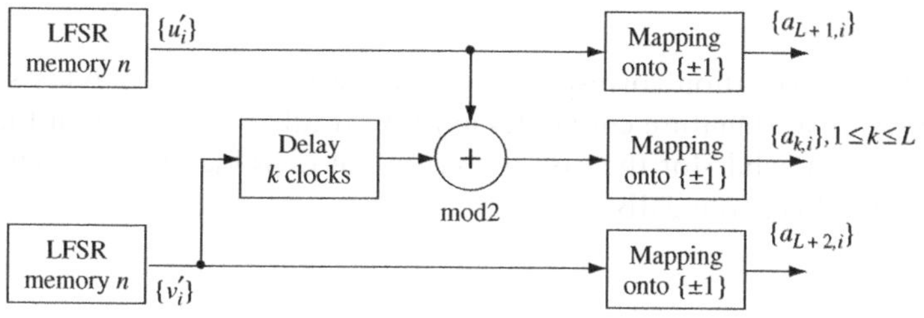

Figure 7.18 Generating Gold sequences

Let us estimate the correlation peak of the Gold ensemble, beginning by calculating correlations of the first L sequences:

$$R_{p,kl}(m) = \sum_{i=0}^{L-1} a_{k,i}a_{l,i-m} = \sum_{i=0}^{L-1} u_i u_{i-m} v_{i-k} v_{i-l-m}$$

It is seen that since the case $m = 0 \bmod L$ and $k = l$ corresponds to the mainlobe of the kth ACF, the situation should be analysed where these equalities are not fulfilled simultaneously. But then either both $u_i u_{i-m}$ and $v_{i-k}v_{i-l-m}$ are just some other shifts of the initial sequences $\{u_i\}, \{v_i\}$, or only one of those products is a sequence consisting of only ones. In the first case we have the CCF of the initial m-sequences $\{u_i\}, \{v_i\}$ taking on only the three values indicated by (7.50) or (7.51), while in the second we have the non-normalized ACF sidelobe of one of the sequences $\{u_i\}, \{v_i\}$, i.e. -1.

Consider now the CCF of $\{a_{k,i}\}, k = 1, 2, \ldots, L$ and $\{a_{l,i}\}, l = L + 1$:

$$R_{p,kl}(m) = \sum_{i=0}^{L-1} u_i u_{i-m} v_{i-k}$$

If $m = 0 \bmod L, u_i u_{i-m} = 1$ and the CCF is simply a constant component of $\{v_i\}$, i.e. -1. Otherwise $u_i u_{i-m} = u_{i-s}$ for some s and we have the CCF of initial m-sequences obeying the restrictions (7.50) or (7.51). The same is true for the CCF of $\{a_{k,i}\}, k = 1, 2, \ldots, L$ and $\{a_{l,i}\}, l = L + 2$.

Finally, the CCF of $\{a_{L+1,i}\}$ and $\{a_{L+2,i}\}$ is directly the CCF of the initial m-sequences, while their autocorrelation functions, like those of m-sequences, have non-normalized ACF sidelobes equalling -1. Collecting all of the results together, we see that the correlation peak (7.44) of the Gold set is determined by the maximal in modulus value of the original CCF (7.50) or (7.51). After normalizing it to the length L we come to the estimation:

$$\rho_{\max}^2 = \begin{cases} \dfrac{\left(\sqrt{2(L+1)}+1\right)^2}{L^2}, n \neq 0 \bmod 2, \\[2ex] \dfrac{\left(2\sqrt{L+1}+1\right)^2}{L^2}, n = 2 \bmod 4 \end{cases} \approx \begin{cases} \dfrac{2}{L}, n \neq 0 \bmod 2 \\[2ex] \dfrac{4}{L}, n = 2 \bmod 4 \end{cases} \quad (7.53)$$

with the last approximation corresponding to large length $L \gg 1$. As is seen, for any odd memory n Gold signature ensembles asymptotically ($L \gg 1$) attain the Sidelnikov lower bound (7.46), while for the case of even n not divisible by four their loss in $\rho_{\max}$ against this bound is about 3 dB.[3]

[3] When $n = 0 \bmod 4$ a Gold ensemble also exists with the same correlation peak as in the case $n = 2 \bmod 4$, but with number of sequences smaller by one [67,70].

Example 7.5.1. Construct Gold sequences of length $L = 2^3 - 1 = 7$. An ensemble of that small length is impractical but useful for elucidating the idea. Let us start with the binary $\{0, 1\}$ m-sequence first met in Example 6.6.1: $\{u_i'\} = \{1, 0, 0, 1, 0, 1, 1\}$. The decimation index $d = 3$ meets the limitation of item 2 above. Then the decimation sequence is $\{v_i'\} = \{1, 1, 1, 0, 1, 0, 0\}$. Symbol-wise summation of $\{u_i'\}$ and $\{v_i'\}$ modulo 2 gives the sequence $\{0, 1, 1, 1, 1, 1, 1\}$, which after mapping to alphabet $\{\pm 1\}$ gives the first Gold sequence $\{a_{1,i}\} = \{+ - - - - - -\}$. Shifting $\{v_i'\}$ to the right by one position and adding modulo 2 to $\{u_i'\}$ gives the sequence $\{1, 1, 1, 0, 0, 0, 1\}$, which after transition to symbols $\{\pm 1\}$ gives the second Gold sequence $\{- - - + + + -\}$. Six more Gold sequences are obtained by further shifts of $\{v_i'\}$, adding modulo 2 to $\{u_i'\}$ and changing symbols into $\{\pm 1\}$. Along with $\{u_i'\}$ and $\{v_i'\}$ transformed into $\{\pm 1\}$ sequences, we obtain $K = 2^3 + 1 = 9$ sequences altogether. Checking the value of the correlation peak in this simplest case makes little sense, since with $L = 7$ no non-normalized periodic correlation, but the ACF mainlobe, may exceed 5, predicted by (7.53). Building Gold ensembles of greater lengths and checking their optimality is the subject of Problem 7.40.

Gold ensembles enjoy great popularity in modern CDMA systems. Suffice it to say that they are employed in the space-based global navigation system GPS for multiplexing satellite signals, and in the 3G mobile radio UMTS standard for scrambling CDMA codes, etc.

7.5.3 *Kasami sets and their extensions*

The idea of constructing Kasami sets is very close to that described above for the Gold scheme. Let us decimate a binary $\{\pm 1\}$ m-sequence $\{u_i\}$ of even memory $n = 2h$ with the decimation index $d = 2^h + 1$. Obviously, this d is not co-prime to the period $L = 2^n - 1 = (2^h - 1)(2^h + 1)$ of the sequence $\{u_i\}$, resulting in a decimation sequence $\{v_i\} = \{u_{di}\}$ of the period being a factor of L. It may be shown that if $\{u_i\}$ is initialized so that $u_0 = -1$ the 'short' sequence $\{v_i\}$ is actually a binary m-sequence of period $L_1 = 2^h - 1$, whose non-normalized periodic CCF with $\{u_i\}$ over the long period L takes only two values [9,70,73]:

$$R_{p,uv}(m) = \pm 2^h - 1 = \pm\sqrt{L+1} - 1, \ m = 0, 1, \ldots, L-1 \tag{7.54}$$

Then L_1 Kasami signatures of length L are formed as symbol-wise products of the initial m-sequence $\{u_i\}$ with L_1 different cyclic replicas of $\{v_i\}$, and one more signature is a 'long' sequence itself:

$$\begin{aligned} a_{k,i} &= u_i v_{i-k}, \ k = 1, 2, \ldots, L_1 \\ a_{L_1+1,i} &= u_i \end{aligned} \tag{7.55}$$

where $i = \ldots, -1, 0, 1, \ldots$. There are $K = L_1 + 1 = 2^h = \sqrt{L+1}$ such signatures of period L in total. Of course, again, multiplication of $\{\pm 1\}$ sequences $\{u_i\}, \{v_i\}$ may be

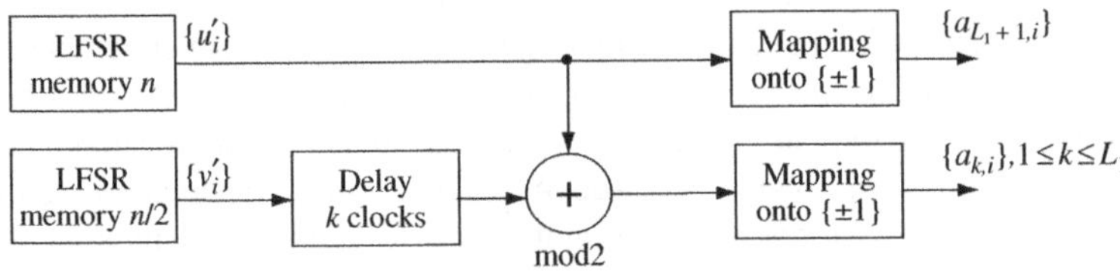

Figure 7.19 Generating Kasami sequences

realized as modulo 2 addition of their $\{0, 1\}$ predecessors $\{u_i'\}$, $\{v_i'\}$, but, unlike the Gold set, to form the 'short' sequence $\{v_i'\}$ the necessary length of LFSR is two times smaller: $h = n/2$ (see Figure 7.19).

Proof of the minimax property of the Kasami set:

$$\rho_{\max}^2 = \frac{\left(\sqrt{L+1}+1\right)^2}{L^2} \approx \frac{1}{L}, L \gg 1 \tag{7.56}$$

is performed based on (7.54), similarly to that of the Gold set, and is left to the reader as an exercise (Problem 7.28). The comparison of the two binary ensembles shows a significant gain (6 dB) of Kasami sets in the correlation peak versus Gold ensembles of the same length in exchange for much smaller ($(L+2)/\sqrt{L+1} \approx \sqrt{L}$ times) number of sequences K.[4]

Example 7.5.2. Construct the Kasami set of length $L = 2^4 - 1 = 15$ ($h = 2, K = \sqrt{L+1} = 4$). Start by building the binary $\{0, 1\}$ *m*-sequence $\{u_i'\}$ of length $L = 15$ based on the primitive polynomial $f(x) = x^4 + x + 1$ and initial loading $u_0' = 1, u_1' = u_2' = u_3' = 0$. We have $\{u_i'\} = \{1, 0, 0, 0, 1, 0, 0, 1, 1, 0, 1, 0, 1, 1, 1\}$. Decimation of this sequence with the index $d = 2^h + 1 = 5$ produces the *m*-sequence of period three $\{v_i'\} = \{1, 0, 1, 1, 0, 1, 1, 0, 1, 1, 0, 1, 1, 0, 1\}$. Modulo 2 sums of $\{u_i'\}$ and three shifted replicas of $\{v_i'\}$ after transferring to the alphabet $\{\pm 1\}$ are the first three Kasami sequences: $\{a_{1,i}\} = \{++------+---+-+\}$, $\{a_{2,i}\} = \{+-+-++-+--++++-\}$ and $\{a_{3,i}\} = \{---++-+++++--++\}$. The fourth one is $\{u_i'\}$ converted into $\{\pm 1\}$ symbols: $a_{4,i} = \{-+++-++--+-+---\}$. Direct calculation shows that all their non-normalized CCF as well as the non-normalized ACF sidelobes of the first three take on only values -5 and 3, so that $\rho_{\max}^2 = 1/9$ in full agreement with (7.56). A Matlab program for building arbitrary Kasami sets and verifying their correlation properties is the subject of Problem 7.41.

The relatively small number of Kasami sequences makes rather remarkable the method found by Kamaletdinov [74] to extend the Kasami set almost two times without

[4] Bounds (7.45) and (7.46) may be slightly improved for binary sets allowing for the non-normalized correlations taking on only integer values. As a result it appears that both Gold sets of odd memory and Kasami sets are strictly (not only asymptotically!) optimal in correlation peak among all binary sets [67,70].

sacrificing the correlation peak. Let n be divisible by 4: $n = 4r$, r integer, so that $L = 2^{4r} - 1 = 16^r - 1 = 15, 255, 4095, \ldots$. Then in addition to the Kasami set another binary ensemble of length L and size $\sqrt{L+1}$ exists called the *bent sequence ensemble* [9,75] and possessing the same minimax property $\rho^2_{max} = (\sqrt{L+1} + 1)^2/L^2 \approx 1/L$. In very general terms constructing bent sequences again consists of symbol-wise multiplication of two initial sequences: a 'long' m-sequence of period $L = 2^{4r} - 1$ and some special sequence based on the so-called bent function. The details of this are tricky enough and will not be discussed here, but the important thing is that any bent sequence has normalized CCF with any of the first L_1 Kasami sequences (7.55) not exceeding by its modulus the correlation peak of both the Kasami and bent sequence ensembles. Therefore, it is possible to arrange a composite ensemble including $L_1 = 2^h - 1 = 2^{2r} - 1 = \sqrt{L+1} - 1$ Kasami and $\sqrt{L+1}$ bent sequences and possessing the former correlation peak $\rho^2_{max} = (L+1)/L^2 \approx 1/L$. The ensemble thus obtained is unique in the sense that among all known binary ensembles with correlation peak $\rho^2_{max} \approx 1/L$ this one has the greatest number of signatures $K = 2\sqrt{L+1} - 1$.

7.5.4 *Kamaletdinov ensembles*

More binary minimax ensembles exist [9,67]; however, some of them differ from the described ones only in a fine structure of sequences but not in length L, size K and correlation peak ρ_{max}. Against this background the ensembles discovered by Kamaletdinov [76] are of a particular interest, covering the range of lengths differing from those of Gold and Kasami sets.

In order to make the idea easier to understand, we describe a somewhat narrowed version of Kamaletdinov sets, although with no loss as to the length range or parameters achievable. To outline the first Kamaletdinov scheme let us take prime odd $p > 3$ of the form $p = 4h + 3 = 3\text{mod}4$ and extend the definition of the binary character $\psi(x)$ given in Section 6.8 to the zero element of $GF(p)$ putting $\psi(0) = 1$ (an alternative $\psi(0) = -1$ will produce the same final result). Let us treat a position number i of the sequence symbol as an element of $GF(p)$, i.e. being reduced modulo p, and form $p+1$ p-ary sequences $d_{k,i}$ over $GF(p)$ (i.e. with elements from this field) as follows:

$$d_{k,i} = \begin{cases} i + \alpha^{i+k} + \alpha^{-i}, & k = 1, 2, \ldots, p-1 \\ i + \alpha^{i}, & k = p \\ i + \alpha^{-i}, & k = p+1 \end{cases} \tag{7.57}$$

where all arithmetic is that of $GF(p)$, α is a primitive element of $GF(p)$ and $i = \ldots, -1, 0, 1, \ldots$. One may see that every sequence in (7.57) is formed as a sum of sequences of co-prime periods p and $p-1$ ($\alpha^{p-1} = \alpha^0 = 1$), and therefore has the period $L = p(p-1)$. Now perform a mapping of sequences (7.57) onto the binary alphabet $\{\pm 1\}$ using the extended binary character:

$$a_{k,i} = \psi(d_{k,i}), \; k = 1, 2, \ldots, p+1, \; i = \ldots, -1, 0, 1, \ldots \tag{7.58}$$

The binary set thus generated has parameters:

$$L = p(p-1), K = p+1, \rho_{\max}^2 = \frac{(p+3)^2}{L^2} \tag{7.59}$$

Length L may be made large enough only by a proper choice of p ($p \gg 1$), in which case $(p+3)^2/L = (p+3)^2/(p^2-p) \approx 1$ and $\rho_{\max}^2 \to 1/L$, showing after comparing with (7.45) at least asymptotical optimality of the ensemble in the correlation peak.

Example 7.5.3. Let $p = 7$. Direct verification confirms that $\alpha = 3$ is a primitive element in $GF(7)$. Then the sequences $\{\alpha^i\}$ and $\{\alpha^{-i}\}$ are both of period $p-1=6$: $\{\ldots, 1, 3, 2, 6, 4, 5, 1, 3, \ldots\}$ and $\{\ldots, 1, 5, 4, 6, 2, 3, 1, 5, \ldots\}$, respectively. Combined modulo 7 with the sequence $\{i\} = \{\ldots, 0, 1, 2, 3, 4, 5, 6, 0, 1, \ldots\}$ of period 7, as prescribed by (7.57), they produce $K = p+1 = 8$ 7-ary sequences of period $L = p(p-1) = 42$. For instance, the first of them is $\{d_{1,i}\} = \{221136110025006614665503554462443351332240\}$. Replacing their 7-ary elements by the extended characters according to the rule $\psi(0) = \psi(1) = \psi(2) = \psi(4) = 1$, and $\psi(3) = \psi(5) = \psi(6) = -1$ converts the sequences into 8 binary ones, e.g. $\{a_{1,i}\} = \{++++--+++++-++--++----+---++-+++---+--++++\}$. It is rather tiresome to compute their ACF and CCF 'by hand', and Problem 7.42 provides Matlab software to support it.

The second Kamaletdinov construction exploits the p-ary ($p = 4h+3 = 3\bmod 4$) linear sequence $\{c_i^k\}$ obtained by decimation with the index $d = p-1$ of $p-1$ shifts $\{d_{i+k}\}, k = 1, 2, \ldots, K = p-1$ of the p-ary m-sequence $\{d_i\}$ having memory $n = 2$, i.e. length $p^2 - 1$. Since d divides p^2-1 the sequence $\{c_i^k\} = \{d_{di+k}\}$ has the period $(p^2-1)/(p-1) = p+1$. Now let us build $p-1$ sequences over $GF(p)$:

$$d_{k,i} = i + c_i^k, \; k = 1, 2, \ldots, p-1, \; i = \ldots, -1, 0, 1, \ldots, \tag{7.60}$$

and map them onto the binary alphabet $\{\pm 1\}$ according to (7.58). This generates the ensemble with parameters:

$$L = p(p+1), \; K = p-1, \; \rho_{\max}^2 = \frac{(p+1)^2}{L^2} = \frac{1}{p^2} \tag{7.61}$$

Again, for the case of long lengths ($p \gg 1$) the ratio $(p+1)^2/L = (p+1)^2/(p^2+p) \to 1$ and $\rho_{\max}^2 \to 1/L$, demonstrating at least asymptotical optimality of the ensemble.

Example 7.5.4. This time there is no exclusion for $p = 3$ and the p-ary m-sequence $\{d_i\}$ of memory $n = 2$ and length $p^2 - 1 = 8$ may be formed by the primitive polynomial over $GF(3)$ of the second degree $f(x) = x^2 + x + 2$, or equivalently, by a recurrent equation $d_i = 2d_{i-1} + d_{i-2}$. The initial loading $d_0 = 1, d_1 = 0$ produces the sequence $\{d_i\} = \{\ldots, 1, 0, 1, 2, 2, 0, 2, 1, 1, 0, \ldots\}$. Its shifts decimated with index $d = p - 1 = 2$ transform into two sequences of period 4: $\{\ldots, 1, 1, 2, 2, 1, 1, 2, 2, \ldots\}$ and $\{\ldots, 0, 2, 0, 1, 0, 2, 0, 1, \ldots\}$. After symbol-wise addition with the sequence $\{i\} = \{\ldots, 0, 1, 2, 0, 1, 2, \ldots\}$ they give $K = p - 1 = 2$ sequences of period $L = p(p+1) = 12$: $\{1, 2, 1, 2, 2, 0, 2, 0, 0, 1, 0, 1\}$ and $\{0, 0, 2, 1, 1, 1, 0, 2, 2, 2, 1, 0\}$. The last step, replacement of their elements by extended characters $\psi(0) = \psi(1) = +1, \psi(2) = -1$, produces the Kamaletdinov set of two binary sequences of length $L = 12$: $\{a_{1,i}\} = \{+-+--+-+++++\}$ and $\{a_{2,i}\} = \{++-++++---++\}$. Their ACF and CCF are not difficult to compute by hand (or with the aid of the program of Problem 7.43), coming to $\rho^2_{max} = 1/9$ in full agreement with (7.61).

Table 7.1 Examples of binary minimax signature sets

Ensemble	Length L	Size K	Squared correlation peak ρ^2_{max}
Gold	$2^n - 1, n \neq 0 \bmod 4$	$L + 2 = 2^n + 1$	$\frac{(\sqrt{2(L+1)}+1)^2}{L^2} \to \frac{2}{L}, n$ odd
	7,31,63,127,511,1023		$\frac{(2\sqrt{(L+1)}+1)^2}{L^2} \to \frac{4}{L}, n$ even
Kasami	$2^n - 1$, n even 15,63,255,1023	$\sqrt{L+1}$	$\frac{(\sqrt{L+1}+1)^2}{L^2} \to \frac{1}{L}$
Union of Kasami and bent sequences	$2^n - 1, n = 0 \bmod 4$ 15,255	$2\sqrt{L+1} - 1$	$\frac{(\sqrt{L+1}+1)^2}{L^2} \to \frac{1}{L}$
Kamaletdinov 1	$p(p-1)$, ($p = 3 \bmod 4$, prime) 42,110,342,506,930	$p + 1 = \frac{\sqrt{4L+1}+3}{2} \to \sqrt{L}$	$\frac{(p+3)^2}{L^2} \to \frac{1}{L}$
Kamaletdinov 2	$p(p+1)$, ($p = 3 \bmod 4$, prime) 12,56,132,380,552,992	$p - 1 = \frac{\sqrt{4L+1}-3}{2} \to \sqrt{L}$	$\frac{(p+1)^2}{L^2} \to \frac{1}{L}$

To prove the propositions on the correlation peak of the ensembles above, the theory of quadratic equations in finite fields is necessary. Leaving this sophisticated issue aside, we refer the interested reader to the original paper [76].

Let us summarize now our knowledge on the binary minimax ensembles in Table 7.1, presenting length (listing all the lengths of existing ensembles within the range $7 \leq L \leq 1023$), number of signatures and squared correlation. The table is expressive enough as regards the significant contribution of Kamaletdinov sets: the number of their lengths in the considered range is 11, compared to 6 for Gold and 4 for Kasami sets.

Problems

7.1. In a DS CDMA system based on periodic binary signatures and BPSK data modulation, a user transmits the signal $\{++---+---+++-+++----+-\}$ covering more than two data bits. What is the signature code of this user (the common sign of all symbols being immaterial) if data bit duration equals signature period?

7.2. In a DS CDMA system based on periodic binary signatures and BPSK data modulation, a user employs the signature code $\{+++--+-\}$, data bit duration being equal to the signature period 7Δ. Due to the failure of timing recovery in the receiver, the despreading reference lags behind the received spreading signal by one chip. What is the result of data demodulation when a stream of zero data bits is transmitted?

7.3. How will the presence of amplitude modulation in an APSK signature affect the structure of a DS spreading receiver? Will the despreading in this case return a data symbol to the form characteristic of a non-spread transmission?

7.4. A DS CDMA system uses QPSK for data transmission at the rate 64 kbps and spreading code with chip rate 1.28 megachips per second (Mcps). Find the spreading factor and bandwidth occupied by the system.

7.5. An FH CDMA system uses a 4-frequency spreading signal of length $N = 4$: $\{1, 4, 2, 3\}$ and 4-FSK data transmission (each couple of bits is transmitted by one of 4 frequencies). The transmitted bit stream is 00101101. Draw a possible time–frequency array of the transmitted signal if one data bit covers two chip durations. What sort of FH is used: fast or slow?

7.6. An FH CDMA system uses a 4-frequency spreading signal of length $N = 4$: $\{1, 4, 2, 3\}$ and 4-FSK data transmission (each couple of bits is transmitted by one of 4 frequencies). The transmitted bit stream is 10110100. Draw a possible time–frequency array of the transmitted signal if one signature chip covers two data bits. What sort of FH is used: fast or slow?

7.7. A fast FH CDMA system uses 16-frequency spreading signal and 4-FSK data modulation. The chip duration is 10 μs. Estimate the minimal bandwidths of spreading and transmitted signals if chips of different frequencies should be orthogonal.

7.8. A synchronous CDMA system with BPSK data transmission at the rate $R = 9.6$ kbps should be organized within an available bandwidth $W_t = 76.8$ kHz. How many users can it accommodate to preserve the optimality of a single-user receiver? Design an appropriate binary signature set. How will the number of users change if BPSK data transmission is replaced by QPSK, 8-PSK or 16-QAM? If any of them increases the number of users, at what cost does this happen?

7.9. A synchronous CDMA system serves 36 users employing orthogonal signatures of equal energy per bit. How many new signatures (bandwidth and user data rate being fixed) of the same bit energy can one add to the existing ones without sacrificing minimum distance between different group signals?

7.10. What is the minimum length of synchronous signatures allowing a no smaller than 33% increase of the number of users in the oversaturation scheme (7.23)?

7.11. Add a supplementary signature to the four Walsh functions of length $N = 4$. Is the supplementary signature binary? If not, can you modify the primary signatures to make the supplementary one binary?

7.12. $K = (4N - 1)/3$ synchronous signatures are built according to the oversaturation scheme (7.23). Is it a good idea to use them for a K-user CDMA, if only a single-user receiver is acceptable?

7.13. Find the minimal length potentially allowing MAI power per signature per conventional receiver in a synchronous oversaturated CDMA no higher than −30 dB relative to the useful signal power, if the number of users is 101.

7.14. Prove that three or more binary sequences of length N cannot be orthogonal to each other unless their length is a multiple of four.

7.15. Can an oversaturated Welch-bound set of $K = 21$ binary signatures exist? What about $K = 22$, 23 or 32?

7.16. Outline the procedure of building a Welch-bound set of $K = 256$ binary sequences of length $N = 100$.

7.17. (Karystinos and Pados [64].) Prove that for an oversaturated set of an odd number K of binary signatures, the Welch bound (7.30) rises to:

$$TSC \geq \frac{K^2}{N} + \frac{N-1}{N}$$

7.18. Build an ensemble of $K = 15$ binary signatures of length $N = 12$ achieving the bound of the previous problem. Generalize the procedure to $K = 2^m - 1$ signatures ($K > N$).

7.19. What is the minimum period of $K = 11$ asynchronous signatures which does not prohibit obtaining average squared correlation between all their cyclic replicas within -20 dB?

7.20. Consider random signatures meeting (7.37). Prove that multiplication of signatures by data symbols (data modulation) does not disturb (7.37), provided data symbols are independent of signature symbols.

7.21. Prove that if two sequences of the same least period L both have perfect periodic ACF, their periodic CCF cannot equal zero for all mutual shifts.

7.22. Find the maximal number of asynchronous signatures of the period $L = 100$ which does not prohibit retaining the correlation peak below -23 dB.

7.23. $K = 50$ users may move freely within a zone of radius $D_c = 15$ km. The maximal delay spread of the channel between a user and central station $\tau_{ds} = 20\,\mu$s. Bandwidth allocated to the system is 2 MHz. Find the minimum lengths of the binary m-sequence and Legendre sequence allowing arranging time-offset signatures for the link 'user–central station'. Find the minimal length of the perfect PACF ternary sequence of memory 3 matching this problem.

7.24. In Section 7.5.1 an m-sequence is used to generate a frequency-offset signature set. Does any other binary minimax sequence (e.g. a Legendre one) allow the same way of obtaining a signature set with squared correlation peak around $1/L$? If not, why?

7.25. A CDMA system operates at carrier wavelength 4 cm with signature chip duration 1 µs. The length of signatures should be $L = 2^{10} - 1 = 1023$. What maximal number of frequency-offset signatures may be arranged, if the user's velocity ranges up to 144 km/h?

7.26. Find all decimation indexes fitting the Gold algorithm for lengths 63, 127, 511, 1023.

7.27. A signature ensemble is necessary to serve $K = 100$ users with correlation peak no greater than 0.064. What is the minimal length of the Gold ensemble meeting these demands?

7.28. Prove the minimax property (7.56) of Kasami sets.

7.29. A signature ensemble is necessary of size no smaller than 31 with correlation peak below -23 dB. Find the ensemble of minimal length among the known binary ones matching this requirement.

7.30. A signature ensemble is necessary of size no smaller than 24 with correlation peak below −25 dB. Find the ensemble of minimal length among the known binary ones matching this requirement.

Matlab-based problems

7.31. Write a program illustrating the principle of DS spreading–despreading (see Figures 7.3 and 7.4).

7.32. Figure 7.20 illustrates Matlab simulation of fast FH spreading using the example of 6 chips per one bit duration, the 4-frequency signature $\{F_i\} = (1, 3, 2, 1, 0, 3)$ and binary FSK data modulation (frequency offset of bit one $f_d = 4$). Write a program simulating fast FH spreading–despreading for a range of parameters.

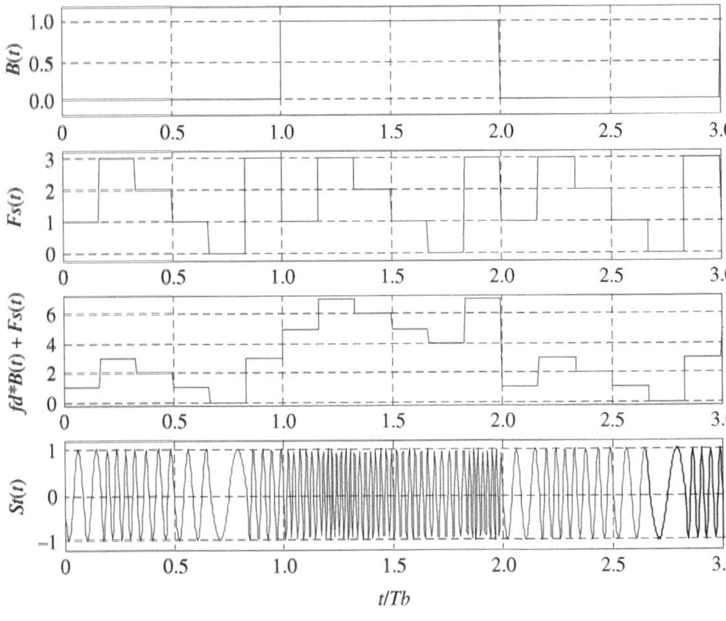

Figure 7.20 Fast FH spreading

7.33. Write a program calculating the squared distance between two group signals (7.11) using equation (7.15) for an arbitrary given synchronous signature set and binary data transmission. Test it for an orthogonal signature set (e.g. Hadamard matrix rows), varying randomly the patterns of differences ε_k.

7.34. Modify the program of the previous problem to calculate the minimum squared distance in a constellation of group signals (7.11) for an arbitrary set of K synchronous signatures and binary data transmission. Up to which set sizes K does the program run in a reasonable time?

7.35. Write a program generating an oversaturated signature set according to the scheme (7.23) for lengths $N = 4^l$ with arbitrary l. Take the lth Kronecker degree of $\mathbf{H}_4$ from Example 7.2.1 as the primary signatures. Make sure that all the supplementary signatures are binary. Use the program of Problem 7.33 for a spot check of the distances between group signals.

7.36. Write a program simulating multiuser reception in oversaturated synchronous CDMA. Steps to be done:

(a) Form $K = 21$ binary signatures as in the previous problem.
(b) Form a random K-dimensional vector of bits transmitted by K users.
(c) Modulate the signatures by bits in the BPSK manner and form a group signal.
(d) Add the Gaussian noise to the group signal, setting the noise standard deviation to something like the signature amplitude.
(e) Plot the observation obtained.
(f) Try all possible 2^K bit patterns, each time forming a candidate group signal and measuring the Euclidean distance from it to the received observation.
(g) Give out the decision on the bit pattern providing the closest candidate group signal to the observation and check whether all bit decisions are correct.
(h) Run the program, increasing the noise level, and comment on the results.

7.37. Write a program computing total squared correlation and average squared correlation per signature pair for an arbitrary synchronous signature set.

7.38. Write a program constructing an oversaturated binary Welch-bound set of $K = 2^n$ signatures of arbitrary length $N < K$. Use the set obtained in simulating K-user synchronous CDMA. Steps to be done:

(a) Construct a Hadamard matrix of size K.
(b) Discard $N - K$ rows in it and use the columns of the truncated matrix as signatures.
(c) Form the K-dimensional vector of users' data, and use it to manipulate signatures and form a group signal.
(d) Simulate K single-user receivers, each calculating the correlation of the received group signal with an appropriate signature and taking the decision according to the polarity of the correlation.
(e) Compare the K-dimensional vector of decisions on the data with the one really transmitted and find the number of erroneous bits.
(f) Repeat items (c)–(e) 1000–10 000 times and find the bit error probability per user.
(g) Run the program for $n = 5, 6, 7, 8$ finding each time the minimal length N (maximal oversaturation K/N) where errors do not still occur.

7.39. Write a program simulating an asynchronous CDMA employing time-offset replicas of the same binary m-sequence. Recommended steps:

(a) Set the maximal number of users $K = 20–25$ and the maximal delay in number of chips $m_{\max} = 80–100$.
(b) Form the binary $\{\pm 1\}$ m-sequence of length $L \geq K(m_{\max} + 1)$.
(c) Take as K signatures K cyclic replicas of the m-sequences, each delayed from the previous by $m_{\max} + 1$.
(d) Pick out K_a first signatures (active users) and manipulate all but the first by random independent bits, bit duration fixed within $N = 160–200$ chips.

(e) Shift all manipulated signatures but the first randomly and independently versus their initial versions within the range $[0, m_{max}]$ and add their sum to the first signature to come to a group signal.
(f) Imitate a single-user receiver for the first user, calculating the correlation within one bit of the first signature and the group signal.
(g) Demodulate the first user bit and compare it to the true one (which is zero).
(h) Varying the number of active users, e.g. $K_a = 5, 10, 20, \ldots$ repeat items (d)–(g) 1000–5000 times for every K_a. Calculate an experimental floor SIR at the correlator output and the bit error probability and compare them with the predictions $q_I^2 = N/(K-1)$, $P_e = Q(q_I)$.
(i) Plot 5–7 overlapped example realizations of a group signal and cumulative sum at the receiver integrator output (Figure 7.21 presents example plots for $K = 25,\ K_a = 20,\ N = 169,\ m_{max} = 80$).

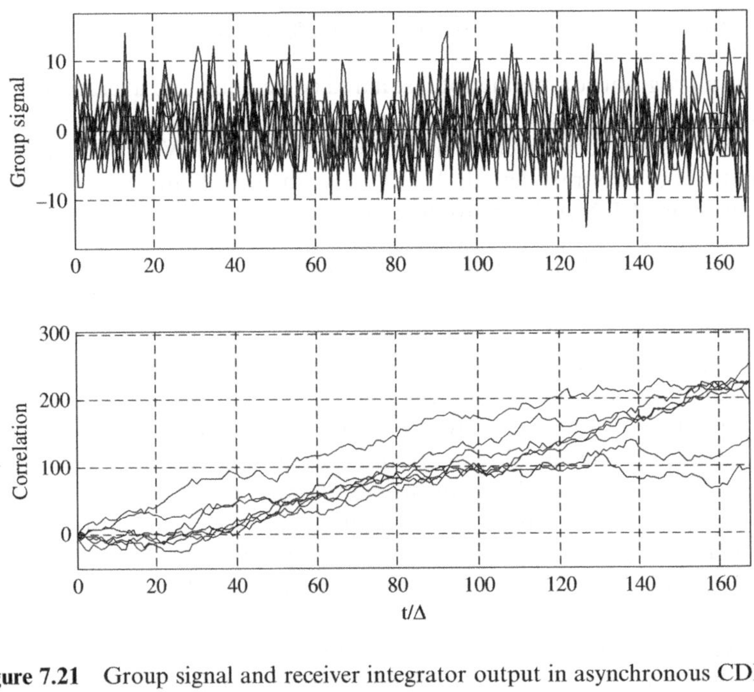

Figure 7.21 Group signal and receiver integrator output in asynchronous CDMA

7.40. Write a program generating the Gold set and outputting the histogram of the values of the periodic ACF and CCF of its members. Figure 7.22 presents example plots for $N = 2^5 - 1$.

(a) Generate a binary m-sequence of a proper memory $n \geq 5$.
(b) Decimate it in an appropriate manner to come to the second initial m-sequence.
(c) Form 15–20 Gold sequences as prescribed by (7.52), including initial m-sequences.

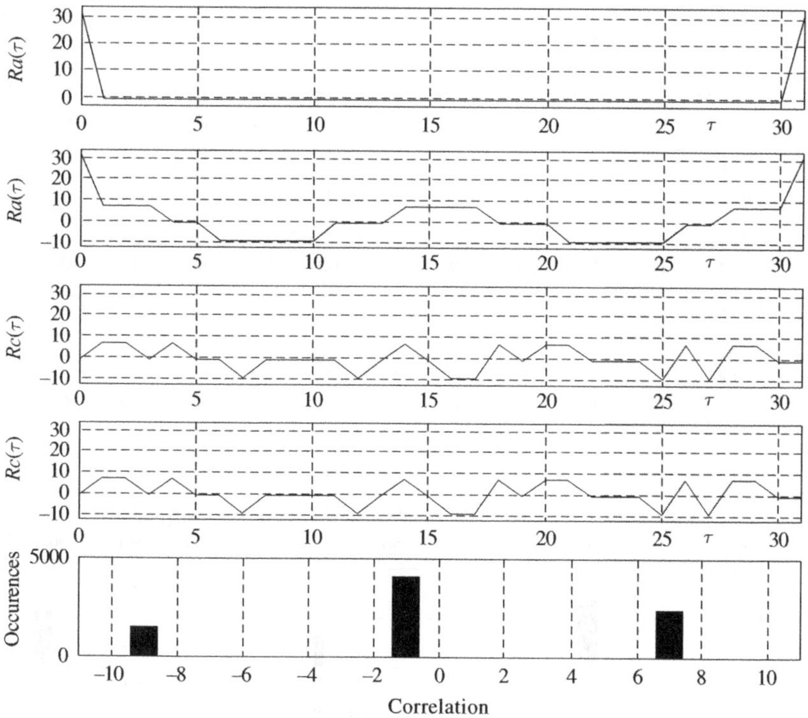

Figure 7.22 Periodic correlations and histogram of correlations for the Gold set of length 31

(d) Calculate their periodic ACF and CCF.
(e) Plot ACF of one of the initial *m*-sequences, ACF of some other Gold sequence, CCF of two initial *m*-sequences, and CCF of another pair out of the Gold sequences obtained.
(f) Plot a histogram of all unwanted correlations (CCF values and ACF sidelobes).
(g) Output the correlation peak and compare it with the theoretically predicted one and the lower bound.
(h) Run the program for relevant odd and even values *n* and interpret the results.

7.41. Write a program to investigate Kasami sets. The procedure may be in general similar to that of the previous problem, but now it makes sense to calculate the correlations between all (not only selected) sequences.

7.42. Write a program to investigate Kamaletdinov sets determined by (7.57)–(7.58). Figure 7.23 present the plots for the case $p = 11$. Recommended steps:

(a) Set $p = 3 \bmod 4$, find a primitive element α in $GF(p)$ and its inverse α^{-1}.
(b) Form $K = p + 1$ sequences over $GF(p)$ as defined by (7.57), all arithmetic being of $GF(p)$.
(c) Map the sequences obtained onto the binary $\{\pm 1\}$ alphabet according to (7.58).

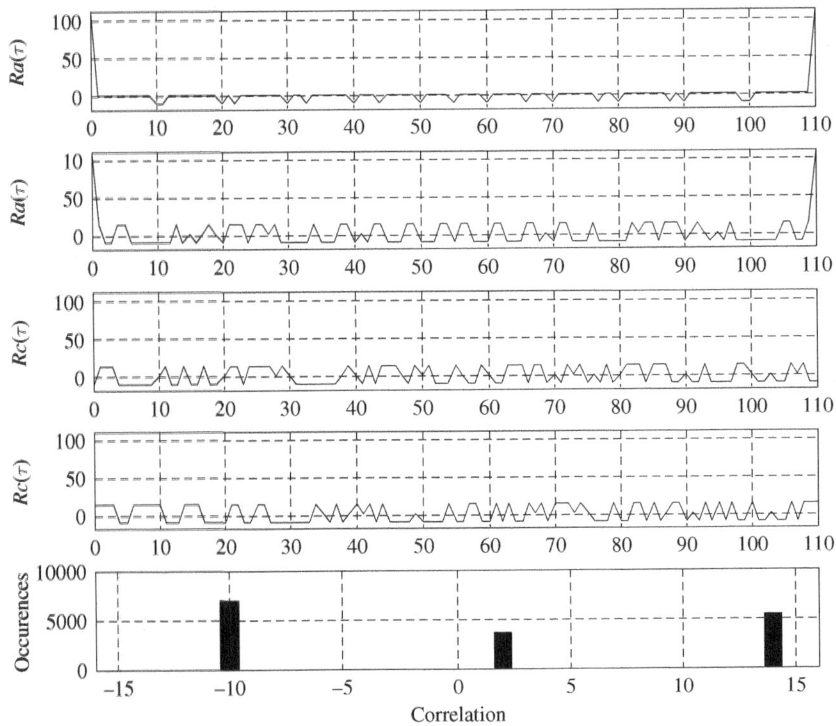

Figure 7.23 Periodic correlations and histogram of correlations for Kamaletdinov set of length 110

(d) Calculate the periodic ACF and CCF of all the sequences of the previous item.
(e) Plot ACF of sequences $\{a_{K,i}\}$ and $\{a_{1,i}\}$.
(f) Plot CCF of this pair and pair $\{a_{K-1,i}\}, \{a_{K,i}\}$.
(g) Plot a histogram of all unwanted correlations (CCF values and ACF side-lobes).
(h) Output the correlation peak and compare it with the theoretically predicted one and the lower bound.
(i) Run the program, varying p, and interpret the results.

7.43. Write a program to investigate Kamaletdinov sets determined by (7.60). Run it varying p, plot selected periodic ACF, CCF, and histograms of unwanted correlations. Register the correlation peak, and compare it to the theoretically predicted one and to the lower bound.

8

DS spread spectrum signal acquisition and tracking

8.1 Acquisition and tracking procedures

One of the most characteristic problems in spread spectrum technology is measuring the time of arrival and frequency of the received signal. In the systems where spread spectrum signals are used for ranging and measurement of object motion parameters (radar, sonar, navigation), time–frequency estimation is the main task. In spread spectrum communications it is the core of the timing recovery procedure. In fact, to correctly demodulate the transmitted data a receiver of any digital communication system should know with sufficient accuracy the borders of symbols, frames etc. in the received datastream. In other words, the local receiver clock should be properly synchronized with the received datastream. In spread spectrum systems a particularly precise synchronism is demanded, since time-mismatch between the received spreading signal and its local despreading replica (the reference) exceeding or equal to chip duration, will completely destroy the despreading and subsequent data demodulation (see Section 7.1). Therefore, the synchronization-related tasks of a receiver include preliminary (to starting the data recovery session) alignment of its own despreading reference with the spreading code of the arriving signal and maintaining rather accurate synchronism between them over the whole subsequent data reception time. Certainly, from the theoretical point of view, the synchronization procedure is not anything new: to align a local reference with the received signal one has just to measure the time–frequency shift of the received signal against the local clock. Then, if necessary, the receiver oscillator may be time–frequency corrected and thereby synchronized with the received signal.

The optimal (ML) strategies of measuring time delay and frequency were thoroughly discussed in Sections 2.12–2.14. In practice, however, their 'pure' realization very often runs into serious obstacles. The initial (e.g. when the receiver is primarily activated) bias

Spread Spectrum and CDMA: Principles and Applications Valery P. Ipatov

of the local clock in time and frequency against the received signal may appear rather large. Among the factors causing such a mismatch are autonomous operation of the transmitter and receiver clocks, the wide range of variations of path length between the transmitter and receiver, Doppler frequency shift due to their relative motion etc. In such circumstances direct implementation of the ML rule sometimes proves to be excessively demanding or even prohibitive in terms of resource consumption, as is demonstrated by the example below.

Example 8.1.1. The ranging C/A signal of the GPS (see Section 11.2) has period containing $L = 1023$ chips or in real time $L\Delta = 1$ ms. To solve the navigation task, one should measure the time position of the signal with accuracy no worse than a split microsecond, for instance 0.1 of chip duration Δ. If the receiver is initialized with no prior knowledge of a local clock mismatch versus the signal, the uncertainty of the signal delay spans one period, i.e. $L\Delta$. Turning to the correlator-bank receiver of Figure 2.18, one can see that its realization means implementing 10 230 parallel correlators. Switching to the matched filter structure (Figure 2.19) does not make the problem easier: such a filter with memory $L\Delta$, digitally realized, would have to operate with at least 10 samples per chip, performing 1023 summations during one sampling interval (smaller than 100 ns). Involving such an enormous hardware or software resource for performing only one of many tasks of the receiver does not look commercially justified, at least considering current technological tendencies.

In order to avoid implementation difficulties, the practical procedures of time–frequency estimation in a wide uncertainty region are often performed in the form of two successive steps. The first, called *acquisition* (code acquisition, search), performs a coarse measuring of the necessary parameters and provides preliminary estimates used by the second step, called *tracking*. This second step, typically performed by special code tracking and frequency tracking loops, delivers fine time–frequency estimations used further immediately by a local reference generator to align the despreading signal with a received spreading code. But in order to capture synchronism (pull-in) and enter the tracking state, the tracking loops need an initial targetting, e.g. knowledge of a received signal timing within one chip duration or so. This, as was already pointed out, is the task of the acquisition stage, reducing the primary uncertainty of signal parameters to that demanded by a tracking loop. Comparatively soft requirements towards the accuracy of estimates at the acquisition step allow cutting down the amount of calculated statistics and simplifying the implementation. To come back to Example 8.1.1, slackening the demands on the necessary precision of time measurement to one chip duration means a ten times smaller number of correlators in the scheme of Figure 2.18 or ten times lower processing speed in the matched filter structure. Yet the main resource-saving technique exploited by acquisition is a partial or complete replacement of parallel computations of the decision statistics by a serial one.

To explain this let us treat unknown delay τ and frequency shift F of the signal as signal coordinates on the time–frequency plane. Suppose that the initial uncertainty ranges of τ and F are τ_u and F_u, respectively, and that as a result of acquisition those ranges should be reduced to $\delta\tau$ and δF. Then, as Figure 8.1 shows, signal position is within one of M $\delta\tau \times \delta F$ rectangular *cells*, where $M = (F_u\tau_u)/(\delta F \delta\tau)$. The acquisition

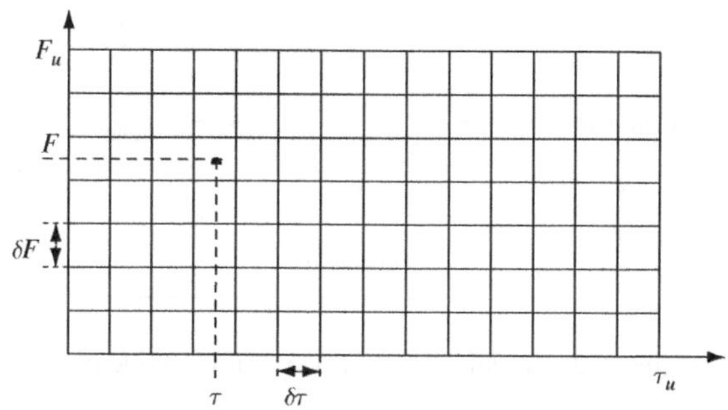

Figure 8.1 Search zone and signal position on the delay–frequency plane

should find out which one of M cells contains the signal, i.e. test M competitive hypotheses (see Section 2.8). If the optimal testing procedure were used, M correlations (2.74) would be computed in parallel for values τ and F corresponding to cell centres, and then the decision made in favour of τ, F corresponding to the highest correlation. Typical acquisition procedures, however, utilize the long presence of signal at the receiver input, which permits calculation of only several (not all M) correlations at a time. If none of them is large enough, the decision is taken that there is no true cell (i.e. containing the signal) among those tested and the search continues by examining another group of cells. The procedure goes on this way until some correlation is recognized as large enough to suggest that the corresponding cell is true. This terminates the acquisition, after which a tracking loop starts working, targetted by the estimations obtained. Remarkable research has been done concerning acquisition algorithms and strategies (see, for example, the bibliography in [77]). Below we will limit ourselves to only very brief discussion, starting with the simplest version of acquisition.

8.2 Serial search

8.2.1 Algorithm model

In a *serial search* only one cell at a time is tested, i.e. only a single correlation is calculated of the observation and a local signal replica, having some specific time–frequency shift. The correlation magnitude is then analysed in order to decide whether the cell is true or false. Various criteria may serve to take the decision. For example, the search may continue until all the cells inside the uncertainty region (see Figure 8.1) are tested, all the time storing in memory the maximal correlation observed up to now along with the values of τ, F corresponding to it. Then, after the last cell is analysed, the cell believed to be true is known automatically by its coordinates kept in memory, and all to be done is just reading them out. This strategy is equivalent to implementing the ML estimation rule, but calculating the necessary correlations not simultaneously but sequentially in time for successively arriving signal segments.

Still more typical of practical receivers is another version of a serial search, where the currently found correlation magnitude is just compared with a threshold [6,9,77]. If the

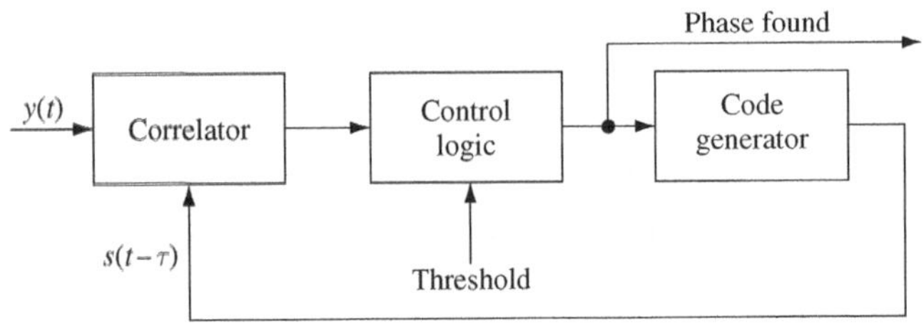

Figure 8.2 Serial search of a spreading code phase

correlation is larger than the threshold the decision is made that the current cell is true and the search finishes. Otherwise the search system examines the next cell and so forth.

From the point of view of performance analysis it does not matter how many parameters are unknown and to be estimated in the course of searching: both time and frequency (or whatever else) or some one of them. The only material thing is the overall number of cells to be checked. Yet to make further deliberations more transparent we will treat them as though an acquisition consists in only measuring the time delay of a received signal, the frequency being known a priori with sufficient precision. Figure 8.2 presents the structure running a serial search in this case. When a spreading code is periodic, the maximal uncertainty time zone may span only one period and all greater delays are reduced to fall within one period. In this light the name 'phase' is also appropriate as a synonym of delay of a periodic code [2,6,9] and transferring from one cell to the next in the uncertainty region means just changing the phase of the local code replica. When a current correlation is below the threshold the search control logic orders the local generator to increment the phase of the code replica $s(t-\tau)$ at its output by one chip or a split chip, and the procedure goes over to examining the next cell. If the current correlation exceeds the threshold the control logic signals that acquisition is finished and the code generator keeps the code phase corresponding to the cell declared to be true. In the following sections we discuss the performance of this algorithm based on the analysis of [78] and referring the reader inclined to learn more to [6,9,77,79–81].

8.2.2 Probability of correct acquisition and average number of steps

To simplify the analysis (still with no compromise of the basic regularities), let us assume that the local code replica is shifted versus the received signal by an integer number of chip durations. Then with L being the code period there are L possible code phases altogether, only one of which is true. To put it differently, there may be at most L search cells and every transition from one cell to the next means incrementing the code phase by one chip duration. Very often checking all of these L cells in the course of a search is needless thanks to reliable prior information shortening the search region to only M of L cells. Let us begin with an assumption that the search starts with the least favourable empty cell inside the uncertainty region, i.e. most distant from the true one. Figure 8.3 illustrates this premise: the first reaching of the true cell (black circle) takes place only after passing safely over $M-1$ empty cells (white circles).

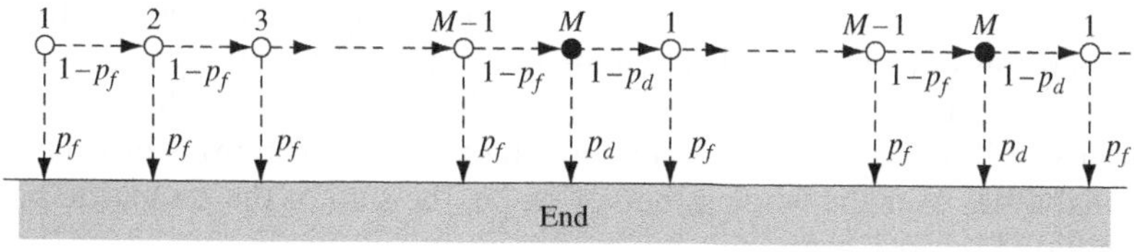

Figure 8.3 A serial search of a code within the uncertainty region of M cells

In order to achieve a reliable distinction between correlation levels in the false and true cells the search system should calculate the correlation in every cell over a sufficient time interval: *dwell time*. Whatever dwell time and threshold are set up, decisions on whether a cell is true or false cannot be absolutely faultless. One sort of possible error is a false alarm (see Section 3.2), i.e. declaring an empty cell true. When that happens, the search procedure finishes in the false cell, and it is natural to try to keep the probability of such an event low enough. On the other hand, dwelling in a true cell may also end up with non-zero probability by the wrong decision: missing the signal and transition to the next (empty) cell, which had been already examined earlier. This makes the search procedure cyclic: if it is not finished after the first scanning of the uncertainty region, a second one is performed and so on, each of the successive restarts initiating a new search *cycle*. Below we accept that the sought signal is present at the receiver input permanently so that the number of possible cycles has no upper limit. It is seen from Figure 8.3 that if the search system comes safely to a true cell there are two possible routes from it: correct decision (and finishing search) with detection probability p_d or missing signal with probability $1 - p_d$. The latter event entails only continuing the search by the next cycle of testing and has no negative consequences except for the extra time spent before the acquisition. Dwelling in an empty cell also has two possible outcomes: correctly declaring it false, accompanied by transferring to the next one with probability $1 - p_f$, or recognizing it true with false alarm probability p_f. In comparison to signal missing, this second error is more catastrophic, since the wrong code phase delivered by the search means that all the operations performed by the receiver afterwards are useless. To secure low risk of this event every decision that the code phase is found is typically rechecked at the cost of additional dwelling (see the next section) in the suspicious cell, but nevertheless some non-zero probability remains of terminating the search in the wrong cell.

Let us call a *search step* every dwelling in some cell ending with the decision to either prolong or stop the search. It is seen directly from Figure 8.3 that when the search starts at cell number one (most remote from the true one) it may stop at the tth false cell ($t = 1, 2, \ldots, M - 1$) after $mM + t$ steps, and in the true cell after $mM + M = (m + 1)M$ steps, where $m = 0, 1, \ldots$ is the number of complete 'idle' search cycles, i.e. passages across the uncertainty region preceding the final ($m + 1$th) cycle, at which the search stops in either a false or true cell. Then reading Figure 8.3 as a flow chart produces the following equation for probability $p(s)$ of finishing the search after s steps:

$$p(s) = \begin{cases} p_f(1 - p_f)^{t-1}\left[(1 - p_d)(1 - p_f)^{M-1}\right]^m, & s = mM + t, t = 1, 2, \ldots, M - 1 \\ p_d(1 - p_f)^{M-1}\left[(1 - p_d)(1 - p_f)^{M-1}\right]^m, & s = mM + M \end{cases} \tag{8.1}$$

where $m = 0, 1, \ldots$. The events (and only those ones) whose probabilities are expressed by the second row of (8.1) all imply finishing the search with a correct estimation of code phase. Therefore, the overall probability P_{c1} (the index '1' indicates that the search starts from the first cell) of correct search outcome (correct acquisition) is just the sum of all these probabilities over the whole range of m, i.e. the sum of the geometric progression with the ratio $(1-p_d)(1-p_f)^{M-1}$:

$$P_{c1} = p_d(1-p_f)^{M-1}\sum_{m=0}^{\infty}\left[(1-p_d)(1-p_f)^{M-1}\right]^m = \frac{p_d(1-p_f)^{M-1}}{1-(1-p_d)(1-p_f)^{M-1}} \tag{8.2}$$

Another important parameter is the average number of steps $\bar{s}_1$ of the search:

$$\bar{s}_1 = \sum_{s=1}^{\infty} sp(s) = \overline{m}M + \bar{l} \tag{8.3}$$

where again index '1' points at the starting cell number one. In equation (8.3) $\overline{m}$ stands for the average number of idle search cycles running before the final one, while $\bar{l}$ denotes the average number of steps within the last cycle finishing in either a false or true cell. The probability of an individual cycle being idle is a product of the probabilities of non-stopping in all $M-1$ empty cells and a unique true cell, i.e. $(1-p_d)(1-p_f)^{M-1}$. Consequently, the probability $p(m)$ of exactly m idle cycles elapsing before the search stops is:

$$p(m) = \left[1-(1-p_d)(1-p_f)^{M-1}\right]\left[(1-p_d)(1-p_f)^{M-1}\right]^m, m = 0, 1, \ldots \tag{8.4}$$

As is seen, probability distribution $p(m)$ obeys the geometric law, whose expectation is well known, yet bearing in mind further needs we show how it is found through the generating function [14,66]:

$$g_m(z) = \overline{z^m} = \sum_{m=0}^{\infty} z^m p(m)$$

The derivatives of the generating function at the point $z = 1$ allow calculating moments of an associated random variable. In particular:

$$\left.\frac{\mathrm{d}g_m(z)}{\mathrm{d}z}\right|_{z=1} = \sum_{m=0}^{\infty} mp(m) = \overline{m}$$

For a generic geometric distribution $p(l) = a^l(1-a), l = 0, 1, \ldots; 0 < a < 1$, the generating function is obtained by summation of a geometric progression:

$$g_l(z) = \sum_{l=0}^{\infty} z^l a^l(1-a) = \frac{1-a}{1-za}$$

so that after differentiation:

$$\bar{l} = \left.\frac{a(1-a)}{(1-za)^2}\right|_{z=1} = \frac{a}{1-a} \tag{8.5}$$

Comparing a generic geometric law with distribution (8.4) shows readily that substitution $a = (1 - p_d)(1 - p_f)^{M-1}$ in (8.5) produces the desired expectation $\overline{m}$:

$$\overline{m} = \frac{(1 - p_d)(1 - p_f)^{M-1}}{1 - (1 - p_d)(1 - p_f)^{M-1}} \tag{8.6}$$

To find the second term $\overline{t}$ in (8.3) note that the probability of terminating a final cycle in an empty cell number t regardless of how many idle cycles preceded may be found by summation in m of all probabilities of the first row in (8.1) (see also Figure 8.3), while the probability of safely reaching the unique (Mth) true cell and stopping in it (again, regardless of the number of preceding idle cycles) is just (8.2). Thus, the probability distribution $p(t)$ of the number of steps t within the final cycle is:

$$p(t) = \begin{cases} p_f(1 - p_f)^{t-1} \sum_{m=0}^{\infty} \left[(1 - p_d)(1 - p_f)^{M-1}\right]^m = \dfrac{p_f(1 - p_f)^{t-1}}{1 - (1 - p_d)(1 - p_f)^{M-1}}, \\ \qquad t = 1, 2, \ldots, M - 1 \\ P_{c1} = \dfrac{p_d(1 - p_f)^{M-1}}{1 - (1 - p_d)(1 - p_f)^{M-1}}, t = M \end{cases}$$

The generating function of this probability distribution is:

$$\begin{aligned} g_t(z) &= \sum_{t=1}^{M} z^t p(t) = \frac{p_f}{1 - (1 - p_d)(1 - p_f)^{M-1}} \sum_{t=1}^{M-1} z^t (1 - p_f)^{t-1} + z^M P_{c1} \\ &= \frac{p_f}{1 - (1 - p_d)(1 - p_f)^{M-1}} \cdot \frac{z - z^M(1 - p_f)^{M-1}}{1 - z(1 - p_f)} + z^M P_{c1} \end{aligned}$$

Differentiating $g_t(z)$ at the point $z = 1$ gives, after some elementary algebra:

$$\overline{t} = \frac{1 - (1 - p_f)^M - Mp_f(1 - p_d)(1 - p_f)^{M-1}}{p_f[1 - (1 - p_d)(1 - p_f)^{M-1}]} \tag{8.7}$$

Then using (8.6) and (8.7) in (8.3) ends in the average number of steps:

$$\overline{s}_1 = \frac{1 - (1 - p_f)^M}{p_f[1 - (1 - p_d)(1 - p_f)^{M-1}]} \tag{8.8}$$

Let us now abandon our initial assumption about a starting cell and consider how beginning the search in a cell number r affects the results above. In this case the partial cycle (attribute it as number zero) arises spanning $M - r$ empty cells plus one true cell. The following events are possible within this cycle: finishing the search at a false tth cell with probability $p_0(t|r) = p_f(1 - p_f)^{t-r}, t = r, r + 1, \ldots, M - 1$, finishing it in the true cell with probability $p_0(t = M|r) = p_d(1 - p_f)^{M-r}$, and, lastly, missing the signal and continuing the search from the first cell with probability $P_{0m}(r) = (1 - p_d)(1 - p_f)^{M-r}$. Then the overall probability of the correct acquisition when starting from the rth cell is:

$$P_{cr} = p_0(t = M|r) + P_{0m}(r)P_{c1} = \frac{p_d(1 - p_f)^{M-r}}{1 - (1 - p_d)(1 - p_f)^{M-1}} \tag{8.9}$$

The average number of steps is recalculated in the same way. When the search terminates at the tth cell of zero cycle the number of steps passed is $t - r + 1$, but if the signal is missed, $\bar{s}_1$ extra steps on average will be added to the $M - r + 1$ passed already. Thus, the average number of steps $\bar{s}_r$, when starting from an arbitrary (rth) cell, is:

$$\bar{s}_r = \sum_{t=r}^{M} (t - r + 1)p_0(t|r) + (M - r + 1 + \bar{s}_1)P_{0m}(r) \tag{8.10}$$

The first sum here, after changing the summation index to $i = t - r + 1$, becomes:

$$p_f \sum_{i=1}^{M-r} i(1 - p_f)^{i-1} + (M - r + 1)p_0(t = M|r) \tag{8.11}$$

where the first term is easily evaluated through the generating function, as was done in the derivation of (8.7):

$$\sum_{i=1}^{M-r} i(1 - p_f)^{i-1} = \frac{1 - (1 - p_f)^{M-r+1} - (M - r + 1)p_f(1 - p_f)^{M-r}}{p_f^2}$$

Using this in (8.11) and (8.10) along with equality $p_0(t = M|r) + P_{0m}(r) = (1 - p_f)^{M-r}$ after a plain algebraic treatment gives:

$$\bar{s}_r = \frac{1 - (1 - p_f)^{M-r+1}}{p_f} + \bar{s}_1 P_{0m}(r) = \frac{1 - P_{cr}}{p_f} + \frac{P_{cr}}{p_d} \tag{8.12}$$

Certainly, under substitution $r = 1$ (8.9) and (8.12) turn into (8.2) and (8.8), respectively.

Now, knowing the prior probability distribution $p_0(r)$ of an initial cell number r we may average (8.9) and (8.12) over all initial cells within the search region. For a uniform a priori distribution $p_0(r) = 1/M, r = 1, 2, \ldots, M$ this operation results in the following overall average probability $\overline{P}_c$ of a correct acquisition and average number of steps $\bar{s}$ of the search:

$$\overline{P}_c = \frac{1}{M}\sum_{r=1}^{M} P_{cr} = \frac{p_d[1 - (1 - p_f)^M]}{Mp_f[1 - (1 - p_d)(1 - p_f)^{M-1}]} \tag{8.13}$$

$$\bar{s} = \frac{1 - \overline{P}_c}{p_f} + \frac{\overline{P}_c}{p_d} \tag{8.14}$$

8.2.3 *Minimizing average acquisition time*

Calculation and comparing with a threshold of correlation for every candidate code phase means dwelling for some finite time at every analysed cell. In general this time may be random, and, moreover, will depend on whether a current cell is true or false. We, however, limit ourselves to considering here only the simplest version of a serial search assuming fixed dwell time T_d. Some other options will be briefly discussed in the next section. Then the average time $\overline{T}_s$ spent by the search system is just a product of the average number of steps and dwell time: $\overline{T}_s = \bar{s}T_d$. It is evident that, signal power fixed, the longer is the dwell time T_d the more reliable may be the decision on whether the cell is true or false, i.e. smaller values of the false alarm (p_f) and signal miss ($1 - p_d$) probabilities per cell may be secured.

Certainly, the reliability of the search characterized by the probability of a correct acquisition $\overline{P}_c$ should not be worse than some predetermined quantity. As (8.13) shows, the same value of $\overline{P}_c$ may be achieved via different combinations of the probabilities p_f, p_d per cell. This fact underlies the opportunity of minimizing average search time by varying one of the parameters p_f or p_d, while the probability $\bar{P}_c$ of the correct acquisition is maintained constant. The physical nature of such optimization is pretty clear. Suppose we impose on the detection probability a strict requirement of being very close to one. This means that the search will almost certainly succeed in only one (initial) cycle; however, to secure high detection probability the dwell time at every cell has to be long, so that the search drifts slowly toward the true cell and average acquisition time $\overline{T}_s$ is large. On the other hand, we may accept a high probability of signal miss in order to shorten T_d, but this will lead to a high probability of repeated cycles, increasing the average number of steps as compared to the previous case, owing to which average acquisition time $\bar{T}_s$ may again appear large. Obviously, some intermediate optimum should exist for the detection probability per cell p_d minimizing the value $\overline{T}_s$.

To solve the task $\overline{T}_s = \min, \overline{P}_c = \text{const}$, one needs to specify explicitly the dependence of dwell time on probabilities p_f, p_d, i.e. equivalently, the channel model. Assuming AWGN channel, we recall that the correlation modulus is physically just a real envelope at the matched filter output, which is a Gaussian noise envelope for an empty cell and an envelope of signal plus noise mixture if the cell is true. It is well known and may be found in any communications handbook (e.g. [2,4,7,8]) that the PDF of the Gaussian noise envelope obeys the Rayleigh law (see Section 3.2 or (3.12)), while the envelope of the sum of signal and Gaussian noise has Rician PDF. In the normalized form convenient here, the latter may be written as:

$$W(Y) = \begin{cases} Y \exp\left(-\dfrac{Y^2 + q_d^2}{2}\right) I_0(q_d Y), \ Y \geq 0 \\ 0, \ Y < 0 \end{cases} \tag{8.15}$$

where Y is the value of the envelope normalized to the noise standard deviation, q_d is voltage SNR accumulated during the dwell time T_d and $I_0(\cdot)$ is the modified zero-order Bessel function of the first kind. Of course, substitution $q_d = 0$, meaning absence of signal, turns (8.15) into the Rayleigh PDF (note that $I_0(0) = 1$).

Remember that the decision on the contents of the cell is done by a comparison of Y with a threshold. If the threshold normalized to the noise standard deviation is Y_t then the decision that the cell is true is taken whenever $Y \geq Y_t$. Then we may write for the probabilities p_f, p_d:

$$p_f = \int_{Y_t}^{\infty} W(Y|H_0)\mathrm{d}Y, p_d = \int_{Y_t}^{\infty} W(Y|H_1)\mathrm{d}Y$$

where PDFs $W(Y|H_0)$ and $W(Y|H_1)$ are versions of (8.15) for the hypotheses H_0 (empty cell, $q_d = 0$) and H_1 (true cell, $q_d > 0$), respectively. The integrals above are:

$$p_f = \exp\left(-\frac{Y_t^2}{2}\right), p_d = Q_M(q_d, Y_t) \tag{8.16}$$

where $Q_M(\cdot, \cdot)$ is just the designation of the integral of Rician PDF, also called the Marcum Q-function [7,8].

Solving the first equation (8.16) for the unknown Y_t, p_f being set up, gives the threshold necessary for retaining the false alarm probability at the predetermined level: $Y_t = \sqrt{2\ln(1/p_f)}$. Substituting it into the second equation (8.16) associates p_d and p_f directly, given accumulated SNR q_d:

$$p_d = Q_M\left(q_d, \sqrt{2\ln\frac{1}{p_f}}\right) \tag{8.17}$$

In its turn, SNR provided by dwelling during time T_d is defined as usually (see Section 3.2): $q_d = \sqrt{2PT_d/N_0}$, with P being the signal power and N_0 being the one-side noise power spectrum density. Now, let $T_d(p_f, p_d)$ and $q_d(p_f, p_d)$ be the dwell time and SNR necessary to secure fixed probabilities p_f, p_d. Then:

$$T_d(p_f, p_d) = \frac{q_d^2(p_f, p_d)}{2P/N_0} \tag{8.18}$$

Eventually, the average search time according to (8.14):

$$\overline{T}_s = \overline{s}T_d = \left(\frac{1 - \overline{P}_c}{p_f} + \frac{\overline{P}_c}{p_d}\right) T_d(p_f, p_d)$$

or in normalized form:

$$\overline{T}_s\left(\frac{2P}{N_0}\right) = \left(\frac{1 - \overline{P}_c}{p_f} + \frac{\overline{P}_c}{p_d}\right) q_d^2(p_f, p_d) \tag{8.19}$$

Now the optimization solution becomes straightforward. Let the probability of correct acquisition $\overline{P}_c$ and size of the search region M be specified.

1. Set some value of false alarm probability per cell p_f within the range $0 < p_f < 2(1 - \overline{P}_c)/M$.
2. Solving (8.13) as an equation for unknown p_d find its value securing the required $\overline{P}_c$ along with given p_f:

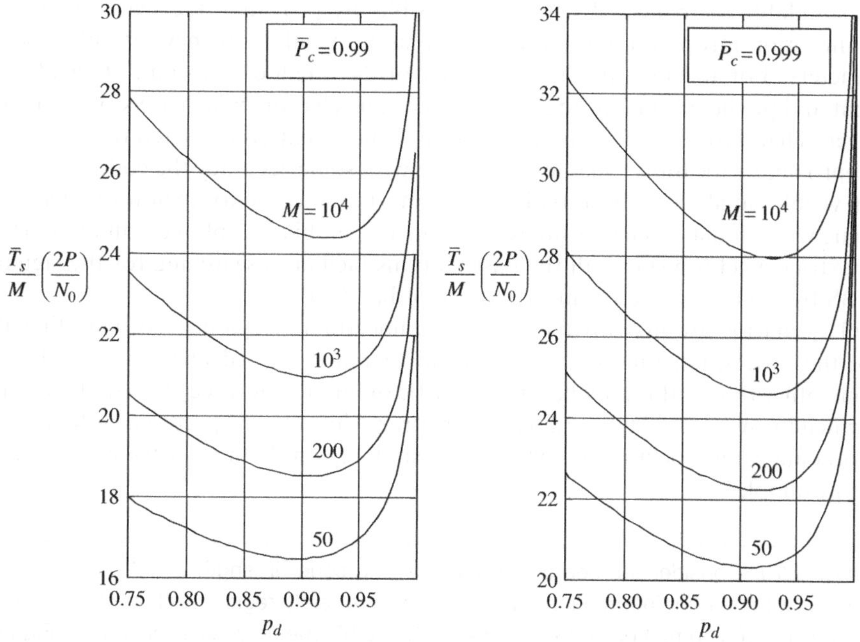

Figure 8.4 Normalized average acquisition time against the detection probability per cell

$$p_d = \frac{M\overline{P}_c p_f[1-(1-p_f)^{M-1}]}{1-(1-p_f)^{M-1}[1+p_f(M\overline{P}_c-1)]}$$

3. Solving (8.17) as an equation for unknown q_d find its value securing the pair p_f, p_d.
4. Substitute this solution into (8.19).
5. Varying p_f build the dependence of average search time on p_d and pick the pair p_d, p_f delivering minimum to the $\overline{T}_s$.

Figure 8.4 presents the curves of the normalized acquisition time $\overline{T}_s(2P/N_0)M^{-1}$ against the detection probability per cell p_d for two probabilities of successful acquisition $\overline{P}_c = 0.99, 0.999$ and four widths of an uncertainty region $M = 50, 200, 10^3$ and 10^4. As it shows, the minimum of average time occurs in the segment $p_d \in [0.9, 0.95]$. Allowing for rather flat behaviour of curves in the vicinity of the minimum value, $p_d = 0.9$ may be accepted as a universal figure for optimal detection probability per cell independently of the uncertainty range M and the probability of a correct acquisition $\overline{P}_c$.

8.3 Acquisition acceleration techniques

8.3.1 Problem statement

Naturally, average acquisition time grows with the number of cells M to be tested, i.e. uncertainty region extension, all other factors kept the same. The tendency is then understandable to reduce M whenever possible, by providing the search system with relevant prior information on the code phase, signal frequency or other scanned parameters. For instance in the GPS each of 24 satellites transmits data about the current and predicted state of the whole space constellation, which are stored in a user's receiver. Thank to this, after the user captures the signal of any satellite, he—knowing his approximate location—may pre-compute with some accuracy the code phases of the other visible satellites, substantially narrowing the uncertainty region of their search. A somewhat similar scenario takes place in the mobile telephone cdmaOne (IS-95), where strict synchronism of all the base stations facilitates searching for the signal of a new BS by the user's receiver in the course of handover.

Still, scenarios are inescapable where the uncertainty region is so wide that it can make the acquisition time of a plain serial search described above intolerably long. Among other things, this may be characteristic of initializing a receiver (first switching it on), when its own clock standard has an arbitrary shift with respect to a system one, and a priori data cannot be used to narrow the search region. Let us illustrate such a case by way of example.

Example 8.3.1. Consider the searching phase of the code of length $L = 2^{15}$ ('short' code of cdmaOne) with no prior information narrowing an uncertainty region. Putting $M = L = 2^{15}$ and the probability of a correct outcome no smaller than 0.99 one may extrapolate from Figure 8.4: $\overline{T}_s \geq 25 \times 2^{15}/(2P/N_0)$. If we accept as appropriate the figure $2P/N_0 = 40$ dB Hz, the average acquisition time will exceed 80 s. The estimation obtained is, however, rather optimistic, being based on the assumption of perfect initial chip synchronism (see beginning of Section 8.2.2). In such an idealized situation, it would be adequate to increment the code phase from step to step by a full chip duration. In practice, such a chip-synchronism often is not available and one-chip incrementing is at risk of hitting the signal ACF at the slope of the mainlobe instead of the peak point (see, e.g., Figure 6.9), which increases the hazard of missing a true cell. To avoid this trouble one should use a smaller increment, typically half of a chip, increasing the number of cells M within the uncertainty region and, automatically, the acquisition time.

The issue of time-consumption may become all the more crucial in applications employing very long spread spectrum codes, like systems of ranging and tracking remote space objects. Let us briefly and without resorting to mathematical subtleties describe the main techniques of accelerating the search operation.

8.3.2 Sequential cell examining

The strategy of constant dwelling time independently of whether the cell is false or true, which was accepted above, might be refined bearing in mind that the majority of cells are empty. Indeed, the potential technique allowing quick recognition of an empty cell at the cost of prolonged examination of the true one should intuitively save acquisition time. Such methods do exist and are covered by the general name of *sequential analysis*. The simplest sequential procedure is a two-dwell one [9,78,82]. Its key idea consists in dividing the examination process into two stages. In the first of them a rather low threshold secures a low probability of signal miss even with dwelling time T_{d1} short enough. At the same time, the false alarm probability appears to be much higher than would be tolerable within the previous (single dwell) method. Due to the short dwelling time T_{d1} false cells are on average examined quickly, but a high percentage of them (up to 10% or even more) are declared as true. In order to sift out false cells mistaken by the first stage for true, the second stage, having much better reliability than the first, is performed. The latter is achieved by a proper parameter choice: longer dwelling time T_{d2} and higher threshold provide total (including the first stage, too) error probabilities per cell satisfactory to meet the required probability of a correct acquisition (8.13). If the stages are performed independently, i.e. correlation at the second is computed ignoring that accumulated at the first, total false alarm and detection probabilities per cell are $p_f = p_{f1}p_{f2}$ and $p_d = p_{d1}p_{d2}$, the second index numerating the stage. The total dwell time per cell now proves to be random since examining any cell either finishes at the first stage or with some probability passes on to the second stage. The false alarm probability of the first stage p_{f1}, being many times higher than the total one p_f, is still much smaller than the detection probability p_{d1}. This implies that the average dwell time for a false cell $\overline{T}_{df} = T_{d1} + p_{f1}T_{d2}$ is smaller than a similar value at the true cell $\overline{T}_{dt} = T_{d1} + p_{d1}T_{d2}$, which is favourable for saving acquisition time. With an optimal choice of p_{d1}, p_{d2} the average acquisition time may be reduced two times or more compared to the search with fixed dwell time [9,78].

Further enhancement is possible with a greater number of stages, where each one rechecks the decisions of the previous [9,78,83–86]. The extreme case of this multiple dwell strategy is the Wald sequential analysis [85], where decision attempts are committed continuously with processing each successive chip. Two thresholds are then used and a cell is declared empty as soon as an accumulated correlation drops below the lower one, while the decision that a cell is true is taken if the upper threshold is crossed. As long as the decision statistic (correlation) remains between the thresholds dwelling in the cell continues by integration of more and more chips [77,78].

Calculation of acquisition time of multiple dwell (or in general sequential) strategies of a serial search may seem more complicated as compared to the case of a fixed-time one, because of the randomness of dwell time per cell. An effective way of simplifying the problem is given in [78], where it is proved that the average acquisition time for the search starting from the least favourable cell may be found as the product of the average number of cycles and average duration of one cycle.

8.3.3 Serial-parallel search

An evident resource of search acceleration involves several parallel correlators, each operating autonomously and scanning a separate part of the uncertainty region. In this case an initial uncertainty region just breaks into n_c sub-regions each covering M/n_c cells, where n_c is the number of parallel channels, and acquisition time accordingly reduces n_c times. In the uttermost case when $n_c = M$ the search becomes fully parallel and does not require serial steps. This opportunity enjoys wide application in real equipment and is especially efficient when exploiting hardware components which are necessarily present in the receiver but would otherwise be idle during the search session.

For example, any modern GPS receiver contains a multitude of correlator channels necessary for parallel tracking of signals of all (or, at least, four) visible satellites. During the search period these channels are free of other tasks and may be used for signal acquisition.

8.3.4 Rapid acquisition sequences

Scenarios are possible where speeding up an acquisition becomes of paramount importance. Imagine, for example, a system of object positioning in remote space. In order to enable unambiguous measurement of a distance in a very wide range, estimated to be maybe hundreds of thousands of kilometres or more, a spread spectrum signal of a suitably large period (e.g. hundreds of thousands or millions of chips) is necessary. Needless to say, the traditional search strategies discussed above will appear prohibitively slow unless the number of correlators goes up to many hundreds. For such cases special code sequences optimized according to the minimum acquisition time criterion may appear an effective option.

Why is a serial search so slow when applied to signals with good (having low sidelobes) ACF? The answer is straightforward: examining and rejecting any current empty cell reduces the uncertainty region by only one cell, and no fewer than M trials are necessary to pass across the whole zone starting at the least favourable cell. Then another question arises: is it not possible to arrange a sequence which would allow halving an initial uncertainty zone after testing one correlation, instead of discarding only one cell? Stiffler found an exhaustive solution of this problem [86], although some effective codes saving acquisition time had been proposed earlier [87].

Stiffler's *rapid acquisition sequences* have a plain structure, being just the sum of strictly synchronized n components. The first is a sequence of chips of alternating polarities $(\cdots + - + - + - + - \cdots)$, i.e. has period $L_1 = 2$ chips. The second is a meander wave $(\cdots + + - - + + - - \cdots)$ with period $L_2 = 4$ chips, etc. up to the nth meander of period $L_n = 2^n$ chips. We may treat the kth component as though it is a sequence of polarity-alternating 'long' chips of duration $2^{k-1}\Delta$, where Δ is chip duration of the first meander. A single correlator search starts by defining a phase of the first component. Under the assumption of chip synchronism existing, there are only two possible values of it, and correlating the observed waveform with the local reference, which is a replica of the first meander, removes this uncertainty: the correlation is positive if the local replica is in-phase with the arriving first meander and negative if they are antipodal. After this step is finished the chip synchronism with the second arriving meander (the chips are now of duration 2Δ!) is secured and only two phases of the second meander are again possible. This uncertainty is resolved in the same way, by correlating with the local replica of the second meander, and so on until at the nth step the phase of the 'slowest' (nth) meander is determined by correlating with its local replica. Thus, the whole search procedure takes only n steps, each reducing two times an initial uncertainty zone $M = 2^n$.

The procedure of sectioning the uncertainty zone into two equal halves is called dichotomy. Stiffler's rapid acquisition sequence is best matched to this procedure.

A straightforward generalization of the idea to the case of n_c-correlator search equipment consists in replacement of meanders by n components of period

$L_1 = n_c + 1$ chips, where the chip of each component is L_1 times longer than the preceding. The complete period of the summary sequence is $L = L_1^n$ initial chips. Each component should have as good a periodic ACF as possible and among binary components minimax ones (see Sections 6.7 and 6.9) are best. At the first step an uncertainty in L_1 possible phases of the 'fastest' component is resolved after a parallel correlating of the received signal with n_c local references, being the differences of $L_1 - 1$ cyclic replicas of the first component with the L_1th one. If among n_c correlations some are non-negative, the reference providing maximal correlation determines the phase, which is declared as synchronized with the first component of the received signal. Otherwise the L_1th cyclic replica is believed to be true. After this the chip bounds of the second component are known and the next step is fulfilled, repeating the same operations as the previous with the second component, etc. The procedure terminates after the nth step, where uncertainty on the phase of the slowest component is resolved in the same way as before. As is seen, each step in this case realizes testing of $L_1 = n_c + 1$ hypotheses and reduces L_1 times the initial uncertainty region $M = L = L_1^n$. Again, the sequences just described are best fit to this L_1-alternative testing: for every receiver complexity (number of parallel correlators n_c) the unique code exists minimizing acquisition time. Due to this, such codes deserve to be called matched [70]. Stiffler's rapid acquisition sequence is one of them, corresponding to a single correlator receiver.

A numerical evaluation shows that the gain in acquisition time accompanying the use of these codes may range far beyond hundreds of times [70, 86] (see also Problem 8.14).

An ordinary sum of binary components is a multilevel rather than a binary sequence, which entails signal amplitude modulation and is often considered objectionable. Clipping the sum and retaining only its sign transforms a matched code into binary, preserving the search procedure unchanged. The only penalty for this is a slight (within 1.5–2 times) reduction in the acquisition time gain.

Note that the two-stage synchronization code of UMTS (see Section 11.4.11) is an example of a similar approach: at the first stage the primary code is found whose period is 15 times smaller than that of the secondary one. Then the second search stage removes the uncertainty on which of $M = 15$ phases of the secondary code is true.

8.4 Code tracking

8.4.1 Delay estimation by tracking

Closed tracking loops are used universally in wireless receivers to instrument continuous and accurate parameter measuring. Depending on the nature of a measured parameter, examples are automatic frequency control, phase-lock loop, automatic gain control, and others. A strict mathematical investigation proving the optimality of tracking loops for the case of time-varying signal parameters is based on the theory of nonlinear estimation [88], but their primary idea follows directly from the ML rule studied in Chapter 2. The specific character of a spread spectrum receiver manifests itself chiefly in a despreading operation (see Section 7.1), appealing to a precise synchronism of a local despreading reference with the arriving signal. Keeping this in mind, we concentrate here on the precise measurement of delay (or code phase) of the arriving signal.

To get the point most quickly let us simplify the problem up to estimating the delay of a baseband signal, removing the effects of random phase. Suppose that τ is an unknown delay of a baseband signal $s(t)$. Then on the strength of the ML rule (2.55) the optimal estimator should form estimation $\hat{\tau}$ of this parameter as its value maximizing the correlation $z(\tau)$ between the reference signal replica $s(t-\tau)$ and observation $y(t)$. One way to implement this is a correlator bank (as in Figure 2.18) directly computing the function $z(\tau)$ at M sample points; the other is a matched filter structure reproducing $z(\tau)$ in real time (Figure 2.19, the envelope detector being unnecessary for a baseband signal). But both schemes may appear infeasible for the case of a spread spectrum signal of a long length: the first due to the necessity for numerous correlators and the second due to a problematic matched filter implementation (see Example 8.1.1).

The tracking loop structure is one more alternative. Note that at the maximum point of $z(\tau)$ its derivative vanishes:

$$z'(\hat{\tau}) = \left.\frac{dz(\tau)}{d\tau}\right|_{\tau=\hat{\tau}} = -\int_0^T y(t)s'(t-\hat{\tau})dt = 0 \tag{8.20}$$

Let us call $e(\hat{\tau}) = z'(\hat{\tau})$ the *error signal*, the reasons for which will be clear soon. As is seen, one may search for $\hat{\tau}$ as an argument making the error signal zero. Assume that the genuine signal delay is τ and a tentative estimate is $\hat{\tau}$, and calculate the mean of the error signal over all noise realizations in $y(t)$:

$$\overline{e(\hat{\tau})} = \overline{z'(\hat{\tau})} = -\int_0^T \overline{y(t)}s'(t-\hat{\tau})dt = -\int_0^T s(t-\tau)s'(t-\hat{\tau})dt = \int_0^T s'(t-\tau)s(t-\hat{\tau})dt \tag{8.21}$$

The last equality here follows after integration by parts, whenever care is taken that the integration interval spans the whole signal 'body' regardless of τ: $s(-\tau) = s(T-\tau) = 0$. Under the same condition the average error signal at $\hat{\tau} = \tau$ (tentative estimate coincides with a true parameter value) is zero:

$$\overline{e(\hat{\tau} = \tau)} = -\int_0^T s(t-\tau)s'(t-\tau)dt = -\left.\frac{[s'(t-\tau)]^2}{2}\right|_0^T = 0$$

but the derivative of $\overline{e(\hat{\tau})}$ at the same point according to the last part of (8.21):

$$\left.\frac{d\overline{e(\hat{\tau})}}{d\hat{\tau}}\right|_{\hat{\tau}=\tau} = \overline{z''(\tau)} = -\int_0^T [s'(t-\tau)]^2 dt$$

is minus energy of the signal derivative, i.e. negative. This means that if $\hat{\tau}$ lies in a sufficiently small vicinity to the left of τ, $\overline{e(\hat{\tau})}$ is positive, while with $\hat{\tau} > \tau$ $\overline{e(\hat{\tau})}$ is negative. This suggests the structure of a *delay-lock loop* (DLL) solving equation (8.20) by way of iterations and shown in Figure 8.5. The local reference generator creates a time-shifted replica of the signal

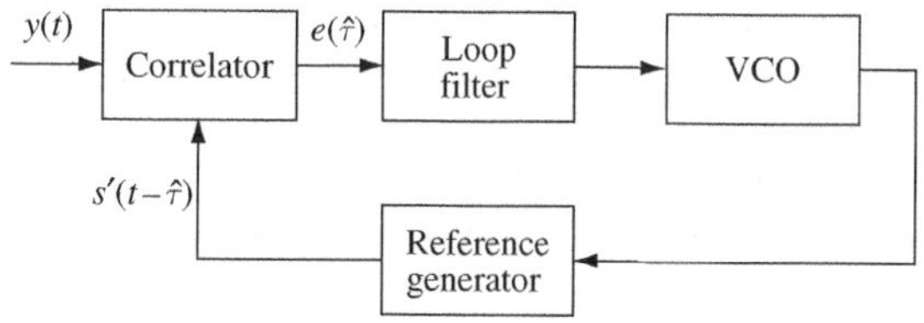

Figure 8.5 General DLL structure

derivative $s'(t-\hat{\tau})$, which is correlated in the correlator with the observation $y(t)$. The resulting error signal is then cleared off the noise by a loop filter to approximate averaging in (8.21). When the smoothed error signal is positive it tells with a high probability that the local reference $s'(t-\hat{\tau})$ is ahead of the signal, and forces the voltage controlled oscillator (VCO) to lower its frequency, i.e. increase the reference (tentative) delay $\hat{\tau}$. On the other hand, a negative smoothed error signal drives VCO[1] to the higher frequency, i.e. reduces reference delay. Clearly, in the steady state DLL maintains the error signal around zero, securing synchronism between the local reference and the arriving signal.

Obviously, to operate adequately the DLL needs an initial targeting, i.e. a starting value of $\hat{\tau}$, which is close enough to a genuine signal delay. On the one hand, this imposes demands on the precision of the acquisition procedure. On the other hand, the reference waveform following from the ML rule is usually not feasible (e.g. it may include delta functions when the signal consists of rectangular chips). Typically some other waveform replaces it, which, being more convenient to implement, preserves the main property: a distinct odd dependence of error signal at the correlator output on estimation error $\hat{\tau}-\tau$. In constructing such a reference the desire to provide pull-in (capturing synchronism) with softer requirements for the initial targeting may play an influential role. We will have more comments on this in Section 8.4.3.

8.4.2 *Early–late DLL discriminators*

The first element of the DLL structure is a *discriminator*, i.e. correlator–reference combination forming an error signal $e(\hat{\tau})$. In one of the classical schemes of DLL discriminator for a baseband signal $s(t)$, the reference $s_r(t)$ is the difference of two time-offset replicas of the signal: the late $s(t-\delta/2)$ and early $s(t+\delta/2)$ ones, δ being their time separation. Then a useful component of the error signal due to mismatch $\varepsilon=\hat{\tau}-\tau$ of reference versus the received signal is:

$$\overline{e(\hat{\tau})} = A\int_0^T s(t-\tau)s_r(t-\hat{\tau})\mathrm{d}t = A\int_0^T s(t-\tau)\left[s\left(t-\hat{\tau}-\frac{\delta}{2}\right)-s\left(t-\hat{\tau}+\frac{\delta}{2}\right)\right]\mathrm{d}t$$
$$= AE[\rho(\varepsilon+\delta/2)-\rho(\varepsilon-\delta/2)] \tag{8.22}$$

[1] In a real implementation input quantity controlling the oscillator frequency may be not a voltage, e.g. in a digitally realized loop input number plays this role. However, to avoid unnecessary multiplicity of terms we use the traditional term VCO as universal.

where E is the energy of a standard signal $s(t)$ over integration time T, A is the amplitude of the received signal scaling it versus $s(t)$, and $\rho(\tau)$ is the normalized signal ACF calculated over the time space $[0, T]$. Take, for instance, a discrete periodic signal (spread spectrum code) of length N with chips of duration Δ, and let integration time span integer number l of periods: $T = lN\Delta$, then $\rho(\tau)$ is periodic ACF. If $s(t)$ is a minimax binary sequence (e.g. m-sequence, Legendre sequence) with rectangular chips, then its normalized periodic ACF $\rho(\tau)$ looks as is shown in Figure 8.6a. The advancing and retarding (with minus) copies of $\rho(\tau)$ entering (8.22) are shown in Figure 8.6b by the dashed lines for an example separation $\delta = 2\Delta$ and their difference $\overline{e(\varepsilon)} = \overline{e(\hat{\tau} = \tau + \varepsilon)}$, called the *discriminator curve*, is given in the same plot by the solid line. It is readily seen that the discriminator under consideration is fully adequate: advancing or lagging of the reference versus an input signal causes positive or negative error signal, respectively, enabling the VCO to change its frequency and move the reference in the proper direction.

Figure 8.7 presents one possible structure of an early–late discriminator for the case of a baseband code. The code generator clocked by VCO forms the early signal replica $s(t + \delta/2)$, which delayed by δ gives a late replica. Their difference enters the correlator as a reference signal. The code replica $s(t)$ synchronized with the input signal and used for despreading may be obtained as delayed by the $\delta/2$ early replica. If the code generator is a shift-register-based one and $\delta = 2\Delta$ no external delays are necessary, since all three code replicas may be read from three successive register flip-flops.

An alternative scheme of the same discriminator involves two correlators first separately correlating the observation with early and late signal replicas and subtracting the results to come to an error signal $e(\hat{\tau})$ [18,77].

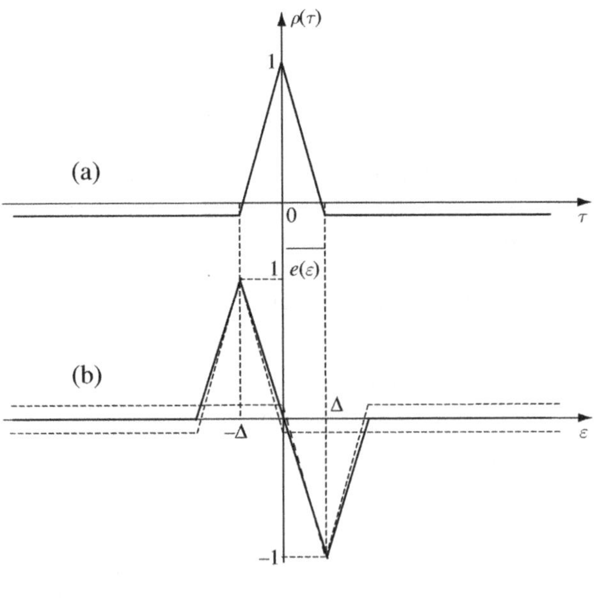

Figure 8.6 Discriminator curve of the early–late DLL

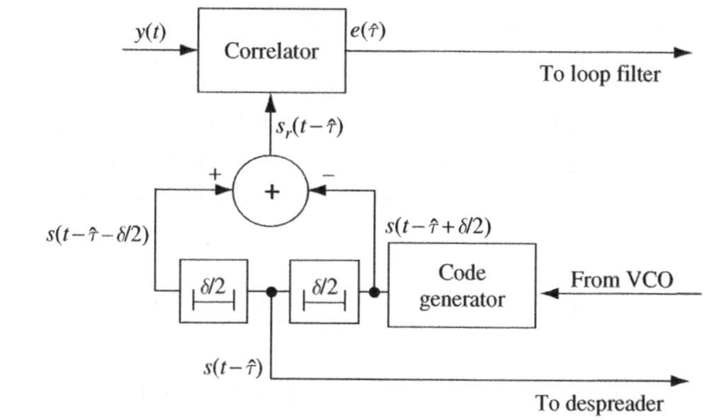

Figure 8.7 Early–late discriminator for a baseband signal

In the case of a bandpass signal the discriminator just considered may be applicable only if the receiver phase recovery loop is previously synchronized with the incoming signal carrier, so that a bandpass signal may be transformed into its real baseband equivalent. For this reason it is often called *coherent*, as is the DLL employing it. If preliminary phase synchronization is not possible, a *noncoherent* discriminator is used based on comparison of squared moduli of two correlations. These are calculated between the observed complex envelope $\dot{Y}(t)$ and the complex envelopes of two references, being early and late replicas of the signal code. Above all, this discriminator proves to be efficient despite the presence of data modulation. Technically it is often implemented as shown in Figure 8.8, where multiplication of complex envelopes is performed through heterodyning (see the comment at the end of Section 7.1.2). Let us inspect how this structure works, ignoring input noise and setting, with no generality violation, $\tau = 0$. Let the references be time-offset replicas of the spreading signal having complex envelope $\dot{S}(t)$ and carrier frequency f_1, which differs from the received one f_0:

$$s_r\left(t - \varepsilon \pm \frac{\delta}{2}\right) = \mathrm{Re}\left[\dot{S}\left(t - \varepsilon \pm \frac{\delta}{2}\right)\exp(j2\pi f_1 t + j\vartheta)\right],$$

where an initial phase ϑ will finally play no role. The incoming signal complex envelope $A\dot{B}(t)\dot{S}(t)\exp(j\phi)$ includes along with the spreading code also signal amplitude A, data

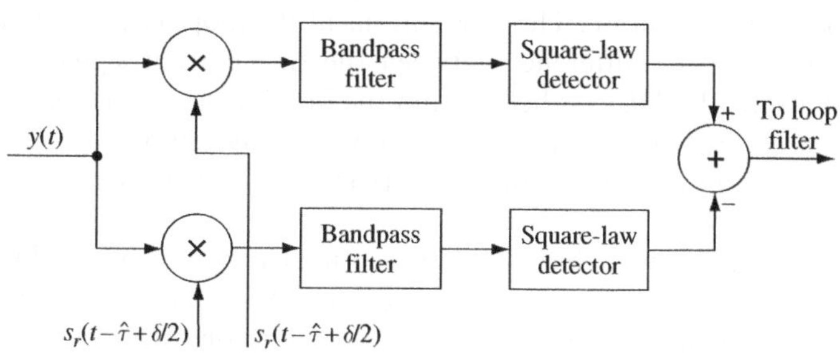

Figure 8.8 Early–late noncoherent discriminator

modulation component $\dot{B}(t)$ and unknown initial phase ϕ. After multiplying the input signal with the references and extracting the low-frequency component the two resulting bandpass signals of difference carrier frequency $f_0 - f_1$ will have complex envelopes $A\dot{B}(t)\dot{S}(t)\dot{S}^*(t - \varepsilon \pm \delta/2)\exp[j(\phi - \vartheta)]$. Suppose that bandpass filters after the multipliers have pulse response with complex envelope $\dot{H}(t)$. Then real envelopes at the filter outputs calculated in terms of the convolution integral (see Section 2.12.1) are:

$$\left| \frac{A}{2} \int_{-\infty}^{\infty} \dot{B}(\theta)\dot{S}(\theta)\dot{S}^*\left(\theta - \varepsilon \pm \frac{\delta}{2}\right)\dot{H}(t-\theta)\,\mathrm{d}\theta \right|$$

If the filter pulse response is rectangular of duration T equal to data symbol duration, and data modulation is PSK, then samples of output real envelopes at the moment $t = T$ are:

$$\left| \frac{A}{2} \int_{0}^{T} \dot{B}(t)\dot{S}(t)\dot{S}^*\left(t - \varepsilon \pm \frac{\delta}{2}\right)\mathrm{d}t \right| = \left| \frac{A}{2} \int_{0}^{T} \dot{S}(t)\dot{S}^*\left(t - \varepsilon \pm \frac{\delta}{2}\right)\mathrm{d}t \right| = AE\left|\dot{\rho}\left(\varepsilon \mp \frac{\delta}{2}\right)\right|$$

being proportional to the modulus of the corresponding value of ACF $\dot{\rho}(\tau)$ of the spreading complex envelope. The difference of squared moduli again gives a discriminator curve of form similar to the one of Figure 8.6 (see Problems 8.7 and 8.15).

Implementation of the scheme of Figure 8.8 may run into trouble in the form of parameter imbalance of the early and late branches. In order to get round it various solutions are known [9,18,77], including the *tau-dither* loop (another name is 'time-shared'), where only a single branch is involved, switching by turns between the early and late references.

8.4.3 DLL noise performance

DLL is just a particular case of a phase-lock loop and exposes the difficulties as to the analysis of its behaviour, which are generic to nonlinear feedback systems [89,90]. Still, one of the most important characteristics of DLL performance–noise error of a steady-state delay tracking–is easily calculated whenever the linear approximation is applicable.

In practice, rather small noise error is usually wanted, meaning good filtering capability of the loop against noise. The fluctuations at the loop output may be considered small if the error, i.e. the difference between the true current signal delay τ and its estimation $\hat{\tau}$ delivered by DLL, is held within the linear zone of the discriminator curve with probability close to one. If this condition is met, one can believe that the discriminator curve is linear in the infinite range of error $\varepsilon = \hat{\tau} - \tau$. This allows linearizing the system model as follows.

Let us limit ourselves to a baseband (or equivalently, coherent) discriminator of DLL and calculate the noise power spectrum $\tilde{N}_d(f)$ at its output. Since the correlator of this discriminator correlates observation with a reference signal $s_r(t) = s(t - \delta/2) - s(t + \delta/2)$, output noise variance according to (2.15) is found as $\sigma^2 = N_0 E_r/2$, where E_r is reference energy over integration time T. With E being, as earlier, standard signal energy, $E_r = 2E[1 - \rho(\delta)]$ and $\sigma^2 = N_0E[1 - \rho(\delta)]$. Integration over interval T may be thought of as low-pass filtering with bandwidth $W_f = 1/T$, meaning that the found noise power is spread over this bandwidth and therefore one-side noise power spectrum at the discriminator output is:

$$\tilde{N}_d(f) = \frac{\sigma^2}{W_f} = N_0ET[1 - \rho(\delta)] \tag{8.23}$$

With a rectangular chip shape and negligible level of sidelobes $\rho(\tau)$ is an isosceles triangle of unit height and base 2Δ, so that $\tilde{N}_d(f) = N_0ET$, whenever $\delta \geq \Delta$.

Replacing the true discriminator curve by an imaginary linear one means just an infinite continuation of a linear segment surrounding zero point with the same slope S_d. The last may be found from (8.22) allowing for evenness of ACF:

$$S_d = \left.\frac{\mathrm{d}e(\varepsilon)}{\mathrm{d}\varepsilon}\right|_{\varepsilon=0} = AE\left[\rho'\left(\frac{\delta}{2}\right) - \rho'\left(-\frac{\delta}{2}\right)\right]$$

Leaning again upon a triangular shape of ACF gives:

$$S_d = \begin{cases} -2AE/\Delta = -2AEW, & \delta < 2\Delta \\ -AE/\Delta = -AEW, & \delta = 2\Delta \end{cases} \tag{8.24}$$

where estimation of discrete signal bandwidth $W = 1/\Delta$ is substituted.

Imagine now that instead of a true noise $n(t)$, which is added to the received signal, a dummy noise $n_\tau(t)$ with power spectrum density $\tilde{N}_\tau(f) = \tilde{N}_d(f)/S_d^2$ is added directly to a measured parameter τ. At the output of a linear discriminator replacing the real one, this fictitious noise will be indistinguishable from the true noise at the real discriminator output, since its power spectrum $S_d^2\tilde{N}_\tau(f) = \tilde{N}_d(f)$ is absolutely the same. We then arrive at the system model of Figure 8.9, whose input is not a signal corrupted by noise, but instead parameter τ itself in a mixture with a fictitious additive noise $n_\tau(t)$. This mixture is processed by a linear closed loop, where estimation $\hat{\tau}$ is subtracted from the input quantity, outputting the error ε. The filter with transfer function $\tilde{h}(f)$ then smoothes error ε scaled by a discriminator slope S_d to produce the output estimation $\hat{\tau}$. Filtering here aggregates operations fulfilled by a loop filter and VCO to convert a real discriminator error signal $e(t)$ into a corresponding shift of a reference signal. This model is

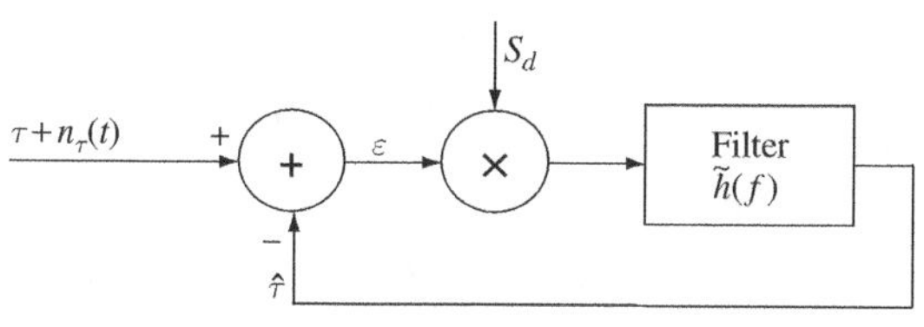

Figure 8.9 Linearized model of DLL

entirely linear, and variance $\mathrm{var}\{\hat{\tau}\}$ of random fluctuations at its output may be found, based on the superposition principle, independently of signal component, as:

$$\mathrm{var}\{\hat{\tau}\} = \int_0^{\infty} \tilde{N}_{\tau}(f)\left|\tilde{h}_l(f)\right|^2 \mathrm{d}f \tag{8.25}$$

where $\tilde{h}_l(f)$ is the transfer function of a closed loop. To find the last quantity it is enough to apply the delta function to the loop input. Then the output spectrum, being exactly $\tilde{h}_l(f)$, obeys the equation $\tilde{h}_l(f) = [1 - \tilde{h}_l(f)]S_d\tilde{h}(f)$ leading to a rule, which is well known in the theory of linear feedback systems [2,7]:

$$\tilde{h}_l(f) = \frac{S_d\tilde{h}(f)}{1 + S_d\tilde{h}(f)} \tag{8.26}$$

A dummy spectrum of delay fluctuations $N_{\tau}(f)$ is spread over the bandwidth $W_f = 1/T$, which is typically much wider as compared to the bandwidth of the closed loop; otherwise the latter could not smooth noise fluctuations effectively. This allows the following version of (8.25):

$$\mathrm{var}\{\hat{\tau}\} = \tilde{N}_{\tau}(f)B_N = \frac{\tilde{N}_d(f)B_N}{S_d^2} \tag{8.27}$$

where the loop *noise bandwidth* B_N is determined as:

$$B_N = \int_0^{\infty} \left|\tilde{h}_l(f)\right|^2 \mathrm{d}f = \int_0^{\infty} \left|\frac{S_d\tilde{h}(f)}{1 + S_d\tilde{h}(f)}\right|^2 \mathrm{d}f$$

Returning now to (8.23) and (8.24) and noting that $\rho(\delta) = 1 - \delta/\Delta, 0 < \delta < \Delta$ we may write (8.27) in the form:

$$\mathrm{var}\{\hat{\tau}\} = \begin{cases} (\delta/\Delta)/(4W^2q_l^2), & 0 < \delta < \Delta \\ 1/(4W^2q_l^2), & \Delta \le \delta < 2\Delta \\ 1/(W^2q_l^2), & \delta = 2\Delta \end{cases} \tag{8.28}$$

where

$$q_l^2 = \frac{A^2E}{N_0B_NT} = \frac{A^2P}{N_0B_N} \tag{8.29}$$

is called SNR in the loop. The reason for such a name is obvious: the numerator of (8.29) contains an actual signal power, since $P = E/T$ is a power of a standard (having amplitude $A = 1$) signal. At the same time, the denominator presents a noise power within the noise bandwidth of the loop.

Equation (2.28) is quite similar to the Woodward formula for potentially achievable accuracy of time measurement met in Section 2.12.2:

$$\mathrm{var}\{\hat{\tau}\} \approx \frac{1}{(2\pi W_{rms})^2 q^2}, q \gg 1$$

stressing that a tracking loop is an adequate means of time measurement.

There are three parameters affecting the steady-state accuracy of DLL: signal level with respect to noise A^2P/N_0, loop noise bandwidth B_N and early–late separation δ. The first is a brute force resource and needs no special comment. Optimization of the second is not a straightforward task, since a mechanical reduction of noise bandwidth without careful design of a loop filter may dramatically deteriorate the dynamic properties of the system like pull-in duration and ability of tracking a varying-delay signal. Choosing the separation δ is a matter of compromise, too: reduction of δ versus chip duration Δ provides higher estimation accuracy, thanks to positive correlation of noises at the outputs of early and late branches. At the same time the smaller is separation δ, the narrower is the discriminator characteristic itself (see Figure 8.6). This imposes more rigid demands on the acquisition precision, since the latter should guarantee falling time mismatch of the local reference and received signal inside an active (non-zero) zone of discriminator characteristic. Another factor to be kept in mind is the risk of loss of synchronism increasing with narrowing discriminator characteristic. A popular way to reconcile the conflicting requirements for the parameters of a tracking loop is adaptation: at the initial stage of pulling-in wider noise bandwidth and larger separation may be used, which after finishing the transient processes are reduced to come to a higher steady-state precision.

Problems

8.1. A serial search should be organized with a constant dwell time $T_d = 2\,\mathrm{ms}$. The discrete signal to be searched occupies bandwidth 1 MHz and has code length $L = 1000$. No prior information on code phase is known and the initial frequency bias of the local clock versus the signal carrier frequency lies in the range $\pm 10\,\mathrm{kHz}$. Estimate roughly the minimal number of cells to be tested.

8.2. Find asymptotic approximations of overall average probability and average number of steps of a serial search if the false alarm probability becomes very small ($Mp_f \ll 1$). Try to explain the results physically.

8.3. A serial search is used with fixed dwell time and threshold, which are optimized for some signal power P. What happens to the overall average probability of acquisition and average number of steps in two limiting cases: $P \to 0$ and $P \to \infty$ if no readjustment of dwell time and threshold is done? Give physical reasoning for the results.

8.4. Find expressions for probabilities of false alarm and detection and then for dwell time per cell necessary to secure given p_f, p_d, if signal amplitude fluctuates according to the Rayleigh law (3.12) and the initial phase is a random constant uniformly distributed over the interval $[-\pi, \pi]$. (Hint: the easiest way to do this is by treating a signal as a Gaussian process independent of noise.)

8.5. Find expressions for probability of correct acquisition and acquisition time for Stiffler's rapid acquisition sequence of length $L = 2^n$.

8.6. Build a discriminator curve of a coherent DLL for the cases $\delta = \Delta, 0 < \delta < \Delta$, and $\Delta < \delta < 2\Delta$. Find an extension of the maximal-slope zone. Why is $\delta > 2\Delta$ irrelevant?

8.7. Build a discriminator curve of a noncoherent DLL with separation $\delta = \Delta$. Why is $\delta = 2\Delta$ irrelevant?

8.8. Prove that a voltage controlled oscillator of DLL operates as an integrator of the error signal. Suppose no noise is present on the input and signal delay is constant. Prove that steady-state error at the DLL output is zero. Is the same true if the signal delay changes linearly and the DLL contains no more integrators?

8.9. Consider the DLL where no additional loop filter is used. Suppose the discriminator slope is 0.5 V/μs and VCO changes its frequency by 100 kHz per one volt. The initial difference of frequencies of VCO and received signal is 10 kHz. Find a steady-state noise-free error of DLL.

8.10. Find noise bandwidth and variance of the output error in terms of discriminator slope and VCO gain of a DLL having no loop filter. Give physical reasoning for the dependences of these quantities on system parameters.

Matlab-based problems

8.11. Write a program evaluating the average acquisition time (8.19) and plotting dependences similar to those of Figure 8.4 for an arbitrary search region, given the overall probability of a correct acquisition. Running the program for various values of $M, \overline{P}_c$, observe and comment on the behaviour of optimal probability of detection per cell.

8.12. Modify this program for the signal model of Problem 8.4. On running the program, observe and try to explain the difference in optimal detection probability and acquisition time against the previous case.

8.13. Write a program illustrating the dynamics of serial search of a baseband m-sequence. Recommended steps:

(a) Generate a $\{\pm 1\}$ m-sequence of memory 7–10 and oversample it two times in order to have two search steps per chip.

(b) Repeat this sequence as a reference with a random cyclic shift.

(c) Form several Gaussian noise realizations with standard deviation about $\sqrt{(2^n - 1)/8}$ higher than the signal amplitude, add them to the original sequence and plot with superposition the observation realizations obtained.

(d) Plot a reference sequence.

(e) Repeat the iteration steps, calculating every time the correlation between reference and an observation, each time updating the noise realization; if a current correlation normalized to the length exceeds the threshold 0.5 break the cycle and declare the search finished, otherwise shift the reference by one position (half a chip) and continue to the next step;

(f) At every step plot the correlation versus a current cell number and current reference to see its motion against the signal; use the operator 'pause' for a better visualization.

(g) Run the program, varying the threshold, and observe events like false alarm, signal miss and repeated cycles.

8.14. Leaning upon the results of Problem 8.5, write a program calculating search time for Stiffler's rapid acquisition sequences, given the probability of correct acquisition. Running the program for various code lengths, estimate the gain in acquisition time versus a serial search of an ordinary binary sequence of the same length. Plot the dependences of both acquisition times on n for different probabilities of correct acquisition.

8.15. Write a program calculating and plotting the discriminator curves of coherent and noncoherent DLL for various chip forms and early–late separations. Run the program for rectangular and half-cosine chips and comment on the results.

8.16. Write a program for simulating and exploring coherent DLL with no loop filter. Recommended steps:

(a) Generate a $\{\pm 1\}$ m-sequence of length $L = 63$ and oversample it 100 times.
(b) Set early–late separation δ within the range $0 < \delta \leq 2\Delta$ and form early and late references as shifted copies of the m-sequence.
(c) Add noise with a standard deviation exceeding signal amplitude by 15 times to the m-sequence of item (a) to obtain observations.
(d) Calculate correlations (normalized to reference energy) with early and late references and error signal as their difference.
(e) Multiply the error signal by gain G and round the result.
(f) Shift references according to the scaled error signal of the previous item.
(g) Plot 'pure' signal, observation and references.
(h) Repeat items (c)–(g) 1000 times and observe the behaviour of the DLL. Use the operator 'pause' inside the cycle to get a proper visualization.
(i) Changing gain in the range 10–40 and varying separation (e.g. $\delta = \Delta/2, \Delta, 2\Delta$) observe and compare with the theoretical prediction the dependence of output variance on these parameters.
(j) Find an experimental estimation of error variance and compare its value with the one found in Problem 8.10.

9

Channel coding in spread spectrum systems

9.1 Preliminary notes and terminology

In the course of transmitting, storing or processing the data need to be presented in some appropriate form. In digital communications the primary message generated by a source may be thought of as a sequence of data bits or a *bit stream*. It is mapping the bit stream onto a sequence of symbols of some predetermined alphabet that is traditionally called *coding*. The goals of coding may be different. For example, the terms *source coding* or *data compression* mean removal of redundancy from a bit stream to represent the source data in the most economical form. Another case of coding is encryption, which is performed to protect data from unintended interception or forging. The subject of this chapter is *channel coding*, aimed at making data transmission over the communication channel as immune as possible to the corrupting effects of unavoidable channel interference. The particular cases considered in Sections 2.3 and 2.5–2.7 show how important it is to find a proper signalling manner for overcoming the degrading influence of channel noise. Along with modulation, channel coding governs reliable data transmission over a noisy channel.

For over five decades of its existence, channel coding theory has been directed and motivated by the fundamental Shannon's capacity theorem mentioned in Chapter 1. According to this theorem, any channel is characterized by the constant C (measured in bits per second) called capacity, which establishes the upper bound of achievable rate R of information transmission over the channel. Whenever $R > C$ no signalling mode can secure an arbitrarily reliable data transmission. On the other hand, when $R < C$ one can always find a code guaranteeing as small a probability of mistaking one message for another at the receiving end as desired (see Figure 1.1). Shannon's capacity theorem, being a pure mathematical existence assertion, does not point to any concrete coding algorithm to achieve the quality tipped by it. Moreover, its proof, based on averaging the error probability over all possible channel codes, shows that almost all codes of

Spread Spectrum and CDMA: Principles and Applications Valery P. Ipatov

sufficient length are good from this angle. And yet finding specific code rules allowing Shannon's limit to be approached remained impenetrable up to the moment of discovering turbo codes in 1993, although lots of important and widely utilized results had been obtained in pursuing this target.

Of course, modern coding theory is too sophisticated to permit pressing even its initial basics into a brief chapter. It looks all the more inappropriate against the background of the key role of coding theory in general information technology, of which spread spectrum communications is only a particular branch. Still, the importance of channel coding in spread spectrum systems is extremely high, since the majority of them are designed to operate in a very noisy environment and, what is more, many themselves create strong intra-system interference (MAI in asynchronous CDMA). The MAI effects, unlike the natural (thermal) noise, cannot be overcome just by brute force, i.e. increasing signal power, since all users have equal rights and gain in SIR for one of them obtained in this way turns into a loss for the others (see Sections 4.5 and 4.6). This leaves the designer with only two resources for withstanding MAI: increasing the spreading factor and involving powerful channel codes. Trying to handle the available space reasonably, we limit this chapter to only coding issues related to the commercial 2G and 3G spread spectrum standards cdmaOne, UMTS and cdma2000. Accordingly, the mathematical tools, designations and description manner below are narrowly adapted to match this particular task in the most economical and fast way. We refer readers interested in a more universal scope to the books on coding theory (e.g. [31,33,91]).

Let us start with some basic classification of channel codes. The first feature to distinguish between them is alphabet size, according to which we talk about binary, ternary etc. codes. Although the range of applications of non-binary (e.g. Reed–Solomon or Ungerboeck) codes is pretty wide nowadays, we concentrate on only binary ones, which are used in the specifications mentioned above. Another form of classification is the way in which information data are mapped onto the *codewords* or *code vectors* (i.e. sequences of code symbols carrying the transmitted message). The point is that any channel coding consists of inserting some redundancy into the message, making the transmitted signals more distant from each other, and thereby reducing the risk of confusion between them. Depending on the way of adding this redundancy, all channel codes are classified into *block* or *tree* (*trellis*) codes. A characteristic of block codes is segmentation of the source bitstream, which is divided into blocks of k information bits, every block being encoded into $n > k$ binary symbols. In so doing, the redundant $n - k$ symbols serve to protect only the k source bits of their own codeword. Codewords of tree (e.g. convolutional) codes have a different structure: a continuous-source bitstream is encoded into an infinite stream of code symbols (*codestream*) with no fragmentation (see details in Section 9.3).

When it arrives at the receiving end the encoded word should be mapped back onto the transmitted data bits. This operation is called *decoding*. Physically, due to modulation, any codeword travels via the channel as some signal. When transmitted over the AWGN (or another state-continuous) channel, the signal gets corrupted by noise whose instantaneous samples are continuous. The optimal (ML) decision strategy of a receiver in the case of Gaussian noise is equivalent to the minimum Euclidean distance rule (see Section 2.1), which means declaring true the signal closest to the observation obtained. This straightforward procedure ends directly in decoded data bits and bears the name (along with its numerous approximations) *soft* decoding. The complexity of soft algorithms is the reason why an alternative *hard* decoding is often used instead. This decoding mode includes two steps: at the first of them decisions are made on all individual code symbols and as a result an observation appears to be demodulated into the vector consisting of symbols belonging to the code alphabet (in the considered case, binary). Some of the symbols of the *binary observation* thus obtained may be erroneous and then the whole demodulated vector is likely to differ from all codewords. The second step finds among all allowed codewords the one having maximal likelihood, meaning the greatest probability to be transformed by the channel to the current demodulated binary observation. Such a decoding procedure, typically ending in declaring some specific code vector as true, is called *error correction*. Alternatively, the goal of decoding may be limited to only checking whether the current binary observation is a true codeword or whether some errors occurred due to the channel corrupting effects. Then, if a binary observation enters the set of allowed codewords, it is mapped back onto the corresponding data bit sequence. Otherwise the fact of unsuccessful transmission is registered and the receiver either requests the transmitter to repeat the message (as in ARQ systems) or tries to restore it by interpolating the previous and subsequent ones. This sort of decoding, called *error detection*, is characteristic of the application of block codes in modern commercial wireless spread spectrum systems, and the following section focuses on coding formats complying with the error detection task.

9.2 Error-detecting block codes

9.2.1 Binary block codes and detection capability

Suppose that $b_0, b_1, \ldots, b_{k-1}$ are k source bits to be encoded in a binary codeword $\mathbf{u} = u_0, u_1, \ldots, u_{n-1}$ of length $n > k$. All 2^k combinations of k source bits are assumed possible, meaning that there are $M = 2^k$ codewords altogether. In every codeword $n - k$ binary symbols are redundant in the sense that only k symbols are necessary for one-to-one mapping of M source messages onto binary vectors. These redundant symbols make codewords more discernible from each other, securing better noise resistance. The set of all $M = 2^k$ words of length n is called (n,k) block code. The hard decoding, i.e. demodulation of continuous observation into a binary one, corresponds to the model of the binary symmetrical channel (BSC), transforming an input symbol $u = 0, 1$ into the opposite output binary one $y \neq u$ with the *crossover* (or *symbol error*) probability p. The attribute 'symmetric' emphasizes the equal probabilities of transitions of '0' into '1' and vice versa (see Figure 9.1).

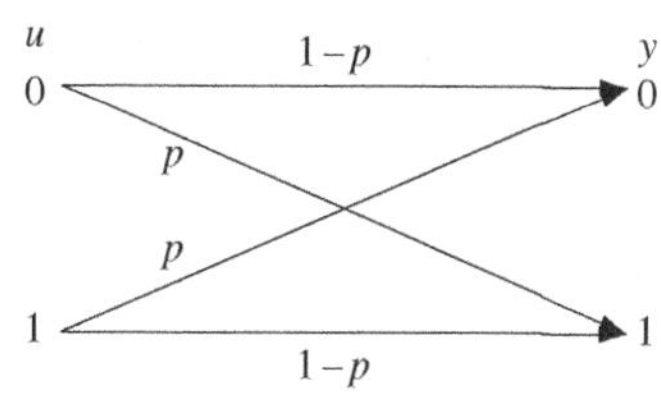

Figure 9.1 Binary symmetric channel model

Suppose that $\mathbf{y} = (y_0, y_1, \ldots, y_{n-1})$ is the binary observation at the BSC output. If $\mathbf{y}$ does not coincide with one of M code vectors $\mathbf{u}$, the receiver sees that erroneous symbols are present (error detection), otherwise k data bits are released corresponding to the estimated codeword $\hat{\mathbf{u}} = \mathbf{y}$. Clearly, if a transmitted word is $\mathbf{u}_0$ but the binary observation coincides with another codeword, i.e. $\hat{\mathbf{u}} = \mathbf{y} \neq \mathbf{u}_0$, an undetected error occurs and the released bits are not the true transmitted ones.

Let us introduce some more definitions. The *Hamming distance* $d_H(\mathbf{f}, \mathbf{g})$ between two vectors $\mathbf{f} = (f_0, f_1, \ldots, f_{n-1})$ and $\mathbf{g} = (g_0, g_1, \ldots, g_{n-1})$ of the same length n is the number of positions where the vectors have different elements $f_i \neq g_i$. The *Hamming weight* $w_H(\mathbf{f})$ of a vector $\mathbf{f}$ is the number of its non-zero components. If, for example, $\mathbf{f} = (01011)$, $\mathbf{g} = (11000)$, then $d_H(\mathbf{f}, \mathbf{g}) = 3$, $w_H(\mathbf{f}) = 3$ and $w_H(\mathbf{g}) = 2$. It is straightforward to make sure that $d_H(\mathbf{f}, \mathbf{g}) = w_H(\mathbf{f} - \mathbf{g})$ and $w_H(\mathbf{f}) = d_H(\mathbf{f}, \mathbf{0})$, with $\mathbf{0}$ being zero vector.

Let us assume as transmitted a codeword $\mathbf{u}$. The BSC will transform it into the different fixed codeword $\mathbf{v}$ (causing thereby an undetected error and releasing untrue data bits corresponding to $\mathbf{v}$), if symbol errors occur in all $d_H(\mathbf{u}, \mathbf{v})$ positions where $\mathbf{u}$ and $\mathbf{v}$ differ with no corruption of $n - d_H(\mathbf{u}, \mathbf{v})$ symbols coinciding in $\mathbf{u}$ and $\mathbf{v}$. For a memoryless BSC, i.e. the one where all symbol errors are independent we may then put the probability $P(\mathbf{y} = \mathbf{v}|\mathbf{u})$ of the event above as:

$$P(\mathbf{y} = \mathbf{v}|\mathbf{u}) = p^{d_H(\mathbf{u},\mathbf{v})}(1 - p)^{n - d_H(\mathbf{u},\mathbf{v})} \tag{9.1}$$

Since $p < 1/2$ (otherwise we just interchange the designations of output '0' and '1'), to reduce the probability of confusion of codewords $\mathbf{u}$ and $\mathbf{v}$ the Hamming distance between them should be as large as possible. Consider now Hamming distances $d_H(\mathbf{u}, \mathbf{v})$ between all different code vectors of a code U. Denote the least among them as d_H and call it the (minimum) *code distance* of a code U:

$$d_H = \min_{\substack{\mathbf{u} \neq \mathbf{v}, \\ \mathbf{u},\mathbf{v} \in U}} d_H(\mathbf{u}, \mathbf{v}) \tag{9.2}$$

Any codeword of U may appear as a transmitted one and to minimize the risk of confusing the two closest code vectors the distance between them, i.e. code distance d_H, has to be maximal. We arrive thereby at the following assertion.

Proposition 9.2.1. Code U is capable of detecting any t_d or fewer symbol errors (up to t_d errors) if and only if its code distance $d_H \geq t_d + 1$.

Indeed, take a code with $d_H \leq t_d$ and pick out a pair of its closest code vectors $\mathbf{u}, \mathbf{v}$. If d_H symbols of $\mathbf{u}$ different from those of $\mathbf{v}$ are corrupted, then $\mathbf{u}$ becomes $\mathbf{v}$, meaning that a pattern exists of no more than t_d errors, which is not detectable. Conversely, if the distance between any code vectors exceeds t_d, no pattern of t_d or fewer errors can transform one codeword into another.

One may prove the following statement in the same way (Problem 9.4).

Proposition 9.2.2. Code U is capable of correcting up to t_c or fewer symbol errors if and only if its code distance $d_H \geq 2t_c + 1$.

9.2.2 Linear codes and their polynomial representation

Let us treat binary code symbols $\{0, 1\}$ as elements of a binary finite field $GF(2)$ (see Section 6.6) and consider symbol-wise linear operations over codewords of a code U obeying the arithmetic of $GF(2)$. Clearly, there is only one non-trivial operation of this sort, namely symbol-wise $GF(2)$ addition:

$$\mathbf{u} = (u_0, u_1, \ldots, u_{n-1}),\ \mathbf{v} = (v_0, v_1, \ldots, v_{n-1}) \quad \Rightarrow \mathbf{u} + \mathbf{v} = (u_0 + v_0, u_1 + v_1, \ldots, u_{n-1} + v_{n-1})$$

For example, if $\mathbf{u} = (100111)$ and $\mathbf{v} = (010110)$, then $\mathbf{u} + \mathbf{v} = (110001)$. Symbol-wise subtraction has no independent role and just repeats addition, since in $GF(2)$ the negative of an element is the element itself. In the same way symbol-wise multiplication by a scalar from $GF(2)$ (i.e. by 0 or 1) of any codeword either makes it zero vector or does not change it at all.

A binary code U is called *linear* if the sum of any of its code vectors is again some code vector belonging to U. The name stems from the fact that such a code is a vector (linear) space over the field $GF(2)$ [31,33,91], although this concept is not involved seriously in our further discussion. Any linear code U of length n contains zero vector (i.e. with n zero components) as a code vector, since the sum of an arbitrary code vector entering U with itself produces exactly zero vector: $\mathbf{u} + \mathbf{u} = \mathbf{0}$. The following statement explains one of the reasons why linear codes are of special interest.

Proposition 9.2.3. The code distance of a linear code U is equal to the minimum Hamming weight over all non-zero codewords:

$$d_H = \min_{\substack{\mathbf{u} \in U \\ \mathbf{u} \neq \mathbf{0}}} w_H(\mathbf{u}) \tag{9.3}$$

To prove this just make a substitution $d_H(\mathbf{u}, \mathbf{v}) = w_H(\mathbf{u} - \mathbf{v})$ in (9.2) and note that $\mathbf{u} - \mathbf{v} = \mathbf{u}'$ is again a codeword of U.

As (9.3) shows, there is no need to test all $M(M-1)/2$ different vector pairs to find code distance of a linear code containing M words. It is enough to 'weigh' $M-1$ non-zero code vectors, i.e. to perform $M/2$ smaller number of tests, which, considering the typically enormous number of words M, offers a highly significant gain.

To get the idea of designing error-detecting codes of 2G and 3G wireless standards a polynomial description of linear codes is a good aid.

Let us associate with a codeword $\mathbf{u} = (u_0, u_1, \ldots, u_{n-1})$ the *code polynomial* $u(z)$ of a dummy variable z arranged as:

$$u(z) = u_{n-1}z^{n-1} + u_{n-2}z^{n-2} + \cdots + u_0$$

where by way of agreement we put $z^0 = 1$. This polynomial representation, which is widely used in coding theory, is just a form of the z-transform, underlying discrete linear system analysis, discrete signal processing, digital filtering etc. [2,7]. One-to-one correspondence between sets of codewords and code polynomials means that the sum of two code polynomials $u(z)$, $v(z)$ of a linear code U is again a code polynomial of the same code. Specifically, if $u(z) = \sum_{i=0}^{n-1} u_i z^i$, $v(z) = \sum_{i=0}^{n-1} v_i z^i$ are code polynomials of words $\mathbf{u}$, $\mathbf{v}$ of a linear code U, $u(z) + v(z) = \sum_{i=0}^{n-1} (u_i + v_i) z^i$ is a code polynomial of the word $\mathbf{u} + \mathbf{v} \in U$, where addition of coefficients follows the rules of the field they belong to, i.e. in our case $GF(2)$.

Polynomial arithmetic used in coding analysis and design includes two more operations: multiplication and division with *remainder*. The rules of these operations are universal regardless of the fields of polynomial coefficients, but dealing now with binary codes we use the terms of binary arithmetic. Consider an arbitrary (not necessarily a code one) binary (i.e. with coefficients from $GF(2)$) polynomial $a(z)$. The highest power of z in this polynomial holding non-zero coefficient is called the *degree* of $a(z)$ with symbolism deg $a(z)$. Let $a(z), b(z)$ be two binary polynomials and deg $a(z) = m$, deg $b(z) = n$. Then their product $a(z)b(z)$ is a polynomial of degree $m + n$ obtained by extending the commutativity ($z^i a = a z^i$) and distributivity laws to operations including a formal variable z and gathering together coefficients of equal degrees of z:

$$\begin{aligned} a(z)b(z) &= (a_m z^m + a_{m-1} z^{m-1} + \cdots + a_0)(b_n z^n + b_{n-1} z^{n-1} + \cdots + b_0) \\ &= a_m b_n z^{m+n} + (a_m b_{n-1} + a_{m-1} b_n) z^{m+n-1} + \cdots + (a_1 b_0 + a_0 b_1) z + a_0 b_0 \\ &= \sum_{k=0}^{m+n} \left(\sum_{i=0}^{k} a_i b_{k-i} \right) z^k \end{aligned}$$

Certainly, we put $z^m z^n = z^{m+n}$, all operations over a_i, b_i are performed in $GF(2)$, and in the last internal sum coefficients a_i, b_i, whose indexes become negative or exceed the polynomial degree should be set equal to zero. Take for example binary polynomials $a(z) = z^4 + z^3 + 1$, $b(z) = z^2 + z + 1$, then their product $a(z)b(z) = z^6 + z^3 + z^2 + z + 1$.

The algorithm of dividing a dividend $a(z)$ by the divisor $b(z)$ with remainder looks as follows:

$$a(z) = q(z)b(z) + r(z) \tag{9.4}$$

where $q(z)$ is a quotient and $r(z)$ is remainder. The uniqueness of $q(z), r(z)$ is secured by relations deg $a(z) =$ deg $q(z) +$ deg $b(z)$ and deg $r(z) <$ deg $b(z)$. The algorithm (9.4) is similar to the 'school' rule of dividing integers with remainder, the polynomial degree replacing a number magnitude. One of the techniques realizing this operation is 'long division', i.e. successive computation of remainder and division of it by a divisor until the remainder degree becomes smaller than the degree of the divisor. At the first iteration $b(z)$ is multiplied by z raised to the power equalizing the degree of the product with that of a dividend $a(z)$. Subtraction (equivalent to addition in $GF(2)$) from $a(z)$ of the product obtained results in the first remainder, which at the second iteration plays the role of dividend, and so forth. Let us illustrate it by an example.

Example 9.2.1. Suppose $a(z) = z^4 + z^3 + 1, b(z) = z^2 + z + 1$ and apply a long division algorithm:

$$\begin{array}{r} z^2+1 \\ z^2+z+1 \overline{) z^4+z^3+1} \\ + \\ \underline{z^4+z^3+z^2} \\ z^2+1 \\ + \\ \underline{z^2+z+1} \\ z \end{array}$$

After two iterations we have $q(z) = z^2 + 1, r(z) = z$ so that division with remainder results in $z^4 + z^3 + 1 = (z^2 + 1)(z^2 + z + 1) + z$.

Similar to integers, we say that $a(z)$ is divisible by $b(z)$ (or $b(z)$ divides $a(z)$) when the remainder is zero, i.e. $a(z) = q(z)b(z)$.

Consider now a linear code U of length n with all code polynomials divisible by the fixed polynomial $g(z)$ of degree $r \geqslant 1$. Any code polynomial of U then has the form $u(z) = b(z)g(z)$, and since there are 2^{n-r} different factors $b(z)$ securing the product degree no greater than $n - 1$, such a code can include 2^{n-r} codewords at most. In fact, it is always in one's power to go the reverse way and build up a code with this maximal number of words, i.e. transmitting $k = n - r$ information bits, polynomial $g(z)$ of a fixed degree r given. For that it is enough to use $k = n - r$ data bits $b_0, b_1, \ldots, b_{k-1}$ as coefficients of a *data polynomial* $b(z) = b_{k-1}z^{k-1} + b_{k-2}z^{k-2} + \cdots + b_0$ and construct the corresponding code polynomial as a product $u(z) = b(z)g(z)$. Then 2^k different data k-bit blocks are in one-to-one correspondence with the same number of code polynomials of degree no greater than $n - 1$. The linearity of this code can be checked readily (Problem 9.11). Polynomial $g(z)$ in such a construction is called the *generator polynomial* of U. Note that the number of redundant or *check* symbols of such a code always equals r, i.e. the degree of the generator polynomial.

When a data polynomial $b(z)$ is multiplied with the generator polynomial $g(z)$ directly the codewords corresponding to polynomials $u(z) = b(z)g(z)$ are non-systematic, i.e. data bits in them are not explicitly seen. To come to a *systematic* code, in which, e.g., the last k binary symbols are information bits themselves and $r = n - k$ first ones are check symbols, some rearrangement of codewords may be done. The complete set of codewords in so doing remains the same and only the mapping of data bits onto codewords alters. Let us multiply a data polynomial $b(z)$ by $z^r = z^{n-k}$, coming to a polynomial $z^{n-k}b(z)$ of degree no greater than $n - 1$. If the remainder $r(z)$ of its division by $g(z)$ is discarded (just added to $z^{n-k}b(z)$ in the binary case), it becomes divisible by a generator polynomial $g(z)$, i.e. becomes a code polynomial. The last one:

$$u(z) = z^{n-k}b(z) + r(z)$$

corresponds to a systematic codeword, since the data bits are exactly k senior coefficients of $z^{n-k}b(z)$, and $r(z)$, with degree smaller than $r = n - k$, cannot affect them.

Example 9.2.2. Let us find a codeword of linear (5, 2) code with the generator polynomial $g(z) = z^3 + z^2 + 1$, if data bits are $b_0 = 1, b_1 = 1$. Then $b(z) = z + 1, z^{n-k}b(z) = z^4 + z^3$, having remainder $r(z) = z$ after division by $g(z)$. Summation of $z^{n-k}b(z)$ with this remainder leads to the code polynomial $u(z) = z^4 + z^3 + z = zg(z)$ corresponding to a systematic word, with the two last symbols being data bits.

9.2.3 Syndrome calculation and error detection

Suppose that word $\mathbf{u}$ of the linear code U is transmitted over the BSC. In an output binary observation $\mathbf{y}$ the elements distorted by the channel will differ from those transmitted, which may be written as:

$$\mathbf{y} = \mathbf{u} + \mathbf{e} \tag{9.5}$$

with $\mathbf{e}$ being the *error vector*, having zeros and ones at the places of undistorted and distorted symbols, respectively. For instance, if the word $\mathbf{u} = (01011)$ from Example 9.2.2 is transformed by a BSC into the observation $\mathbf{y} = (11110)$, the error vector $\mathbf{e} = (10101)$. In the same way as codewords, observation $\mathbf{y} = (y_0, y_1, \ldots, y_{n-1})$ and error vector $\mathbf{e} = (e_0, e_1, \ldots, e_{n-1})$ can be presented in the form of polynomials $y(z) = y_{n-1}z^{n-1} + y_{n-2}z^{n-2} + \cdots + y_0$ and $e(z) = e_{n-1}z^{n-1} + e_{n-2}z^{n-2} + \cdots + e_0$. Then (9.5) takes the form:

$$y(z) = u(z) + e(z)$$

Let $g(z)$ be the generator polynomial of U. Remainder $s(z)$ after division of observation polynomial $y(z)$ by $g(z)$ is called the *syndrome*. Since any code polynomial is divisible by $g(z)$, the syndrome repeats the remainder of dividing the error vector $e(z)$ by $g(z)$. Therefore, a non-zero syndrome always reports about the presence of errors in the observation $\mathbf{y}$, and error detection may be implemented as calculating a syndrome for the observation $\mathbf{y}$ and deciding that error occurred whenever it is non-zero.

Certainly, not all error patterns are detectable, and any undetectable error vector is always some code vector. Indeed, if $e(z)$ is a code polynomial it is divisible by $g(z)$ and the syndrome is zero. Conversely, zero remainder tells only that $e(z)$ is divisible by $g(z)$, but any polynomial of degree up to $n - 1$ divisible by $g(z)$ is a code polynomial.

Example 9.2.3. Suppose that the transmitted codeword $\mathbf{u} = (01011)$ of the (5,2) linear code is distorted by BSC into $\mathbf{y} = (11001)$, i.e. two symbol errors occurred. Dividing $y(z) = z^4 + z + 1$ by $g(z) = z^3 + z^2 + 1$ results in a non-zero syndrome $s(z) = z^2$, signalling the presence of errors. On the contrary, if the observation were $\mathbf{y} = (11101)$, corresponding to three corrupted symbols, the syndrome would appear zero: $y(z) = z^4 + z^2 + z + 1 = (z + 1)g(z)$, failing to detect such an error pattern.

Sometimes the ability of a linear code to detect errors is characterized by the share of undetectable error patterns among all error patterns. Since there are 2^n possible error vectors altogether and only those repeating 2^k code vectors are undetectable, the share above is $2^{-(n-k)} = 2^{-r}$.

The linear codes just discussed and built on the basis of generator polynomials are known as cyclic codes or shortened cyclic codes. When used to only detect (not to correct) errors, they are often called *cyclic redundancy codes* (CRC).

9.2.4 Choice of generator polynomials for CRC

The share of undetected errors drops exponentially with the number of check symbols, which may motivate using generator polynomials of high degrees. It should be remembered, however, that check symbols present an overhead and increasing their number above some reasonable level may be wasteful. CRC are typically used on higher system protocol layers for checking the quality of data frames retrieved on the physical layer, i.e. after resources of much more powerful error-correction codes have already been utilized. Thanks to that, symbol errors met by CRC are rather rare and it is quite unlikely that in the codeword with hundreds of symbols more than a few errors happen. Thus, detection of up to three errors per codeword may often be considered satisfactory. Let us describe the procedure of designing CRC meeting this requirement.

Take a binary primitive polynomial $g_1(z)$ of degree m (see Section 6.6). An important fact proved in algebra of extended fields is that a primitive polynomial of degree m never divides any binomial $z^l + 1$ of non-zero degree $l < 2^m - 1$ [30,32]. Then we may prove the following statement.

Proposition 9.2.4. A linear code U with the generator polynomial $g(z) = (z+1)g_1(z)$ of degree $m+1$ detects up to three errors whenever its code length $n \leq 2^m - 1$.

Because of the linearity of U (see (9.3)) and Proposition 9.2.1 it is only necessary to prove that the minimal weight of a non-zero word of U is no smaller than four. Every code polynomial $u(z)$ being divisible by $z+1$ may be written as $u(z) = q(z)(z+1)$. Although z is a formal variable, the latter equality should hold after substituting $z = 1$ in both its parts, resulting in $u(1) = u_{n-1} + u_{n-2} + \cdots + u_0 = 0$, which means evenness of the number of non-zero elements of a word, i.e. evenness of its weight. Assuming the existence of a word of weight two with non-zero ith and jth symbols ($j > i$), we have a code polynomial $u(z) = z^j + z^i = z^i(1 + z^{j-i})$ not divisible by $g_1(z)$, since the latter, being irreducible, cannot have z as a factor and does not divide $1 + z^{j-i}$ ($j - i < n \leq 2^m - 1$) due to primitivity. Hence, the smallest weight of a non-zero word of U is four.

The choice of appropriate CRC generator polynomial is now straightforward. If the desirable code length is n, then one should just find a primitive polynomial $g_1(z)$ of degree $m \geq \lceil \log_2(n+1) \rceil$ and the generator polynomial $g(z) = (z+1)g_1(z)$ is ready.

CRC of this type are nothing other than quite popular Hamming codes (shortened whenever $n < 2^m - 1$) with all odd-weight words removed. In alternative applications they are used to correct any single and detect any double errors.

Example 9.2.4. Several CRC are used in 2G and 3G specifications [18, 69, 92]. The exemplary one entering all three standards (cdmaOne, UMTS and cdma2000) has the generator polynomial $g(z) = z^{16} + z^{12} + z^5 + 1 = (z + 1)g_1(z)$, where $g_1(z) = z^{15} + z^{14} + z^{13} + z^{12} + z^4 + z^3 + z^2 + z + 1$ and is primitive. Similarly, generator polynomials of other CRC of those standards (of degrees 30, 24 etc.) are factored into the binomial $z + 1$ and a primitive polynomial.

9.3 Convolutional codes

Convolutional codes are in common use in modern telecommunications as an effective tool for securing reliable data transmission over noisy channels. Belonging to a more general class of tree codes, they are distinguished inside it by the linearity of the encoding algorithm. The difference between convolutional and block codes is rather fuzzy: any convolutional code may be thought of as a block code of a properly large length. It is more satisfactory to see the peculiarity and the reason for the extreme popularity of convolutional codes in their recurrent nature, which allows a much more feasible error-correction decoding procedure (Viterbi algorithm) as compared to the others.

9.3.1 Convolutional encoder

The idea of a convolutional encoding may in general terms be described as follows. Take a block (vector) of ν_c consecutive bits of a source and convert it linearly into $n > 1$ output binary code symbols occupying time space of one source bit. The linearity as applied to vectors with components from $GF(2)$ means just modulo 2 summations of selected components. After this, update a block of source bits, inserting one new bit and discarding the oldest. We again have a block of ν_c source bits, this time lagging the initial one by one bit (and containing $\nu_c - 1$ former bits and a new one), which is encoded to the new n code symbols. These steps are continuously executed one by one, each time involving a new bit and dropping the oldest. Figure 9.2, where $\nu_c = 3, n = 3$, illustrates the procedure: the encoder watches a source bit stream through the sliding window of width ν_c and encodes all bits currently seen into n code symbols. After every step the window moves by one source bit and the next step is accomplished. The number of source bits ν_c determining the

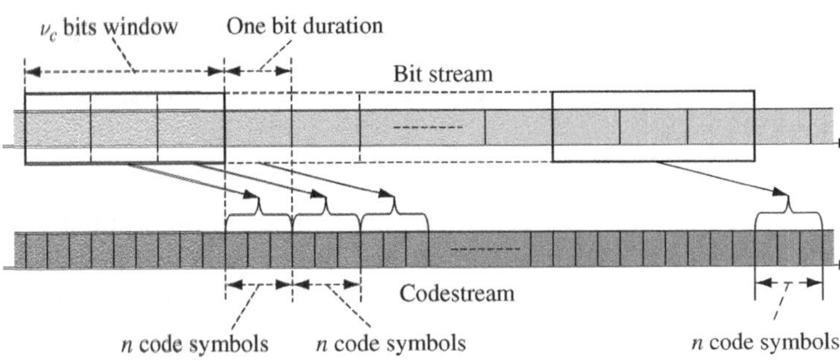

Figure 9.2 Convolutional encoding

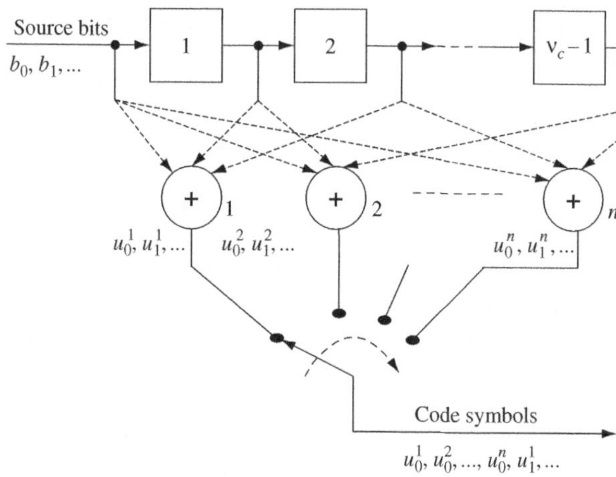

Figure 9.3 Convolutional encoder

code symbols at one step is called the *constraint length*. The principle described can be implemented as shown in Figure 9.3, where the shift register containing $\nu_c - 1$ flip-flops stores $\nu_c - 1$ previous source bits. Along with the incoming bit, they are fed into the linear logic circuit consisting of n modulo 2 adders. The output of every flip-flop and encoder input may or may not be connected to each adder (that is why the links are marked as dashed lines), the scheme of connections determining the concrete dependence of output code symbols on ν_c source bits, i.e. the encoding rule. With a current source bit b_i arriving at the input n code symbols $u_i^1, u_i^2, \ldots, u_i^n$ appear in parallel at the adders' outputs. After every clocking, the bit pattern in the register shifts to the right by one cell, preparing the circuit to generate the next n code symbols. The output switch running through n positions during one bit space converts the code symbols from a parallel to a serial pattern, creating an output codestream $u_0^1, u_0^2, \ldots, u_0^n, u_1^1, u_1^2, \ldots, u_1^n, \ldots$. It is seen that the encoder of Figure 9.3 in the steady state responds to every new source bit by n code symbols (see also Figure 9.2), so that its code rate R_c measured in bits per code symbol is $1/n$. The example below helps to explain better the principle of convolutional encoding.

Example 9.3.1. Figure 9.4 illustrates the implementation of the convolutional encoder with the constraint length $\nu_c = 3$ and rate $R_c = 1/2$. Input bits $b_0, b_1, \ldots$ produce two streams of code

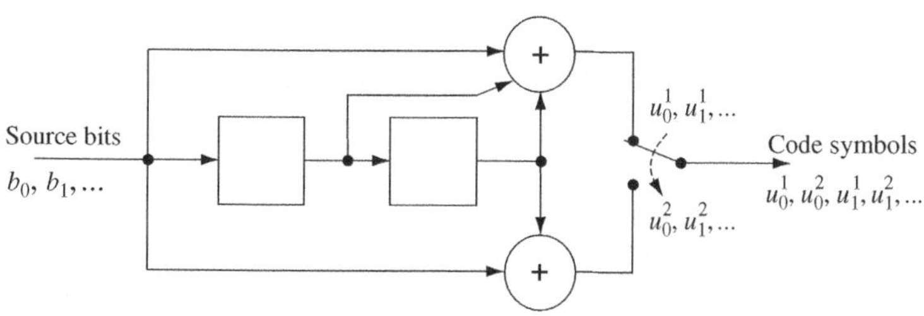

Figure 9.4 Convolutional encoder of rate 1/2

symbols $u_0^1, u_1^1, \ldots$ and $u_0^2, u_1^2, \ldots$, which are then multiplexed so that u_i^1 and u_i^2 occupy, respectively, even and odd positions in a common codestream. For instance, a bit stream $\{b_i\} = 10100100\ldots$ produces sequences $\{u_i^1\} = 11011111\ldots$ and $\{u_i^2\} = 10001101\ldots$ which are then multiplexed into a codestream $\{(u_i^1, u_i^2)\} = 1110001011111011\ldots$.

The following question sounds natural: are rates R_c other than $1/n$, i.e. equal to k/n, where $1 < k < n$, possible with convolutional encoding? There are two classical ways of solving this task. The first of them just generalizes the principle above: at every step $k\nu_c$ bits rather than ν_c are linearly transformed into n code symbols, after which k oldest bits (instead of one) are replaced by k new ones and the encoder proceeds to the next step. The second way, called *puncturing*, uses rate $1/n$ code as the raw material and removes some of its code symbols according to a pre-assigned pattern. The puncturing, when arranged properly, reduces the number of code symbols per data bit, providing rate k/n. For reasons of implementation feasibility puncturing is frequently considered preferable, accentuating further a dominant interest towards the codes of rate $1/n$. We will follow this line and focus further only on the convolutional codes of rate $1/n$.

Clearly, a circuit including a shift register and an individual modulo 2 adder with all its connections is nothing but an FIR filter (see Figure 6.20) outputting the convolution of the input bit stream with the filter pulse response, which explains the name of the codes under study. This also underlies one of the convenient ways to describe formally a convolutional encoder. The convolution relating a code symbol u_i^l (i.e. appearing at the lth adder output when a bit b_i arrives) to the input bit stream is:

$$u_i^l = \sum_{t=0}^{\nu_c - 1} b_{i-t} g_t^l, \quad i = 0, 1, \ldots; \ l = 1, 2, \ldots, n \tag{9.6}$$

where $g_t^l = 1$ if the lth adder is connected to the tth flip-flop ($t = 0$ corresponds to the encoder input) and $g_t^l = 0$ otherwise; $b_i = 0$ whenever $i < 0$. An appropriate frequency-domain instrument for discrete systems is z-transform, which was mentioned in the previous section. A convolution corresponds in the z-domain to a product of z-transforms, so that (9.6) takes an equivalent form:

$$u^l(z) = \sum_{i=0}^{\infty} u_i^l z^i = b(z) g_l(z), \ l = 1, 2, \ldots, n \tag{9.7}$$

where $b(z) = \sum_{i=0}^{\infty} b_i z^i$ is a z-transform of an input bit stream and:

$$g_l(z) = g_0^l + g_1^l z + \ldots g_{\nu_c - 1}^l z^{\nu_c - 1}, \ l = 1, 2, \ldots, n \tag{9.8}$$

is a transfer function of the lth FIR filter (i.e. forming the lth code symbol) called also the lth *generator polynomial* of a convolutional code. The set of n generator polynomials determines a convolutional code exhaustively, since their non-zero coefficients specify connections of adders with a shift register.

Example 9.3.2. The encoder shown in Figure 9.4 has generator polynomials $g_1(z) = 1 + z + z^2$ and $g_2(z) = 1 + z^2$. Make sure that sequences of code symbols cited in Example 9.3.1 may be obtained from (9.7).

Putting $g_1(z) = 1$ leads to a *systematic* convolutional code, in which bits of source data are directly seen on definite positions. The trouble, however, is that within the fixed structure of Figure 9.3 systematic codes are usually not the best ones as regards error correction capability (see Problem 9.15). This underlies modifying the shift register of a convolutional encoder to a feedback structure when the systematic property is critical. We will have to revisit this matter in more detail on familiarizing ourselves with turbo codes in Section 9.4.1.

9.3.2 Trellis diagram, free distance and asymptotic coding gain

The shift register of a convolutional encoder has 2^{ν_c-1} possible states and there are only two states where it can pass from a current state after clocking. It is an incoming source bit b_i that selects one of these two paths. When a state of the register on the ith clock interval is $(b_{i-1}, b_{i-2}, \ldots, b_{i-\nu_c+1})$, then the next state will be either $(0, b_{i-1}, \ldots, b_{i-\nu_c+2})$, if the incoming source bit $b_i = 0$, or $(1, b_{i-1}, \ldots, b_{i-\nu_c+2})$, if $b_i = 1$. Similarly, the register comes to the state $(b_i, b_{i-1}, \ldots, b_{i-\nu_c+2})$ if the previous state was either $(b_{i-1}, b_{i-2}, \ldots, b_{i-\nu_c+2}, 0)$ or $(b_{i-1}, b_{i-2}, \ldots, b_{i-\nu_c+2}, 1)$. To depict schematically all these details of register behaviour the *trellis* is an appropriate tool. It includes two columns of 2^{ν_c-1} nodes, the left column for the current state and the right for the next one. The branches (arrows) go out from each node of the left column to two nodes of the right, solid and dashed branches showing paths selected by incoming bit zero and one, respectively. In the same way, two branches enter every node of the right column, being both either solid (zero input bit) or dashed (input bit equalling one). Every branch is labelled by an n-tuple which is nothing but a group of n code symbols issued by the encoder when an input source bit directs it from one state to another.

The example below clarifies the method of building the trellis for the convolutional encoder of Figure 9.4.

Example 9.3.3. The four states of a two-cell register starting with the left flip-flop are: (00), (10), (11) and (01). Figure 9.5 presents the trellis built as described. For instance, the branch from

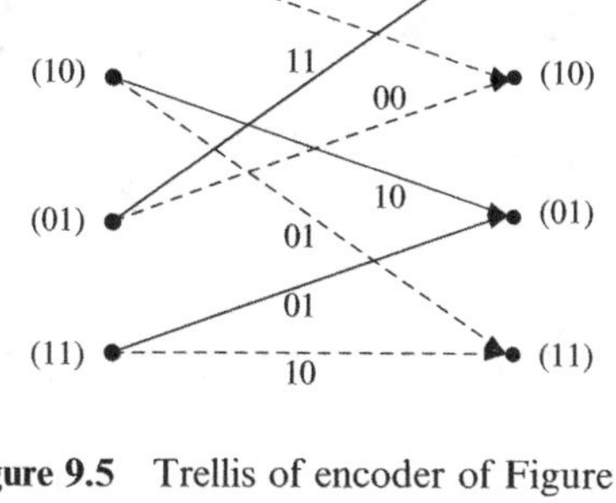

Figure 9.5 Trellis of encoder of Figure 9.4

(10) to (01) is solid, while the one to (11) is dashed, code symbols 01 label the branch leading from (11) to (01), since with ones in both cells and zero input bit the upper adder outputs zero, while the lower outputs one, etc.

During every clock interval the encoder moves along some branch of a trellis issuing code symbols labelling the branch. Tracing this process in a diagram presented as in Figure 9.5, one has to jump from the right column to the same node in the left column at every next step. To escape this, let us just repeat the trellis as many times as is wished using the right column of the current step as the left one for the following step. Then encoding will be identified with a motion along the obtained *trellis diagram*, the current input bit directing the encoder along a solid or dashed branch depending on whether it is zero or one. To illustrate this, let us turn to Figure 9.6, which presents a trellis diagram for the code of Example 9.3.1. Every candidate sequence of input data bits selects a specific path on the trellis diagram, which we may trace, e.g. for the sequence $\{b_i\} = 10100100\ldots$. Its first bit is 1, which directs the encoder from the node (00) to the node (10), issuing output code symbols 11; the second bit 0 transfers the encoder from (10) to (01), generating code symbols 10; the third changes the state from (01) to (10), outputting 00 and so forth. The thick line shows the resulting codeword: 1110001011111011.

It was already noted that one might treat a convolutional code as the block one of a properly long length. The codewords of this latter block code are just different paths on the trellis diagram, the minimum Hamming distance over all pairs of them giving a code distance. In its turn, the code linearity simplifies the task of finding code distance: on the strength of Proposition 9.2.3, the minimum distance between paths is the least Hamming weight over all non-zero words. Suppose now that an encoded bit stream is terminated after some large enough (no smaller than ν_c) number of bits and padded by $\nu_c - 1$ tail zeros to set the register to the all-zero state. Realized practically (one example is cdmaOne), this padding would not insert a material overhead whenever the length of an encoded bit stream is many times larger than the constraint length. On the other hand, the padding makes all the paths converge to the all-zero register state, as Figure 9.6 shows for the code of Example 9.3.1. If we take a padded bit stream starting

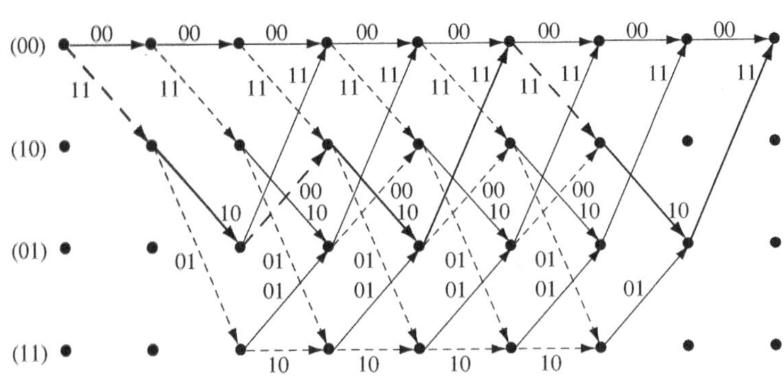

Figure 9.6 Trellis diagram of encoder of Figure 9.4

with some number n_0 of zeros followed by bit '1' and replace these n_0 initial zeros to the end of the stream, we just move a corresponding path on the diagram by n_0 steps to the left with no change of its weight. After such a shift the path diverges from the zero path at the very first step and merges with it no later than n_0 steps before the last tail zero. Along this path several returns to zero and further diverging may happen (see Figure 9.7), each of them only increasing the path weight. Since the objective is to find the minimum weight, any deviations from the zero path but the first one should be ignored. Summarizing, we conclude that to find the distance of a convolutional code one ought to investigate only the paths deflecting from the zero path at the origin of the trellis diagram and having no deviation from the zero path after the first merging with it. In the theory of convolutional codes this entity is traditionally called the *free distance*. Denoting it d_f we see, for example, that among all the paths in Figure 9.6 with only a single deviation from the zero path the codeword 11101100 ... encoding a bit stream 100... has minimal weight, so that $d_f = 5$. Certainly, free distance d_f guarantees correcting any $\left\lfloor \frac{d_f - 1}{2} \right\rfloor$ symbol errors (see Proposition 9.2.2); however, typically many patterns with a greater number of errors are corrected, too. There are only a few examples of effective algebraic rules for convolutional encoding. The majority of known convolutional codes with good correction capability are found by a computer search [31,33,93].

Due to the specificity of the encoding algorithm, finding all possible weights of words (weight spectrum) of an arbitrary convolutional code proves to be not so analytically difficult as it is for many linear block-codes. In particular, directly from the trellis (or, equivalently, state diagram) a system of linear equations is made up, the solution of which leads to an explicit expression for a weight spectrum [2,7,93].

Coding gain, which shows how many times signal energy per bit or signal power can be reduced as a result of encoding, error probability fixed, is a universal measure to characterize the efficiency of one or other code. We discussed this notion in Section 2.6 applied to orthogonal signalling and noted that the asymptotic coding gain for the case of the AWGN channel is the gain in Euclidean distance. With BPSK transmission every discrepancy in symbols of two signals adds to their squared Euclidean distance $4E_s$, where E_s is symbol energy. There is a pair of words of a convolutional code having d_f different symbols and no pair with smaller discrepancy (Hamming distance). Therefore, the minimum squared Euclidean distance between BPSK-mapped convolutional codewords $d^2_{\min, cc} = 4d_f E_s$. At the same time (see Section 2.6) a similar quantity for uncoded transmission $d^2_{\min, u} = 4E_b$, resulting in asymptotic coding gain of a convolutional code:

$$G_a = \frac{d^2_{\min,\ cc}}{d^2_{\min,\ u}} = \frac{d_f E_s}{E_b} = d_f R_c$$

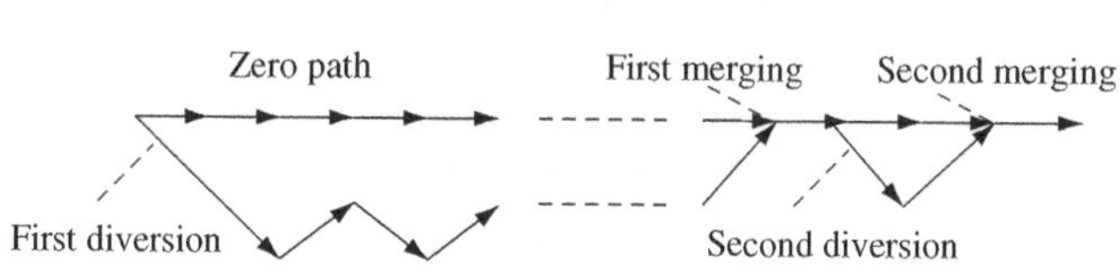

Figure 9.7 Diversion and merging of paths

code rate R_c being measured in bits per symbol. For the code of Example 9.3.1 $R_c = 1/2, d_f = 5$, so that $G_a = 2.5$ or about 4 dB. We remember once again that G_a is derived for the AWGN channel (not BSC!); in other words, for the case of soft decoding. Hard decoding degrades this figure by 2–3 dB depending on code parameters and symbol SNR [31,33].

9.3.3 The Viterbi decoding algorithm

As was mentioned, among the reasons for the great popularity of convolutional codes the existence of a feasible decoding algorithm is of special importance. Let us begin with the following statement.

Proposition 9.3.1. The ML hard error-correction decoding of a binary code is equivalent to the minimum Hamming distance rule:

$$d_H(\hat{\mathbf{u}}, \mathbf{y}) = \min_{\mathbf{u} \in U} d_H(\mathbf{u}, \mathbf{y}) \Rightarrow \hat{\mathbf{u}} \text{ is declared the received word} \tag{9.9}$$

As one can see, the rule is very similar to (2.3) with a single change: in the case of BSC the Hamming distance replaces the Euclidean one, adequate for the AWGN channel. To prove (9.9) it is enough to note that (9.1) rewritten as:

$$p(\mathbf{y}|\mathbf{u}) = p^{d_H(\mathbf{u},\mathbf{y})}(1-p)^{n-d_H(\mathbf{u},\mathbf{y})} = \left(\frac{p}{1-p}\right)^{d_H(\mathbf{u},\mathbf{y})}(1-p)^n$$

is nothing but the BSC transition probability, i.e. the probability for the sent code vector $\mathbf{u}$ of length n being transformed by BSC into a binary observation $\mathbf{y}$. Since the crossover probability of BSC $p < 0.5$, the transition probability $p(\mathbf{y}|\mathbf{u})$ is a decreasing function of the Hamming distance from an observation $\mathbf{y}$ at the BSC output to a code vector $\mathbf{u}$, and, therefore, the ML codeword $\hat{\mathbf{u}}$ is the one closest to $\mathbf{y}$ by the Hamming distance.

A direct realization of (9.9) for an arbitrary code appeals to comparison of M Hamming distances from $\mathbf{y}$ to all codewords. Since M is typically quite large such a solution might appear infeasible. On the contrary, because of the beneficial structure of convolutional codes the ML decoding is not a big technological challenge, at least if the constraint length is moderate.

The Viterbi decoding procedure implements the ML strategy in a recurrent, step-by-step form of searching a path on the trellis diagram closest to the binary observation $\mathbf{y}$. Every new decoding step starts with the arrival of the next group of n observation symbols. At the ith step the decoder calculates the distance of n incoming observation symbols from every branch of a trellis diagram and increments the distances for all paths calculated over the $i-1$ previous steps. One might work out the distances for an arbitrary code similarly, as new observation symbols arrive, but it is the recurrent nature of a convolutional code that makes this routine computationally economical due to the immediate discarding of many paths at each step.

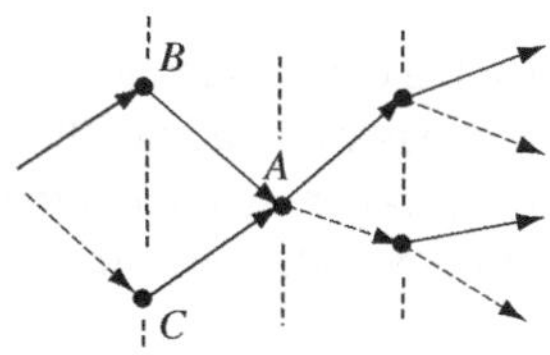

Figure 9.8 Paths going through node A at the ith step

Consider all paths passing through a fixed node A at the ith step, as shown in Figure 9.8. Continuation of any path after the ith step does not depend on the route of arriving at A, so that different paths reaching A may further merge with each other. But this means that of all the paths going through A and continuing the same way after the ith step the one having minimum distance from **y** up to the ith step will remain closest to **y** forever, since a common continuation adds an equal contribution to all distances! Why then go on with computing distances for the rest as soon as it is clear that they have no chance of finally appearing closest to the observation? Instead we discard them, retaining only the one reaching the node A with minimum distance. This latter is called a *survivor* and for the time being we believe that there is only a single survivor for every node of the trellis diagram at the ith step (see comments below). The current, i.e. calculated over all observed symbols up to the ith step, distance of the node A survivor from **y** is called the *node metric* of A.

Now recall that there are only two branches entering any node; Figure 9.8 shows this for some node A. Two branches enter it from two nodes B, C of the previous step and, therefore, continue the survivors of B and C. We may calculate the distances of two paths reaching A by just measuring the *branch metrics*, i.e. the distances of branches from the arriving group of n observed symbols, and adding them to the node metrics of B and C. The path with smaller distance is declared the survivor of A and is recorded in memory along with its distance (node metric) while the other is discarded. On performing these operations for all nodes of the trellis diagram the decoder proceeds to the next step.

To summarize briefly, the Viterbi decoder calculates branch metrics at every step, adds them to all the node metrics accumulated before and then sifts out the more distant of two paths leading to every node. Since there are 2^{ν_c-1} nodes (i.e. register states) altogether, the complexity of the decoder is determined only by the constraint length ν_c and remains fixed independently of the theoretically unlimited number of codewords (paths).

Coming back to our assumption on the uniqueness of a survivor for every node, note that since Hamming distance is integer, the probability always exists that two paths leading to the same node have equal current distances from **y**. Different strategies are possible to solve ambiguity of this sort. One of them is just a random choice: assigning tails of a fair coin to one of the paths and declaring this one the survivor if tails really falls after flipping the coin. This certainly violates the decoding optimality, although the accompanying energy loss is typically insignificant. Alternatively, both competitive paths may be declared survivors and recorded in memory until further steps remove the ambiguity. The latter option preserves decoding optimality at the cost of involving extra memory.

Example 9.3.4. Consider decoding the observation $\mathbf{y} = 0100110000\ldots$ for the code of Example 9.3.1. Figure 9.9 illustrates the process with node metrics placed immediately near the nodes, frames containing the pairs of observation symbols arriving at the current step. The decoder starts the process assuming zero (i.e. (00)) initial state of the encoder register. The starting

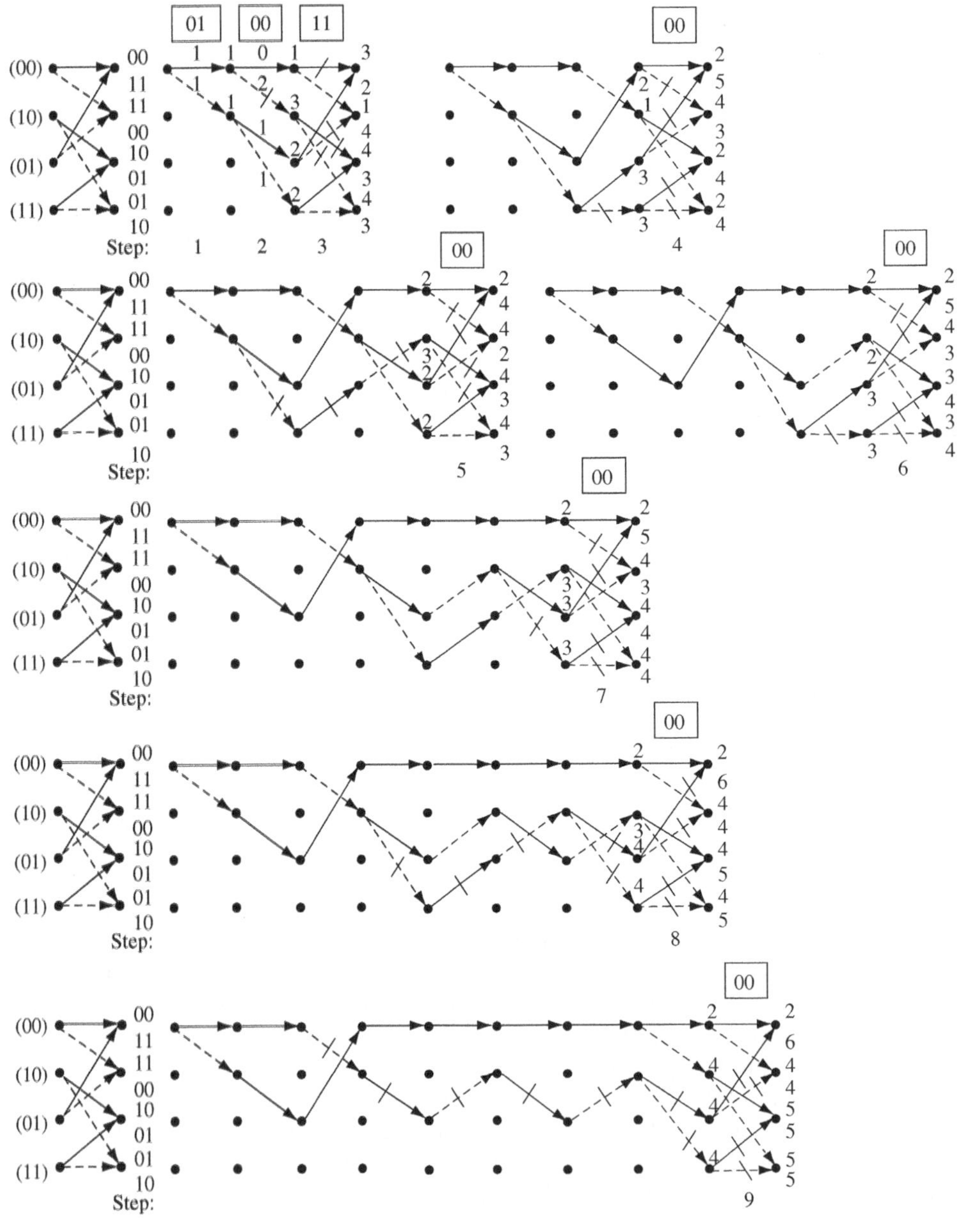

Figure 9.9 Dynamics of decoding for the convolutional code of Example 9.3.1

$\nu_c - 1 = 2$ steps correspond to transient behaviour of the encoder register, when only one branch enters every state (see Figure 9.6) and so all paths are survivors. At the first step the decoder compares the first group of $n = 2$ observation symbols with two branches emanating from the state (00). According to their Hamming distances from the observed symbol group 01, both the solid and dashed lines obtain metric 1, shown near the branches. Consequently, the node metrics of the two nodes, at which the branches arrive, both become equal to one. At the next step the distance is measured between the second group of observation symbols 00 and two pairs of branches emanating from nodes (00) and (10), resulting in the metrics labelling the branches. Added to the node metrics of the previous step, they update the metrics of nodes (00) and (10) and produce the metrics of two more nodes (01) and (11). Starting with the third step two branches enter any node of a trellis diagram of Figure 9.6, meaning that the decoder must decide which of them is a survivor. We do not show in Figure 9.9 the metrics of the branches beginning with this step in order to avoid clutter. As is seen, there are two paths leading to the node (00) at the third step. Their distances from the observed symbols 010011 are 3 and 2, respectively. The first is not a survivor and the decoder discards it along with its metric, so it is crossed out and not present at the plot of the next step. The second is a survivor, and is recorded in memory with its metric until the next step. In the same way, the decoder finds the survivors for the rest of the nodes. The decoding proceeds similarly at further steps, storing in memory only $2^{\nu_c - 1} = 4$ survivors, and every plot of Figure 9.9 depicts only the paths declared survivors at the previous step.

At step 7 the decoder first runs into the problem of ambiguity: two paths arrive at the node (01) with equal distances, and the same occurs with the node (11). The choice of survivors illustrated by the figure reflects some realization of flipping a fair coin. The same events happen at steps 8 and 9. The reader is encouraged (Problem 9.19) to make sure that any alternative resolving of the ambiguity will not change the final result of decoding except for the step number when the decoded bits are first released (see below).

The situation after the ninth step is very important: all paths appear to merge with each other up to the seventh symbol group. Whatever happens afterwards, this part of all merging paths will remain common forever, meaning that the data bits corresponding to it may be released right away as decoded ones. Hence, the decoder produces the decoded data 1000000. Comparing the codeword $\mathbf{u} = 11101100000000\ldots$ corresponding to it with the observation $\mathbf{y} = 0100110000\ldots$ we may note that the Hamming distance between them is 2, and if a transmitted word was really the one declared by the decoder, two errors were corrected, in full consistence with free distance $d_f = 5$. Similar situations will arise at further steps, allowing the decoder to release decoded bits in the course of processing an observation symbol stream.

Certainly, outputting the decoded data at random moments of the merging of survivors, as in the example above, looks impractical, and a more regularly arranged procedure is desirable. It has been repeatedly verified by experiments and simulation that during the ith decoding step the merging part of all survivors almost never ends after the data bit number $i - 5\nu_c$, so the decision on every bit may be regularly output with delay $5\nu_c$ [94].

A very significant feature of convolutional codes making them even more attractive is the comparative simplicity of implementing soft decoding. In a general case of a block code with M words the soft decoding means direct calculation of M Euclidean distances or correlations, no algebraic tricks like syndrome decoding being available. With a gigantic M typical of many applications this often makes the task completely infeasible. At the same time the penalty for simplifying the receiver using a hard decoder is energy loss varying for the AWGN channel around 2 dB: a figure today considered pretty significant. Turn now to the Viterbi algorithm and replace the Hamming distance in it by a (squared) Euclidean one. Obviously, it converts the decoding into a soft one, optimal for the AWGN channel. The branch and node metrics then become just the corresponding Euclidean distances (or correlations). Such a modification by no means affects the implementation advantages of the Viterbi algorithm. Indeed, the node metrics are calculated in a recursive way as before through step-by-step incrementing by branch metrics, and the path entering a node with worse metric than the other may again be discarded at every step as a non-survivor.

Certainly, digital realization of decoders is mostly appropriate, presuming quantizing input observation. It is a conventional trend to classify decoders of binary codes involving quantization with more than two levels as soft ones. A profound analysis shows that in the majority of cases 3-bit (8-level) quantization is enough to achieve almost the potential (characteristic of continuous processing) performance [94].

9.3.4 Applications

Many efficient convolutional codes are now known and the range of their telecommunication involvement is extremely vast. In particular the 2G cdmaOne and 3G UMTS standards exploit codes of constraint length $\nu_c = 9$ and rates $R_c = 1/2$, $R_c = 1/3$ providing asymptotic coding gain about 7.8 dB [18,69,92]. The 3G cdma2000 standard, in addition to those above, employs the code with parameters $\nu_c = 9$, $R_c = 1/4$. Beyond their own significant value convolutional codes constitute the basis for turbo codes, which near the Shannon limit on data transmission reliability. These are discussed briefly in the following section.

9.4 Turbo codes

As we mentioned in Section 9.1, despite a very strong endeavour for decades following the advent of Shannon's information theory the attempts to find regular coding rules securing reliable data transmission near channel capacity remained unsuccessful. In this light, the discovery of *turbo codes* in 1993 was a fascinating breakthrough, first met by the communications community with understandable disbelief. Nowadays, however, turbo codes are widely recognized as a very efficient means of high-quality communications especially at low SNR per data bit. An accurate and compact theory of this class of codes remains something of a holy grail; intuition combined with extensive computer search has played a significant role in obtaining many results up to now.

9.4.1 Turbo encoders

The other name of turbo codes is parallel concatenation convolutional codes, which reflects the core idea of the encoding algorithm: two parallel (component) convolutional encoders encode the same source data bit stream [95,96]. The component encoders are as a rule identical, i.e. have the same constraint lengths and sets of generator polynomials. The first of them encodes the data directly, and before arriving at the second the data bit stream undergoes *interleaving*. This operation just permutes the data bits in a pseudo-random fashion within a block of fixed length I.

As was noted in Section 9.3.1, among the convolutional codes based on the FIR structure of Figure 9.3, there is no equivalence between systematic and non-systematic codes, non-systematic ones being typically more powerful as to the distance properties. At the same time, the decoding principle associated with turbo codes and imparting special attractiveness to them appeals to systematic component encoders. Exploiting infinite impulse response (IIR) in place of the FIR structure opens the way to forming a systematic convolutional code containing the same words, and therefore having the same distance properties as the non-systematic code. To explain it, start with the encoder of Figure 9.3 described by a set of generator polynomials $g_l(z), l = 1, 2, \ldots, n$ and establish a one-to-one mapping between two bit streams $b(z)$ and $b_1(z)$ as $b(z) = b_1(z)g_1(z)$. Turn then to equation (9.7) and see that for an input bit stream $b_1(z)$ it may be written in the form:

$$u_l(z) = b_1(z)g_l(z) = \frac{b(z)}{g_1(z)}g_l(z),\ l = 1, 2, \ldots, n \tag{9.10}$$

showing that the structure capable of dividing source bit stream $b(z)$ by $g_1(z)$ before inputting the FIR scheme of Figure 9.3 would encode $b(z)$ into the same convolutional codewords as the FIR encoder itself. The only difference is that the codeword assigned earlier to $b_1(z)$ is reassigned now to $b(z)$, which—considering the equal rights of any source bit streams—is of no significance as soon as the decoder knows this new correspondence order. But after such an operation all the codewords become systematic, since $u_1(z) = b(z)$! The feedback register in Figure 9.10 implements the division by $g_1(z) = g_r^1 z^r + g_{r-1}^1 z^{r-1} + \cdots + 1$, where $r = \nu_c - 1$. Indeed, according to the general rule

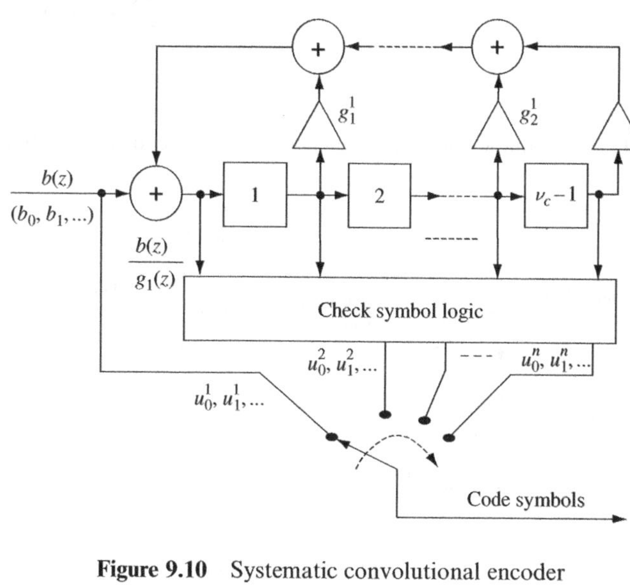

Figure 9.10 Systematic convolutional encoder

applicable to any linear system (one example of it is (8.26)), the z-domain transfer function $h_f(z)$ of the closed feedback loop is expressed (using equivalence of addition and subtraction in binary arithmetic) as [2,7]:

$$h_f(z) = \frac{h(z)}{1 + \beta(z)h(z)}$$

$h(z)$ and $\beta(z)$ being the transfer functions of the open loop and feedback circuit, respectively. Considering the output of the leftmost adder of Figure 9.10 as the output of a closed loop, we have $h(z) = 1$, $\beta(z) = g_1^1 z + g_2^1 z^2 + \cdots + g_r^1 z^r$ so that $h_f(z) = 1/g_1(z)$. Since the divided bit stream $b(z)/g_1(z)$ is present directly at the register input, the check symbol logic fixed by polynomials $g_l(z)$, $l = 2, \ldots, n$ may be connected to the same register for a further ordinary convolutional encoding. Summarizing, in a systematic feedback encoder a source bit stream gives directly a stream of data bits $u_1(z)$, while check symbol streams $u_l(z)$, $l = 2, \ldots, n$ are formed by the same logic as in Figure 9.3 but connected to the feedback register. Figure 9.11 presents the example of transforming the non-systematic encoder of Figure 9.4 to a systematic form.

The complete structure of a turbo encoder includes two such systematic encoders and an interleaver, as is shown in Figure 9.12 for the component encoders of rate $R_c = 1/2$. The 'direct' data bits from the second component encoder are discarded and only its check bits

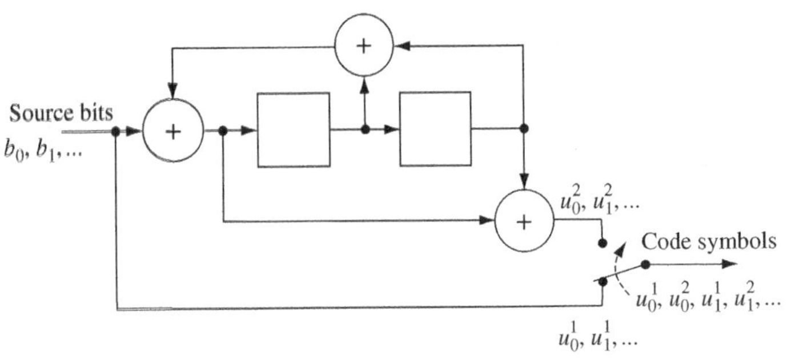

Figure 9.11 Systematic convolutional encoder for the code of Example 9.3.1

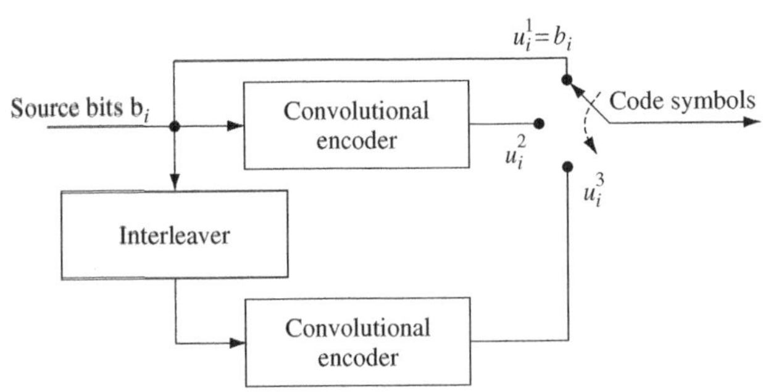

Figure 9.12 Turbo encoder

are used. Thus, if the component convolutional codes are of rate $R_c = 1/2$, the resulting turbo code contains two check symbols u_i^2, u_i^3 per data bit b_i, i.e. has the rate $R_c = 1/3$. If desired, its rate may be increased to 1/2 by transmitting only one check symbol per data bit taken alternately from the component encoders. The interleaver permutes data bits within the block of length I. Typically $\nu_c - 1$ tail bits are used to set the first component encoder to zero and a turbo code is treated as the block code carrying I information bits.

9.4.2 Iterative decoding

Although a turbo code consists of two convolutional codes its optimal decoding cannot be realized as two independent Viterbi procedures, since the paths on the component trellis diagrams are related to each other via the same (though interleaved) encoded data. The pioneers of turbo coding [95,96] proposed the use of an iterative version of the *maximum a posteriori probability* (MAP) rule applied not to a codeword but instead to every information (not check!) bit. Its description below presents only a general idea; the reader curious about details may consult [96,97].

A posteriori probabilities $p(b_i = 0|\mathbf{y})$ and $p(b_i = 1|\mathbf{y})$ show the likelihood of one or the other value of the ith data bit calculated on the basis of observation vector $\mathbf{y}$. They concentrate in themselves complete knowledge about b_i delivered by $\mathbf{y}$ and extracted from it. In the case of reliable communication, one of these probabilities is close to unity, while the other is almost zero. Naturally, a MAP decoder will output as an estimation $\hat{b}_i$ of the ith data bit the value with greater a posteriori probability, following the rule:

$$\Lambda_i = \frac{p(b_i = 0|\mathbf{y})}{p(b_i = 1|\mathbf{y})} \underset{\hat{b}_i=1}{\overset{\hat{b}_i=0}{\gtrless}} 1 \tag{9.11}$$

The recurrent algorithm for calculating bit a posteriori probability was developed in [98]. In the case under study the data bit b_i is physically present in observations due to systematic encoding. On the way forward the recurrent MAP algorithm computes a posteriori probability of b_i using all observations before and including this bit. After all the observation samples are received it goes back and refines the results, incorporating information extracted from the observations arriving later than b_i. Thus, after passing forwards and backwards, the a posteriori probabilities for all data bits may be found. To run this algorithm one needs the trellis of the code, channel transition probability and a priori probability distributions $q(b_i)$ for every data bit. Certainly, there would be no problem in applying it individually to each of the two component convolutional codes constituting a turbo code. The core of the turbo code, however, is encoding the same permuted data by the two component codes, so that information about b_i could be retrieved from both of them jointly. The iteration process arranged as shown in Figure 9.13 meets this demand.

The complete observation vector at the receiving side may be split into three vectors: $\mathbf{y}_l = \mathbf{u}_l + \mathbf{n}_l$, $l = 1, 2, 3$, where $\mathbf{u}_1 = (u_0^1, u_1^1, \ldots u_{M-1}^1), u_i^1 = b_i$ is the vector of data bits b_i directly present in any codeword due to the systematic character of turbo codes, $\mathbf{u}_2, \mathbf{u}_3$ are vectors of redundant symbols of the first and second component encoders, respectively, and $\mathbf{n}_l$ are vectors of independent noise samples.

At the first step the decoder of the first component code calculates a posteriori probabilities $p_1(b_i|\mathbf{y}_1, \mathbf{y}_2)$ of all I data bits b_i using observations $\mathbf{y}_1, \mathbf{y}_2$ associated with the data

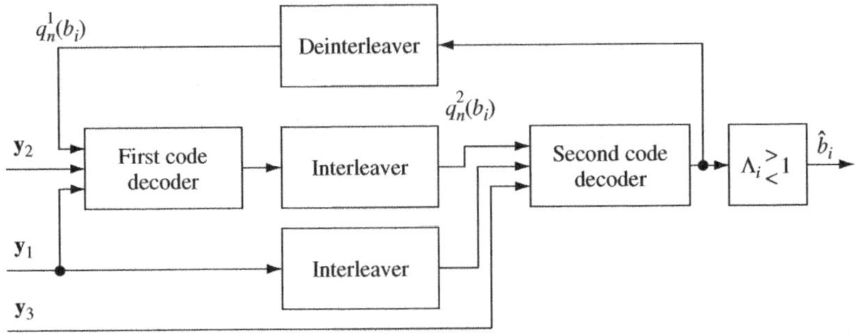

Figure 9.13 Iterative turbo decoder

and check symbols of this code. As initial information it uses a uniform a priori distribution $q_1^1(b_i) = 1/2, b_i = 0, 1$, since setting all data bit patterns a priori equiprobable is natural. After this, the decoder of the second code computing a posteriori probabilities $p(b_i|\mathbf{y}_1, \mathbf{y}_3)$ may rest not only on appropriate observations $\mathbf{y}_1, \mathbf{y}_3$ but also on the information delivered by the first decoder, using its a posteriori distribution $p_1(b_i|\mathbf{y}_1, \mathbf{y}_2)$ as the a priori one: $q_1^2(b_i) = p_1(b_i|\mathbf{y}_1, \mathbf{y}_2)$. The result is the first approximation $p_1(b_i|\mathbf{y})$ of a posteriori distribution $p(b_i|\mathbf{y})$. Since at this step the first decoder was not supported by the information of the second one, this is done at the second iteration, where the first decoder again decodes the first component code but using a priori distribution $q_2^1(b_i) = p_1(b_i|\mathbf{y})$. Going on in this way, after the nth step the next approximation $p_n(b_i|\mathbf{y})$ of $p(b_i|\mathbf{y})$ is formed by the second decoder and used by the first one as the next a priori distribution $q_{n+1}^1(b_i)$ to output $p_{n+1}(b_i|\mathbf{y}_1, \mathbf{y}_2)$. This latter in its turn is used by the second decoder as the next a priori distribution $q_{n+1}^2(b_i)$ to produce the next approximation of the desired a posteriori probabilities $p_{n+1}(b_i|\mathbf{y})$, etc. Since the interleaver permutes data bits before inputting the second encoder the interleavers permute in the same way observations $\mathbf{y}_1$ and a priori distribution $q_n^2(b_i) = p_n(b_i|\mathbf{y}_1, \mathbf{y}_2)$ entering the second decoder. Similarly, the deinterleaver restores the original order of bits in a feedback passing on $p_n(b_i|\mathbf{y})$ from the second decoder output to the first decoder input. With these rearrangements, all data processed by both decoders are aligned properly.

A vast simulation has confirmed the convergence of these iterations experimentally, although theoretical justification remains a matter of question.

9.4.3 Performance

As mentioned before, turbo codes were the first regular codes to provide reliable data transmission over the band-limited channel at near-capacity rates and low energies per bit. To illustrate it by examples, let us first examine some fundamental limits on BPSK data transmission. On the strength of the sampling theorem any bandpass signal of bandwidth W is a vector of dimension $2WT$ (see Section 2.3). In the case of BPSK each component of such a vector may take on only two values, implying that within the time–frequency resource WT the number M of BPSK signals obeys the bound $M \leq 2^{2WT}$, or equivalently, no more than $2WT$ data bits may be transmitted. This restricts the transmission rate achievable with BPSK within the bandwidth W as $R = (\log M)/T \leq 2W$, imposing, in its turn, the bound on the rate per Hertz:

$R/W \leq 2$ bps/Hz. Now consider a binary code with rate $R_c = 1/2$, meaning that only every second component of a signal vector carries data, the rest being assigned to check symbols (two signal samples are spent per data bit), so that the ratio between data rate and bandwidth $R/W = 1$ bps/Hz. Turning to Shannon's bound (1.2), we may see that the minimum bit energy normalized to noise E_b/N_0 (half of power bit SNR) necessary to provide errorless transmission over an AWGN channel at such a rate equals 0 dB. We are, however, now dealing not with an arbitrary Gaussian channel but with one whose input symbols are limited to the BPSK alphabet (Gaussian channel with binary input). This restriction increases the minimal E_b/N_0 corresponding to $R/W = 1$ bps/Hz up to 0.19 dB [97]. The turbo code of constraint length 5 and block length 65 536 proposed in [95,96] secures bit error probability $P_b < 10^{-5}$ (this figure is often referred to as the practical criterion of wireless errorless operation) at $E_b/N_0 \approx 0.7$ dB, i.e. yielding only about 0.5 dB to Shannon's limit. At the moment of their publication these results seemed fantastic, since a long unsuccessful history had made many experts believe that finding deterministic coding rules allowing operation near Shannon's bound was a hopeless task. Following the original works [95,96], other efficient turbo as well as serial concatenation codes have been found (see bibliography in [97]).

It should be noted that the asymptotic ($E_b/N_0 \to \infty$) behaviour of turbo codes is not better than that of convolutional codes of the same rate and memory, since they do not possess any advantage in minimum distance. Turning to (2.23), we may see that asymptotically (with growth of SNR) the effect of multiplicity $n_{\min}$, i.e. the number of signals with minimum Euclidean distance $d_{\min}$ from a transmitted one, plays a secondary role against $d_{\min}$ itself due to the exponential drop of the Q-function with SNR (take the logarithm of P_e to make sure of it). For this reason dependence of P_e on E_b/N_0 sooner or later achieves a 'floor' character determined by $d_{\min}$ and analogous to the one typical of other codes with the same minimum distance. This, however, occurs at SNR values securing very small bit error probabilities, falling far beyond the range of practical needs. The explanation of why turbo codes guarantee such excellent operational quality at low bit SNR is not in their large minimum distance, but rather in the relatively small number of words lying from each other at small distances, in particular the small multiplicity $n_{\min}$ in (2.23). This redistribution of distances towards a greater number of bigger ones versus convolutional codes happens due to the pseudorandom interleaving of data bits encoded by the second component encoder. If the data bit pattern is unlucky to generate a small-weight word of the first component code, its permutation may appear different enough to produce the second component word of a remarkably higher weight.

9.4.4 Applications

Despite their short history, turbo codes are now in widespread use and enter the specifications of many systems. The most interesting in our context is their involvement in 3G mobile radio standards. The UMTS specification includes turbo codes of rate 1/3 based on two component convolutional codes of constraint length 4 and interleaver of variable length in the range from 40 to 5114 [92,97]. The cdma2000 standard also contains two-component turbo codes of constraint length 4 with interleaver size ranging from 250 to 4090. Appropriate puncturing allows rates of 1/2, 1/3, 1/4 or 1/5 to be obtained [69,97].

9.5 Channel interleaving

The analysis above rested on the memoryless model, where corrupting effects inflicted on the code symbols by channel interference were independent of each other. In real wireless channels with their shadowing and fading phenomena (see Section 3.5), these additive memoryless effects are supplemented by multiplicative ones: comparatively slow and sporadic drops of a received signal level spanning many code symbols. The technique outlined below is universally relevant regardless of soft or hard decoding mode, but to make the discussion more transparent let us specify a hard decoding procedure. When AWGN is the only channel interference all symbol errors are independent and are randomly scattered along the codewords. Signal drops due to fading cause grouping of symbol errors into packets or bursts. Certainly, if length B of the burst is no greater than the code correction (detection) capability t_c (t_d), the decoder will with no difficulty correct (detect) B errors, since the specific pattern of errors within the code error control capability is of no matter. The fading nature, however, often leads to rare but rather long error bursts, so that meeting the condition $B \leq t_c$ would require codes with a large check symbol overhead, i.e. small rate and wasteful bandwidth utilization. An efficient and feasible alternative is the very popular *channel interleaving*, under which symbols of the codestream are permuted before transmission to disperse as far as possible those positioned close to each other and, on the other hand, make closer those positioned far apart. At the receiving side deinterleaving takes place, returning all the code symbols to their initial positions. If a burst of errors of length B occurs in the channel, corrupted symbols after deinterleaving will prove to be far from each other as though the errors were independent. If block codes of appropriate length and distance are used these errors will with high probability fall into different words to be corrected by the decoder. With convolutional codes the chance of their correction will again be good, since such codes correct many patterns of errors counting beyond the free distance, unless they gather into too dense packets. The simplest implementation of this technique is a block interleaver writing symbols into a square matrix row by row and then reading them out column by column. Of course, the deinterleaver rearranges the symbols in the reverse manner.

Interleaving is an integral part of the majority of modern digital wireless communication systems, including all 2G and 3G mobile standards.

Problems

9.1. A binary block code of length $n = 9$ is used for transmitting $M = 32$ messages. How many check symbols does it have? What is its rate? What is the number of redundant binary observation vectors?

9.2. A binary block code has minimum Hamming distance $d_H = 7$. What is its minimum Euclidean distance if binary symbols are transmitted in FSK mode by non-overlapping pulses of energy E_s?

9.3. The binary block code $U = \{10101, 00011, 11000, 01110\}$ is used for data transmission over BSC. The observation vector is $\mathbf{Y} = (00110)$. Into what code vector will it be decoded with error correction? What would be the answer for $\mathbf{Y} = (11011)$? What are the correction and detection capabilities of the code? Is this code linear?

9.4. Prove Proposition 9.2.2.

9.5. What should the minimum distance of a binary code be to correct up to t_c errors and above it to detect up to $t_d > t_c$ errors?

9.6. Find the number of binary vectors having weights no greater than t_d (volume of a binary sphere of radius t_d).

9.7. Using the result of Problem 9.6 prove the Gilbert bound: a binary block code detecting up to t_d errors always exists whenever:

$$(M-1)\sum_{i=0}^{t_d} C_n^i < 2^n$$

where M, n are the number of codewords and code length, respectively.

9.8. Calculate the result of the following operations over binary polynomials:

$$f(z) = (z^3 + z^2 + 1)(z^4 + 1) - (z^2 + 1)^2(z^3 - 1) + z^6 + z^2$$

9.9. Open the brackets in the binary polynomial $(z+1)^{2^i}$ where i is a positive integer.

9.10. Find the remainder of dividing $z^5 + z^3 + 1$ by $z^2 + z + 1$ over $GF(2)$.

9.11. Let $g(z)$ be a binary polynomial of degree r. Prove that the binary block code of length n using this generator polynomial (having code polynomials $u(z) = b(z)g(z)$, data polynomials $b(z)$ being arbitrary binary polynomials of degree no greater than $n - r - 1$) is linear.

9.12. The binary polynomial $g_1(z) = z^4 + z + 1$ is primitive. What is the largest length of CRC detecting up to three errors based on this polynomial? Will the observation polynomial $y(z) = z^8 + z^5 + z + 1$ be declared erroneous or not? What about the polynomial $y(z) = z^8 + z^4 + z + 1$?

9.13. The generator polynomials of a convolutional code are $g_1(z) = 1$, $g_2(z) = 1 + z$. What are its rate and constraint length? Sketch the encoder scheme, draw the trellis diagram and find the free distance of the code.

9.14. The generator polynomials of a convolutional code are $g_1(z) = 1 + z + z^2$, $g_2(z) = 1 + z + z^2$, $g_3(z) = 1 + z^2$. What are its rate and constraint length? Sketch the encoder scheme, draw the trellis diagram and find the free distance of the code.

9.15. One of two generator polynomials of the convolutional code is $g_1(z) = 1 + z + z^2$. Which of the polynomials $g_2(z) = 1$, $g_2(z) = 1 + z$ or $g_2(z) = 1 + z^2$ is better as the second generator polynomial to maximize the asymptotic coding gain? What is this maximal gain? What sort of implication follows from this problem about comparison of systematic and non-systematic codes?

9.16. Generator polynomials of the convolutional code are $g_1(z) = 1 + z + z^2$, $g_2(z) = 1 + z^3$. Encode the bit stream 110110110110110... and explain why this code enters the class of so-called *catastrophic* codes, which are not recommended for practical use.

9.17. Decode as many data bits as possible if the observation is 100101100011000 and the generator polynomials of the convolutional code are $g_1(z) = 1$, $g_2(z) = 1 + z$ and $g_3(z) = 1 + z$.

9.18. Decode the observation 111111100000001111, if generator polynomials of the convolutional code are as in Problem 9.14 and it is known that after encoding four data bits the encoder is forced to a zero state by tail bits. How far is the decoded word from the observation? If the decoding result is true, how many errors did the decoder correct?

9.19. Revise the decoding procedure of Example 9.3.4, storing all paths entering a node with equal metrics as survivors. Continue after the seventh step up to the first step where decoded data bits may be released. At what step does it occur? How many bits are released?

9.20. The bit stream is encoded by a binary convolutional code with generator polynomials $g_1(z) = 1, g_2(z) = 1 + z$. Binary symbols are transmitted by BPSK $(0 \rightarrow +1, 1 \rightarrow -1)$. The communication channel is Gaussian and output observation samples are $\mathbf{y} = (-0.5, -0.5, -3, -4, -6, 2, -4, 5, 3, -2)$. Decode this observation using the hard and soft (based on correlation of the observation with trellis diagram paths) Viterbi algorithm. Assume that the last two code symbols correspond to a tail bit setting the encoder to zero. Explain the difference (if any) of the results of the two procedures. Which of them is more trustworthy?

9.21. Someone wants to build a turbo code starting with the convolutional code described by the generator polynomials $g_1(z) = 1 + z^3 + z^4$ and $g_2(z) = 1 + z + z^2 + z^4$. Sketch a component encoder of the turbo code.

Matlab-based problems

9.22. Write and run a program to calculate sum, product, quotient and remainder for two arbitrary binary polynomials.

9.23. Write a program that finds a generator polynomial for binary CRC of given length detecting up to three errors.

9.24. Write a program illustrating error detecting by CRC. Recommended steps:

(a) Using the program of Problem 9.23, find an appropriate generator polynomial $g(z)$ for CRC of a given length n detecting up to three errors.

(b) Take any binary error vector of length n and weight 1, 2 or 3 and divide it by the generator polynomial; make sure that for any such vector but zero the syndrome is non-zero.

(c) Take any decimal number smaller than 2^{n-r}, where $r = \deg g(z)$, and convert it into a binary vector (the function 'de2bi' is a good aid for this); multiply the corresponding polynomial by the generator polynomial.

(d) Using the polynomial of the previous item as the error one, check that this error pattern coinciding with some codeword cannot be detected.

(e) Run the program 1000–10 000 times for CRC of length $n = 40–200$ with independent random error vectors and calculate syndromes. How often do undetected errors happen? Compare the result with the theoretical prediction.

9.25. Write a program running convolutional encoding, given the set of generator polynomials. In Matlab, as is customary, generator polynomials are given in an octal notation. The binary vector of length ν_c of polynomial coefficients in a power-ascending order is appended if necessary by left zeros to make the length a multiple of three. Then every binary triple is put as an octal figure, the rightmost bit being the least significant, and the set of n polynomials in octal notation is presented as an n-tuple. For instance polynomials $g_1(z) = 1 + z$ and $g_2(z) = 1 + z + z^2$ have octal notation (6,7). Run the program for codes with polynomials (7,5), (15,17), (53,75), (561,753), (5,7,7), (25,33,37) and (133,145,175) trying to get a non-zero codeword of minimum weight. What is the trend in the behaviour of free distance versus constraint length and rate?

9.26. Develop the program of Problem 9.25 to find the free distance of a convolutional code. Run the program for codes with polynomials (7,5), (15,17), (23,35), (53,75), (133,171), (247,371), (561,753), (1167,1545), (5,7,7), (13,15,17), (25,33,37), (47,53,75), (133,145,175), (225,331,367), (557,663,711) and (1117,1365,1633).

9.27. Using operators of the Matlab Communication Toolbox, write a program for convolutional hard-decision decoding. Assuming that the transmitted message is $10\nu_c$ zeros, run the program for the code with generator polynomials (23,35). Enter the error vectors of weight 2–3 and make sure that errors are corrected whenever the parameter 'tblen' is around $(4-5)\nu_c$. Try to find the error of weight 4 which is not corrected, and explain why it may not be easy despite the free distance of the code being 7. Increasing the error weight, observe the decoder behaviour. Run the program 1000 times for independent random error patterns of a given weight and calculate bit error probability depending on error weight. Do the same for the codes (47,53,75) and (133,145,175) (free distances 13 and 15, respectively).

10

Some advancements in spread spectrum systems development

10.1 Multiuser reception and suppressing MAI

In Section 4.1 we met two options for making decisions on the data in a K-user CDMA system. One of them runs the optimal (ML) procedure realized by the so-called multiuser receiver, while the other involves a single-user or conventional procedure. The conventional receiver treats MAI as no more than an additional random noise, fully ignoring the deterministic nature of signatures and correlations between them. On the other hand, multiuser algorithms utilize a priori knowledge about signature codes or, at least, their ensemble correlation properties. In this section we are going to discuss briefly the ideas underlying multiuser reception, starting with the simplest case of synchronous CDMA.

10.1.1 Optimal (ML) multiuser rule for synchronous CDMA

In order to make the discussion free of secondary details let us consider the plainest, yet general enough, model of K-user DS CDMA involving real signatures and BPSK data transmission. The model covers, among others, any system with BPSK signature and data modulation. As in Section 7.2, within this subsection we consider a fully synchronous case when both chips and borders of data symbols (bits) of all users are strictly aligned in time. This, along with the assumption of the independence of consecutive data bits of any user, permits limiting the observation interval to a single bit duration: $T = T_b$. Then the group signal of K users:

$$s(t;\mathbf{b}) = \sum_{k=1}^{K} A_k b_k s_k(t) \tag{10.1}$$

Spread Spectrum and CDMA: Principles and Applications Valery P. Ipatov

where, similarly to (4.1), $A_k > 0$ is the real amplitude of the kth user signal, $\mathbf{b} = (b_1, b_2, \ldots, b_K)$ is the vector of data bits of K users (bit pattern) and $s_k(t)$ is the kth user's signature.

As was mentioned in Section 4.1, the globally optimal (ML) procedure involves searching the estimate $\hat{\mathbf{b}} = (\hat{b}_1, \hat{b}_2, \ldots, \hat{b}_K)$ of the K user data bits pattern $\mathbf{b}$ as the value of $\mathbf{b}$ minimizing the Euclidean distance (or its square $d^2(\mathbf{s}, \mathbf{y})$) between the observation $y(t)$ and group signal (10.1). Calculating $d^2(\mathbf{s}, \mathbf{y})$ in the same way as in (4.3) results in:

$$d^2(\mathbf{s}, \mathbf{y}) = \int_0^T [y(t) - s(t; \mathbf{b})]^2 \, dt = \|\mathbf{y}\|^2 - 2\sum_{k=1}^{K} A_k b_k z_k + \sum_{k=1}^{K}\sum_{l=1}^{K} A_k A_l b_k b_l \rho_{kl} \qquad (10.2)$$

where

$$z_k = \int_0^T y(t) s_k(t) \, dt \qquad (10.3)$$

is, as usual, correlation of the observation $y(t)$ with the kth signature, ρ_{kl} is the correlation coefficient of the kth and lth signatures, and the presence of amplitudes A_k allows use of a convenient normalization of signatures:

$$E_k = \|\mathbf{s}_k\|^2 = \int_0^T s_k^2(t) \, dt = 1, \; k = 1, 2, \ldots K$$

Let us introduce two matrices: $\mathbf{G} = \mathrm{diag}(A_1, A_2, \ldots, A_K)$, a diagonal $K \times K$ matrix of users' amplitudes, and $\mathbf{C} = [\rho_{kl}]$, $k, l = 1, 2, \ldots, K$, the correlation matrix of signatures. With designation $\mathbf{z} = (z_1, z_2, \ldots, z_K)$ for the vector of correlations (10.3), the squared distance (10.2) becomes (superscript T symbolizing vector–matrix transpose):

$$d^2(\mathbf{s}, \mathbf{y}) = \|\mathbf{y}\|^2 - 2\mathbf{bGz}^T + \mathbf{bGCGb}^T \qquad (10.4)$$

The first term in the right-hand side of (10.4) is fixed with a current observation $y(t)$ and consequently the ML estimate $\hat{\mathbf{b}}$ may be found as the value of $\mathbf{b}$ maximizing the difference of two other terms:

$$2\hat{\mathbf{b}}\mathbf{Gz}^T - \hat{\mathbf{b}}\mathbf{GCG}\hat{\mathbf{b}}^T = \max_{\mathbf{b}}(2\mathbf{bGz}^T - \mathbf{bGCGb}^T) \qquad (10.5)$$

When a CDMA system is not oversaturated ($K \le N$; see Section 7.2), all signatures are allowed to be orthogonal, so that $\rho_{kl} = \delta_{kl}$ and $\mathbf{C} = \mathbf{I}_K$, where $\mathbf{I}_K$ is the Kth order identity matrix. Then $\mathbf{bGCGb}^T = \sum_{k=1}^{K} A_k^2$ does not depend on the user's bit pattern $\mathbf{b}$. As was noted in Section 4.1, multiuser detecting then degenerates into a conventional one, where the sign of the correlation z_k defines the estimate $\hat{b}_k$ of the kth user's bit b_k. In the case of non-orthogonal signatures (e.g. oversaturated CDMA, $K > N$) the conventional receiver yields to the ML one, but the latter may appear prohibitively complex. Indeed, the data vector $\mathbf{b}$ is strictly restrained by the BPSK alphabet limitation $b_k = \pm 1$, and no procedure exists that is more computationally efficient than just trying all 2^K possible bit patterns and comparing the results of their substitution to the right-hand side of (10.5). Therefore, ML multiuser detection according to (10.5) has exponential complexity versus the number of users (see numerical example in Section 4.1). On the other hand, the motivation to involve multiuser detection is often strongest when the number of users is so great that conventional detection fails due to the high level of MAI. This explains the interest in various quasi-optimal multiuser algorithms, some of which are surveyed in the rest of the section.

10.1.2 Decorrelating algorithm

Let us start with a conventional (i.e. correlation-based) receiver of user number one's data. According to (10.1) the observation:

$$y(t) = s(t; \mathbf{b}) + n(t) = \sum_{k=1}^{K} A_k b_k s_k(t) + n(t) \tag{10.6}$$

This, after substitution to (10.3), where $k = 1$, leads to:

$$z_1 = A_1 b_1 + \sum_{k=2}^{K} A_k b_k \rho_{k1} + n_1 \tag{10.7}$$

where $n_1 = \int_0^T n(t) s_1(t)\,\mathrm{d}t$ is a noise sample at the first correlator output. The second term of (10.7) is MAI, and the question is whether it may be suppressed to zero by means of some linear transform of the input observation. Whatever this linear operation is, finally we should have some MAI-free substitute ς_1 of z_1, i.e. the scalar, producing the decision on the first user's current bit as:

$$\hat{b}_1 = \operatorname{sign}(\varsigma_1) \tag{10.8}$$

Any linear operation transforming $y(t)$ into a scalar may be described as correlation:

$$\varsigma_1 = \int_0^T y(t) u(t)\,\mathrm{d}t \tag{10.9}$$

differing from (10.3) by only a reference signal $u(t)$. Therefore, we are going to suppress MAI declining matched reference $s_1(t)$ in favour of mismatched one $u(t)$, i.e. at the cost of loss in SNR with respect to a thermal noise. We have already resorted to this trick when seeking zero-forcing filters to remove autocorrelation sidelobes (see Section 6.12). Using (10.6) in (10.9) replaces (10.7) by:

$$\varsigma_1 = A_1 b_1 \rho_{1u} + \sum_{k=2}^{K} A_k b_k \rho_{ku} + n_1' \tag{10.10}$$

where ρ_{ku} is the correlation coefficient of the kth signature with $u(t)$, normalization of reference $u(t)$ is the same as for signatures, and $n_1' = \int_0^T n(t)u(t)\,dt$ is a noise contribution in ς_1.

Let us put signatures and a reference signal $u(t)$ in the form (2.50) typical of DS CDMA with real-valued signatures:

$$s_k(t) = \sum_{i=0}^{N-1} a_{k,i}s_0(t - i\Delta), \;\; u(t) = \sum_{i=0}^{N-1} u_i s_0(t - i\Delta) \tag{10.11}$$

where $u_i, i = 0, 1, \ldots, N-1$ is a real code sequence of the reference $u(t)$. Using the vector notation of code sequences $\mathbf{a}_k = (a_{k,0}, a_{k,1}, \ldots, a_{k,N-1}), \mathbf{u} = (u_0, u_1, \ldots, u_{N-1})$ (see Section 7.2) and setting with no lost of generality chip energy $E_0 = 1$, we come to equations $\rho_{kl} = (\mathbf{a}_k, \mathbf{a}_l) = \mathbf{a}_k\mathbf{a}_l^T$, $\rho_{ku} = (\mathbf{a}_k, \mathbf{u}) = \mathbf{a}_k\mathbf{u}^T$. To remove the MAI term in (10.10) independently of amplitudes and bits of interfering users we need to fulfil $K-1$ conditions: $\rho_{ku} = 0, k \geq 2$. In other words, the reference code $\mathbf{u}$ should be a solution of the linear equation set $\mathbf{a}_k\mathbf{u}^T = \mathbf{u}\mathbf{a}_k^T = 0$, $k = 2, 3, \ldots K$. Preserving the non-zero useful effect ρ_{1u} has to be non-zero, hence $\mathbf{u}$ is just a properly scaled solution $\mathbf{v}$ of the equation:

$$\mathbf{vA} = \mathbf{e}_1 \tag{10.12}$$

where columns of the $N \times K$ signature matrix $\mathbf{A}$ are signature code vectors: $\mathbf{A} = (\mathbf{a}_1^T, \mathbf{a}_2^T, \ldots, \mathbf{a}_K^T)$ and $\mathbf{e}_1$ is a K-dimensional vector of the view $\mathbf{e}_1 = (1, 0, 0, \ldots, 0)$. When all signature vectors are linearly independent, the system (10.12) may have a set of solutions, but among all vectors $\mathbf{v}$ satisfying (10.12) we choose the one that is a linear combination of signatures, i.e. rows of $\mathbf{A}^T$: $\mathbf{v} = \mathbf{xA}^T$, where $\mathbf{x}$ is an unknown K-dimensional row vector. The reason for such a choice is that inclusion in $\mathbf{v}$ of any component orthogonal to the space of the signature vectors will only increase the norm of $\mathbf{v}$, i.e. the noise component at the correlator output, with no increase of the useful first term in (10.10). With this substitution (10.12) becomes:

$$\mathbf{xA}^T\mathbf{A} = \mathbf{xC} = \mathbf{e}_1$$

Linear independence of signatures (columns of $\mathbf{A}$) means rank K of the correlation $K \times K$ matrix $\mathbf{C} = \mathbf{A}^T\mathbf{A}$, i.e. its invertibility and uniqueness of solution of the equation above: $\mathbf{x} = \mathbf{e}_1\mathbf{C}^{-1}$. Then:

$$\mathbf{v} = \mathbf{xA}^T = \mathbf{e}_1\mathbf{C}^{-1}\mathbf{A}^T = \mathbf{e}_1(\mathbf{A}^T\mathbf{A})^{-1}\mathbf{A}^T \tag{10.13}$$

is a desired solution of (10.12), whose scaling $\mathbf{u} = \mathbf{v}/\|\mathbf{v}\|$ results in the normalized decorrelation reference $\mathbf{u}$, so that $\mathbf{ua}_1^T = \rho_{1u}$. This normalization is practically unnecessary, having no effect on the sign of ς_1 in the decision rule (10.8).

Physically the reference vector (10.13) is just orthogonal to all signatures but the first, entirely eliminating MAI at the output of a correlator tuned to the first user's signal. We would find the reference signal for the kth user's receiver in the same way, replacing $\mathbf{e}_1$ in (10.13) by $\mathbf{e}_k$, whose unique non-zero component is the kth one.

The main drawback of the described *decorrelating* receiver is its working capacity with only linearly independent signatures. If this demand is not observed, any attempt to force MAI to zero will inevitably make a useful effect (the first term) in (10.10) vanish too. At the same time, linear independence means that $K \leq N$, in which case the most adequate choice of signatures is an orthogonal set (see previous subsection) entailing optimality of a single-user receiver and automatically rejecting MAI with no SNR loss and no special decorrelation processing. In the case of oversaturation ($K > N$) linear independence of signatures is impossible and the decorrelating algorithm cannot be used.

10.1.3 Minimum mean-square error detection

Let us again exploit the idea of mismatched processing in a correlator tuned to the first user's signal, but this time, instead of forcing MAI to zero, we will try to minimize the overall corrupting effect of MAI and noise. Coming back to (10.7), we may note that only the term $A_1 b_1$ in it is a useful component, the other two presenting an overall interference (MAI plus noise). In this light it is natural to look for a linear operation (10.9), imitating a useful contribution with *minimum mean-square error* (MMSE). To formalize the problem we first rewrite (10.9) in a vector form, substituting $u(t)$ from (10.11):

$$\varsigma_1 = \sum_{i=0}^{N-1} u_i y_i = \mathbf{u}\tilde{\mathbf{y}}^T \tag{10.14}$$

where $\tilde{\mathbf{y}} = (y_0, y_1, \ldots, y_{N-1})$ and $y_i = \int_0^T y(t)s_0(t - i\Delta)\,\mathrm{d}t, i = 0, 1, \ldots, N-1$. Technically y_i may be obtained as a sample at the output of a chip matched filter taken at the appropriate moment (see (2.68)), allowing to look at $\tilde{\mathbf{y}}$ as a vector of observations after the chip matched filtering. Our task now is to minimize mean square deviation $\overline{\varepsilon^2}$ of ς_1 from $A_1 b_1$ by an appropriate choice of a reference code vector $\mathbf{u}$:

$$\overline{\varepsilon^2} = \overline{|A_1 b_1 - \varsigma_1|^2} = \overline{|A_1 b_1 - \mathbf{u}\tilde{\mathbf{y}}^T|^2} = \min_{\mathbf{u}}$$

Note that no a priori normalization of the reference $\mathbf{u}$ is necessary. After squaring and term-wise averaging, the mean-square error takes the form:

$$\overline{\varepsilon^2} = \overline{(A_1 b_1)^2} - 2A_1 \mathbf{u}\overline{\tilde{\mathbf{y}}^T b_1} + \overline{(\mathbf{u}\tilde{\mathbf{y}}^T)^2} = \overline{(A_1 b_1)^2} - 2A_1 \mathbf{u}\overline{\tilde{\mathbf{y}}^T b_1} + \mathbf{u}\overline{\tilde{\mathbf{y}}^T\tilde{\mathbf{y}}}\mathbf{u}^T \tag{10.15}$$

where we use elementary matrix algebra (commutativity of multiplication by a scalar and associativity of vector–matrix multiplication, commutativity of a scalar product $\mathbf{u}\tilde{\mathbf{y}}^T = \tilde{\mathbf{y}}\mathbf{u}^T$) and non-randomness of $\mathbf{u}$. The ith component of $\tilde{\mathbf{y}}$ after substitution of (10.6) and then (10.11) becomes:

$$y_i = \int_0^T y(t)s_0(t - i\Delta)\,\mathrm{d}t = \int_0^T \left[\sum_{k=1}^{K} A_k b_k s_k(t) + n(t)\right] s_0(t - i\Delta)\,\mathrm{d}t$$

$$= \sum_{k=1}^{K} A_k b_k \sum_{j=0}^{N-1} a_{k,j} \int_0^T s_0(t - j\Delta)s_0(t - i\Delta)\,\mathrm{d}t + \nu_i = \sum_{k=1}^{K} A_k b_k a_{k,i} + \nu_i \tag{10.16}$$

where $\nu_i = \int_0^T n(t)s_0(t - i\Delta)\,dt$ is the ith noise sample at the chip matched filter output and a natural assumption is used that the chips time-spaced by a non-zero integer number of repetition periods Δ are orthogonal (e.g. if chip duration is no longer than Δ, those chips do not overlap). Now we see that $\overline{y_i b_1} = A_1 a_{1,i}$, because bits of different users are independent of each other ($\overline{b_k b_l} = \delta_{kl}$) and of noise ($\overline{\nu_i b_k} = \overline{\nu_i} \cdot \overline{b_k} = 0$), and hence:

$$A_1 \overline{\tilde{\mathbf{y}}^T b_1} = A_1^2 \mathbf{a}_1^T \tag{10.17}$$

In a similar manner we calculate the matrix $\overline{\tilde{\mathbf{y}}^T \tilde{\mathbf{y}}}$, whose elements are simply correlation moments $\overline{y_i y_j}$ of samples y_i. Then, according to (10.16) and allowing for non-correlatedness of noise samples after the chip matched filter:

$$\overline{y_i y_j} = \sum_{k=1}^{K}\sum_{l=1}^{K} A_k A_l \overline{b_k b_l} a_{k,i} a_{l,j} + \overline{\nu_i \nu_j} = \sum_{k=1}^{K} A_k^2 a_{k,i} a_{k,j} + \sigma^2 \delta_{ij}$$

σ^2 being the variance of a noise component of y_i. Thus, the $N \times N$ correlation matrix $\mathbf{R}$ of the observation vector $\tilde{\mathbf{y}}$ is:

$$\mathbf{R} = \overline{\tilde{\mathbf{y}}^T \mathbf{y}} = [\overline{y_i y_j}] = \mathbf{A}\mathbf{G}^2\mathbf{A}^T + \sigma^2 \mathbf{I}_N \tag{10.18}$$

where $\mathbf{I}_N$ is the Nth order identity matrix. Substituting (10.17) in (10.15) after discarding the first term independent of $\mathbf{u}$ gives the following scalar function to minimize by adjusting $\mathbf{u}$:

$$f(\mathbf{u}) = \mathbf{u}\mathbf{R}\mathbf{u}^T - 2A_1^2 \mathbf{a}_1 \mathbf{u}^T \tag{10.19}$$

At the point $\mathbf{u}$ of extremum of $f(\mathbf{u})$ the gradient of $f(\mathbf{u})$, i.e. vector, whose components are derivatives of $f(\mathbf{u})$ with respect to every component of the vector $\mathbf{u}$, should be a zero vector. The gradient of $f(\mathbf{u})$ is readily found (see Problem 10.3) as $2(\mathbf{u}\mathbf{R} - A_1^2 \mathbf{a}_1)$. Thus, with invertible matrix $\mathbf{R}$ the vector $\mathbf{u}$ delivering an extremum to $f(\mathbf{u})$ is defined by the equation:

$$\mathbf{u} = A_1^2 \mathbf{a}_1 \mathbf{R}^{-1} \tag{10.20}$$

where $\mathbf{R}$ is given by (10.18). The reader is challenged to check that the extremum just found is really a minimum of (10.19) (Problem 10.3).

Clearly, this algorithm does not rest on the invertibility of the signature correlation matrix $\mathbf{C} = \mathbf{A}^T\mathbf{A}$; just the observation correlation matrix (10.18) should be invertible, which is practically always true. Hence, the solution (10.20), in contrast to (10.13), is universal regardless of the relation between K and N. At the same time, at least in one important particular case the solution (10.20) degenerates to the single-user algorithm. Let the signature set be a Welch-bound one, meaning that the rows of the signature matrix $\mathbf{A}$ are orthogonal (see Section 7.2.2), i.e. $\mathbf{A}\mathbf{A}^T = \mathbf{I}_N$. If all signals have the same intensity A, $\mathbf{G}^2 = A^2\mathbf{I}_K$, and the observation correlation matrix (10.18) becomes the simplest, $\mathbf{R} = (A^2 + \sigma^2)\mathbf{I}_N$, resulting in $\mathbf{u} = [A^2/(A^2 + \sigma^2)]\mathbf{a}_1$, which reproduces a scaled first signature, i.e. the reference of a conventional receiver. Thus, no special MMSE processing exists for the Welch-bound signatures of equal power. This fact is rather trivial if $K \le N$, since then such signatures are orthogonal and a conventional receiver eliminates MAI fully with best noise filtering, but for the oversaturation scenario ($K > N$) the statement is not that predictable.

In the literature the result (10.20) is often given in another form including explicitly the signature correlation matrix $\mathbf{C} = \mathbf{A}^T\mathbf{A}$ [99–101]. Deriving it is possible, for instance, through the matrix inversion lemma given here in the form fitted to the context:

$$\mathbf{R}^{-1} = (\mathbf{A}\mathbf{G}^2\mathbf{A}^T + \sigma^2\mathbf{I}_N)^{-1} = \frac{1}{\sigma^2}\mathbf{I}_N - \frac{1}{\sigma^2}\mathbf{A}(\mathbf{A}^T\mathbf{A} + \sigma^2\mathbf{G}^{-2})^{-1}\mathbf{A}^T \tag{10.21}$$

Proof of this result consists in a direct check (Problem 10.4). Note that it works whenever the matrix $\mathbf{G}$ is invertible, which is observed automatically if all users' amplitudes are non-zero. Making use of (10.21) and equation $A_1^2\mathbf{a}_1 = \mathbf{e}_1\mathbf{G}^2\mathbf{A}^T$ in (10.20) results in:

$$\mathbf{u} = \mathbf{e}_1\mathbf{G}^2\mathbf{A}^T\mathbf{R}^{-1} = \frac{1}{\sigma^2}\mathbf{e}_1\mathbf{G}^2\left[\mathbf{A}^T - \mathbf{A}^T\mathbf{A}(\mathbf{A}^T\mathbf{A} + \sigma^2\mathbf{G}^{-2})^{-1}\mathbf{A}^T\right]$$
$$= \frac{1}{\sigma^2}\mathbf{e}_1\mathbf{G}^2\left[\mathbf{A}^T - (\mathbf{A}^T\mathbf{A} + \sigma^2\mathbf{G}^{-2})(\mathbf{A}^T\mathbf{A} + \sigma^2\mathbf{G}^{-2})^{-1}\mathbf{A}^T\right] + \mathbf{e}_1(\mathbf{A}^T\mathbf{A} + \sigma^2\mathbf{G}^{-2})^{-1}\mathbf{A}^T$$

and eventually:

$$\mathbf{u} = \mathbf{e}_1(\mathbf{C} + \sigma^2\mathbf{G}^{-2})^{-1}\mathbf{A}^T \tag{10.22}$$

Returning back to (10.14), we write the final form of the decision rule on the first user's bit as:

$$\hat{b}_1 = \text{sign}(\varsigma_1) = \text{sign}[\mathbf{e}_1(\mathbf{C} + \sigma^2\mathbf{G}^{-2})^{-1}\mathbf{A}^T\tilde{\mathbf{y}}^T] \tag{10.23}$$

Spreading this rule to the receiver of the kth user's data is again immediate: $\mathbf{e}_k$ has to replace $\mathbf{e}_1$.

Emphasizing again that the rule under study is universal independently of signature correlation matrix invertibility, it is nevertheless noteworthy that if $\mathbf{C}$ is non-singular ($K \leq N$ is a necessary condition of it) and thermal noise diminishes, the MMSE detector converges asymptotically to the decorrelating one:

$$\mathbf{u} = \mathbf{e}_1(\mathbf{C} + \sigma^2\mathbf{G}^{-2})^{-1}\mathbf{A}^T \underset{\sigma^2 \to 0}{\to} \mathbf{e}_1\mathbf{C}^{-1}\mathbf{A}^T.$$

To demonstrate the efficiency of MMSE it is appropriate to compare the signal-to-interference-plus-noise ratio (SINR) for the receiver effect ς_1 in cases of reference (10.22) and that of a conventional receiver $\mathbf{u} = \mathbf{a}_1$. The contribution of the kth signal in ς_1 as seen from (10.14) is $A_k\mathbf{u}\mathbf{a}_k^T$, so that the useful power (created by the first useful signal) is $A_1^2(\mathbf{u}\mathbf{a}_1^T)^2$ and MAI power is $\sum_{k=2}^K A_k^2(\mathbf{u}\mathbf{a}_k^T)^2$. Uncorrelatedness of noise samples ν_i after a chip matched filter means that their powers are added after weighting by u_i in (10.14), so that the total power of the noise component of ς_1 is $\sigma^2\sum_{i=0}^{N-1}u_i^2 = \sigma^2\|\mathbf{u}\|^2$. Combining these results produces SINR:

$$q_I^2 = \frac{A_1^2(\mathbf{u}\mathbf{a}_1^T)^2}{\sum\limits_{k=2}^{K} A_k^2(\mathbf{u}\mathbf{a}_k^T)^2 + \sigma^2\|\mathbf{u}\|^2} = \frac{q_{1b}^2(\mathbf{u}\mathbf{a}_1^T)^2}{\sum\limits_{k=2}^{K} q_{kb}^2(\mathbf{u}\mathbf{a}_k^T)^2 + \|\mathbf{u}\|^2} \tag{10.24}$$

where $q_{kb}^2 = A_k^2/\sigma^2$ is power SNR per data bit for the kth user.

Example 10.1.1. Consider an oversaturated synchronous CDMA with Welch-bound signatures. Binary Welch-bound ensembles exist for any $K > N$ allowing the existence of a $K \times K$ Hadamard matrix. The signatures then are just K columns of this matrix after discarding any $K - N$ rows. In the light of the aforesaid, the case of equal powers will not display any advantages of MMSE against conventional detection. For the values $K = 64, N = 48$ a random pattern of users' amplitudes was taken as 64 samples of Rayleigh variable to imitate the Rayleigh channel. Figure 10.1 presents the dependence of SINR (10.24) for MMSE and conventional detectors on bit SNR under some 'benign' (well-scattered) amplitude pattern. The curves show that the gain of MMSE sometimes appears significant (in Figure 10.1 up to about 10 dB). Still, it should be remembered that such a profit is just a matter of chance: for some amplitude patterns it may appear even bigger, but the more uniform the amplitude pattern, the smaller is the difference in SINR versus the conventional receiver. One more remark is that the MMSE detector has much better resistance to the scatter of users' intensities (if signatures are non-orthogonal, of course) in comparison with the conventional receiver, making it especially attractive wherever the power control is not perfect.

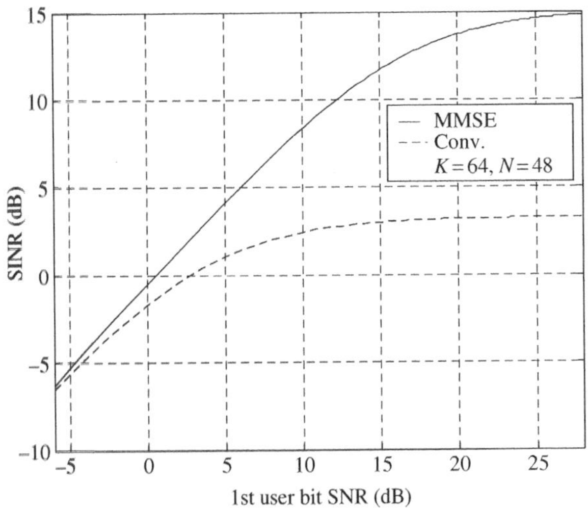

Figure 10.1 Example SINR curves for MMSE and single-user receivers

10.1.4 Blind MMSE detector

Although the computational complexity of the MMSE algorithm (as well as the decorrelating one) is not at all practically prohibitive there is still one implementation issue motivating further research. As (10.20) shows, the key operation of the MMSE algorithm is inversion of the observation correlation matrix $\mathbf{R}$ defined by (10.18). To perform it the receiver of the kth user should know, along with its own signature, also the signatures of all the other users. In some situations such a requirement looks fairly excessive. One of the most typical involvements of synchronous CDMA is the downlink of mobile radio, and making every mobile know all the signatures currently utilized by the other users would complicate the system dramatically. Fortunately, the correlation matrix of the observed process may be estimated experimentally from the observation itself, provided the observation period is long enough. This is the core idea of *blind* multiuser algorithms. Suppose that $\tilde{\mathbf{y}}_i$ is an N-dimensional row vector of samples at the chip matched filter output corresponding to the data bit number $i = 0, 1, \ldots$. Then estimate $\hat{\mathbf{R}}_i$ of the observation correlation matrix may be found as:

$$\hat{\mathbf{R}}_i = \frac{1}{i+1}\sum_{t=0}^{i}\tilde{\mathbf{y}}_t^T\tilde{\mathbf{y}}_t = \frac{i}{i+1}\cdot\frac{1}{i}\sum_{t=0}^{i-1}\tilde{\mathbf{y}}_t^T\tilde{\mathbf{y}}_t + \frac{1}{i+1}\tilde{\mathbf{y}}_i^T\tilde{\mathbf{y}}_i = \frac{i}{i+1}\hat{\mathbf{R}}_{i-1} + \frac{1}{i+1}\tilde{\mathbf{y}}_i^T\tilde{\mathbf{y}}_i$$

Therefore, it is possible to compute $\hat{\mathbf{R}}_i$ in a recurrent way: as soon as the portion of observation samples spanning the new bit arrives, it is used to update the estimate $\hat{\mathbf{R}}_i$. After this the result is substituted in (10.20) to find the current reference as $\mathbf{u}_i = A_1^2\mathbf{a}_1\hat{\mathbf{R}}_i^{-1}$, calculating from (10.14) $\varsigma_1 = \mathbf{u}_i\tilde{\mathbf{y}}_i^T$, and the decision on the ith bit according to (10.23). Variations and advancements of blind multiuser algorithms are plentiful and may be found in the literature (see [19,100,101] and their bibliographies).

10.1.5 Interference cancellation

One can construe the low-complexity of both decorrelating and MMSE detectors by the fact that they exploit a single-user philosophy, i.e. a linear operation of multiplying the observation vector $\tilde{\mathbf{y}}$ by a mismatched reference vector $\mathbf{u}$. The interference cancellation strategy is again based on a conventional receiver, supplemented by a loop of subtraction of MAI terms from the output effect (10.7). Suppose that the first user receiver knows the signatures and amplitudes of all users and one way or another has obtained estimates $\hat{b}_2, \hat{b}_3, \ldots, \hat{b}_K$ of data bits of the side users. Then one is capable of regenerating all side user signals, subtracting their sum from the observation $y(t)$ and utilizing the result as an input (presumably free of MAI) to the conventional receiver. Certainly, the efficiency of such a detector will dramatically depend on the reliability of knowledge of side signal amplitudes and the accuracy of estimates of side user bits. Among others, the multistage procedure is widely discussed in the literature [19,102]. Its first stage involves successive estimates of user bits transferring from the stronger to weaker signals, and using the already estimated $k-1$ user bits to remove corresponding MAI terms when estimating the kth user's bit. After all K bits are estimated by so doing the procedure runs the next stage, where all the same operations are repeated. This time, however, the receiver's knowledge on MAI is richer compared to the previous stage, and subtraction of the recreated MAI starts from the very beginning, i.e. estimating a bit of the strongest user. Stages like this are iterated as many times as is wished, each starting with an updated MAI recreation and continuing to refine it during the course of user bit pattern estimation. When one or another terminating criterion is met, the procedure outputs the final estimate of the bit of the user of interest.

10.1.6 Asynchronous multiuser detectors

Although the multiuser detection algorithms briefly discussed above are tailored to synchronous CDMA, they are, after an appropriate modification, extendable to the asynchronous case, too. Let us come back to (10.6) and adapt it to the asynchronous situation illustrated by Figure 7.15, assuming alignment of chip boundaries of all users and operating with only baseband real signatures (i.e. ignoring mutual phase shifts of bandpass signatures). Assume also that all signatures are periodic with period equal to bit duration $T = T_b$ and consider as before the first user receiver ($\tau_1 = 0$) at the observation interval of data bit number zero. Then:

$$y(t) = \begin{cases} A_1 b_{1,0} s_1(t) + \sum_{k=2}^{K} A_k b_{k,-1} s_k(t - \tau_k) + n(t), 0 < t \leq \tau_1 \\ A_1 b_{1,0} s_1(t) + \sum_{k=2}^{K} A_k b_{k,0} s_k(t - \tau_k) + n(t), \tau_1 < t \leq T \end{cases} \quad (10.25)$$

As (10.25) shows, the character of MAI created by the kth user depends now on whether the kth user's data bit changes or not during the observation interval, the kth signature altering its polarity at the moment $t = \tau_k$ in the second case. Suppose that the first user receiver knows the timings τ_k of all users. In this case the number of potentially harmful MAI vectors created by $K - 1$ side users is $2(K - 1)$, unlike the synchronous case, where a similar number was only $K - 1$. Together with the first signature vector we have $2K - 1$ vectors in total. If they are all linearly independent, a modified version of the decorrelating algorithm (10.13) may be used to eliminate MAI entirely regardless of the bits transmitted by users. Clearly, within the dimension (spreading factor) N, the maximal number of users K allowing such a processing is limited to $2K - 1 \leq N$, i.e. cannot be greater than $\lceil N/2 \rceil$. The necessary alteration of (10.13) is replacement of the 'pure' $N \times K$ signature matrix $\mathbf{A}$ by an extended $N \times (2K - 1)$ one, whose columns are the first signature, $K - 1$ properly time-shifted other signatures and the latter with polarity hops, starting at chip number zero (see Figure 7.15).

The MMSE detector interprets MAI statistically, i.e. as a random process described by its correlation matrix. The same interpretation is, certainly, applicable to the asynchronous case and again some revision is necessary to allow for changing the observation correlation matrix [19].

Since in an asynchronous system every side user bit may cover two consecutive bits of the user of interest, MAI correlation extends beyond the duration of one bit. That is the reason why the performance of asynchronous multiuser detectors typically improves when the observation interval spans several data bits.

We cannot go deeper into discussing multiuser detecting due to space limitations. This area is among the most challenging and rapidly advancing currently, and readers eager to get better insight into its current state as well as to learn about the prospects of its application in 3G and beyond are recommended to consult [19,100,102–104] and their references.

10.2 Multicarrier modulation and OFDM

In recent years the transmission mode called *multicarrier* (MC) modulation has gained great popularity in telecommunications. In general terms, MC modulation means no more than involving a multitude of parallel *subcarriers* to transmit symbols of the same datastream. The specific form of utilizing the subcarriers may vary depending on the goal pursued. In our context the issue of primary interest is the relation of MC modulation with the spread spectrum concept. It follows from the previous contents that all advantages inherent to spread spectrum have their origin in a high time–frequency product (processing gain, spreading factor) WT. Various ways exist to widen the signal spectrum, given the duration given. DS spread spectrum solves this task, replacing conventional 'slow' symbol pulses by 'fast' signals of some convenient shape, i.e. controlling signal bandwidth via the time domain. MC modulation from this angle may be referred to as an alternative method of spreading the spectrum through its direct shaping in the frequency domain.

10.2.1 Multicarrier DS CDMA

The simplest version of incorporating the MC technique into CDMA is an ordinary frequency multiplexing of data of the same user. To realize it M_c subcarriers transmit demultiplexed and DS spread data with non-overlapping spectra of subcarrier signals. Figure 10.2 explains this transmission mode, which is called multicarrier DS CDMA

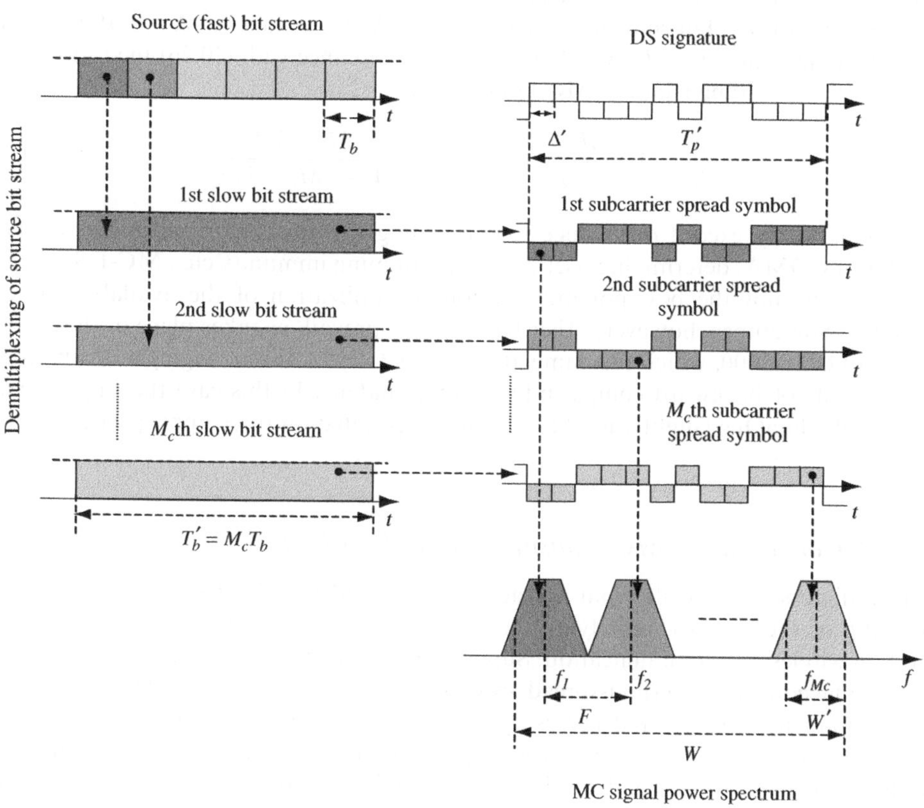

Figure 10.2 Explanation of MC-DS-CDMA

(MC-DS-CDMA). The source 'fast' bit stream of rate $R = 1/T_b$ is demultiplexed (split) into M_c parallel 'slow' bit streams of rate R/M_c (or bit duration $T'_b = M_c T_b$) each. The ith slow bit stream is transmitted on its specific subcarrier of frequency $f_i, i = 1, 2, \ldots, M_c$ using DS spreading and some data modulation mode (BPSK, QPSK or other mode). With spreading chip duration Δ' and subcarrier data symbol duration T'_p the bandwidth and spreading factor of one subcarrier signal are found as $W' \approx 1/\Delta'$ and $N' = T'_p/\Delta' \approx W'T'_p$, respectively. Spacing F of the subcarrier frequencies should exclude spectra overlapping, i.e. meet the restriction $F = W' + F_g$, where F_g is a guard interval allowing for out-of-band spectra sidelobes. Therefore the total bandwidth occupied by the system:

$$W = (M_c - 1)F + W' = M_c W' + (M_c - 1)F_g. \tag{10.26}$$

The receiver of MC-DS-CDMA consists of M_c parallel identical receivers, each tuned on its own subcarrier and processing it independently of the others. Demodulated slow bit streams from their outputs are then multiplexed to restore the transmitted source fast bit stream.

Comparison of this system with DS CDMA of the same total bandwidth W and data rate R shows that due to the necessity for guard intervals MC-DS-CDMA yields to DS-CDMA in spreading factor. If both systems use the same modulation mode then symbol duration T_p for DS CDMA should be M_c times shorter than the MC-DS-CDMA symbol duration T'_p, but chip duration in DS CDMA may be taken as $\Delta \approx 1/W$ providing the spreading factor $N = T_p/\Delta \approx WT'_p/M_c$. In accordance with (10.26) the ratio of this entity to the spreading factor of MC-DS-CDMA:

$$\frac{N}{N'} = 1 + \frac{M_c - 1}{M_c} \cdot \frac{F_g}{W'} = 1 + \frac{(M_c - 1)F_g}{W - (M_c - 1)F_g} \tag{10.27}$$

may remarkably exceed one. Since the spreading factor is the most important parameter of CDMA, determining user capacity, jamming immunity etc., MC-DS-CDMA proves to be not the best potential option for utilization of the available spectral resource. Sometimes, however, other factors may prevail, as takes place in the specification of cdma2000, which recommends MC-DS-CDMA for arranging a downlink on the grounds of backward compatibility with cdmaOne. In this case the single-carrier format of cdmaOne (slightly modified) is just replicated on three appropriately spaced carriers.

10.2.2 Conventional MC transmission and OFDM

Let us digress for a while from spread spectrum and CDMA to get a better understanding of the reasons underlying the considerable interest in the MC technique in modern wireless telecommunication. Suppose one wants to transmit a source bit stream using some conventional (non-spread-spectrum) modulation mode (BPSK, QPSK etc.). With M-ary modulation and necessary data transmission rate R, the duration of the data symbol pulse is $T_p = (\log_2 M)/R$. Suppose that the channel coherence bandwidth B_c (see Section 3.6) is significantly narrower than the bandwidth of data symbols ($B_c \ll W \approx 1/T_p = R/\log_2 M$), or, putting it differently, the delay spread τ_{ds} exceeds symbol duration remarkably. Then under the 'direct' transmission (see Figure 10.3a),

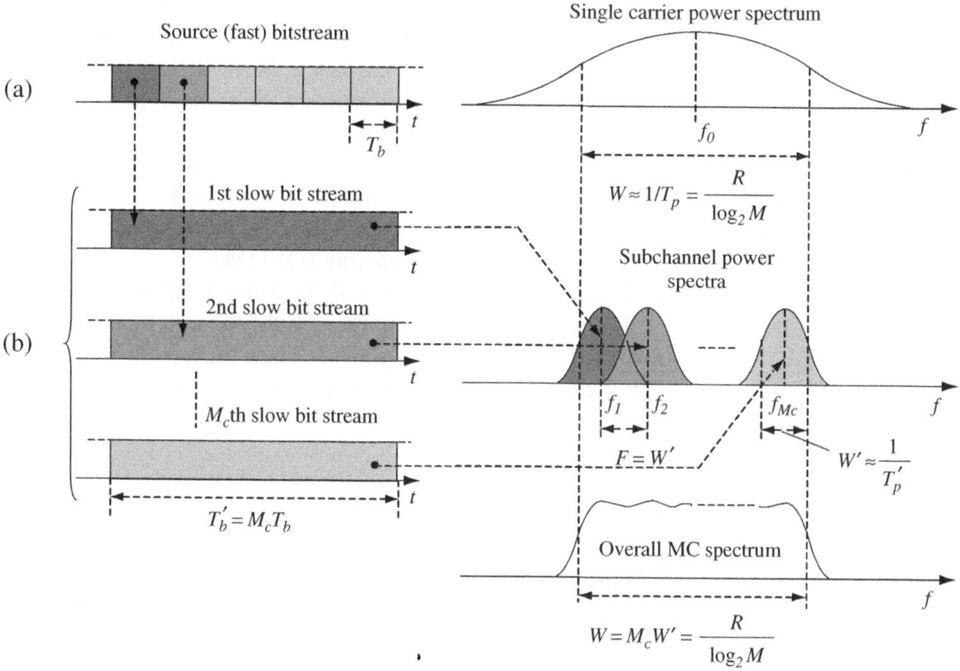

Figure 10.3 Single-carrier (a) and MC (b) data transmission

a deep ISI (see Section 3.5.4) will be present, distorting the data symbols following the current one. To counter it the receiver will have to involve a rather complex equalizer with a long memory, typically realized as an adaptive FIR filter, i.e. tapped delay-line with adjustable tap weights.

MC transmission offers an alternative solution (Figure 10.3b) avoiding the need for complex equalizing. Let us again demultiplex the 'fast' source bit stream of rate R to $M_c \geq W/B_c$ parallel 'slow' bit streams having rate R/M_c each. Certainly, the overall rate provided by all slow bit streams is equal to the original one, i.e. R. Now let us take M_c subcarriers $f_1, f_2, \ldots, f_{M_c}$ spaced uniformly with interval $F = W' = W/M_c$ and use each of them to transmit one of M_c slow bit streams in the same modulation mode as before. Every individual subcarrier forms a separate subchannel operating regardless of the others and transmitting a slow bit stream by longer pulses (symbols) of duration $T'_p = M_c T_p$, i.e. occupying M_c times narrower bandwidth $W' = W/M_c$ than before. This means that within a subchannel the fading is no longer frequency-selective, since $W' = W/M_c \leq B_c$. For a flat fading the delay spread does not go far beyond a single pulse, and ISI is less dramatic than it was initially and may be countered by comparatively simple equalizers. The total bandwidth occupied by the MC system is around $W \approx M_c/T'_p = 1/T_p = R/\log_2 M$, i.e. equalling that of the single-carrier transmission. In fact, spectral efficiency of the MC system appears to be even better, since the shape of its real spectrum is closer to a rectangle.

We do not use any frequency guard intervals now, and, moreover, allow subchannel spectra to overlap, since a traditional estimate of subchannel bandwidth $W' \approx 1/T'_p$ usually leaves remarkable out-of-band spectrum sidelobes. Nevertheless, the mutual interference between subchannels may be entirely suppressed. Suppose that the symbol pulse is rectangular and by agreement its bandwidth is measured as $W' = 1/T'_p$. Then frequency spacing between adjacent subcarriers $F = W' = 1/T'_p$ guarantees the orthogonality of subchannel signals, i.e. complete elimination of mutual interference between the MC subchannels. That is why this version of the MC modulation technique bears the name *orthogonal frequency division multiplexing* (OFDM).

Let the modulation symbols (complex amplitudes) of M_c subcarriers of OFDM be $b_i = A_i \exp(j\phi_i), i = 1, 2, \ldots, M_c$, where A_i and ϕ_i are real amplitude and phase, respectively, and symbols are transmitted by rectangular pulses. Then the resulting signal in complex notation, which is physically a complex envelope to be up-converted further to some central carrier f_0, is:

$$\dot{S}(t) = \sum_{i=1}^{M_c} b_i \exp(j2\pi f_i t) = \sum_{i=1}^{M_c} b_i \exp\left[\frac{j2\pi(i-1)t}{T'_p}\right] \tag{10.28}$$

where f_1 is set equal to zero, and therefore $f_i = (i-1)F = (i-1)/T'_p$. The latter assumption leads to no loss of generality, since the eventual value of the central frequency is afterwards set up by an up-conversion. Signal (10.28) is OFDM symbol, the number of different OFDM symbols (OFDM alphabet size M_{OFDM}) being determined by the size of the alphabet of modulation symbols M and the number of frequencies: $M_{\mathrm{OFDM}} = M^{M_c}$. For the example of BPSK OFDM $M_{\mathrm{OFDM}} = 2^{M_c}$.

Sampling (10.28) with interval $T_s = T'_p/M_c$ produces the sequence:

$$\dot{S}_l = \dot{S}(lT'_p/M_c) = \sum_{i=1}^{M_c} b_i \exp\left[\frac{j2\pi(i-1)l}{M_c}\right], \quad l = 0, 1, \ldots, M_c - 1 \tag{10.29}$$

replicating (except for an immaterial constant coefficient) the inverse discrete Fourier transform (IDFT) of the sequence of modulation symbols $\{b_i\}$. This uncovers the key reason for the popularity of OFDM: to implement this MC mode there is no need to modulate parallel generators of M_c subcarriers and sum the results. The same output effect is obtained with the aid of IDFT of modulation symbols. Thus, the typical structure of an OFDM transmitter (Figure 10.4) includes demultiplexer, and an IDFT unit outputting the IDFT vector (10.29), which is then converted from parallel to serial form of sequential samples and interpolated to produce a continuous OFDM symbol

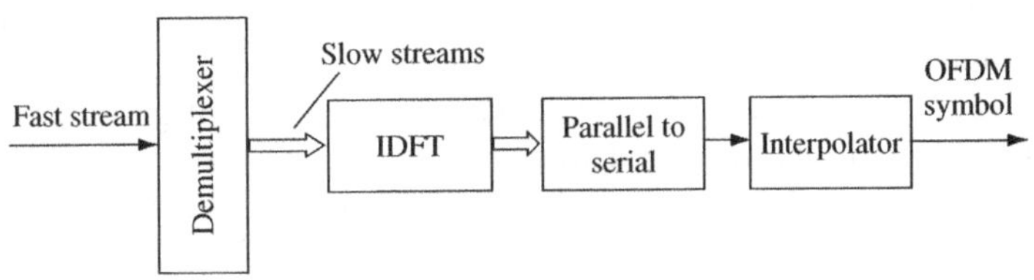

Figure 10.4 Generation of OFDM symbol

(10.28). This latter (after introducing a prefix; see below) is up-converted and transmitted on the desired central frequency f_0.

At the receiving end there is also no need to use M_c parallel receivers each tuned to its individual subcarrier, since one may extract $\{b_i\}$ from (10.29) by the direct discrete Fourier transform (DDFT):

$$\sum_{l=0}^{M_c-1} \dot{S}_l \exp\left[-\frac{j2\pi(i-1)l}{M_c}\right] = M_c b_i, \; i = 1, 2, \ldots, M_c \tag{10.30}$$

This shows that a DDFT unit is an appropriate device to split a received OFDM symbol into M_c subchannel effects necessary to retrieve transmitted data. Still, at the real channel output the receiver does not have at its disposal a 'pure' OFDM symbol. Instead it observes a complex envelope $\dot{Y}(t)$, containing an OFDM symbol distorted by noise and ISI. As follows from its principle, the MC technique, increasing symbol duration limits the depth of ISI propagation to the symbol following the current one. To exclude this residual ISI, too, one may insert a guard interval of duration $T_g \geq \tau_{ds}$ between adjacent MC symbols. This interval should not compulsorily be empty. Moreover, filling it with a cyclic prefix of the OFDM symbol remarkably facilitates neutralization of the channel multipath effects. Appending a cyclic prefix serves to convert a convolution of the transmitted signal with the channel pulse response into the cyclic one, corresponding to the product of DFT images. Denote $\nu = \lfloor \tau_{ds}/T_s \rfloor$ an integer number of sampling intervals in maximal channel delay and append v last samples of (10.29) as a prefix to the transmitted OFDM symbol. The receiver will drop the first ν samples of $\dot{S}_l, l = -\nu, -\nu + 1, \ldots, M_c - 1$, so that multipath replicas of the previous OFDM symbol will not affect M_c samples left. The delayed replicas of the current symbol itself will be influential, however, on the latter samples due to the channel multipath propagation. If the channel pulse response samples are $\dot{H}_0, \dot{H}_1, \ldots, \dot{H}_\nu$, then each observation sample $\dot{Y}_l$, $l = 0, 1, \ldots, M_c - 1$ (noise neglected) is found as the convolution:

$$\dot{Y}_l = \sum_{m=0}^{\nu} \dot{S}_{l-m} \dot{H}_m, l = 0, 1, \ldots, M_c - 1 \tag{10.31}$$

Due to the cyclic prefix the sequence $\dot{S}_{l-m}, l = 0, 1, \ldots, M_c - 1$ is a cyclic shift of the sequence $\dot{S}_l$ for any $m = 0, 1, \ldots, \nu$, therefore (10.31) is a cyclic convolution, and its DFT spectrum is a product of DFTs $\tilde{\dot{S}}_i$ and $\tilde{\dot{H}}_i$ of sequences $\dot{S}_l$ and $\dot{H}_l$. DFT of the former is just a scaled sequence of modulation symbols $\{b_i\}$ (see (10.30)), while

$$\tilde{\dot{H}}_i = \sum_{l=0}^{\nu} \dot{H}_l \exp\left(-\frac{j2\pi(i-1)l}{M_c}\right)$$

is the channel transfer function at the frequencies $f_i = f_1, f_2, \ldots, f_{M_c}$, so that:

$$\tilde{\dot{Y}}_i = M_c b_i \tilde{\dot{H}}_i, i = 1, 2, \ldots, M_c \tag{10.32}$$

This result shows that to remove the channel influence on the OFDM signal, i.e. perform equalizing, it is enough to simply divide each sample at the output of the receiver DDFT

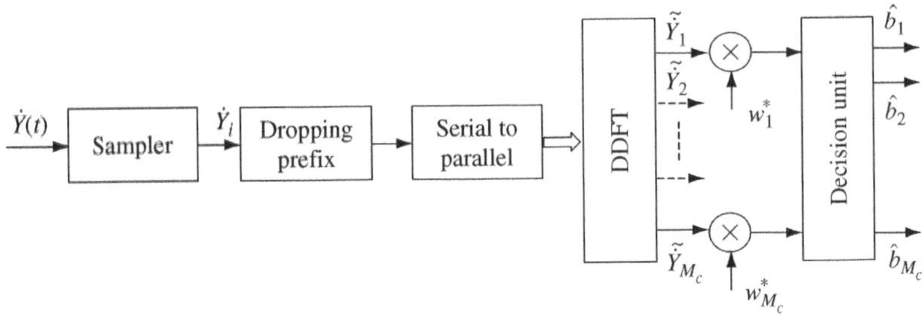

Figure 10.5 OFDM receiver structure

unit by the channel transfer function $\tilde{\dot{H}}_i$ at the corresponding frequencies.[1] To learn the current channel state, i.e. $\dot{H}_i$, some special training procedures are typically used. Note also that since the guard interval creates a sort of overhead reducing the data transmission rate, it is often reasonable to increase M_c, making the guard interval a small fraction of OFDM symbol duration.

When data are transmitted with PSK and subchannel signals are processed separately, i.e. the ith component of DDFT (10.32) is used independently of the others to demodulate the modulation symbol b_i, the equalization above may be simplified to just compensation of the channel phase shift, since subcarrier amplitude is redundant for a decision. If joined subchannel processing is necessary, however, e.g. in the case of MC-based CDMA considered below, the amplitudes of $\dot{H}_i$ are of serious importance, and the ultimate equalizing may appear preferable.

Summarizing, we may present the OFDM receiver structure in the form of Figure 10.5. The sampler provides samples $\dot{Y}_l$, from which the prefix ones are then discarded. The sample sequence is then transformed into a parallel form. The DDFT unit outputs DFT spectral components $\dot{Y}_i$, which are data symbols b_i distorted by noise and channel effects. Therefore they, after equalizing (just multiplying by an appropriate weight coefficients w_i^*), may serve to elaborate data symbol estimates $\hat{b}_i$ in the same manner as for BPSK, QPSK, QAM or other modulation mode.

10.2.3 Multicarrier CDMA

The MC modulation scheme is easily adaptable to the multiuser environment to provide code division multiplexing. Unlike DS CDMA, where an appropriate signature shaping in the time domain provides separation of user signals, in multicarrier CDMA (MC-CDMA) signatures are formed in the frequency domain, by controlling the amplitudes and phases of subcarriers in a user-specific manner. One way of explaining MC-CDMA is linking DS CDMA with MC transmission. Let us refer back to Figure 10.3 and imagine that instead of the source fast bit stream we have the kth user data symbol

[1] Such evening of the channel transfer function is known as zero-forcing equalizing and was mentioned earlier in Section 6.12.

stream spread by the kth user DS signature of length (spreading factor) N. By this we have N DS chips of duration Δ per kth user data symbol of duration T_p. Let us demultiplex this fast DS spread stream into $M_c = N$ slow streams, unfolding every data symbol into $M_c = N$ parallel chips of long duration $\Delta' = T_p = N\Delta$. Each of these parallel slow chip streams is further transmitted in an MC (OFDM) manner, so that each user has his specific law of modulating subcarriers.

Let us describe the same more directly. Let b_k be a current data symbol of the kth user. Let $\mathbf{a}_k = (a_{k,0}, a_{k,1}, \ldots, a_{k,N-1})$ be the kth user signature vector, now used in the frequency domain. To form the MC-CDMA signal N components of vector $b_k\mathbf{a}_k$ manipulate in parallel amplitudes and phases of $M_c = N$ subcarriers $f_1, f_2, \ldots, f_N$ during the pulse of duration T_p. Summation of all manipulated subcarriers produces the ith MC-CDMA symbol of duration T_p transmitted by the kth user. Certainly, with $F = 1/T_p$ implementing MC-CDMA is more feasible in a typical OFDM DFT-based form, but the direct way of generating the MC-CDMA signal shown in Figure 10.6 for the case of real (e.g. BPSK) alphabets of data symbols and signatures is more transparent as an illustration of the idea. Its generalization to complex alphabets is straightforward. In the OFDM implementation the IDFT unit replaces the multi-channel structure of Figure 10.6.

In synchronous non-oversaturated systems, like mobile radio downlink, any set of $K \leq N$ orthogonal signature vectors (Walsh functions etc.) might provide MAI-free separation of OFDM MC-CDMA user signals, since the orthogonality of the DFT spectra guarantees the orthogonality of OFDM symbols. Selection of signatures in asynchronous systems (e.g. mobile radio uplinks) is not that straightforward, although some minimax signature ensembles characteristic of asynchronous DS CDMA (see Section 7.5) may be of interest for MC-CDMA, too [105–107]. There is one more complication related to designing MC-CDMA signatures which is especially topical for mobile uplinks: the real envelope of the MC-CDMA signal, in contrast to that of DS CDMA, has significant variations, making the peak-factor perceptibly greater than

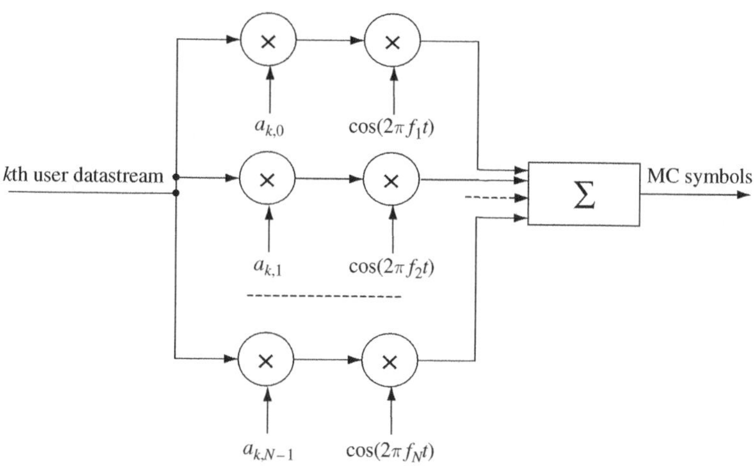

Figure 10.6 Generation of kth user MC-CDMA signal

one. This issue is to be taken into consideration and, all other factors being the same, of the many candidates the signature set providing the smallest peak-factor should be preferred.

Figure 10.7 presents a generic (non-DFT-based) structure of the kth user's MC-CDMA receiver. It consists of $N = M_c$ channels, each tuned to its own frequency and realized as a complex correlator processing the observation complex envelope $\dot{Y}(t)$. To counter flat fading within each frequency subchannel, the complex value $\dot{z}_{k,i}$ from the ith correlator output is weighted by a complex coefficient w_i^* and multiplied by the conjugated signature symbol $a_{k,i}^*$. The last operation is nothing but despreading in the frequency domain. Summation of such products over all subchannels produces the statistic $\dot{z}_k$ to be used in estimating the kth user's current data symbol b_k. Again, in a DFT realization the DDFT unit replaces the set of correlators.

Let us briefly touch upon the issue of choosing weight coefficients w_i^*, $i = 1, 2, \ldots, N$. In the scenario of MC-CDMA this task is somewhat more complicated than in conventional MC transmission, due to the necessity of controlling MAI level. Suppose that all user signals pass through the same channel, as is the case, e.g., for a mobile downlink. Since subcarrier spacing F is no smaller than the channel coherence bandwidth, values of the channel transfer function $\dot{\tilde{H}}_i$, $i = 1, 2, \ldots, N$ at the subcarrier frequencies may disperse independently in a wide range. Normalizing the channel transfer function as $N^{-1}\sum_{i=1}^{N}\left|\dot{\tilde{H}}_i\right|^2 = 1$ lets us characterize the kth signal undistorted intensity by amplitude A_k. Then the powers P_k, P_{nk}, P_l created by the useful signal, noise and the lth MAI, respectively, at the kth user output are calculated as:

$$P_k = A_k^2\left|\sum_{i=1}^{N}|a_{k,i-1}|^2\dot{\tilde{H}}_i w_i^*\right|^2, \quad P_{nk} = \sigma^2\sum_{i=1}^{N}|a_{k,i-1}|^2|w_i|^2, \quad P_l = A_l^2\left|\sum_{i=1}^{N}a_{l,i-1}a_{k,i-1}^*\dot{\tilde{H}}_i w_i^*\right|^2$$

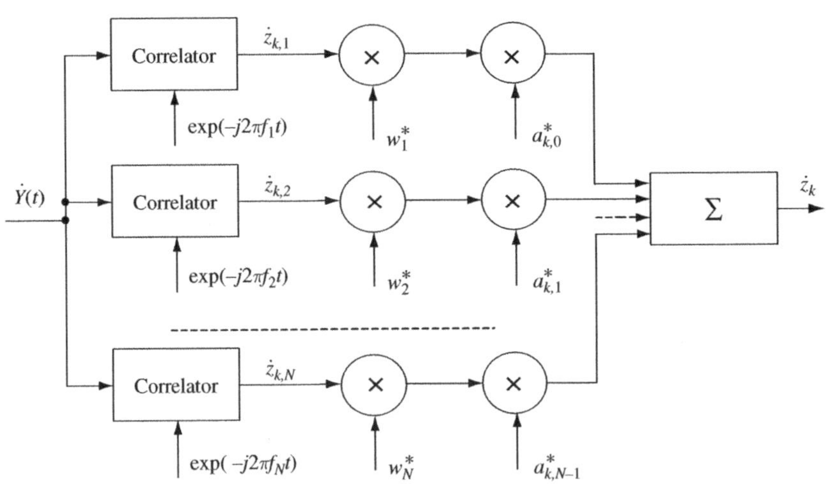

Figure 10.7 Generic scheme of MC-CDMA receiver

where σ^2 is noise subchannel power. It is seen now that even if signatures are originally orthonormal:

$$\sum_{i=0}^{N-1} a_{l,i}a_{k,i}^* = \delta_{kl}$$

the frequency selectivity over the subchannels may destroy orthogonality, amplifying some and suppressing other subcarriers. As a result MAI emerges so that not all P_l, $l \neq k$ are zeros. To preserve signature orthogonality at the receiving end independently of the channel current state one should select $w_i^* = 1/\dot{\tilde{H}}_i$, $i = 1, 2, \ldots, N$, i.e. realize zero-forcing equalizing entirely compensating for the channel effects. This, however, is a mismatched processing whenever the channel amplitude transfer function is non-uniform, so that the penalty for complete suppression of MAI is loss in SNR $q_{k,zf}^2$ corresponding to zero-forcing combining:

$$q_{k,zf}^2 = \frac{P_k}{P_{nk}} = \frac{A_k^2}{\sigma^2} \frac{\left|\sum_{i=1}^{N} |a_{k,i-1}|^2\right|^2}{\sum_{i=1}^{N} |a_{k,i-1}|^2 \left|\dot{\tilde{H}}_i\right|^{-2}} \leq q_{k,mf}^2 = \frac{A_k^2}{\sigma^2} \sum_{i=1}^{N} |a_{k,i-1}|^2 \left|\dot{\tilde{H}}_i\right|^2$$

where $q_{k,mf}^2$ is power SNR achieved with the matched processing, i.e. maximal ratio combining (Section 3.6.1) $w_i^* = \dot{\tilde{H}}_i^*$. For the case of originally equal intensities of all subcarrier components $|a_{k,i}| = 1/\sqrt{N}, i = 1, 2, \ldots, N$ SNR loss $\gamma = q_{k,mf}^2/q_{k,zf}^2$ of zero-forcing combining versus the maximal ratio one (cf. (6.42)):

$$\gamma = \frac{1}{N} \sum_{i=1}^{N} \left|\dot{\tilde{H}}_i\right|^{-2} \tag{10.33}$$

It is seen from (10.33) that when nonuniformity of the channel amplitude frequency distortion is remarkable (some $\left|\dot{\tilde{H}}_i\right|$ are close to zero), SNR loss, i.e. the penalty for radical elimination of MAI, may appear intolerable, and it is more reasonable to seek for a compromise between the levels of unsuppressed MAI and noise. One of the approaches of this sort leads to MMSE equalization, the idea of which is similar to that discussed in the previous section applied to multiuser detection. Details of this technique, as well as further insight into the spread spectrum MC philosophy, can be found in [105,106] and the numerous references listed there.

In conclusion, we again stress that no hard barrier exists between DS and MC CDMA. They are just parallel technical ways of getting the same result: the spread spectrum signature. The latter may always be synthesized either as a superposition of harmonics in the frequency domain (MC) or by direct shaping in the time domain (DS).

10.2.4 Applications

The penetration of the MC technique into digital telecommunication is presently very wide. Among examples of its practical application are the standards of digital audio and video broadcasting DAB, DVB-T etc. The positive experience accumulated to date promises a remarkable attractiveness of MC-based versions of CDMA. In particular, MC-CDMA is currently considered as one of the most plausible platforms for 4G air interfaces.

10.3 Transmit diversity and space–time coding in CDMA systems

10.3.1 Transmit diversity and the space–time coding problem

From the brief discussion of Section 3.6, we see that involving multiple receive and transmit antennas is a resourceful way of arranging the diversity branches necessary to oppose the destructive fading effects. Typically antenna arrays are employed for this purpose, consisting of elements spaced by several wavelengths to secure independence of their multipath patterns. The term *multiple input multiple output* (MIMO) serves to define the system jointly processing signals captured by several receive antennas from several transmitting antennas. Figure 10.8 gives a general description of a MIMO channel containing n_T receive and n_R transmit antennas. The ith transmit and jth receive antennas form a subchannel, whose current state is characterized by a complex fading coefficient $\dot{H}_{ij}$, $i = 1, 2, \ldots, n_T$, $j = 1, 2, \ldots, n_R$, which in the general case may be both time and frequency dependent. Usually, thanks to a special piloting, the receiver knows the channel state and is able to use coefficients $\dot{H}_{ij}$ for an efficient joint processing (i.e. combining) of subchannel signals. As for the transmitter, it may be aware of the channel state and capable of adjusting the signal to current propagation conditions only if a reliable data feedback 'receiver–transmitter' is organized (closed loop transmit diversity).

Suppose that there are only one transmit and n_R receive antennas, and hence n_R subchannels with fading coefficients $\dot{H}_1, \dot{H}_2, \ldots, \dot{H}_{n_R}$. Then the utilization of the potential of the receive diversity is in general terms just a proper combining of signals received by parallel antennas. It is not a big problem, at least in principle, since the receiver has signals of different antennas at separate outputs and (knowing the states of all diversity branches $\dot{H}_j$) may process them in the best possible manner. The situation changes critically with the involvement of the transmit diversity, too. When parallel transmit antennas operate simultaneously, the receiver runs into the problem of separating their signals, which are superimposed on each other in every receive antenna, to further utilize the knowledge of subchannel states $\dot{H}_{ij}$ and combine subchannel signals in

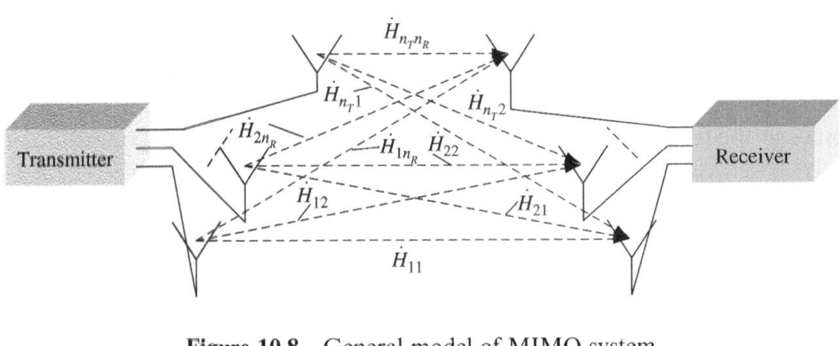

Figure 10.8 General model of MIMO system

an appropriate fashion. To provide a chance of such separation, the data transmission through n_T parallel transmit antennas should be arranged carefully, and ways of doing this constitute a subject of the problem called *space–time coding*. The name reflects the fact that the group of transmitted data bits is mapped one-to-one onto the two-dimensional $n_T \times n$ codeword $[u_t^i]$. The i, t entry u_t^i of the array is a code symbol transmitted by the ith antenna at the tth time moment, n being the code length. Note that in some cases the receive diversity may appear infeasible, e.g. in the mobile radio downlink, where the small dimensions of a handset does not give enough space for several receive antennas. In these scenarios the transmit diversity and, hence, an adequate space–time coding become especially valuable. In what follows we assume that only a single receive antenna is used to concentrate only on investigating the efficiency of the transmit diversity. This allows us to simplify designations of the channel fading coefficients, retaining only a single subscript pointing at the transmit antenna: $\dot{H}_{i1} = \dot{H}_i$.

10.3.2 Efficiency of transmit diversity

A lot of research has been undertaken to evaluate the Shannon capacity, i.e. the potential rate of error-free data transmission, of MIMO channels, and the profitable role of antenna multiplicity has been proved for the basic fading models [108–110]. There is no wonder in the benefits of receive diversity, since extra receive antennas utilize signal energies from extra space points which would be lost irrevocably with a single antenna. With the maximal ratio combining of n_d identical receive diversity branches, average power SNR grows n_d times (see Section 3.6.1), and although this factor is not central in the improvement of error probability and channel capacity, it still makes this improvement readily predictable. Unlike this, the nature of gaining capacity or reducing error probability through transmit diversity is not that obvious, considering the division of the limited power resource between multiple transmit antennas. In fact, no gain in average SNR takes place under the maximal ratio combining of identical diversity branches with fixed overall power, unless the transmitter knows the channel state and is able to coordinate transmitting through different branches so that subchannel signals are summed coherently in the receive antenna. Indeed, let channel state information be unavailable to the transmitter and the total power P be equally divided between n_d identical diversity branches (antennas, frequency channels etc.). Then the average power SNR per branch is q^2/n_d, where q^2 is the average power SNR, which would exist at the receiver with no diversity. Clearly, the maximal ratio combining would only increase average power SNR per branch n_d times, making it equal to the one with no diversity.

Accordingly, there are two contradictory trends in the transmit (as well as frequency etc.) diversity. On the one hand, increasing the number of branches, given the total power, provides a greater number of independent subchannels, which, supporting each other, secure higher probability that at least some of them are not poor. On the other hand, conditions (branch SNR) in each of the diversity subchannels become poorer with growth of n_d. An accurate theoretical analysis shows that the first of these factors overweighs the second. We put aside mathematical derivations concerning the channel capacity, which may be found in the literature (e.g. [108–110]), but Problems 10.11 and 10.12 contain the plainest examples illustrating the issue. As for the positive effect of transmit diversity on the error probability, it becomes obvious from the following consideration of BPSK data transmission over the Rayleigh fading channel.

Let $A_i = |\dot{H}_i|$ be an amplitude-fading coefficient of the ith diversity branch with the average square normalized as $\overline{A_i^2} = 1$. Then the maximal-ratio combined current receiver SNR (see (3.15)) $q_r^2 = \sum_{i=1}^{n_d} A_i^2 (q^2/n_d)$, and, according to (2.19), conditional bit error probability $P_e(A_1, A_2, \ldots, A_{n_d})$, with the subchannel states A_i, $i = 1, 2, \ldots, n_d$, fixed, is:

$$P_e(A_1, A_2, \ldots, A_{n_d}) = Q(q_r) \tag{10.34}$$

To come to the unconditional bit error probability P_e we have to average (10.34) in all subchannel amplitudes A_i, $i = 1, 2, \ldots, n_d$ using their joint PDF $W(A_1, A_2, \ldots, A_{n_d})$. Due to the independence of branches this PDF is just a product of n_d one-dimensional PDFs of all amplitudes, and:

$$P_e = \int_0^\infty \int_0^\infty \ldots \int_0^\infty Q(q_r) \left[\prod_{i=1}^{n_d} W(A_i) \right] \mathrm{d}A_1 \, \mathrm{d}A_2 \ldots \mathrm{d}A_{n_d} \tag{10.35}$$

To make the integration variables of (10.35) separable, let us approximate the complementary error function $Q(x)$ by its upper bound (see Problem 10.13) $Q(x) \le (1/2)\exp(-x^2/2), x \ge 0$, coming to:

$$P_e \le \frac{1}{2} \prod_{i=1}^{n_d} \int_0^\infty \exp\left(-\frac{A_i^2 q^2}{2n_d} \right) W(A_i) \, \mathrm{d}A_i$$

For the channel with Rayleigh fluctuations PDF $W(A_i)$ obeys the law (3.12), which leads to:

$$P_e \le \frac{1}{2} \prod_{i=1}^{n_d} \int_0^\infty 2A_i \exp\left[-A_i^2 \left(1 + \frac{q^2}{2n_d} \right) \right] \mathrm{d}A_i = \frac{1}{2} \left(\frac{2n_d}{q^2 + 2n_d} \right)^{n_d} \tag{10.36}$$

Figure 10.9 demonstrates the behaviour of the bit error probability depending on the total average SNR q for 1, 2, 3, 4, 5 and 6 diversity branches. It is of great importance that two diversity branches provide a significant energy gain, error probability preassigned. For example, if a tolerable bit error probability is no greater than 10^{-4}, two diversity branches cut down the necessary transmitted energy by more than 15 dB. At the same time, with further addition of diversity branches the energy gain grows at a dropping rate, and, say, transition from 5 to 6 branches promises a saving of only around 1 dB of emitted energy. This explains why in many practical systems (e.g. mobile radio downlinks) two transmit antennas are chosen as a good balance between the diversity gain and equipment complexity.

Note that when the number of branches tends to infinity the right-hand side of (10.36) turns into $(1/2)\exp(-q^2/2)$ (see Problem 10.14), i.e. an upper bound (dashed line in Figure 10.9) of the bit error probability for the case of a non-fading Gaussian channel. In other words, by increasing the number of transmit diversity branches one can in the limit (at least theoretically) completely eliminate the harmful effect of multipath propagation.

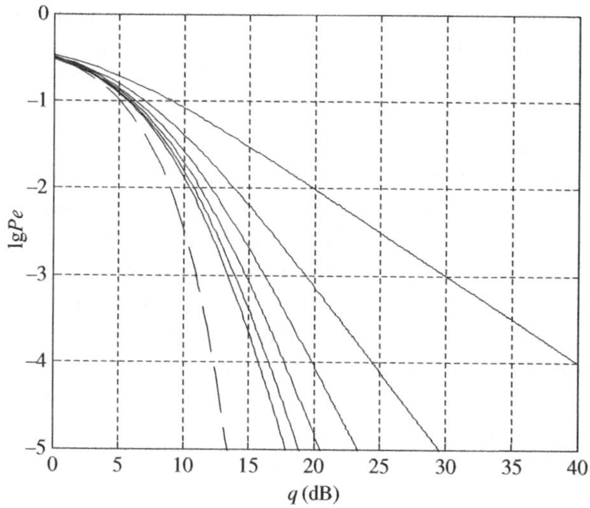

Figure 10.9 Bit error probability versus overall SNR in the transmit diversity scheme for 1, 2, 3, 4, 5 and 6 diversity branches

10.3.3 Time-switched space–time code

There may be different approaches to designing space–time codes depending on the fading model. Fast fading (see Section 3.5) implies such rapid fluctuations of the multipath pattern that values of the fading coefficient of the same subchannel at two adjacent symbol intervals are independent. In what follows we are dealing with the opposite case of slow fading, assuming that subchannel fading coefficients remain constant over all codeword duration.

To accentuate the non-trivial character of the problem of designing codes securing separability of the signals of different transmit antennas at the receiving side, let us start with the simplest example.

Example 10.3.1. Let the fixed power resource P be divided equally between $n_T = 2$ transmit antennas sending simultaneously the same data symbol with no measures allowing separation of subchannel signals at the receiver (repetition space–time code). Let a single receive antenna be used and the fading coefficients $\dot{H}_1, \dot{H}_2$ of two subchannels be independent Gaussian complex numbers with zero means and equal variances. This is exactly the case of Rayleigh fading, since magnitudes $A_i = \left|\dot{H}_i\right|$, $i = 1, 2$ are subject to the Rayleigh PDF. With a sent complex envelope $\dot{S}(t)$, the received one is $(\dot{H}_1 + \dot{H}_2)\dot{S}(t)/\sqrt{2}$, where the square root of two is responsible for power splitting. It is now evident that two subchannels form the resultant channel with an overall fading coefficient $\dot{H} = (\dot{H}_1 + \dot{H}_2)/\sqrt{2}$, which is again Gaussian with zero mean, and the same variance as any of $\dot{H}_i$. Therefore, magnitude $A = \left|\dot{H}\right|$ is Rayleigh and the resultant channel is again a Rayleigh fading channel. If the mean squares of A_i are normalized to one, $\overline{A^2} = 1$, too. Therefore, the resultant channel is absolutely identical to each of the subchannels, and using two antennas in this case cannot give any benefit as compared to a single antenna. In other words, the repetition code is a degenerated one, providing no real diversity.

As the example shows, the number of effective diversity branches may appear smaller than the number of transmit antennas. One of the most important parameters of any space–time code is the *diversity gain*, i.e. the number of really contributing diversity branches secured by the coding scheme (see Problem 10.15).

The obvious way to provide separation of signals of different transmit antennas in the receiver is to emit the same data symbol by n_T antennas by turns, i.e. with no time overlapping. In other words, one and only one of the transmit antennas emits the signal at each time, employing the total power resource. Then the situation is identical to that of multiuser TDMA communications, i.e. the signal of each antenna is identified by its time position and is orthogonal to others due to non-overlapping in the time domain. Thereby, the receiver observes all the subchannel signals not in the mixture but following successively in time with no mutual interference. Then, knowing the current fading coefficients $\dot{H}_i$, the receiver is entirely certain of which of these signals are more or less reliable, and able to combine them in any appropriate way, e.g. maximize SNR by using maximal ratio weighting. This simplest coding scheme corresponds to the *time-switched space–time code*.

Example 10.3.2. Consider again the case of two antennas ($n_T = 2$) repeating transmission of the same current data symbol, this time operating in an intermittent manner: when the first emits energy, the second is inactive, and vice versa. The receiver observes the signals, one after another, passing through the subchannels with fading coefficients $\dot{H}_1$ and $\dot{H}_2$, both distorted by additive noise of power σ^2. To realize the maximal ratio combining, these observations are summed with weights $\dot{H}_i^*$, $i = 1, 2$, respectively, resulting in power SNR (see (3.15)) $q_r^2 = (A_1^2 + A_2^2)q_s^2/2$, where q_s^2 is an overall 'non-fading' power SNR per one transmitted data symbol and halving arises due to splitting the fixed symbol energy between two antennas. For identical subchannels with magnitudes normalized to one ($\overline{A_i^2} = 1$) average power SNR is, of course, again the same as in the case of transmitting the whole symbol energy through a single subchannel: $\overline{q_r^2} = 2q_s^2/2 = q_s^2$. However, if the combiner is treated as the output of the resultant channel, the latter is no longer Rayleigh and has error probability smaller than a Rayleigh one under the same average SNR (see Figure 10.9). Thus, two diversity branches really exist and provide a predicted gain, which is achieved in return for two times smaller transmission rate per bandwidth unit. Indeed, the transmission rate in bit/s fixed, each antenna now transmits every data symbol over duration $T_p/2$, i.e. occupying a doubled bandwidth. One more instructive comparison is with the two-branch receive diversity system ($n_T = 1$, $n_R = 2$). It is easy to see that if a single transmit antenna uses the full interval T_p to transmit a current symbol, the receiver maximal ratio combiner provides average power SNR $\overline{q_r^2} = 2q_s^2$, i.e. 3 dB higher versus that in the transmit diversity scheme, energies per symbol equal. The roots of the energy loss of the transmit diversity against the receive diversity have been repeatedly pointed out: splitting total fixed energy between antennas in the transmit diversity scheme.

Time-switched space–time codes are very simple but they realize maximal diversity gain n_T in exchange for widening bandwidth (rate fixed) and discontinuity of transmission, i.e. increasing signal peak-factor. This strongly motivates the search for space–time codes allowing separation of signals from different antennas despite their overlapping in time. The simplest, but very important, example of such codes is introduced in the next subsection.

10.3.4 Alamouti space–time code

The coding scheme proposed in [111] exploits two transmit antennas, operates with no extra bandwidth and offers maximal possible diversity gain for two antennas $n_d = n_T = 2$. Let b_0 and b_1 be two successive data symbols standing for even and odd time positions, respectively, and belonging to some fixed modulation alphabet (PSK, QAM etc.). Codewords of the Alamouti space–time code are 2×2 arrays of the form:

$$\mathbf{u} = \begin{bmatrix} b_0 & -b_1^* \\ b_1 & b_0^* \end{bmatrix} = \begin{bmatrix} \mathbf{u}_1 \\ \mathbf{u}_2 \end{bmatrix} \tag{10.37}$$

meaning that code length $n = 2$. As is seen, at the even symbol interval two antennas simultaneously transmit code symbols $u_0^1 = b_0$ (first antenna) and $u_0^2 = b_1$ (second antenna), while at the odd interval the transmitted symbols are $u_1^1 = -b_1^*$ (first antenna) and $u_1^2 = b_0^*$ (second antenna). To put it another way, the antennas simultaneously transmit the length 2 sequences $\mathbf{u}_1 = (u_0^1, u_1^1) = (b_0, -b_1^*)$ (first antenna) and $\mathbf{u}_2 = (u_0^2, u_1^2) = (b_1, b_0^*)$ (second antenna). This arrangement makes the sequences transmitted by the two antennas, i.e. vectors $\mathbf{u}_1$ and $\mathbf{u}_2$, orthogonal: $(\mathbf{u}_1, \mathbf{u}_2) = b_0 b_1^* - b_1^* b_0 = 0$, securing separability of superimposed signals of different subchannels in the receiver. Actually, however, there is no need to fulfil the separation of subchannels as a special procedure, since the optimal (ML) detection of data symbols b_0 and b_1 automatically includes it, as well as maximal ratio combining. For a clear reason we assume that a single code symbol transmitted currently by one antenna utilizes on average half of the total average symbol energy E_s. Let $\dot{\mathbf{Y}} = (\dot{Y}_0, \dot{Y}_1)$ be an observation vector whose components $\dot{Y}_t$, $t = 0, 1$ are samples of the complex envelope at the symbol matched filter output for even and odd positions, respectively, normalized for convenience by the divisor $E_s/\sqrt{2}$. Then:

$$\dot{\mathbf{Y}} = \dot{H}_1 \mathbf{u}_1 + \dot{H}_2 \mathbf{u}_2 + \mathbf{n} \tag{10.38}$$

where $\mathbf{n}$ is a two-dimensional vector of independent complex Gaussian noise samples with zero means and equal variances. Then the ML rule (see Chapter 2) gives out $\hat{b}_0$ and $\hat{b}_1$ as estimations of data symbols b_0 and b_1 if they minimize the Euclidean (squared) distance between the observation $\dot{\mathbf{Y}}$ and the useful component $\dot{H}_1 \mathbf{u}_1 + \dot{H}_2 \mathbf{u}_2$:

$$d^2(\dot{H}_1 \mathbf{u}_1 + \dot{H}_2 \mathbf{u}_2, \dot{\mathbf{Y}}) = \left\| \dot{\mathbf{Y}} - \dot{H}_1 \mathbf{u}_1 - \dot{H}_2 \mathbf{u}_2 \right\|^2 = (\dot{\mathbf{Y}} - \dot{H}_1 \mathbf{u}_1 - \dot{H}_2 \mathbf{u}_2, \dot{\mathbf{Y}} - \dot{H}_1 \mathbf{u}_1 - \dot{H}_2 \mathbf{u}_2)$$

Distributivity and symmetry ($(\mathbf{u}, \mathbf{v}) = (\mathbf{v}, \mathbf{u})^*$) axioms of the inner product along with orthogonality of $\mathbf{u}_1, \mathbf{u}_2$ allow getting:

$$d^2 = \left\| \dot{\mathbf{Y}} \right\|^2 - 2\mathrm{Re}\left[\dot{H}_1^* (\dot{\mathbf{Y}}, \mathbf{u}_1)\right] - 2\mathrm{Re}\left[\dot{H}_2^* (\dot{\mathbf{Y}}, \mathbf{u}_2)\right] + \left|\dot{H}_1\right|^2 \|\mathbf{u}_1\|^2 + \left|\dot{H}_2\right|^2 \|\mathbf{u}_2\|^2$$

where d^2 is a shortened designation for the squared distance in question, or after substituting $\mathbf{u}_1, \mathbf{u}_2$ from (10.37):

$$\begin{aligned} d^2 = &\left\| \dot{\mathbf{Y}} \right\|^2 - 2\mathrm{Re}\left[b_0^* (\dot{H}_1^* \dot{Y}_0 + \dot{H}_2 \dot{Y}_1^*)\right] - 2\mathrm{Re}\left[b_1^* (\dot{H}_2^* \dot{Y}_0 - \dot{H}_1 \dot{Y}_1^*)\right] \\ &+ \left(\left|\dot{H}_1\right|^2 + \left|\dot{H}_2\right|^2\right)\left(|b_0|^2 + |b_1|^2\right) \end{aligned}$$

The transformed observation samples:

$$\dot{z}_0 = \dot{H}_1^*\dot{Y}_0 + \dot{H}_2\dot{Y}_1^*, \dot{z}_1 = \dot{H}_2^*\dot{Y}_0 - \dot{H}_1\dot{Y}_1^* \tag{10.39}$$

as well as the norm of the observation vector do not depend on variables b_0, b_1, with respect to which d^2 has to be minimized. Therefore, in the equation above we are allowed to replace $\|\dot{\mathbf{Y}}\|^2$ with $|\dot{z}_0|^2+|\dot{z}_1|^2$, coming to:

$$d^2 = |\dot{z}_0|^2 - 2\mathrm{Re}(b_0^*\dot{z}_0) + |b_0|^2 + \left(|\dot{H}_1|^2+|\dot{H}_2|^2-1\right)|b_0|^2$$
$$+|\dot{z}_1|^2 - 2\mathrm{Re}(b_1^*\dot{z}_1) + |b_1|^2 + \left(|\dot{H}_1|^2+|\dot{H}_2|^2-1\right)|b_1|^2$$

It is evident now that minimizing d^2 in b_0, b_1 breaks into a separate minimization of two functions of one variable $d^2(b_0) = |\dot{z}_0 - b_0|^2+H^2|b_0|^2$ and $d^2(b_1) = |\dot{z}_1 - b_1|^2+H^2|b_1|^2$, where $H^2 = |\dot{H}_1|^2+|\dot{H}_2|^2-1$, with respect to b_0 and b_1. Thus, the estimates of the data symbols b_0, b_1 are found as:

$$\hat{b}_l = \arg\max_{b_l}\left(|\dot{z}_l - b_l|^2+H^2|b_l|^2\right), \ l = 0, 1 \tag{10.40}$$

where $\dot{z}_l$ is defined by (10.39) and minimization is done over all values of b_l within the given data symbol alphabet. In the particular case of PSK data modulation $|b_l|^2 = 1$ and the only component of the first term of (10.40) dependent on b_l is $\mathrm{Re}(b_l^*\dot{z}_l)$, which turns the decision rule into the ordinary form of PSK demodulation (see Section 7.1.2), but based on the modified matched filter statistics $\dot{z}_l$:

$$\hat{b}_l = \arg\max_{b_l}[\mathrm{Re}(b_l^*\dot{z}_l)], \ l = 0, 1 \tag{10.41}$$

Let us fix transmitted symbols b_l, $l = 0, 1$ and fading coefficients $\dot{H}_i$, $i = 1, 2$. Then useful components of the decision statistics $\dot{z}_l$, $l = 0, 1$ can be evaluated by averaging $\dot{Y}_0$ and $\dot{Y}_1$ in (10.39) with respect to an additive noise. Denoting this operation by $E_n\{\cdot\}$, we obtain (see (10.38)) $E_n\{\dot{Y}_0\} = \dot{H}_1 b_0 + \dot{H}_2 b_1$, $E_n\{\dot{Y}_1\} = -\dot{H}_1 b_1^* + \dot{H}_2 b_0^*$, so that:

$$E_n\{\dot{z}_0\} = (A_1^2 + A_2^2)b_0, \ E_n\{\dot{z}_1\} = (A_1^2 + A_2^2)b_1$$

where, as before, $A_i = |\dot{H}_i|$, $i = 1, 2$. It may be seen now that the useful component of z_l is formed as though no separation problem existed and each symbol were transmitted over two independent diversity branches, further maximal-ratio combined (see Section 3.6.1). In the same way, the variance σ_z^2 of a real or imaginary part of the additive noise entering $\dot{z}_l$ is $\sigma_z^2 = (A_1^2 + A_2^2)\sigma^2$ where σ^2 is the variance of the real part of a noise sample in (10.38). Then power SNR q_{zl}^2 for each of the statistics $\dot{z}_l$, $l = 0, 1$ is:

$$q_{zl}^2 = \frac{|E_n\{\dot{z}_l\}|^2}{\sigma_z^2} = \frac{(A_1^2 + A_2^2)|b_l|^2}{\sigma^2}$$

Now take into consideration the randomness of A_i, $i = 1, 2$, b_l, $l = 0, 1$, and average q_{zl}^2 with respect to all random factors to come to the mean SNR $\overline{q_{zl}^2}$ for each of the statistics $\dot{z}_l$, $l = 0, 1$. Under a natural normalization of fading coefficients and modulation alphabet $\overline{A_i^2} = 1$, $i = 1, 2$, $\overline{|b_l|^2} = 1$, $l = 0, 1$, $\overline{q_{zl}^2} = 2/\sigma^2$, and since $\mathbf{n}$ in (10.38) was normalized by $E_s/\sqrt{2}$, and noise variance at the symbol matched filter output is $N_0E_s/2$ (see (2.15)), $\sigma^2 = N_0/E_s$. The resulting equation is then $\overline{q_{zl}^2} = 2E_s/N_0 = q_s^2$, showing that the Alamouti scheme, as well as the time-switched code, preserves the same average symbol SNR as the no-diversity scheme, providing diversity gain $n_d = n_T = 2$. We stress again that the advantage of the Alamouti code against the time-switched one is the absence of pauses in emission, entailing better peak-factor and higher spectral efficiency.

The Alamouti code is a full-rate one, meaning that two independent modulation symbols are transmitted over two-symbol duration. In general, a space–time code of length n, which allows transmitting k independent data modulation symbols, has rate $R = k/n$. Full-rate codes are preferable, since they involve no extra bandwidth compared to a single-antenna transmission. The existence of more full-rate codes securing maximal diversity gain $n_d = n_T$ strongly depends on the modulation alphabet. For real modulation alphabets (e.g. BPSK) full-rate codes exist for several values of the number n_T of transmit antennas, while for complex modulation symbols (QPSK, QAM etc.) the Alamouti code is unique[2] [112]. At the same time, several interesting constructions of space–time block codes become available for both complex and real alphabets if the full-rate restriction is removed [110,112] (see also Problems 10.16 and 10.17). We refer the reader wishing to gain more information on this issue and become familiar with other aspects of space–time coding to works [110–113] and the papers cited in [110].

10.3.5 Transmit diversity in spread spectrum applications

Seemingly the spread spectrum concept offers a very direct and easy way of providing transmit diversity. Indeed, the main bottleneck of the transmit diversity is separation of signals emitted simultaneously by different transmit antennas at the receiver side. One may get round this stumbling block by spreading the signals of different antennas using different (orthogonal) spreading codes. This at first sight obvious tool of the transmit diversity is, however, far from universal. As a matter of fact, orthogonal sequences in CDMA systems are a deficit resource, since their number determines the potential number of users. Thus, in a saturated ($K = N$) or, more so, oversaturated ($K > N$) CDMA downlink there are no spare orthogonal sequences for arranging transmit diversity, which makes the space–time bandwidth saving codes equally valuable in CDMA applications, too.

The way of incorporating the Alamouti code into the DS CDMA downlink is straightforward and does not require an extra signature resource. Let $\dot{S}_k(t)$ be the complex envelope of the kth user signature (treated as a signal of the same duration T_p as the data symbol) and $b_{k,0}, b_{k,1}$ be even and odd data modulation symbols sent to the kth user. Then it is enough only to use in the array (10.37) the DS spread symbols $b_{k,0}\dot{S}_k(t)$ and $b_{k,1}\dot{S}_k(t)$ in place of b_0, b_1, respectively, to arrange the transmit diversity on the basis of a fixed signature $\dot{S}_k(t)$.

---

[2] We do not consider as different trivial modifications of (10.37) preserving row orthogonality and row norms, like a common conjugation or/and multiplication of rows by -1 as well as by any fixed complex number of magnitude one.

The first and second antennas then transmit the signals $b_{k,0}\dot{S}_k(t) - b^*_{k,1}\dot{S}^*_k(t - T_p)$ and $b_{k,1}\dot{S}_k(t) + b^*_{k,0}\dot{S}^*_k(t - T_p)$, respectively, over two consecutive symbol intervals. Despreading these signals by the reference $\dot{S}^*_k(t) + \dot{S}_k(t - T_p)$ removes the signature and turns the entire problem into the one discussed in the previous subsection. This principle, with a slight modification, is used in the UMTS downlink for arranging an open loop (without user-BS feedback) transmit diversity [114,115]. To be fair, processing synchronization signals (time-delay measurement) encoded by Alamouti code in a user's terminal would appear much more complicated as compared to data demodulation. For this reason the transmit diversity mode employed in the synchronization channel is the time-switched coding touched upon in Section 10.3.3 [115]. One more interesting detail is using closed-loop diversity in the dedicated (i.e. assigned to a specific user) physical channels. Based on the feedback MS-BS data, the BS knows the current state of the channel linking the BS with the specific user and adjusts the phases of the signals of two transmit antennas to make them sum coherently at the terminal input. The amplitudes of the transmitted signals may also be adjusted to realize the maximal ratio combining in the receive antenna and bring the efficiency of transmit diversity nearer to that of the receive one. The cdma2000 downlink specification includes some similar solutions concerning transmit diversity.

Problems

10.1. Let the first signature $\mathbf{a}_1$ be a linear combination of the other signatures. Prove that the linear system (10.12) has no solution, i.e. suppression of MAI automatically removes the useful effect too.

10.2. There is a three-user synchronous DS CDMA system. The signatures are binary of spreading factor $N = 3$: $\mathbf{a}_1 = (+ - -)$, $\mathbf{a}_2 = (- + -)$, $\mathbf{a}_3 = (- - +)$. Find the reference of the first user's decorrelating receiver, demonstrate elimination of MAI, and evaluate SNR loss of the decorrelating algorithm to the matched filtering.

10.3. Derive the gradient of the function (10.19) and prove that $\mathbf{u}$ given by (10.20) is the point of minimum of this function.

10.4. Prove the matrix inverse lemma in the form (10.21).

10.5. Find the reference vector for the MMSE detector for the conditions of Problem 10.2, setting all signature amplitudes and noise variance equal to one. Calculate SINR at the MMSE detector output, compare it with those of the decorrelating detector and matched filtering (Problem 10.2), and explain the results.

10.6. A synchronous CDMA system accommodates 128 users within the spreading factor $N = 96$. The signatures are columns of the 128th order Hadamard matrix, in which 32 rows are discarded. Find MMSE reference vectors for all users, if their signals have equal intensities.

10.7. The MC-DS-CDMA downlink is realized using three subcarriers. Data is transmitted at each subcarrier by BPSK at the rate 32 kbps with spreading factor 64. The guard frequency interval equals $0.5/\Delta$, where Δ is signature chip duration. How would the potential number of users change if DS-CDMA replaced MC-DS-CDMA?

10.8. Data should be transmitted using QPSK at the rate 2.88 Mbps over a channel whose coherence bandwidth $B_c = 50$ kHz. Find the minimal number of subcarriers necessary for MC transmission. What is the minimum length of DFT in the OFDM scheme, if the overhead due to the guard intervals should not exceed 10%?

10.9. A synchronous MC-CDMA downlink in the OFDM version transmits data using QPSK at the rate 40 kbps over the channel with delay spread $\tau_{\max} = 10\,\mu s$. How many users can it serve if all undistorted signatures are orthogonal and the overall bandwidth is 5 MHz?

10.10. Would it be reasonable to use zero-forcing combining in MC-CDMA operating on the Rayleigh subchannels?

10.11. Suppose that n_R antennas receive in parallel the signal transmitted by a single antenna, intensities of all received signals are the same, as well as of independent Gaussian noises corrupting the signals. How does the Shannon capacity of such a channel differ from that corresponding to a no-diversity case, if the receiver knows the path length differences of all signals?

10.12. Suppose that a transmitter is capable of transmitting data involving n_d independent identical diversity branches, total transmitted power being fixed. Suppose that the intensities of all received signals are the same, as well as the independent Gaussian noises corrupting the signals, and the receiver (but not the transmitter!) knows the path length differences of all signals. What is better from the angle of Shannon capacity: to transmit the same or different datastreams over n_d branches?

10.13. Prove the upper bound on the complementary error function: $Q(x) \leq (1/2)\exp(-x^2/2)$ for any $x \geq 0$.

10.14. Prove convergence of the right-hand side of (10.36) to the upper bound of the error probability for a single non-fading branch: $(1/2)\exp(-q^2/2)$, when the number of branches grows without limit.

10.15. According to the strict definition [110,112,113] the diversity gain is the minimal rank among all pairwise differences of distinct space–time codewords, i.e. $n_T \times n$ arrays $[u_t^i]$. Prove that for the Alamouti code this diversity gain equals 2.

10.16. Find the rate (in data symbols per code symbol) and diversity gain of the space–time code with real symbol codewords [112]:

$$\mathbf{u} = \begin{bmatrix} b_0 & -b_1 & -b_2 & -b_3 \\ b_1 & b_0 & b_3 & -b_2 \\ b_2 & -b_3 & b_0 & b_1 \end{bmatrix}$$

10.17. Find the rate (in data symbols per code symbol) and diversity gain of the space–time code with complex symbol codewords [116]:

$$\mathbf{u} = \begin{bmatrix} b_0 & b_1^* & b_2^* & 0 \\ -b_1 & b_0^* & 0 & -b_2^* \\ -b_2 & 0 & b_0^* & b_1^* \end{bmatrix}$$

Matlab-based problems

10.18. Write and run a program simulating conventional, decorrelating and MMSE detectors for arbitrary synchronous signature set and random user signal intensities. Recommended steps:

(a) Form the $N \times K$ matrix of K normalized signatures of length N.
(b) Take the first user's amplitude equal to one and all the rest random, obeying the Rayleigh law with unit square mean.
(c) Add Gaussian noise to the amplitude-scaled signatures with variance corresponding to a pre-assigned bit SNR q_b.
(d) Find the references and calculate SINR for all three types of the first user detector (the decorrelating one does not exist for linearly dependent signatures).
(e) Varying the noise level, signature intensities remaining fixed, build the SINR curves in dependence on bit SNR q_b.
(f) Study the cases of the following signature sets: orthogonal sequences ($K \leq N$), cyclically shifted m-sequences ($K \leq N$) and Welch-bound sequences ($K > N$), and comment on the results.

10.19. Write and run a program illustrating the principle of OFDM modulation and demodulation (see Figure 10.10). Recommended steps:

(a) Set the number of DFT points (frequencies) M_c.
(b) Form and plot a random pattern of M_c source bits.

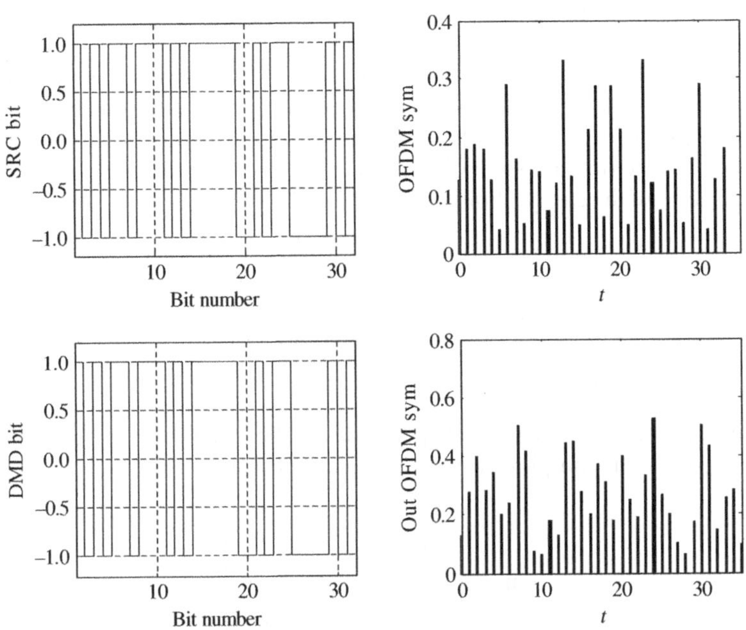

Figure 10.10 Simulating OFDM for $M_c = 32$

(c) Calculate the IDFT of the bit pattern.
(d) Attach a cyclic prefix and plot the OFDM symbol obtained.
(e) Set a random channel delay profile, i.e. integer delays, amplitudes and phases of multiple paths; take the delay spread within 4–6, Rayleigh amplitudes and uniformly distributed over $[-\pi, \pi]$ phases, all independent of each other.
(f) Calculate and plot the OFDM symbol distorted by the channel.
(g) Discard the prefix and tail (due to a channel delay) samples, and calculate the DFT of the vector obtained.
(h) Calculate the channel transfer function and divide the DFT by it.
(i) Demodulate the samples obtained into bits.
(j) Plot the demodulated bit pattern and compare it with the transmitted one.

10.20. Write a program to analyse the effect of choice of orthogonal MC-CDMA signatures on the peak-factor of multicarrier symbols. Run the program for the Walsh-function signature set and complex signatures being cyclic shifts of the polyphase codes of Section 6.11.2. Explain the discrepancy in peak-factor for these two signature ensembles.

10.21. Write a program demonstrating the gain of Alamouti space–time coding versus the no-diversity system for BPSK data transmission over the Rayleigh channel. Recommended steps:

(a) Take a stream of $L_{bs} = 10^4 - 10^5$ random independent bits.
(b) Set their amplitudes independently according to the Rayleigh fading model with the average square equal to one.
(c) Add Gaussian noise of a suitable variance to have pre-assigned average bit power SNR q_b^2.
(d) Demodulate the observation obtained and calculate the empirical bit error rate.
(e) Split the original bit stream into pairs of even and odd bits.
(f) Encode every pair of even and odd bits by rule (10.37), forming two new streams of length L_{bs} corresponding to two antennas.
(g) Form two subchannel amplitude vectors, consisting respectively of even and odd elements of the set of item (b); assign to every element of these vectors random independent phase uniformly distributed over the interval $[-\pi, \pi]$.
(h) Use the first of vectors of item (g) to imitate Rayleigh fading in the first subchannel, assigning its elements to every pair of consecutive bits of the first antenna bit stream as complex amplitudes; do the same for the second antenna bit stream using the second vector of the previous item.
(i) Sum the vectors obtained, dividing the result by $\sqrt{2}$, and add the same noise as in item (c).
(j) Demodulate the observation obtained according to rule (10.41), which for BPSK takes the simplest form: $\hat{b}_l = \text{sign}[\text{Re}(z_l)]$, $l = 0, 1$.
(k) Calculate the empirical bit error rate and compare it to that of item (d).
(l) Run all previous steps, varying bit SNR, and build empirical dependences of the bit error rate on SNR for both transmission modes; compare the results with those predicted from Figure 10.9 and explain the discrepancy, if any.

11

Examples of operational wireless spread spectrum systems

11.1 Preliminary remarks

As was repeatedly underscored in previous chapters, spread spectrum is the basic philosophy of numerous up-to-date and forward-looking wireless systems, ranging from radar and navigation to mobile radio and local area networks. The commercial benefits of spread spectrum technology are brightly confirmed by the impressive market success of 2G cdmaOne (IS-95) mobile telephone, as well as by the consolidation of the international telecommunication community in approving of CDMA as the primary 3G-and-beyond platform. There are a good deal of candidates to illustrate real applications of spread spectrum and CDMA, but we will dwell on only three of them in this chapter, chosen for their outstanding importance both today and in the years ahead, along with their high educational instructiveness. The systems under discussion have already been referred to in the text exemplifying practical implementation of some or other particular principle or idea.

11.2 Global positioning system

It is quite natural for an object moving on the earth's surface or in space to be interested in knowing its current position. The problem of positioning is the basic aim of navigation. Modern navigation equipment is required to inform a user about his instantaneous coordinates, velocities along all coordinate axes, current precise time, predicted position at one or another moment, etc. The most advanced and universal system providing a solution to these tasks is the Global Positioning System (GPS), the operational capability of which was officially declared by the USA in 1993.

Spread Spectrum and CDMA: Principles and Applications Valery P. Ipatov

11.2.1 General system principles and architecture

Three coordinates may describe the position of an object on or above the earth's surface, i.e. latitude, longitude and altitude. In trying to determine them, the object may measure distances (ranges) to three fixed points (beacons) whose coordinates are known in advance with high accuracy. This results in three equations whose three unknowns are exactly the coordinates of the object. In solving them the object learns its own position.

In radio navigation distances are measured via the propagation delay of signals transmitted by the beacons: the distance is just the product of the propagation delay and the speed of light. All beacon transmitters are typically strictly synchronized to each other, operating in a unified *system time*. Unlike this, the clock of any individual user is usually biased relative to system time, and due to frequency drift and instability the user-to-system clock offset proves to be an additional (fourth) unknown. Fixing the time of arrival of a beacon signal versus his local clock, the user will not learn the true propagation delay, since the time-offset is added to it. The easiest way to eliminate this unknown contribution would be to measure the distance by a two-way ranging method, where a beacon sends its signal in response to the user's ranging signal. This, however, implies that users are active, i.e. emitting some energy. The necessity for a beacon to distinguish users' request signals and respond to each of them individually puts a strict limit on the system throughput, i.e. the number of users served. Serving an unlimited number of users is possible only with one-way-ranging, where all the users are passive, i.e. receiving only, which is often desirable for security reasons, too. Then the user receiver can get around the problem of the user-to-system time-offset by measuring times of arrival of signals of four beacons instead of three. Each time of arrival will then contain a 'genuine' propagation delay plus the same unknown clock offset. Multiplied by the speed of light these measurements give *pseudo-ranges*: true ranges plus the product of the clock offset and speed of light. Thus, the user now has four equations with four unknowns, solution of which produces the user position as well as an estimation of the time-offset. This allows the user to fulfil simultaneous positioning and timing.

The GPS is a space-based system meaning that satellites bearing navigation transmitters serve as beacons. Placing transmitters onboard the satellites opens the way to employing UHF radio waves for transmitting the navigation signals. Compared to the longer waves characteristic of earlier ground-based navigation systems (Loran, Omega etc.), UHF waves propagate only along a straight line, with no diffraction, so that trying to employ them in terrestrial transmitters would make positioning possible only within the horizon zone around a beacon. At the same time, the UHF band is much more favourable than longer waves with regard to the dimensions and mass of transmit and receive antennas as well as many other equipment components. The way of reconciling the desires to use the UHF band and to achieve a large coverage area of a navigation transmitter is now rather clear: just put a UHF transmitter onboard a satellite. If the satellite elevation above the earth is high enough, the transmitter will 'illuminate' a large spot on the earth's surface, i.e. meeting the requirement for the size of the navigation coverage zone.

The architecture of the GPS includes three basic segments. The *space segment* contains 24 main satellites (plus several reserve ones) deployed in 6 nearly circular orbits with 4 vehicles in each. The orbits are spaced 60° from each other in longitude, 55° inclined to the equator and have about 12 hours sidereal period. Such a space constellation makes no fewer than four satellites observable simultaneously above the 10°-elevation at any time of day and at any point on the globe. For most of the time, however, the number of observable satellites is greater, sometimes as high as 10. The higher the number of simultaneously visible space vehicles, the better is positioning and timing accuracy. Every satellite bears an atomic (rubidium or caesium) clock having extremely high stability: a day's drift is around or less than 10^{-13} of the nominal frequency. Weather conditions have rather a small effect on the propagation of waves of the GPS band, so that the space segment provides all-time, all-weather positioning all over the globe.

As was pointed out, a user is capable of calculating his position from the measured distances if and only if he is aware of the beacon coordinates. Since satellites revolve on their orbits, their coordinates change constantly and their instantaneous values should be available to a user at any time when he wants to fix his position. The instantaneous location of a satellite on an orbit is not absolutely deterministic due to random disturbances of one or another nature. Similarly, satellite clocks, although very stable, will sooner or later accumulate relative time-offsets, destroying positioning accuracy. Hence, it is necessary to monitor satellite positions and clock behaviour and keep users informed of them. To facilitate this procedure there is a *control segment* of GPS consisting of a main control station in Colorado Springs, Colorado, and five monitoring stations located in Colorado Springs, Hawaii, Kwajalein, Diego Garcia and Ascension Island. The monitoring stations run continuous tracking of all satellites, measuring their clock parameters, testing satellite state, etc. and via special communication links transmit the data to the master control station. The latter computes the current and predicted satellite locations, time-offsets and other relevant parameters, and transmits the fresh data to one of three ground control stations located at the same sites as the monitoring stations (Kwajalein, Diego Garcia and Ascension Island). The ground control stations upload the data to the GPS satellites, employing a dedicated S-band uplink.

The *user segment* covers all users equipped with GPS receivers. There are a lot (at least hundreds) of receiver models currently on the market ranging from the simplest and cheapest handheld devices (used for sport or recreation) to very sophisticated and expensive models, designed for military purposes, surveying etc.

11.2.2 GPS ranging signals

The master atomic oscillator of every GPS satellite outputs the frequency 10.23 MHz. Coherent frequency multiplication by 154 and 120 times produces two L-band frequencies L1 = 1575.42 MHz and L2 = 1227.60 MHz, further modulated by GPS ranging codes. There are two ranging codes: C/A-code (coarse acquisition or clear access code) and P-code (precision or protected code). Each satellite is assigned its specific pair of C/A- and P-code sequences. The upper carrier L1 is a basic one and its signal is available to any GPS receiver, while the lower frequency L2 is planned to be used by only authorized users for high-precision fixing.

Explanation of why two carriers help in accurate positioning lies in the propagation characteristics of the ionosphere, an upper atmosphere layer over 50 km above the earth. Having a high concentration of free ions and electrons due to the ionizing effect of the solar ultraviolet and X-ray radiation, the ionosphere is a dispersive medium, in which the ranging signal propagates at a speed different from that in vacuum, acquiring some additional delay. Unfortunately, this extra delay cannot be precisely pre-calculated due to the fast random fluctuations of medium parameters. On the other hand, the general character of dependence of an ionospheric delay on frequency is well approximated by an inverse square dependence with unknown proportionality coefficient. The latter may be found as the only unknown of the equation obtained by measuring the delays of identical signals on two properly spaced frequencies and subtracting the results. On completing this, the ionospheric correction can be found and removed from the measured pseudo-range.

A C/A-code sequence serves to arrange a scale for positioning with moderate accuracy. It is a Gold one (see Section 7.5.2) of length $N = 1023$ with chip duration (smaller than 1 μs) resulting in the signal period equal to 1 ms sharp. Certainly, the number (1025) of existing Gold sequences of length $N = 1023$ is many times more than necessary for all GPS satellites, so that only some of these sequences are utilized. Periodically repeated Gold sequences modulate frequency L1 only and are continuously transmitted by satellites. Despite all satellites being strictly synchronized, the distances from them to a user change in the course of their revolution, so that the signals of different satellites arrive at the receive antenna with mutual time-offsets varying significantly. Since every satellite is identified by its specific Gold sequence and all these signals have low asynchronous cross-correlations, the user receiver is capable of separating each individual satellite signal from the others. In other words, the C/A-code scale of the GPS space segment employs Gold-ensemble-based asynchronous CDMA. Transmission of satellite coordinates (ephemeris) along with other relevant data (clock error versus the GPS time, corrections for propagation effects, satellite state etc.) is arranged in a DS spreading manner (see Section 7.1): the data bit stream with rate 50 bps BPSK directly modulates the Gold sequence before modulation of the L1 carrier. Thus, every data bit spans 20 periods of a Gold sequence, theoretically neutralizing corruption of correlation properties due to the presence of data modulation (see Section 7.3). The minimum package of data (frame) necessary for positioning occupies 1500 bits (30 s) organized into 5 subframes, each containing 300 bits. The first three subframes are repeated unchanged (except for reloading fresh data) in each frame, while the contents of the rest vary during the superframe covering 25 frames. For better transmission reliability the data stream is encoded by a (32,26) extended Hamming code. One important sort of data transmitted in the 4th and 5th subframes is the almanac: raw ephemeris of other satellites of the system. On capturing the signal of any one satellite the user gets the rough positions of the others and may use this information to accelerate the search for signals of the other satellites (see Section 8.3.1).

The C/A-code provides standard precision of positioning and is available to any user equipped with a GPS receiver on a free-of-charge basis. The P-code is intended for higher precision positioning and so has chip duration 10 times smaller against the C/A-code (below 100 ns) or 10 times wider bandwidth (see Section 2.12.2 to recollect the relation between bandwidth and ranging accuracy). To realize the two-frequency ionospheric error compensation described above, the P-code (DS-modulated by the data bit stream similarly to C/A-code) is transmitted on both frequencies L1 and L2, quadrature multiplexing of C/A and P signals being used on L1 with 3 dB stronger C/A-signal. In its turn, the L2 intensity is 3 dB lower than that of L1. The structure of the P-code is described in open GPS documents. It is formed as a symbol-wise modulo 2 sum of two very long binary sequences differing in length by 37 chips. The resulting period of the sequence thus formed is around 266 days. The non-overlapping 7-day (6.187104×10^{12} chips) segments of this sequence are used as P-codes for different satellites. The USA Department of Defense commissioned the designers of GPS to make provision for strict limitation of access to the P-code, reckoning that unauthorized usage of it may be hazardous to national security. Encryption of the P-code is realized by its modulo 2 summation with a masking or key W-code, whose structure is secret. The resulting Y-code possesses excellent cracking resistance (see Example 3.3.1).

11.2.3 Signal processing

The basic operations of a single-frequency (L1) GPS receiver are very conventional for any DS spread spectrum system. After a coarse acquisition of a satellite C/A-signal (see Sections 8.2 and 8.3), aided when possible by a priori knowledge of satellite locations, the code delay-lock loop (Section 8.4) is locked and starts to output a sequence of estimations of a satellite pseudo-range. Typically, modern GPS receivers include a set of channels processing in parallel the C/A-signals of all visible satellites. On finishing the search for the last used satellite, the receiver is ready to produce the user's coordinates, which is a steady-state process lasting for as long as the user wishes.

The authorized receiver repeats the same operations for the P-codes of both carriers, spending only a little time on searching the signals, since the data frame available from the L1 signal contains a special handover word which facilitates setting the local generator of the P-code to an appropriate initial state.

In many modern GPS receivers these basic operations are supplemented or replaced by a variety of others pursuing improvements of accuracy, speeding up of the initial fixing time, consumer convenience etc. For example, additional accuracy may be gained by measuring pseudo-ranges via integration of the carrier frequency of the received signal. The instantaneous Doppler frequency shift is proportional to the radial speed of the satellite relative to a user. Hence, the integral of the Doppler frequency over some period is proportional to variation of the satellite–user distance over this time interval. Having started from the point with precisely known coordinates, the receiver may further position itself via integrals of instantaneous frequencies of visible satellites, i.e. their current accumulated ranges. Moreover, methods of ambiguity resolution exist, making possible positioning through frequency integrals even without initialization at a known point [117,118].

Another hugely popular operational technique is so-called *differential* or relative positioning, the idea of which is as follows. Let one GPS receiver be set up at the reference site (base) with precisely known coordinates. Then comparing pre-computed satellite ranges with the measured ones, the base receiver can find systematic errors (biases) inserted by system imperfection. Let another receiver be placed at a remote point with unknown coordinates. If the baseline, i.e. the distance between the base and remote receivers, is not very long (e.g. within tens of kilometres) the systematic errors at the base and remote sites are strongly correlated, so the remote receiver may subtract biases estimated by a base receiver from measured ranges, improving their accuracy. Of course, such system modification should contain a communication link providing delivery of base-receiver data to the remote receivers. A vast number of reference sites are now arranged all over the world, transmitting differential corrections via FM stations, broadcasting satellites, radio beacons, cellular radio, Internet etc. [117,118].

11.2.4 Accuracy

The originally planned precision of C/A-code GPS positioning was set up around 100 m in the horizontal and 156 m in the vertical directions, the probability of keeping errors within these limits being 95%. Analogous figures for the P-code fixing were 16 m and 23 m, respectively. However, numerous advanced receiver structures developed by manufacturers have exhibited much better precision even without involving the P-code. This became a matter of anxiety for the US institutions responsible for national security, and in 1990 a selective availability mode was introduced, distorting satellite-transmitted ephemeris and timing and thereby deliberately corrupting positioning accuracy. During the subsequent decade, however, differential navigation, which eliminates these types of errors almost entirely, gained great popularity, so the selective availability mode turned out to be pointless in practice and was terminated in 2000. Nowadays a wide spectrum of offers is characteristic of the GPS equipment market, with proclaimed accuracies ranging from tens of metres to several millimetres and better.

11.2.5 GLONASS and GNSS

The Russian space-based navigation system GLONASS has many common features with GPS. Its space segment consists of 24 satellites located in 3 nearly circular orbits with nominal sidereal period 11 hours 15 minutes and 64.8° inclination to the equator. Again, two frequencies L1 and L2 (respectively in the 1.5 and 1.2 GHz bands) are used to provide ionospheric correction, with C/A-code transmitted on L1 and P-code transmitted on both carriers. Current ephemeris and other relevant data encoded by Hamming code and properly arranged into subframes and frames are superimposed onto ranging codes in a DS manner and transmitted by satellites at the rate 50 bps. A control segment provides continuous monitoring of satellites, computation/prediction of their orbit parameters and uploading them to the satellite onboard memory.

The substantial difference between GLONASS and GPS is that all satellites transmit the same C/A-code, which is a binary m-sequence of length $N = 511$ with real-time period 1 ms. Distinguishing individual satellite signals is possible due to the small mutual carrier offsets between them, transforming the common C/A code into an ensemble of frequency-offset replicas of the m-sequence, as described in Section 7.5.1. In order to save bandwidth antipodal satellites of the same orbit (which are never seen by a user simultaneously) employ the same frequency offset.

The accuracy of GLONASS is of the same level as that of GPS. Both systems are now considered as cooperative, entering the integrated Global Navigation Satellite System (GNSS). It has already been emphasized that increasing the number of processed satellite signals improves the positioning precision, so joint use of both constellations is obviously profitable. In addition, scenarios are not rare, where some satellites over the horizon are obstructed (e.g. by an aircraft wing), so that the total number of available ranging signals within only GPS or GLONASS is not sufficient for positioning. Then again, joint processing of GPS and GLONASS signals may give a considerable gain in positioning integrity. A great number of receiver models presently on the market or in development are capable of combined processing of signals of both systems.

11.2.6 Applications

The role of satellite-based global navigation systems in the modern and future world can hardly be overestimated. Just simply naming the areas of their involvement forms rather a long list, including traditional navigation of ships, aircraft and terrestrial moving objects (cars, trucks etc.), transit systems, mapping utilities (e.g. pipelines), monitoring forestry and natural resources, farming, civil engineering, geodetic surveying, seismic forecasting, airborne mapping, seafloor investigations and many more. Not being able to go deeper into this fascinating topic, we direct the interested reader to the sources [117–119] and their references.

11.3 Air interfaces cdmaOne (IS-95) and cdma2000

11.3.1 Introductory remarks

The first interim specifications of the 2G CDMA cellular telephone of standard IS-95 (presently referred to as cdmaOne, too) were published in 1993–1995, and the operational phase of IS-95 networks started in 1996. Nowadays networks of this standard cover huge territories serving tens of millions of consumers. Its impressive commercial success, widely recognized high quality of service and openness to further modernizations were among the decisive factors favouring the CDMA philosophy as the basic platform for the next generations of mobile radio (3G and beyond).

Initially IS-95 was meant to gradually replace (maintaining compatibility with) an American analog standard, AMPS, operating in the 800 MHz range. The IS-95 documents set up frequency division separation of forward (869–894 MHz) and reverse (824–849 MHz) links,[1] while no limitation on frequency reuse in neighbouring cells or sectors was stipulated. The nominal bandwidth of the IS-95 signal is about 1.25 MHz, so that within the total assigned 25 MHz band an operator has remarkable freedom in carrier selection and frequency planning of the network. All the BSs entering a network are strictly synchronized via GPS to operate in a unified time scale, allowing MS easier

[1] The terms 'forward' and 'reverse' links are synonyms of downlink (BS to MS) and uplink (MS to BS) adopted in the cdmaOne and cdma2000 specifications.

switching from one BS to another (handover). IS-95 and its 3G evolution cdma2000 are typical DS spread spectrum systems, which clearly manifest all the benefits of this technology. They also possess very high educational value, since they demonstrate in a lucid form practical ways of realizing many ideas studied above. In the text to follow we are going to dwell on only the most general principles of spreading, channelization, coding and modulation in the IS-95 and cdma2000 air interfaces. Readers who wish to acquire deeper knowledge may consult the sources [18,69,83,120,121] and many others.

11.3.2 Spreading codes of IS-95

The spreading sequences used in the IS-95 standard were partly mentioned in examples earlier. They are designed to provide CDMA separation of physical channels, distinguishability of signals of different BS arriving at the MS receiver and privacy of the transmitted data. Synchronous CDMA multiplexing of physical channels of the forward link served by a fixed BS is realized on the basis of Walsh sequences (see Section 2.7.3) of length $N = 64$. The orthogonality of Walsh sequences allows separating the corresponding 64 physical channels theoretically with no MAI. The duration of a chip of Walsh sequences is nearly 0.81 μs and the chip rate is 1.2288 Mcps (megachip per second), resulting in the abovementioned bandwidth of 1.25 MHz. Certainly, the number of forward-link physical channels thus implemented is 64 and, consistent with the CDMA principle, they occupy the same common bandwidth with no frequency or time offset. All of the base stations use the same set of 64 Walsh functions, and the spreading by so-called *short codes* makes the signals of different base stations separable from each other in the MS receiver. There are two different basic binary short-code pseudo-noise sequences, PN-I and PN-Q, used in the in-phase and quadrature-phase branches of the BS modulator, respectively. They are primarily generated as two m-sequences whose LFSR generators (see Section 6.6) contain 15 flip-flops and are defined by the primitive polynomials $f_I(x) = x^{15} + x^{13} + x^9 + x^8 + x^7 + x^5 + 1$ for PN-I and $f_Q(x) = x^{15} + x^{12} + x^{11} + x^{10} + x^6 + x^5 + x^4 + x^3 + 1$ for PN-Q. The sequences obtained have length $L = 2^{15} - 1$, but to come to the short codes PN-I and PN-Q they are extended by one more zero symbol following after 14 consecutive zeros. This brings the lengths to $N = L + 1 = 2^{15} = 32\,768$ chips, and with the same chip rate as for Walsh codes there are 37.5 periods of the short codes per second or 75 periods over two seconds. To discriminate between different base stations every one of them employs its BS-specific time-offset replica of the basic short-code sequences. There are 512 such pairs of replicas, every pair being shifted compared to the previous one by 64 chips or about 52 μs. The network planning should assign short-code pairs to the base stations in a way guaranteeing low risk of any MS receiving a signal from an unintended BS, whose timing due to propagation delay is about the same as that of the desired signal and whose strength is sufficient for mixing them up. It should be stressed that relative time-offsets between the base stations entering a network, once set up, remain constant forever, since all the BS use GPS receivers to synchronize their clock oscillators with each other.

One more spreading code is a *long code*, generated primarily as a binary m-sequence of memory 42. According to the specification a primitive polynomial of the LSFR

42-stage generator of the long code prescribes feedback taps from flip-flops with numbers (counting left to right) 7, 9, 11, 15, 16, 17, 20, 21, 23, 24, 25, 26, 32, 35, 36, 37, 39, 40, 41 and 42. The m-sequence thus obtained is again extended by inserting one extra zero after a run of 41 consecutive zeros to come to the length $N = 2^{42}$. Using the same chip rate as before, the long code has period over 42 days. Different user-specific time-offsets (masks) of the long code are used in both the forward link for data protection and the reverse link for CDMA separation of MS signals along with simultaneous data protection (see below).

11.3.3 Forward link channels of IS-95

Logically, i.e. by their information content, there are four types of forward channels:

- pilot channel
- synchronization channel
- paging channels
- traffic channels.

11.3.3.1 Pilot channel

The pilot channel is unique for an individual BS and transmits 'pure' short code but no other data. It looks as though a data sequence of only zeros is DS spread by a short code. Being received by the MS it provides synchronization of its clock with the system (i.e. BS) clock up to the accuracy sufficient for forming a receiver coherent reference to despread and demodulate the received signal. Since the short-code period contains an integer number (512) of periods of Walsh sequences, these latter generated locally are automatically aligned with the spreading Walsh codes of the received signal after a receiver searches and DLL-captures the pilot signal.

The Walsh function, being the first row of a Hadamard matrix of size 64 and consisting of only positive ones, physically separates (channelizes) the pilot channel from others. It is convenient to generate Walsh functions digitally, using binary $\{0, 1\}$ logic, and then replace the zeros and ones by plus and minus ones, respectively, as (6.15) does. Figure 11.1a shows a simplified structure of the pilot channel where W_0 symbolizes binary $\{0, 1\}$ Walsh function number zero consisting of only zeros. As long as spreading by a short code is done for all forward channels commonly in the modulator, i.e. after summing signals of all channels, this operation is shown in the figure to come later. Trivially speaking, the pilot channel shown in Figure 11.1a feeds the modulator by a sequence of only plus ones.

11.3.3.2 Synchronization channel

The synchronization channel is also unique for any individual BS. Figure 11.1b illustrates its general structure. Along with some other information, it transmits the data enabling the receiver to get a user-specific long-code mask and thereby synchronize its local long-code replica with the one used by the BS for data encryption and placing power control

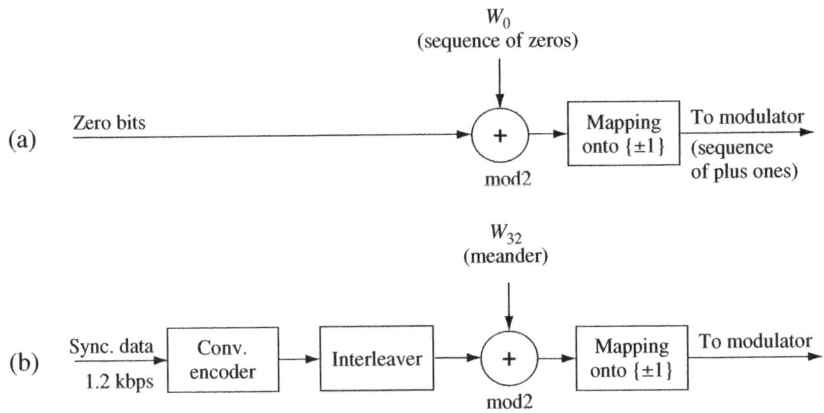

Figure 11.1 Simplified structures of pilot (a) and synchronization (b) channels

bits (see below). The raw data has rate 1.2 kbps and is arranged in frames whose duration coincides with the period of the short code (26.67 ms), each three frames combined in a superframe of 96 bits. The superframes are encapsulated in a message capsule, including 30 CRC symbols (see Example 9.2.4) to realize a message quality indicator. A convolutional code of rate 1/2 and constraint length 9 (see Section 9.3) provides strong protection for the data against channel interference. No tail bits are inserted, i.e. coder and decoder are not reset to the zero state at the end of every frame. The encoded stream rate is 2.4 kbps, but every symbol is treated as though the two-times shorter one is doubled (symbol repetition), so that the input rate of the subsequent interleaver is 4.8 kbps. The interleaver serves to decorrelate error bursts, thereby improving the error correction capability of the convolutional code against long fading drops of signal strength or other correlated degradations (see Section 9.5). Interleaving covers one frame of 128 code symbols (26.67 ms) and uses 16×8 matrix memory. The encoded stream is written column-wise into it and then read out according to the standard-defined pattern.

The Walsh sequence that is the 33rd row of the Hadamard matrix of size 64 physically channelizes the synchronization data. Figure 11.1b shows again primary forming of the binary $\{0, 1\}$ prototype W_{32}, which is simply a meander where 32 zero symbols are followed by 32 ones. The technique of DS spreading illustrated by the figure is a conventional one, based on the one-to-one correspondence between multiplication and modulo 2 summation within real $\{\pm 1\}$ and logical $\{0, 1\}$ alphabets, respectively (see Section 7.5.2). Accordingly $\{0, 1\}$ symbols of the encoded and interleaved stream are symbol-wise modulo 2 added with W_{32} and the resulting stream is then mapped to the $\{\pm 1\}$ alphabet.

11.3.3.3 Paging channels

There are up to seven paging channels intended to notify a subscriber about calls coming from the network, to answer users' requests initiating calls and to transmit other access-related information. The raw data is divided into slots of duration 80 ms consisting of

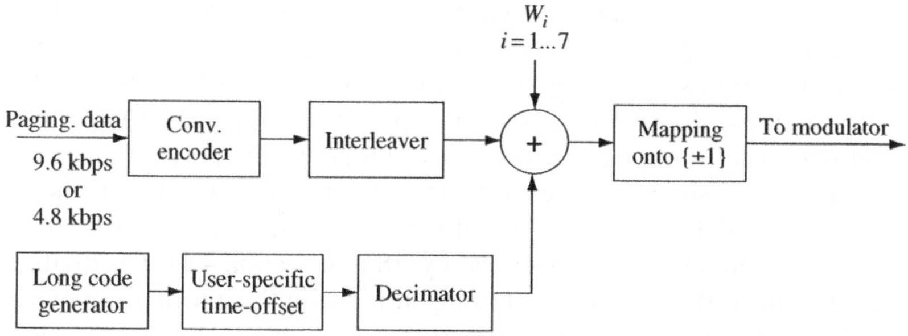

Figure 11.2 Simplified structure of a paging channel

four 20 ms frames. Frames or their parts enter message capsules, each containing up to 1184 bits, including 30 CRC symbols. There are two possible rates of paging data: 9.6 or 4.8 kbps. Figure 11.2 presents a simplified structure of a paging channel. Its first unit is a convolutional encoder of rate 1/2 and constraint length 9. Similarly to the synchronization channel, tail bits are not introduced so there is no resetting of coder and decoder to the zero state at the end of frame. At the input rate 4.8 kbps the encoded stream rate is 9.6 kbps, but every symbol is treated as a doubled shorter one (symbol repetition), so that regardless of the raw data rate the encoded stream is of rate 19.2 kbps. The interleaver operates within 20 ms frames employing 24×16 matrix memory where encoded symbols are written column-wise. The order of reading from the memory is set up by the standard.

The next operation over the encoded and interleaved stream is encryption, which is realized as scrambling, i.e. modulo 2 symbol-wise sum with a pseudorandom binary sequence. The latter is produced by decimation of the user-specific offset replica of the long code with index 64, i.e. taking every 64th symbol. In this way, the rate 1.2288 Mcps of the long code is divided 64 times, coming to $1228.8/64 = 19.2$ kcps. As such, the encryption is specific for any user, preventing an unauthorized interceptor from monitoring the access data sent by the BS.

The Walsh functions with numbers 1 to 7 (second to eighth rows of a Hadamard matrix) are used for the channelization of paging channels. The primary paging channel formed by W_1 is always available, while the rest may either be not activated or used as traffic channels.

11.3.3.4 Traffic channels

The channels considered up to this point perform service functions necessary to start and maintain conveying basic messages to the user. The traffic channels are responsible for delivery of basic information: digitized voice, computer or multimedia data etc. To be specific, let us limit ourselves to speech transmission. The preliminary operation, then, is encoding of speech run by the unit called *vocoder* (voice coder). Leaving aside discussion of the rather complex principles of this device, just note that several types of vocoder are presently used in IS-95 equipment outputting speech data stream with nominal rates 8.6 or 13.3 kbps. The nominal rate corresponds to the highest speech activity (when a speaker is talking unceasingly), while there are three smaller rates for lower activity. For the upper rate 8.6 kbps they are 4.0 kbps, 2.0 kbps and 0.8 kbps. The vocoder constantly traces the energy of analog speech over 20 ms frames and compares it with three adaptive thresholds to set an appropriate rate of digitized output. Rate reduction at periods of low speech activity is accompanied by proportional reduction of emitted signal power, so that the energy per bit remains constant. In its turn lower power means reduced level of MAI to other users inside[2] and outside the cell, and eventually a greater number of users served by BS (see Section 4.6). The digitized speech stream is packed into 20 ms frames containing along with the information bits also CRC symbols and 8 tail bits resetting the convolutional encoder to zero. As a result, the set of rates of the 8.6 kbps nominal rate transforms at the channel input into the set RC1 of 'raw' rates 9.6, 4.8, 2.4 and 1.2 kbps.[3]

Let us turn to the block-diagram of a traffic channel in Figure 11.3. Input data at one of the four rates above arrives at the convolutional encoder of rate 1/2 and constraint length 9. Regardless of the input rate, the output rate due to symbol repetition is always 19.2 kbps, e.g. the stream of rate 1.2 kbps is encoded into one of true rate 2.4 kbps, but in this output stream each symbol is treated as 8 consecutive 8 times shorter symbols. The interleaver operates over 20 ms (384 code bits) permuting symbols of the codestream according to the pattern defined by the specification. The interleaver output is then scrambled in the same way as in paging channels to provide privacy of a transmitted message.

In Sections 4.5 and 4.6 we stated that powers of MS signals should be effectively controlled to overcome the near–far problem and maintain MAI at the BS site below the destructive level. A closed power control loop is one of the arrangements running this task in IS-95. A BS monitors permanently the intensity of each of the received MS signals and sends a command ordering the MS to either increase or decrease its emitted power. The command is just a *power control bit* whose values zero and one dictate

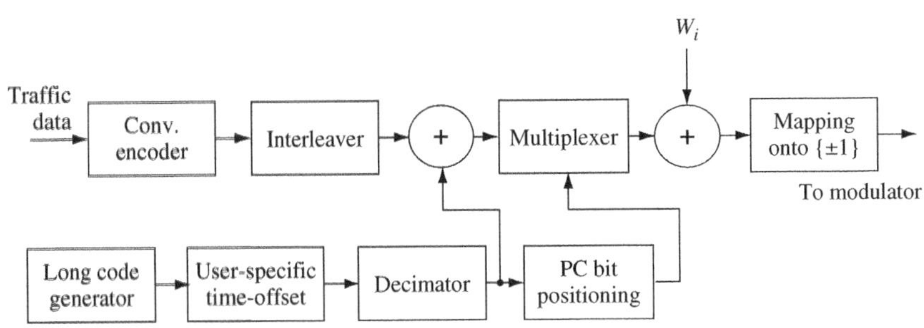

Figure 11.3 Simplified structure of IS-95 forward traffic channel

[2] Despite the forward channel being synchronous and not oversaturated, it is in practice not free of MAI: mutual multipath delays make different Walsh sequences non-orthogonal.

[3] Apart from RC1, IS-95 documents also establish the set of raw rates RC2 for the nominal rate 13.3 kbps: 14.4, 7.2, 3.6 and 1.8 kbps; however, the codestream after the convolutional encoder again has the same rate of 28.8 kbps due to puncturing (see Section 9.3.1).

increase or reduction of MS power, respectively. To insert the commands into the forward link signal every 20 ms frame after the interleaver is divided into 16 power control groups (PCG), each spanning 1.25 ms or $19.2 \cdot 10^3 \times 1.25 \cdot 10^{-3} = 24$ code symbols of the 19.2 kbps codestream. In every PCG a single power control bit overwrites two code symbols. The MS receiver, knowing the positions of overwritten symbols (i.e. power control bits), excludes them from the decoding procedure as having nothing to do with the message contents. This is entirely equivalent to replacing an original convolutional code by a punctured one (see Section 9.3), and its negative effect on the code correction capability is believed to be partly mitigated by random positioning of the power control bits within PCG. The pseudorandom sequence at the output of the first decimator of Figure 11.3 has the same rate as the codestream, i.e. 19.2 kcps. During one 1.25 ms PCG there are 24 chips of this sequence. The last four of them are read as a binary number with the 24th chip giving the most significant bit. This number, ranging from 0 to 15, is used as a position number of the power control bit in the next but one group after the current one. Thus, the power control bit may take randomly any position out of the first 16 in every PCG. In Figure 11.3 the units implementing positioning and inserting power control bits are denoted as 'PC bit positioning' and 'Multiplexer'.

11.3.3.5 Forward link modulation

Figure 11.4 presents the block-diagram of the forward link modulator. Output voltages of all physical channels of BS are first weighted by appropriate gains to realize forward link power control. Every MS periodically informs the BS about the reliability of data received, and the BS properly adjusts the power level of the signal in the traffic channel assigned to this specific MS to maintain the data reception quality above the predetermined threshold. The weighted channel signals are then summed in the adder and fed in

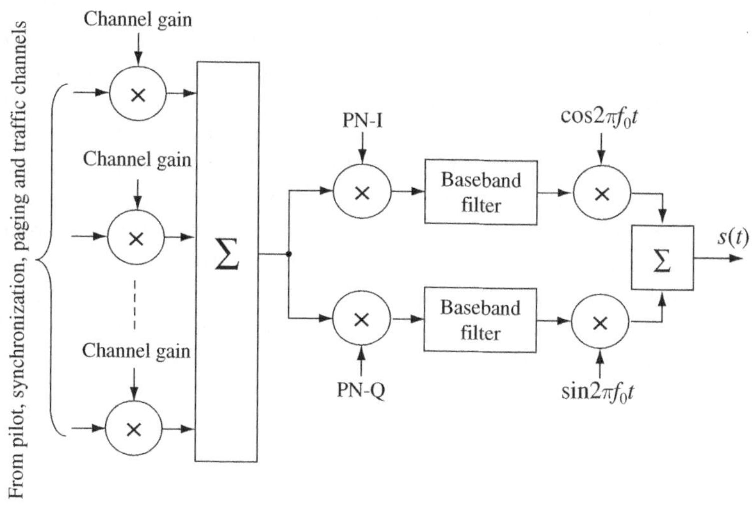

Figure 11.4 IS-95 forward link modulator

parallel to the in-phase and quadrature branches of the modulator to be multiplied by the binary PN-I, PN-Q codes (see Section 11.3.2) and shaped in the frequency domain by the baseband filters. The multiplication of the in-phase and quadrature signals by cosine and sine CW components of frequency f_0 with their subsequent summation performs up-conversion of the BS signal and finishes the modulation process. As is seen, the input baseband signal in both branches is the same. As such, it is a sum of multiple binary voltages, i.e. is multilevel real. Assume for a while that there is only a single physical baseband channel fed immediately to the modulator branches without summation with other channels. We may postulate that each physical channel is processed this way, i.e. there are as many modulator branch pairs as channels and the outputs of all these parallel modulators are added up coherently. Since the scheme of Figure 11.4 is linear, and hence the superposition principle is valid, its output effect is identical to that of the hypothetical scheme above, i.e. with individually modulated channels. That is why we may say that in the forward link of IS-95 DS spreading is used where the binary datastream (channelized by a Walsh function) modulates the QPSK spreading code (see Section 7.1). Since the rate of the codestream at the modulator input is 19.2 kbps, each symbol has duration covering 64 short code chips. Hence, the forward link spreading factor is 64.

It is worth noting that the long code plays no role in DS spreading of the forward link signal, taking part only in data encryption and power control bit positioning. It is often said that forward link spreading is done by both Walsh codes and short PN-codes. Yet, conceptually classifying Walsh functions as channelizing and PN-codes as spreading may look more convenient.

11.3.3.6 MS processing of forward link signal

Signal processing in the MS receiver rests on the classic procedures discussed in depth in the previous chapters. On successful acquisition of a pilot signal, the receiver DLL pulls in and continuously tracks the short code of the contacted BS. The local replica of the short code produced by DLL serves for despreading the received signal. The outcome of the despread pilot channel is a 'pure' CW carrier down-converted to appropriate intermediate frequency. A phase-locked loop tunes the local crystal oscillator to be coherent with this CW signal, providing thereby the reference for coherent data demodulation. After demodulation and deinterleaving the data transmitted over synchronization, paging and traffic channels are separated from each other in correlators using Walsh-sequence references, decoded by the Viterbi algorithm and used according to their destination. For example, a digital-to-analog converter transforms speech data of the traffic channel into voltage, which becomes audible with the aid of an earphone.

Every MS receiver contains several (four or more) parallel channels capable of searching and tracking the pilot signal. One goal of it is arranging the RAKE receiver, which realizes the multipath diversity benefit of spread spectrum (see Section 3.7). Typically at least three such channels are used to implement RAKE fingers. Another procedure requiring autonomous pilot signal channels in the MS receiver is handover. A reserve correlator (or set of them) performs permanent scanning of the time domain, trying to determine if signals of other BSs are present, possibly more intense and preferable for contact. In the latter case the network may order MS switching to another BS, which is done easily, since the receiver is already tracking its signal (*soft handover*).

11.3.4 Reverse link of IS-95

According to the logical content of the data transmitted over the reverse channel every MS operates on one of two types of channels:

- traffic channel
- access channel.

11.3.4.1 Reverse link traffic channel

Figure 11.5 presents a simplified structure of the reverse traffic channel. The MS-transmitted bit stream (digitized speech from a vocoder, computer data etc.) with inserted CRC symbols is divided into 20 ms frames, where 8 tail bits are then inserted for zero-resetting of a convolutional encoder at the start of every encoded frame. As a result the nominal rate of data at the encoder input is 9.6 kbps, but for reduced voice activity three lower rates (4.8, 2.4 and 1.2 kbps) are also employed, in the same way as in the forward link. Due to the asynchronous nature of the reverse link, MAI—unlike in the forward link—would exist even in the hypothetical absence of multipath effects (see footnote 2). This qualitatively justifies the greater strain interference condition of the IS-95 reverse link, explaining why it exploits a convolutional code with higher error correction capability in combination with subsequent 64-ary orthogonal modulation. Since increasing the constraint length ν_c would entail undesirable codec complications, its accepted value is the same as in the forward link ($\nu_c = 9$), reduction of the code rate to 1/3 being a payment for better distance properties. With such a code rate the output codestream rate is 28.8 kbps independently of input bit stream rate: the symbol repetition explained earlier for a forward channel is administered in the reverse link as well.

An individual 20 ms frame of a codestream (576 bits) is divided into 16 power control groups of 36 bits (1.25 ms) each. An interleaver operating over the frame uses a 32×18 matrix, where the codestream is written column-wise. The reading runs row-wise, every pair of odd and the next even rows forming one PCG. However, consecutive pairs of

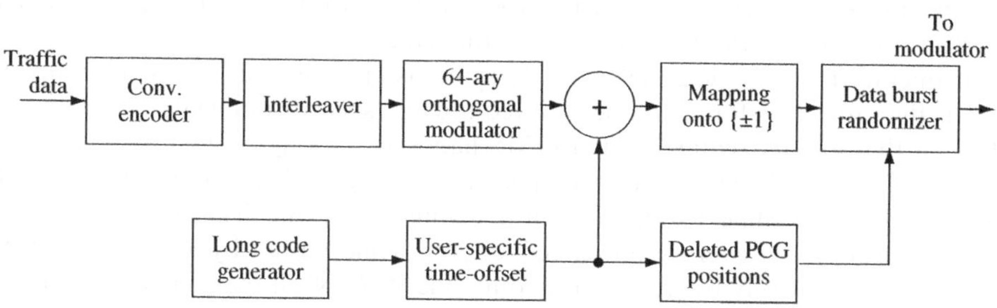

Figure 11.5 Simplified structure of IS-95 reverse traffic channel

rows are read according to a pattern providing adjacency of PCG repeating each other (due to encoded symbol repetition) whenever the raw datastream rates are smaller than 9.6 kbps (i.e. 4.8, 2.4 and 1.2 kbps). For example, when the raw rate is 4.8 kbps every even PCG contains repetition of the same interleaved code symbols as the previous; for the datastream rate 2.4 kbps the groups with numbers $4i + 2, 4i + 3, 4i + 4$ are replicas of the group number $4i + 1$, etc. Such an order is convenient for further lowering of the average transmitted power proportionally to the rate reduction, which is realized in the reverse link slightly differently as compared to the forward link (see below).

The codestream from the interleaver is fed to the 64-ary orthogonal modulator, meaning that every 6-symbol block treated as a binary 6-digit number selects one of 64 orthogonal signals (a Walsh function of this number). This gives an extra coding gain (up to three times asymptotically; see Section 2.6) above that of convolutional coding. Since every 6 input binary symbols are now replaced by 64 binary symbols, the stream of chips at the orthogonal modulator output becomes $64/6 = 32/3$ times faster (307.2 kcps). Let us stress that Walsh functions in the reverse link bear no channelization functions and only implement spread spectrum orthogonal coding for data transmission, as discussed in Section 2.7.3.

The next step in forming a traffic channel is spreading, i.e. modulo 2 summation of the binary symbol stream after the orthogonal modulator with the offset long-code sequence. Since the long code is a chip stream with rate 1.2288 Mcps, there are 4 long-code chips per Walsh symbol or 256 long-code chips per Walsh signal at the orthogonal modulator output. The BS receiver after despreading uses a 64-channel correlator bank, each channel being tuned to one of the Walsh signals, and decides in favour of the Walsh signal referencing the correlator with the strongest response. As such, a processing interval covers the whole Walsh function duration, i.e. 256 long-code chips, and the spreading factor of the reverse link appears to be 256. A long code is unique, strictly specified by the standard, and it is user-specific masks (time-offsets of the code) that secure CDMA separation of different users. Thus, we encounter here the strategy of asynchronous CDMA discussed in Section 7.4. Of course, these offsets should be properly assigned to have no risk of synchronous arrival at the BS of signals from two MSs migrating freely over the whole coverage zone. The value of offset is a current MS identity rendered to it by the network similarly to the channel carrier in FDMA. In parallel with the CDMA channelization, the user-specific mask provides the encryption (scrambling) of the stream after orthogonal modulator. Due to the enormous length of the long code, it is not an easy task for an unauthorized interceptor, who does not know the user's mask, to synchronize a local long-code generator to the intercepted signal and despread (thereby descramble and decrypt) the data.

The operation of mapping logical $\{0, 1\}$ symbols onto the real modulation alphabet $\{\pm 1\}$ is the same as before and does not require any special comment.

One of the primary requirements for MS handset is a long enough battery lifetime. From this angle a linear power amplifier consuming higher average power is less attractive than a nonlinear (operating in a keying mode) one. That is the reason why in the MS transmitter average power reduction is achieved not by lowering an instant power, but alternatively by transmitting only one PCG of all those that replicate each other. For example, with a raw rate 2.4 kbps there are quadruplets of identical PCGs and only one of them is transmitted, while the transmitter is cut off during the three

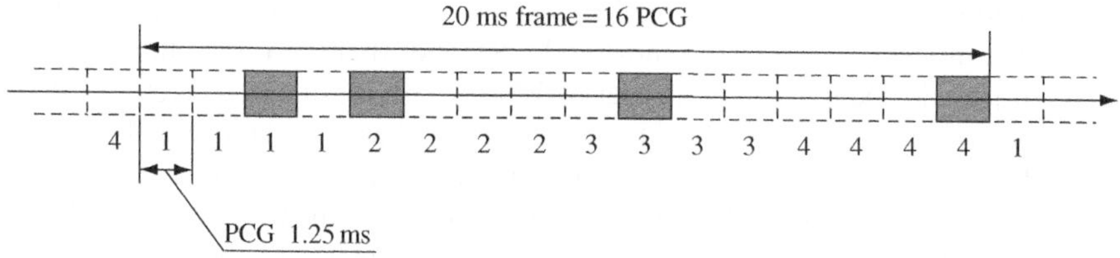

Figure 11.6 Example MS transmission pattern for the raw rate 4.8 kbps

others. Certainly, this makes MS emission discontinuous. Randomization of the positions of discarded PCGs (data burst randomization) enables better averaging of discontinuous MAI at the BS. The pseudorandom pattern of gating off the PCG inside any frame is determined by the last 14 chips of a user's mask, i.e. offset long-code replica, at the end of the previous frame. The rule defined in the specification for every particular raw rate recalculates values of these chips (as binary digits) into the positions of erased PCGs. Figure 11.6, where identical numbers mark replicated PCGs, and shaded and dashed rectangles correspond to transmitted and erased PCGs, shows an example transmission pattern for raw rate 4.8 kbps. We may treat the sequence after the data burst randomizer as ternary with symbols $\{\pm 1\}$ and zero corresponding to active and pausing transmitter, respectively.

11.3.4.2 Access channel

MS uses an access channel when responding to a notification about an incoming call in the idle state and when it either needs to register on the network or initiate a call. The procedures of framing, convolutional encoding, orthogonal modulation, interleaving and long-code spreading in the access channel are basically similar to those of the reverse traffic channel. Of course, no voice data is transmitted through this channel, so no rate/power control according to voice activity is performed. One of the specific features of the access channel relates to initiating access by MS. Not knowing precisely the propagation conditions in the reverse link, the MS starts by sending probe signals of low strength, gradually increasing the signal level with every next attempt until it obtains the BS confirmation that the connection is established. Probe signals are sent in a burst mode with randomized intervals to reduce the probability of colliding requests from several users, because it is not impossible that at the access stage different MSs have the same long code masks.

11.3.4.3 Reverse link modulation

MS can never use both traffic and access channels simultaneously, therefore there is no need for channel signal summation at the modulator input as there was in the forward channel. Traffic or access channel output is immediately fed in parallel into in-phase/quadrature branches, differing from those of Figure 11.4 in the following details. First, offsetting short codes PN-1 and PN-2 to identify BS is now needless (every mobile has its own unique identity—long code mask—all over the coverage zone), and even inconvenient for simultaneous reception of the same MS signal by several BSs during a soft handover. Thus, all MSs use zero offset of the short code. Second, there is a half-chip delay inserted into a quadrature branch after a multiplier by PN-Q. This is done to convert QPSK into its version called offset QPSK (O-QPSK). The latter is believed to be preferable for battery lifetime. The conventional QPSK modulator in Figure 11.4 may be thought of as two BPSK modulators, operating independently with cosine and sine CW carriers. When binary symbols in both branches change into the opposite simultaneously, the phase of the QPSK signal hops by 180°, and the transmitter power amplifier should have linear dynamic range around the doubled signal amplitude. In O-QPSK, due to half-chip time-offset of symbol bounds, when one of two modulating binary streams changes, the other remains constant. Therefore, the maximal hop of the resulting signal phase is only 90°, softening the demand on the amplifier linear dynamic range and helping to make the battery lifetime longer.

11.3.5 Evolution of air interface cdmaOne to cdma2000

One of the main stimuli for promoting 3G standards was the extension of functionality of a mobile handset from just a telephone device to a terminal capable of high-speed data exchange with the network, receiving and outputting multimedia information, access to Internet etc. All of these novelties demand much higher transmission rates as compared to 2G systems, and the ultimate rate associated with the 3G philosophy is 2 Mbps. Such a dramatic rate increase with no compromise of quality of service and number of users served calls for a broader system bandwidth. It is characteristic of cdma2000 projects that, unlike the UMTS concept (see Section 11.4), they consider IS-95 as the starting point and allow for backward compatibility with it. Of all the technologies united by the common name cdma2000, the one based on the MC-DS-CDMA principle (see Section 10.2.1) of the forward link looks most likely to be finally adopted. We cannot touch upon the details of this proposal here (whose extensive description is the subject of [69]), due to their complexity and possibly non-final state, so we give only a very short survey of the main ideas below.

With three carriers (in future this number may increase) the cdma2000 forward link just repeats three times the spectrum of IS-95 occupying total bandwidth of about 3.75 MHz and using in each of three 1.25 MHz subbands spreading and modulation techniques somewhat different from those of IS-95. Unlike IS-95, where BPSK data modulates the QPSK spreading sequence, in cdma2000 both data and spreading code are of QPSK type. This allows doubling the duration of the codestream symbols, given the data rate, thereby increasing by two times (up to 128) the spreading factor per subband within the same chip rate. Certainly, greater spreading factor means doubled number of forward physical channels, and consequently potentially greater number of users served. But even more importantly, the number of parallel channels granted by the network to the same user may permit transmission of data at the number-of-channel times higher total rate. This *multicode* mode inherited from the IS-95B specification is one of the main resources for approaching the rates pre-assigned by the 3G concept.

Another advancement of the cdma2000 forward link is employing transmit diversity, in particular a technique close to that discussed in Section 10.3.5. Since signals from different transmit antennas propagate along different paths, some extra pilot channels above the one of IS-95 are arranged, transmitting antenna-specific pilot signals from two BS antennas involved in diversity. This enables the MS to separate pilots of different antennas and have strictly synchronized references for the diversity signals to demodulate, decode and properly combine diversity branches.

More complex organization of cdma2000 requires stronger systematization of logical forward channels. Instead of only four in IS-95, the hierarchy of logical channels in cdma2000 contains 10 only on the first layer, adding common control channel, common power control channel, broadcast control channel, quick paging channels etc. As for the traffic channels, they include, among others, the fundamental ones (primarily used) and supplemental channels involved in multicode transmission when very high rates not achievable with only fundamental channels are on request. The supplemental channels may use either convolutional or turbo codes (see Section 9.4).

The reverse link of cdma2000 does not use the MC-DS-CDMA technique, being implemented as conventional DS CDMA with three times broader bandwidth (3.75 MHz). This corresponds to three times higher chip rate (3.6864 Mcps) of the long spreading code. With such high DS spreading rate, a number of solutions of IS-95 were revised and replaced by more relevant ones. First of all, a convolutional encoding rate 1/4 is used instead of 1/3, meaning that the raw data rate, say 9.6 kbps, after encoding becomes 38.4 kbps. Using QPSK data modulation and spreading such a codestream by a long code gives a spreading factor $2 \times (3.6864/38.4) \times 10^3 = 192$, which is not a critical loss as compared to 256 in IS-95. At the same time, the reduction of rate of a convolutional code of constraint length 9 to 1/4 returns some extra (non-asymptotic) coding gain. On this ground the designers decided to reject the orthogonal 64-ary modulation and bring the reverse traffic channel structure closer to that of the forward link.

Another distinction is the long list of logical channels, including the reverse pilot channel, which now becomes obligatory, since in the absence of orthogonal modulation the BS should have a local coherent reference to demodulate the QPSK data of MS. There are other new logical channels, some of them running simultaneously. Being channelized by the Walsh signals, they are then linearly summed, meaning that the transmitter power amplifier should be linear so involving O-QPSK is needless. Thus, an ordinary QPSK data modulator is used, where the QPSK-mapped codestream is first multiplied by the QPSK spreading code and then the real and imaginary parts of the product modulate the cosine and sine CW carrier components, as explained in Section 7.1. Note that, similarly to the forward link, reverse supplemental channels may support a reverse fundamental (traffic) channel (one per MS), providing multicode transmission at the high rates unattainable with only fundamental channels. The supplemental channels may again use either convolutional or turbo codes.

11.4 Air interface UMTS

11.4.1 Preliminaries

The UMTS (acronym of *Universal Mobile Telecommunication System*) is a 3G wideband CDMA standard whose development was pioneered by the European telecommunication community. Currently UMTS, along with cdma2000, enters the so-called IMT-2000 family, i.e. the list of standards declared by the International Telecommunication

Union (ITU) to be basic for 3G systems. There are two versions of UMTS: frequency division duplex (FDD) and time division duplex (TDD). As is immediately seen from the names, these systems differ from each other in the method of separation between downlink and uplink. In FDD-UMTS downlink and uplink occupy non-overlapping frequency bands, while in TDD-UMTS they employ different time slots. For brevity we limit the discussion below to only FDD-UMTS, keeping in mind that many solutions adopted in it are common to both systems.

International regulatory documents allocate for FDD-UMTS in Europe frequency bands 1920–1980 MHz (uplink) and 2110–2170 MHz (downlink) with a limitation of 5 MHz on the link bandwidth. DS spreading is a fundamental technique securing separation of physical channels, in particular multiple access (DS CDMA). The standard sets a universal and constant chip rate 3.84 Mcps in full agreement with the bandwidth limitation. Like cdma2000, UMTS is a system in which the data transmission rate may vary in a very wide range. As a consequence of this, along with chip rate invariance, the spreading factor changes with transmission rate.

In what follows we concentrate again on the physical layer of the system, i.e. solutions concerning spreading, channelization and modulation. Note that BSs of UMTS do not rely upon GPS support and operate with their own autonomous non-synchronized clocks. Although this architecture saves on BS equipment costs, it does, however, make cell search and handover procedures more complicated and is responsible for many distinctions between the physical layers of UMTS and cdma2000.

Similarly to Section 11.3, we can hardly present here anything more than a very short sketch, but many recently published books dedicated to the UMTS standard [92,104,114,115,120–122] will help the interested reader to expand his/her knowledge about this system, which is likely to become dominant among mobile telecommunications in the near future.

11.4.2 Types of UMTS channels

According to the terminology of the UMTS specifications there are *logical*, *transport* and *physical* channels. On the layers higher than physical the data is distributed between the logical channels on the basis of information content, but before arriving at the physical layer it is restructured into the transport channels. The criterion to distinguish transport channels is mode and format of data representation, while physical channels (which are just signals bearing messages), as in any CDMA system, are distinguished by their specific codes. The physical layer consists of two sublayers. The transmitted data arrives at the first physical sublayer from the upper layers packed into transport channels according to information content. The first sublayer includes among other things attachment of CRC for data block protection, channel coding and interleaving. Channel coding is either convolutional (constraint length 9, rates 1/2 or 1/3) or turbo coding (constituent encoders of memory 3, rate 1/3). For very high data rate uncoded transmission is possible, too. The second sublayer, i.e. radio link, covers mapping the transport channels to the physical ones (signals), and transmitting signals over the propagation medium. The signal received by MS (downlink) or BS (uplink) then undergoes all necessary reciprocal operations of the two sublayers (demodulation, deinterleaving, decoding etc.), reverse mapping when necessary, and passing the data over to the higher layers.

Another classification applicable to both transport and physical channels discriminates between *common* and *dedicated* ones. The first of them contain data relating to the whole system and commonly used by all users, whereas the second serve to establish a connection between BS and individual MS.

The time structure of all channels is strictly predetermined by the standard. All of them consist of frames of duration 10 ms (38 400 chips) and every frame is in its turn divided into 15 slots of duration 666.66... μs (2560 chips).

11.4.3 Dedicated physical uplink channels

A dedicated physical channel is the one which the network grants to the MS to utilize exclusively, i.e. two mobiles within the same cell can never use the same dedicated channel. There are two types of dedicated uplink channels: *dedicated physical data channel* (DPDCH)[4] and *dedicated physical control channel* (DPCCH), whose names are consistent with their functionalities. The DPCCH transmits to the BS service information: the pilot signal necessary to estimate propagation parameters and synchronize BS coherent reference, message about the format of data in DPDCH, feedback information used in handoff and power control commands. There is always only one DPCCH per MS, and the transmitted user's data arrives at the physical layer packed into one dedicated transport channel per MS. But this unique transport channel may be mapped to several (up to six) DPDCH, if the necessary rate surpasses the maximum provided by a single DPDCH.

Multiplexing of dedicated channels is performed as follows (see Figure 11.7). The DPCCH is always transmitted through a quadrature branch of the QPSK modulator, while the first of the DPDCH is fed into the in-phase branch. The rest of the DPDCH, if any, are distributed between branches as equally as possible. Thus, despite the QPSK modulator involvement data modulation in the uplink is BPSK. The transmission rate over the DPCCH is always constant, equal to 15 kbps or 10 bits per slot. Then each bit occupies $2560/10 = 256$ chips and the spreading factor of DPCCH is always 256. The spreading factor of DPDCH, on the other hand, is variable depending on the necessary data rate and may change from 256 (minimal rate) to 4 (maximal rate). Thus the minimal data transmission rate is 15 kbps, while the maximal per DPDCH is 64 times higher, i.e. 960 kbps. Using up to six parallel DPDCH in the multicode transmission mode described in the previous section allows in principle a maximal rate of 5760 kbps. Certainly, the so-called gross rate (i.e. of codestream after a channel encoder) is meant, the raw datastream rate being lower proportionally to the channel code rate.

Every uplink physical channel has its specific channelization code multiplied with encoded data. Figure 11.7 presents a scenario where all DPDCH are active, using six different channelization codes CC_1–CC_6 with designation CC_c reserved for the channelization code of DPCCH. After multiplication by the channelization code all physical

[4] The names and abbreviations of the UMTS channels reproduce exactly those of the UMTS documents.

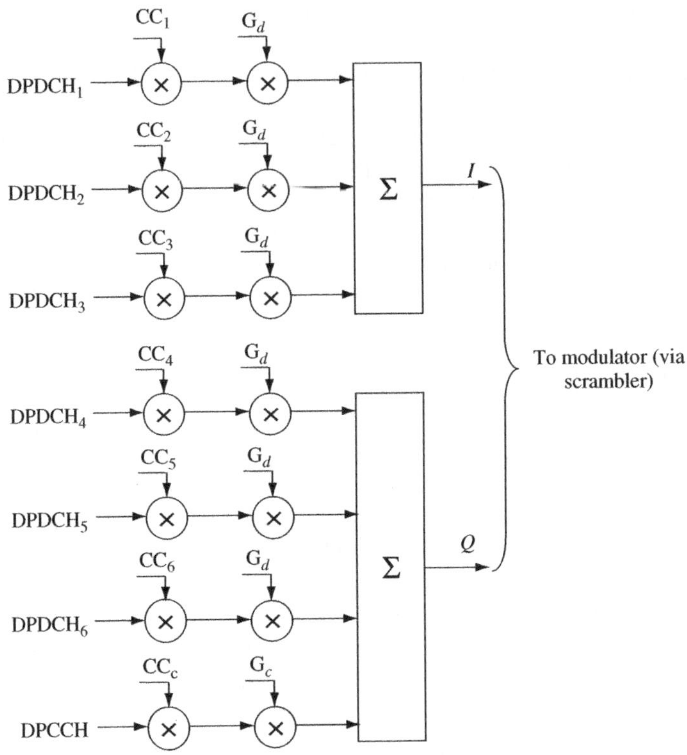

Figure 11.7 Multiplexing of dedicated uplink channels

channels are weighted by their gains, implementing the reverse link power control. Actually, only two different gains are involved: whatever the number of DPDCH, all of them have the same gain G_d, while the gain of DPCCH is G_c. The maximal gain value is one, zero gain cuts off a channel, and the step of variations of gain value is 1/15.

11.4.4 Common physical uplink channels

Common physical channels represent a resource that is at the common disposal of all mobiles. There are two types of common uplink channels: *random access channel* (RACH) and *common packet channel* (CPCH), the terms being used for both the transport and corresponding physical channels. The MS utilizes RACH to initiate a contact (e.g. call) with the network and to transmit short packet messages, CPCH being the main resource for MS packet transmission.

Physical RACH (PRACH) contains a preamble of 4096 chips, which is 16-chip MS identifier repeated 256 times. The MS can start transmission over PRACH at the beginning of any of 15 access slots occupying together two frames, i.e. 5120 chips or 20 ms. The diagram of access slots is set up in the BS transmission format using a special transport downlink channel (*broadcasting channel*, BCH). At the initial stage the MS has no reliable knowledge about signal attenuation in the uplink, and the first transmission of preamble goes at low power. As long as no message comes from the BS confirming connection, the MS randomly selects new access slots and makes new attempts, each time increasing the signal power. After receiving BS confirmation, the MS transmits the message itself covering one or two frames (10 or 20 ms). There is no closed power control loop in PRACH, since the connection sessions on it are quite short.

The structures of the physical CPCH (PCPCH) and PRACH are in general similar; however, the message segment of the PCPCH may occupy a larger number of frames and the preamble segment is followed by one more part: the collision detection preamble, which helps to recognize simultaneous attempts of several MSs to use PCPCH. Due to the longer duration of PCPCH packets the presence of a closed power control loop in it appears to be reasonable in contrast to PRACH.

In the same way as in the dedicated channels, quadrature multiplexing is used in the common channels to combine information and service components of transmitted data. Separation of common and dedicated physical channels is performed by channelization codes, which are considered in the next subsection.

11.4.5 Uplink channelization codes

As stated above, an individual MS uses several types of channels to be separated in the BS receiver. Since the same clock controls all the physical channels of the MS, synchronous code division is a good option to arrange separation between physical channels (multiple DPDCH, DPCCH, PRACH, PCPCH). Note that we are now discussing only the separation of channels of an isolated MS, signals of different mobiles remaining separated by asynchronous CDMA, as in cdmaOne and cdma2000.

In the UMTS documents the channelization coding format is described by the binary tree establishing an iteration procedure. At every iteration any word of the preceding iteration generates two new words of double length by appending either itself or its negation. Let $\mathbf{c}_k$ be some code vector obtained at the kth iteration. Then two code vectors of double dimension $\mathbf{c}_{k+1} = (\mathbf{c}_k, \mathbf{c}_k)$ and $\mathbf{c}'_{k+1} = (\mathbf{c}_k, \ -\mathbf{c}_k)$ are its descendants at the next iteration. In this way, starting with a trivial word (1) of length one leads after k iterations to 2^k channelization codewords of length $N = 2^k$ (Figure 11.8 shows this for the case $k = 3$). One can easily see that this algorithm differs from the Sylvester rule of constructing the Hadamard matrix (see Section 2.7.3) only in reordering the resulting matrix rows, so that the codewords obtained are nothing but Walsh functions. Nevertheless they appear in the UMTS specifications under a special name, *orthogonal variable spreading factor* (OVSF) codes.

The uplink OVSF tree is built using $k = 8$ iterations, so that codewords have maximal length $N = 256$ chips. The word of this length consisting of all ones is allocated to the dedicated control channel DPCCH, whereby a subsequent scrambling secures DPCCH spreading factor 256, mentioned in Section 11.4.3. Choice of channelization code of the same length for DPDCH (spreading factor 256) allows transmitting data at the rate 15 kbps. If a greater rate is on demand, the data symbol becomes shorter, i.e. the spreading factor drops (the chip rate never changes, remaining 3.84 Mcps!). Appropriate DPDCH channelization words are then taken from the intermediate (kth, $SF = 2^k$ being

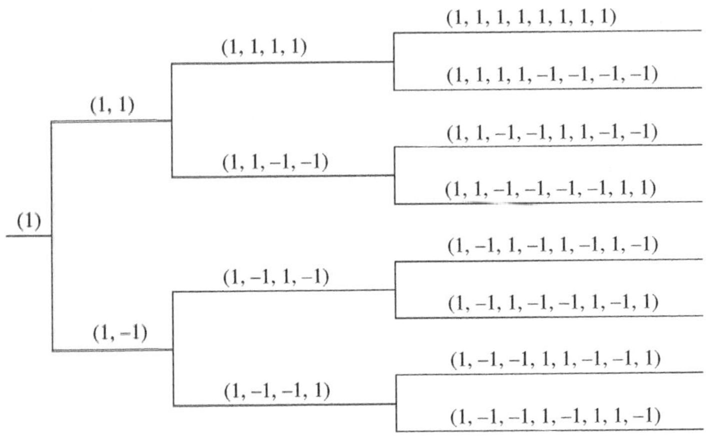

Figure 11.8 Tree of OVSF codes of length 8

the required spreading factor) iteration of the OVSF code tree. If a necessary rate is attainable within a single DPDCH its channelization word number is fixed by the specification as $SF/4 = 2^{k-2}$, counting top-down on the tree and numbering the uppermost word by zero. Such a choice always preserves orthogonality between DPDCH and DPCCH independently of DPDCH spreading factor. When a single DPDCH (with spreading factor 4) cannot provide the necessary data rate, the multicode transmission mode involves several DPDCH, all having always the same minimum spreading factor 4. There are three codewords of length 4 orthogonal to each other and to the word of DPCCH, so, allowing for the chance of using the same word for two DPDCH in quadrature multiplexing (Figure 11.7), up to 6 multicode DPDCH may be arranged.

The technique of allocation of codewords to PRACH and PCPCH set up by the specification also provides their orthogonality to the dedicated channels over the whole range of DPDCH data rates.

11.4.6 Uplink scrambling

The final step in forming the MS signal is its DS spreading by a user-specific signature to realize asynchronous CDMA separation of signals of different MSs in the BS receiver. According to the UMTS thesaurus, this operation, as well as a similar operation in the downlink, is called *scrambling*. This usage of the term is slightly different to that in the cdmaOne and cdma2000 documents (see Sections 11.3.3.3 and 11.3.3.4). There are two types of uplink scrambling codes predetermined by the standard: either long or short codes apply to scramble dedicated channels.

The Gold sequences (see Section 7.5.2) of length $L = 2^{25} - 1$ truncated to one frame period, i.e. 38 400 chips, serve to build up the long scrambling codes. The LFSR generators with feedback polynomials $f_1(x) = x^{25} + x^3 + 1$ and $f_2(x) = x^{25} + x^3 + x^2 + x + 1$ form two m-sequences, of which the second may also be obtained by decimation of the first with the index relevant for Gold codes. The modulo 2 addition of the offset second m-sequence to the first gives a Gold sequence which is then truncated and mapped to the $\{\pm 1\}$ alphabet. The 'pure' truncated Gold sequence $\{c_i\}$ determines the real component of the MS QPSK signature. To obtain the imaginary part the original (before truncation) Gold sequence is first offset by 16 777 232 chips and then truncated and mapped to produce the $\{\pm 1\}$ sequence $\{c'_i\}$. After this, the negative of every even element of the latter sequence replaces the following odd element and the result is chip-wise multiplied with $\{c_i\}$. The last operation halves the number of elements in the resulting QPSK sequence whose polarities are opposite to the preceding. Indeed, the equation describing a QPSK signature a_i thus obtained is:

$$a_i = c(i) + j(-1)^i c(i)c'(2\lfloor i/2 \rfloor), \tag{11.1}$$

where $c(i)$ and $c'(i)$ are just more convenient designations for the elements c_i and c'_i. One can observe now that in passing from an even position $i = 2l$ to the next odd $i = 2l + 1$ the real and imaginary part cannot change simultaneously, meaning that maximal phase jumps are only $\pm 90°$, never $180°$, so a chip polarity may change into the opposite only during transitions from odd to even chips. Reduced frequency of the opposite transitions is typically desirable for battery lifetime, since power amplifier during them dissipates higher energy.

The final step of QPSK scrambling consists in multiplication of the MS multiplexed signal taken from the branches of the scheme of Figure 11.7 with scrambling sequence. This operation, which is performed by an ordinary QPSK modulator, was discussed at length in Section 7.1.2.

A scrambling sequence is precisely synchronized to the MS clock beginning in any frame from the same symbol (chip).

Note that the key advantage of Gold codes—minimax periodic correlation properties—cannot justify their choice for uplink scrambling, because truncation corrupts correlations drastically. An alternative reasoning is ease of generation of many (no fewer than $2^{25} + 1$) pseudorandom sequences.

Short scrambling sequences have length 256 and are planned for employment when a BS receiver is capable of running multiuser algorithms (see Section 10.1). In the case of asynchronous CDMA, the complexity of this type of receiver typically grows with the length of spreading code. The rule of forming short scrambling codes set by the specification includes generating a linear recurrent quaternary sequence of length 255, modulo 4 summation of it with two binary recurrent sequences of lengths 51 and 85, extension of the resulting sequence by one element up to length 256, mapping quaternary symbols onto the QPSK alphabet, and transformation of the obtained complex sequence by rule (11.1).

11.4.7 Mapping downlink transport channels to physical channels

Information conveyed by the network to a specific MS at the transport layer is organized as a single dedicated channel, which is then mapped to two downlink physical channels similar to those in the uplink: data (DPDCH) and control (DPCCH) dedicated channels.

The list of downlink common transport channels is much wider compared to the uplink. It contains, specifically, the already mentioned broadcast channel BCH transmitting parameters of the network or cell used by all MS, the forward access channel (FACH) utilized by the BS to send command data to the MS with known location, and the paging channel (PCH), over which the BS sends commands to the MS with unknown location, etc. At the physical layer the primary common control physical channel (P-CCPCH) transmits BCH data, FACH is mapped to the secondary common control physical channel (S-CCPCH), and transmission of PCH is performed again partly by S-CCPCH and partly by the synchronization channel (SCH). There is also a common pilot channel (CPICH) transmitting data-unmodulated signal to be used by the receiver for estimating the channel parameters. A complete scheme of correspondence between the downlink transport and physical channels set up by the specification may be found in [92,121].

11.4.8 Downlink physical channels format

The mechanism of multiplexing the downlink dedicated data and control channels DPDCH and DPCCH is different from that of the uplink: every slot is split into parts and every part is allocated either to DPDCH or to DPCCH. Under this arrangement the BS transmitter operates at times in a discontinuous manner, due to the pauses in the datastream during periods of low voice activity. The discontinuous emission being used in the MS transmitter might cause interference, badly affecting nearby electronic devices (e.g. hearing aid). This phenomenon has been repeatedly reported for GSM handsets. However, as for the BS, its discontinuous radiation cannot be that harmful, since the devices mentioned above are hardly likely to be close enough to the BS antenna.

Another distinction of the downlink physical channel format is QPSK (unlike BPSK in uplink) data modulation. First, the demultiplexer (DMUX) splits the bit stream after the first physical sublayer (channel encoding, interleaving etc.) into two streams, bearing even and odd original bits, respectively. Both streams mapped to the $\{\pm 1\}$ alphabet are multiplied by a binary channelization code (CC), giving real and imaginary parts of a complex channelized data signal, as Figure 11.9 shows. The final step is multiplication of this QPSK data signal with the QPSK spreading (scrambling) code and QPSK carrier modulation, performed the same way as in the uplink.

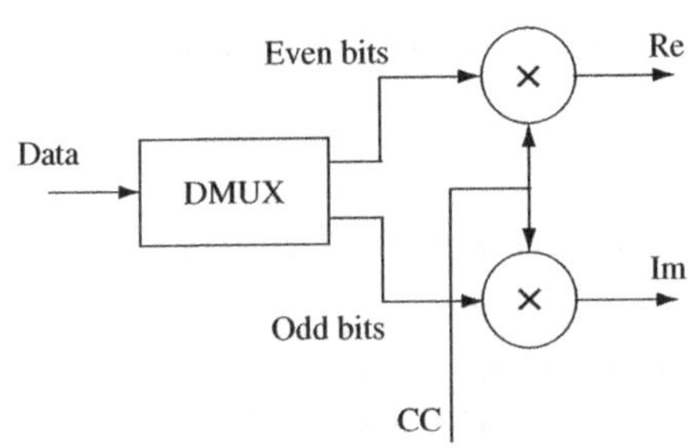

Figure 11.9 Data demultiplexing and channelization in UMTS uplink

11.4.9 Downlink channelization codes

Time multiplexing of data and control channels discussed in the previous subsection means that only one dedicated physical channel (DPCH) is necessary to transmit both these information streams. When a single such channel meets the requirement for the data rate the BS forms it by means of a user-specific channelization code sequence, which, of course, is unique over the cell (or sector) and cannot be utilized to contact any other MS. The scenario of multicode transmission arises when the only physical channel is incapable of transmitting data at the demanded rate. Then the BS involves several parallel physical channels to contact the same MS. These channels always operate with the same spreading factor, but with different channelization codes, which cannot be reused by the same BS in connections with other MSs. There is no need to repeat command messages in all multicode channels, so control information is sent to the MS over only one of them.

As such, the downlink and uplink channelization codes are of the same type. The Walsh functions family or OVSF tree of Figure 11.8 is used with spreading factor in the range from 4 to 512. Some channelization sequences are not available for DPCH, being allocated to common channels, like CPICH. For example, under the minimal spreading factor only three code sequences may serve DPCH, providing maximal gross data rate 2.88 Mbps. Assuming channel code rate 1/2 and unavoidable overhead (control commands etc.), this gross rate does not comply with 3G demands on the 'pure' data rate (up to 2 Mbps), which is one of the reasons why the specification stipulates high-speed uncoded transmission, too.

When many users are operating at different rates their channelization codes should preserve orthogonality despite the difference in values of the spreading factors, or, which is the same, in code lengths. As one may conclude on inspecting Figure 11.8, two Walsh functions of different lengths are orthogonal over the minimal length interval if and only if neither of them is a descendant of the other. It follows, then, that—unlike the uplink, where every MS, being isolated from the others by its unique scrambling law, has the whole set of channelization codes at its exclusive disposal—the management of the downlink channelization is much more complicated. Indeed, different MSs should be assigned different subsets of Walsh sequences, containing no descendants of sequences currently serving other MSs. This problem of dynamic resource coordination is being solved at the upper layers of the network protocol stack.

11.4.10 Downlink scrambling codes

As was mentioned earlier, downlink scrambling codes secure separation of signals of different BSs. Each scrambling sequence has in its basis a Gold sequence of length $L = 2^{18} - 1 = 262\,143$. Two LFSR with the feedback polynomials $f_1(x) = x^{18} + x^7 + 1$ and $f_2(x) = x^{18} + x^{10} + x^7 + x^5 + 1$ generate initial m-sequences related to each other as required to produce a Gold ensemble (see Section 7.5.2). Although there are $2^{18} + 1 = 262\,145$ Gold sequences of the length above, the specification limits the number of those utilized to $2^{13} = 8192$. Two segments of length 38 400 are cut from every allowed sequence: the initial one and the one offset by $2^{17} = 131\,072$ chips, converted then to the $\{\pm 1\}$ alphabet in the usual fashion (6.15). The resulting binary sequences are, respectively, the real and imaginary parts of a downlink QPSK scrambling code.

The specification establishes a strict hierarchy of the scrambling sequences. The set of all scrambling codes is divided into 512 subsets, each consisting of 1 primary and 15 secondary codes. All 512 primary codes are in turn divided into 64 groups, 8 code sequences in each one. Only one unique primary code is allocated to a specific BS. Some physical channels are allowed to use only the primary code, while others may use either primary or secondary codes.

11.4.11 Synchronization channel

11.4.11.1 General structure

The synchronization channel (SCH) possesses quite an important role in the network structure, providing initial cell search and aligning the MS clock diagram with the boundaries of frames and the slots of the received BS signal. Neither channelization nor scrambling codes participate in forming the synchronization channel, since its signal should be found and processed before the MS is aware of the scrambling code of a contacted BS. Note again that BS clocks over the UMTS network are not synchronized to each other, and the initial cell search has to be fulfilled wherever an MS switches to another BS (e.g. in the course of handover). This proved to be one of the key factors in designing the SCH architecture to fit a two-stage search procedure, which is potentially capable of saving acquisition time in comparison with an ordinary serial search (see Sections 8.2 and 8.3). As a matter of fact, the SCH is a pair of primary and secondary synchronization channels, whose signals—primary and secondary synchronization codes—are exploited at the first and second search stages, respectively. Both signals occupy 256 initial chips of every slot. The primary synchronization code sequence is identical not only over all slots but also over all BSs of the network. This makes it impossible for an MS to capture the primary synchronization signal of a predetermined BS; to which of them the MS got connected after the first search stage will become clear only as a result of the second stage. After finding the primary synchronization signal, the MS knows the boundaries of slots but not of frames. At the second stage the MS should remove this ambiguity by testing all 15 (number of slots within a frame) possible values of mismatch between the MS clock and the received BS signal for all possible versions of the secondary synchronization code. To match this task the secondary synchronization code has one frame (15 slots or 38 400 chips) period and a code structure, which is specific for a given BS. Besides, the secondary code sequence is strictly bound to the scrambling code group (one of 64) assigned to the base station. Thus, having completed the second search stage, the MS recognizes the scrambling code group of the captured BS signal, and finally, after testing eight possible primary scrambling code sequences, finds out which of them is used by the contacted BS.

11.4.11.2 Primary synchronization code

The primary synchronization code (PSC) is defined in the specification as a 16-element sequence $\mathbf{a} = (1, 1, 1, 1, 1, 1, -1, -1, 1, -1, 1, -1, 1, -1, -1, 1)$, which is repeated 16 times with or without polarity change to produce the 256-element binary sequence:

$$\mathbf{c}_{ps} = (\mathbf{a}, \mathbf{a}, \mathbf{a}, -\mathbf{a}, -\mathbf{a}, \mathbf{a}, -\mathbf{a}, -\mathbf{a}, \mathbf{a}, \mathbf{a}, \mathbf{a}, -\mathbf{a}, \mathbf{a}, -\mathbf{a}, \mathbf{a}, \mathbf{a})$$

Every BS directly transmits the same sequence $\mathbf{c}_{ps}$ implementing PSC and enabling the MS to perform the first stage of cell search. Since PSC is meant for time measurement in the multipath environment, it should have relevant autocorrelation properties (see Section 6.1). Despite PSC periodicity, most of its period is empty, so aperiodic ACF sidelobe level is an adequate criterion of PSC quality. As pointed out in Example 6.10.2 (see also Figure 6.16), aperiodic ACF of the UMTS PSC is far from prominent, which is probably a penalty for the hardware simplifications pursued by designers.

11.4.11.3 Secondary synchronization code

The secondary synchronization codes (SSC) are built on the basis of a 16-element sequence $\mathbf{b}$ replicating the sequence $\mathbf{a}$ above in the first eight elements and the negative of $\mathbf{a}$ (i.e. $-\mathbf{a}$) in the rest: $\mathbf{b} = (1, 1, 1, 1, 1, 1, -1, -1, -1, 1, -1, 1, -1, 1, 1, -1)$. This sequence is repeated 16 times with or without polarity change to produce the 256-element sequence:

$$\mathbf{z} = (\mathbf{b}, \mathbf{b}, \mathbf{b}, -\mathbf{b}, \mathbf{b}, \mathbf{b}, -\mathbf{b}, -\mathbf{b}, \mathbf{b}, -\mathbf{b}, \mathbf{b}, -\mathbf{b}, -\mathbf{b}, -\mathbf{b}, -\mathbf{b}, -\mathbf{b})$$

The sequence $\mathbf{z}$ is then symbol-wise multiplied with every 16th row of a Sylvester-type Hadamard matrix (see Section 2.7.3) of size 256 to generate 16 orthogonal 256-ary alphabet symbols used further to generate codewords of SSC. The SSC encoding algorithm should secure low levels of all incorrect correlations, i.e. between all 15 cyclic replicas of any one word as well as between any time-shifts of different words. In the UMTS specification all 64 SSC codewords are presented as a table, but an immediate check indicates that these words are just taken from (15,3) 256-ary Reed–Solomon code. The minimum Hamming distance of any Reed–Solomon code is greater by one than the number of check symbols [31,33], i.e. in our case equals 13. This guarantees that there are no more than 2 coinciding 256-ary symbols in any SSC word and any of its 14 cyclic replicas shifted to each other by an integer number of slots. In other words, the normalized periodic ACF of any SSC codeword at such shifts has a level no higher than 2/15. The same is true for any two codewords assigned to different BSs, when only shifts by an integer number of slots are analysed. At the same time, since the BSs of the network do not rely on a common clock, synchronization codes transmitted by them slide against each other and low cross-correlation values should be preserved under arbitrary time-shifts, not only those equal to an integer number of slots. The Reed–Solomon code distance optimality cannot secure a low level of correlations under arbitrary shifts of SSC sequences, and this issue calls for further investigation.

As a common conclusion to Sections 11.3 and 11.4, we again stress that our sketchy excursion into 3G systems and standards dealt only with those solutions relevant to the context of this book, i.e. concerning practical implementation of spread spectrum and CDMA ideas. A lot of books and papers already exist, and many more are anticipated, explaining and clarifying various aspects of 3G and beyond, and covering all the layers of the system protocol stack. In particular, we refer the curious reader to [19,69,92,104,114,115,120–122] and to the 3G specifications directly.

References

[1] Carlson, A. B. *Communication Systems*, McGraw-Hill, New York, 1986.

[2] Sklar, B. *Digital Communications*, Prentice-Hall, Upper Saddle River, NJ, 2001.

[3] Dixon, R. C. *Spread Spectrum Systems with Commercial Applications*, John Wiley & Sons, Chichester, 1994.

[4] Haykin, S. *Communication Systems*, John Wiley & Sons, Chichester, 2001.

[5] Ziemer, R. E. and Peterson, R. L. *Introduction to Digital Communication*, Prentice-Hall, Upper Saddle River, NJ, 2001.

[6] Ziemer, R. E., Peterson, R. L. and Borth, D. E. *Introduction to Spread Spectrum Communications*, Prentice-Hall, Englewood Cliffs, NJ, 1995.

[7] Proakis, J. G. *Digital Communications*, McGraw-Hill, New York, 2001.

[8] Proakis, J. and Salehi, M. *Communication Systems Engineering*, Prentice-Hall, Upper Saddle River, NJ, 2002.

[9] Simon M. K., Omura, J. K., Scholtz, R. A. and Levitt, B. K. *Spread Spectrum Communication Handbook*, McGraw-Hill, New York, 1994.

[10] Scholtz, R. A. The origins of spread spectrum communications, *IEEE Trans. Commun.*, **30**, 822–854, 1982.

[11] Conway, J. H. and Sloane, N. J. A. *Sphere Packings, Lattices and Groups*, Springer-Verlag, New York, 1998.

[12] Sloane, N. J. A. Spherical codes, nice arrangements of points on a sphere in various dimensions, http://www.research.att.com/~njas/packings/index.html.

[13] Adams, R. A. *Calculus: A Complete Course*, 4th edn, Addison-Wesley Longman, Don mills, Ontario, 1999.

[14] Leon-Garcia, A. *Probability and Random Processes for Electrical Engineering*, Addison-Wesley, Reading, MA, 1994.

[15] Freeman, R. L. *Radio System Design for Telecommunications*, John Wiley & Sons, Chichester, 1997.

[16] Berg, O., Berg, T., Haavik, S., Hjelmstad, J. and Skaug, R. *Spread Spectrum in Mobile Communication*, IEE, London, 1998.

[17] Lee, W. C. Y. *Mobile Communications Engineering*, McGraw-Hill, New York, 1997.

[18] Lee, J. S. and Miller, L. E. *CDMA Systems Engineering Handbook*, Artech House, Boston, MA, 1998.

[19] Glisic, S. *Adaptive WCDMA: Theory and Practice*, John Wiley & Sons, Chichester, 2003.

[20] Kim, K. I. CDMA cellular engineering issues, *IEEE Trans. Veh. Tech.*, **42**, 345–350, 1993.

[21] Gradstein, I. and Ryzhik, I. *Table of Integrals, Series, and Products*, Academic Press, New York, 1980.

[22] Eliahou, S. and Kervaire, M. Barker sequences and difference sets, *L'Enseignement Mathématique*, **38**, 345–382, 1992.

[23] Friese, M. Polyphase Barker sequences up to length 36, *IEEE Trans. Inform. Theory*, **42**, 1248–1250, 1996.

[24] Brenner, A. F. Polyphase Barker sequences up to length 45 with small alphabets, *Electron. Lett.*, **34**, 1576–1577, 1998.

Spread Spectrum and CDMA: Principles and Applications Valery P. Ipatov

[25] Lindner, J. Binary sequences up to lengths 40 with best possible autocorrelation function, *Electron. Lett.*, **11**, 507, 1975.
[26] Cohen, M. N., Fox, M. R. and Baden, J. M. Minimum peak sidelobe pulse compression codes, in *Proceedings of the IEEE International Radar Conference*, 7–10 May 1990, Arlington, VA, pp. 633–639, IEEE, 1990.
[27] Deng, X. and Fan, P. New binary sequences with good aperiodic autocorrelation obtained by evolutionary algorithm, *IEEE Commun. Lett.*, **3**, 288–290, 1999.
[28] Baumert, L. D. *Cyclic Difference Sets*, Springer-Verlag, Berlin, 1971.
[29] Schmidt, B. Cyclotomic integers and finite geometry, *J. Am. Math. Soc.*, **12**, 929–952, 1999.
[30] Gilbert, W. J. and Nicholson, W. K. *Modern Algebra with Applications*, John Wiley & Sons, Chichester, 2004.
[31] Bossert, M. *Channel Coding for Telecommunications*, John Wiley & Sons, Chichester, 1999.
[32] Lidl, R. and Niederreiter, H. *Introduction to Finite Fields and their Applications*, Cambridge University Press, Cambridge, 1994.
[33] Blahut, R. E. *Algebraic Codes for Data Transmission*, Cambridge University Press, Cambridge, 2003.
[34] Sverdlick, M. B. *Optimal Discrete Signals*, Sov. Radio, Moscow, 1975 (in Russian).
[35] Lüke, H. D., Schotten, H. D. and Hadinejad-Mahram, H. Binary and quadriphase sequences with optimal autocorrelation properties: a survey, *IEEE Trans. Inform. Theory*, **49**, 3271–3282, 2003.
[36] Lüke, H. D. *Korrelationsignale*, Springer-Verlag, Berlin, 1992 (in German).
[37] Fan, P. and Darnell, M. *Sequence Design for Communication Applications*, John Wiley & Sons, Chichester, 1996.
[38] Amiantov, I. N. *Selected Issues of Statistical Communications Theory*, Sov. Radio, Moscow, 1971 (in Russian).
[39] Levanon, N. and Mozeson, E. *Radar Signals*, John Wiley & Sons, Chichester, 2004.
[40] Ipatov, V. P. Ternary sequences with ideal periodic autocorrelation properties, *Radio Eng. Elect. Physics*, **24**(10), 75–79, 1979.
[41] Ipatov, V. P. Contributory to the theory of ternary sequences with perfect periodic autocorrelation properties, *Radio Eng. Elect. Physics*, **25**(4), 31–34, 1980.
[42] Zierler, N. Linear recurring sequences, *J. Soc. Appl. Math.*, **7**, 31–48, 1959.
[43] Hoholdt, T. and Justesen, J. Ternary sequences with perfect periodic autocorrelation, *IEEE Trans. Inform. Theory*, **29**, 597–600, 1983.
[44] Ackroyd, M. H. and Ghani, F. Optimal mismatched filter for sidelobe suppression, *IEEE Trans. Aerosp. Electron. Syst.*, **9**, 214–218, 1973.
[45] Rihaczek, A. W. and Golden, R. M. Range sidelobe suppression for Barker codes, *IEEE Trans. Aerosp. Electron. Syst.*, **7**, 1087–1092, 1971.
[46] Ipatov, V. P. Total suppression of sidelobes of periodic correlation functions of phase manipulated signals, *Radio Eng. Elect. Physics*, **22**(8), 42–47, 1977.
[47] Ipatov, V. P. Choice of periodical PSK signal and filter combination, *Radioelectron. a. Commun. Syst.* (*Radioelektronika*), **21**(4), 49–55, 1978.
[48] Ipatov, V. P. Synthesis of a binary periodic signal-filter pair, *Radioelectron. a. Commun. Syst.* (*Radioelektronika*), **23**(4), 46–51, 1980.
[49] Ipatov, V. P. Binary periodic sequences with low sidelobe suppression loss, *Radioelectron. a. Commun. Syst.* (*Radioelektronika*), **23**(1), 15–19, 1980.
[50] Ipatov, V. P. and Fedorov, B. V. Regular binary sequences with small losses in suppressing sidelobes, *Radioelectron. a. Commun. Syst.* (*Radioelektronika*), **27**(3), 29–33, 1984.
[51] Blokhuis, A. and Tiersma, H. J. Bounds for the size of radar arrays, *IEEE Trans. Inform. Theory*, **34**, 164–167, 1988.
[52] Hamkins, J. and Zeger, K. Improved bounds on maximum size binary radar array, *IEEE Trans. Inform. Theory*, **43**, 997–1000, 1997.
[53] Golomb, S. W. and Taylor, H. Two-dimensional synchronization patterns for minimum ambiguity, *IEEE Trans. Inform. Theory*, **28**, 600–604, 1982.
[54] Golomb, S. W. and Taylor, H. Construction and properties of Costas arrays, *Proc. IEEE*, **72**, 1143–1163, 1984.
[55] Chatschik, B. An overview of Bluetooth wireless technology, *IEEE Commun. Magazine*, **39**, 86–94, 2001.
[56] Ross, J. A. F. and Taylor, D. P. Vector assignment scheme for $N + M$ users in N-dimensional global additive channel, *Electron. Lett.*, **28**, 1634–1636, 1992.

[57] Paavola, J. and Ipatov, V. P. Binary CDMA signatures for $N + M$ users in N-dimensional global signal space, *Electron. Lett.*, **39**, 738–740, 2003.

[58] Djonin, D. and Bhargava V. New results on low complexity detectors for oversaturated CDMA systems, in *Proceedings of GLOBECOM*, 25–29 November 2001, San Antonio, TX, vol. 2, pp. 846–850, IEEE, 2001.

[59] Paavola, J. and Ipatov, V. P. Oversaturating synchronous CDMA systems on the signature per user basis, in *Proceedings of the 5th European Personal Mobile Communications Conference*, 22–25 April 2003, Glasgow, UK, pp. 427–430, IEEE, 2003.

[60] Learned, R. E., Willsky, A. S. and Boroson, D. M. Low complexity optimal joint detection for oversaturated multiple access communications, *IEEE Trans. Signal Processing*, **45**, 113–123, 1997.

[61] Shi, Z. and Schlegel, C. Spreading code construction for CDMA, *IEEE Commun. Lett.*, **7**, 4–6, 2003.

[62] Welch, L. R. Lower bound on the maximum cross-correlation of signals, *IEEE Trans. Inform. Theory*, **20**, 397–399, 1974.

[63] Rupf, M. and Massey, J. L. Optimum sequence multisets for synchronous code-division multiple-access channels, *IEEE Trans. Inform. Theory*, **40**, 1261–1266, 1994.

[64] Karystinos, G. N. and Pados, D. A. New bounds on the total squared correlation and optimum design of DS-CDMA binary signature sets, *IEEE Trans. Commun.*, **51**, 48–51, 2003.

[65] Ipatov, V. P. On the Karystinos–Pados bounds and optimal binary DS-CDMA signature ensembles, *IEEE Commun. Lett.*, **8**, 81–83, 2004.

[66] Stark, H. and Woods, J. W. *Probability and Random Processes with Applications to Signal Processing*, Prentice-Hall, Upper Saddle River, NJ, 2002.

[67] Sarwate, D. V. and Pursley, M. B. Crosscorrelation properties of pseudorandom and related sequences, *Proc. IEEE*, **68**, 593–619, 1980.

[68] Sidelnikov, V. M. On the mutual correlation of sequences, *Soviet Math. Dokl.*, **12**, 197–201, 1971.

[69] Korowajczuk L., Xavier, B. S. A., Filho, A. M. F., Ribeiro, L. Z., Korowajczuk, C. and DaSilva, L. A. *Designing cdma2000 Systems*, John Wiley & Sons, Chichester, 2004.

[70] Ipatov, V. P. *Periodic Discrete Signals with Optimal Correlation Properties*, Radio I Sviaz, Moscow, 1992 (in Russian).

[71] Goldberg, B.-G. Code division multiplexing by frequency shifted biphase modulated M-sequences, *IEEE Trans. Aerosp. Electron. Syst.*, **17**, 303–304, 1981.

[72] Gold, R. Optimal binary sequences for spread spectrum multiplexing, *IEEE Trans. Inform. Theory*, **13**, 619–621, 1967.

[73] Kasami, T. Weight distribution formula for some class of cyclic codes, Technical Report R-285, Coordinated Science Laboratory, University of Illinois, Urbana, April 1966.

[74] Kamaletdinov, B. Zh. An optimal ensemble of binary sequences based on the union of the ensembles of Kasami and bent-function sequences, *Problems of Inform. Transmission*, **24**, 167–169, 1988.

[75] Olzen, J. D., Scholtz, R. A. and Welch, L. R. Bent-function sequences, *IEEE Trans. Inform. Theory*, **28**, 858–864, 1982.

[76] Kamaletdinov, B. Zh. Optimal sets of binary sequences, *Problems of Inform. Transmission*, **32**, 171–175, 1996.

[77] Glisic, S. and Vucetic, B. *Spread Spectrum CDMA Systems for Wireless Communications*, Artech House, Boston, MA, 1997.

[78] Ipatov V. P., Kazarinov, Yu. M. (ed.), Kolomensky, Yu. A. and Uljanitsky, Yu. D. *Acquisition, Detection and Parameter Measuring of Signals in Radionavigation Systems*, Sov. Radio, Moscow, 1975 (in Russian).

[79] Holmes, J. K. and Chen, C. C. Acquisition time performance of PN spread spectrum systems, *IEEE Trans. Commun.*, **25**, 778–784, 1977.

[80] Polydoros, A. and Weber, C. L. A unified approach to serial search spread-spectrum code acquisition—Part I: General theory, *IEEE Trans. Commun.*, **32**, 542–549, 1984.

[81] Polydoros, A. and Weber, C. L. A unified approach to serial search spread-spectrum code acquisition—Part II: A matched filter receiver, *IEEE Trans. Commun.*, **32**, 550–560, 1984.

[82] Di Carlo, D. M. and Weber, C. L. Multiple dwell serial search: performance and application in direct sequence code acquisition, *IEEE Trans. Commun.*, **31**, 650–659, 1983.

[83] Viterbi, A. J. *CDMA: Principles of Spread Spectrum Communication*, Addison-Wesley, Reading, MA, 1995.

[84] Zigangirov, K. Sh. *Theory of Code Division Multiple Access Communication*, John Wiley & Sons, Chichester, 2004.

[85] Wald, A. *Sequential Analysis*, John Wiley & Sons, New York, 1947.

[86] Stiffler, J. J. *Theory of Synchronous Communications*, Prentice-Hall, Englewood Cliffs, NJ, 1971.

[87] Golomb, S. W. (ed.) *Digital Communications with Space Applications*, Prentice-Hall, Englewood Cliffs, NJ, 1964.

[88] Van Trees, H. L. *Detection, Estimation and Modulation Theory, Part II, Nonlinear Modulation Theory*, John Wiley & Sons, Chichester, 2002.

[89] Lindsey, W. C. *Synchronization Systems in Communication and Control*, Prentice-Hall, Englewood Cliffs, NJ, 1972.

[90] Viterbi, A. J. *Principles of Coherent Communications*, McGraw-Hill, New York, 1966.

[91] Clark G. C. and Cain, J. B. *Error-Correcting Coding for Digital Communications*, Plenum Press, New York, 1988.

[92] Castro, J. P. *The UMTS Network and Radio Access Technology. An Interface Techniques for Future Mobile Systems*, John Wiley & Sons, Chichester, 2001.

[93] Viterbi, A. J. and Omura, J. K. *Principles of Digital Communications and Coding*, McGraw-Hill, New York, 1979.

[94] Heller, J. A. and Jacobs, I. W. Viterbi decoding for satellite and space communications, *IEEE Trans. Commun. Technol.*, **19**, 835–848, 1971.

[95] Berrou, C., Glavieux, A. and Thitimajshima, P. Near Shannon limit error-correcting coding and decoding: turbo codes, in *Proceedings of the IEEE International Conference on Communications, ICC '93*, 23–26 May 1993 Geneva, Switzerland, vol. 2, pp. 1064–1070, IEEE, 1993.

[96] Berrou, C. and Glavieux, A. Near optimum error correcting coding and decoding: turbo codes, *IEEE Trans. Commun.*, **44**, 1261–1271, 1996.

[97] Schlegel, C. B. and Pérez, L. C. *Trellis and Turbo Coding*, John Wiley & Sons, Chichester, 2004.

[98] Bahl, L. R., Cocke, J., Jelinek, F. and Raviv, J. Optimal decoding of linear codes for minimizing symbol error rate, *IEEE Trans. Inform. Theory*, **20**, 284–287, 1974.

[99] Madhow, U. and Honig, M. L. MMSE interference suppression for direct-sequence spread spectrum CDMA, *IEEE Trans. Commun.*, **42**, 3178–3188, 1994.

[100] Wang, X. and Poor, V. H. *Wireless Communication Systems. Advanced Techniques for Signal Reception*, Prentice-Hall, Upper Saddle River, NJ, 2004.

[101] Honig, M. and Tatsanis, M. K. Adaptive techniques for multiuser CDMA receivers, *IEEE Signal Process. Magazine*, **17**, 49–61, 2000.

[102] Verdú, S. *Multiuser Detection*, Cambridge University Press, Cambridge, 1998.

[103] Lu, W. W. (ed.) *Broadband Wireless Mobile. 3G and Beyond*, John Wiley & Sons, Chichester, 2002.

[104] Tanner, R. and Woodward, J. *WCDMA—Requirements and Practical Design*, John Wiley & Sons, Chichester, 2004.

[105] Fazel, K. and Kaiser, S. *Multi-Carrier and Spread Spectrum Systems*, John Wiley & Sons, Chichester, 2003.

[106] Hara, S. and Prasad, R. Overview of multicarrier CDMA, *IEEE Commun. Magazine*, **35**, 126–133, 1997.

[107] Popović, B. M. Spreading sequences for multicarrier CDMA systems, *IEEE Trans. Commun.*, **47**, 918–926, 1999.

[108] Telatar, E. Capacity of multi-antenna Gaussian channels, *Eur. Trans. Telecommun.*, **10**, 585–595, 1999.

[109] Foschini, G. J. and Gans, M. J. On limits of wireless communications in a fading environment when using multiple antennas, *Wireless Personal Commun.*, **6**, 311–335, 1998.

[110] Vucetic, B. and Yuan, J. *Space–Time Coding*, John Wiley & Sons, Chichester, 2003.

[111] Alamouti, S. M. A simple transmit diversity technique for wireless communications, *IEEE J. Select. Areas. Commun.*, **16**, 1451–1458, 1998.

[112] Tarokh, V., Jafarkhani, H. and Calderbank, A. R. Space–time block codes from orthogonal designs, *IEEE Trans. Inform. Theory*, **45**, 1456–1467, 1999.

[113] Tarokh, V., Naguib, A., Seshardi, N. and Calderbank, A. R. Space–time codes for high data rate wireless communication: performance criteria in the presence of channel estimation errors, mobility, and multiple paths, *IEEE Trans. Commun.*, **47**, 199–207, 1996.

[114] Holma, A. and Toskala, A. (eds) *WCDMA for UMTS. Radio Access for Third Generation Mobile Communications*, John Wiley & Sons, Chichester, 2001.

[115] Korhonen, J. *Introduction to 3G Mobile Communications*, Artech House, Boston, MA, 2001.

[116] Hockwald, B., Marzetta, T. L. and Papadias, C. B. A transmitter diversity scheme for wideband CDMA systems based on space-time spreading, *IEEE J. Select. Areas. Commun.*, **19**, 48–60, 2001.

[117] Hoffman-Wellenhoff, B., Lichtenegger, H. and Collins, J. *Global Positioning System: Theory and Practice*, Springer-Verlag, New York, 2001.
[118] El-Rabbani, A. *Introduction to GPS: The Global Positioning System*, Artech House, Boston, MA, 2002.
[119] Farrel, J. and Barth, M. *The Global Positioning System and Inertial Navigation*, McGraw-Hill, New York, 1999.
[120] Steele, R., Lee, Ch. and Gould, P. *GSM, cdmaOne and 3G Systems*, John Wiley & Sons, Chichester, 2001.
[121] Karim, M. R. and Sarraf, R. *W-CDMA and cdma2000 for 3G Mobile Networks*, McGraw-Hill, New York, 2002.
[122] Walke, B., Seidenberg, P. and Althoff, M. P. *UMTS: The Fundamentals*, John Wiley & Sons, Chichester, 2003.

Index

Spread Spectrum and CDMA: Principles and Applications Valery P. Ipatov

Printed and bound by CPI Group (UK) Ltd, Croydon, CR0 4YY
16/09/2026
14963479-0005